DESIGNING GENERAL LINEAR MODELS TO TEST RESEARCH HYPOTHESES

Keith McNeil
Isadore Newman
John W. Fraas

University Press of America,® Inc.
Lanham · Boulder · New York · Toronto · Plymouth, UK

Copyright © 2012 by
University Press of America,® Inc.
4501 Forbes Boulevard
Suite 200
Lanham, Maryland 20706
UPA Acquisitions Department (301) 459-3366

Estover Road
Plymouth PL6 7PY
United Kingdom

Library of Congress Control Number: 2011940399
ISBN: 978-0-7618-5768-6 (paperback : alk. paper)
eISBN: 978-0-7618-5769-3

Dedications

We dedicate this book to our wives; Betti Fraas, Carole Newman, and Maria Lulisa Gonzalez. Their patience, understanding, and unquestioning support have facilitated the completion of our work.

We also want to dedicate the book to Francis "Jack" Kelly. Over 40 years ago he inspired us to explore the utility of the General Liner Model. Although not a statistician, he had a unique intuitive understanding and appreciation for the technique.

Contents

Preface

The General Linear Model (GLM) is a flexible and encompassing statistical technique that can be used to test hypotheses that otherwise could be tested with either correlational techniques or analysis of variance techniques. Understanding of GLM frees the researcher to ask a multitude of research questions without wondering if the particular statistical technique is available to test any one of them. If the researcher is interested in one criterion, then the GLM is applicable to any hypothesis of interest. How many predictor variables are of interest, and in what form (linear, non-linear, or interacting) is irrelevant. Also irrelevant is a special formula for the statistical test and for the degrees of freedom. Thus the researcher is free of those worries, and can focus on the research questions.

We have also chosen to use the term *general linear model (GLM),* although in some places the terms *multiple linear regression technique, multiple linear regression approach,* and *multiple regression* are used.

The first three sections of chapter 1 provide a conceptual, research, and statistical orientation to the entire text. The remainder of chapter 1 provides the rationale for the utility of a conceptual model of behavior, along with one such model that can be used to identify predictor variables. We strongly suggest that readers familiar with the GLM read these three sections before delving into the more advance material. The final chapter presents a discussion of the strategy of research as viewed from the GLM approach. Specifically, we discuss the GLM approach as it relates to several goals of research: (a) predictability, (b) parsimony, (c) replication, and (d) validity.

We have found that the computer examples can provide additional insight and reinforcement for the concepts presented in this book. The reader is provided detailed instructions on how to obtain computer printouts of the regression models used to test various types of research hypotheses. Specifically, instructions are provided for the SPSS and Microsoft Excel computer software, which are two readily available computer software packages. In addition to these instructions, we provide detailed descriptions regarding how to interpret the computer outputs, which are used to test the various research hypotheses. It is our hope that the inclusions of computer instructions and output interpretations will encourage researchers and graduate students to apply the techniques presented in this book.

CHAPTER 1

INTRODUCTION TO THE
GENERAL LINEAR MODEL

CONCEPTUAL ORIENTATION OF THE TEXT

Limited Mathematical Approach

The approach we use in this text relies on using vectors instead of matrices. Few formulae are presented; we focus instead on conceptual understanding. With the ready availability of computers, researchers do not have to calculate their statistics, although they do need to be able to interact with computers. Therefore, the emphasis in this text is on developing one's own statistical models based on the Research Hypothesis, which in turn is based on the researcher's question of interest, and then having the confidence to interact with a computer program to obtain the results of that test. The text provides instructions regarding how to test Research Hypotheses with the statistical computer software found in SPSS for Windows and Microsoft Excel. In addition, the text emphasizes the interpretation of the results obtained from these computer software programs.

Research Guiding the Statistical Effort

It is our firm belief that statistical procedures should be used to answer researchers' questions. Most statistical texts present a vast array of statistical techniques but do not facilitate the understanding of the advantages and problems of the various techniques. We take the position that the Research Hypothesis should be stated first and that the Research Hypothesis dictates all the remaining statistical, computer, and interpretational effort.

Directional and Nondirectional Hypotheses

Most statistical texts discuss directional and nondirectional hypotheses, but few emphasize the appropriate interpretations and necessary adjustments when using the *F* test. In this text, we emphasize the role of each of the two hypotheses and encourage the directional test whenever there is reason to use it.

RESEARCH ORIENTATION OF THE TEXT

Research Hypothesis as the Basis for Inferential Statistics

The reason inferential statistics exist is to answer Research Hypotheses. Inferential statistics are of no other value, and that is why we always capitalize the term *Research Hypothesis*. The Research Hypothesis must come first, and it should not be limited by the statistics that the researcher knows. The most compelling reason for using the General Linear Model (GLM) approach is that the vast majority of Research Hypotheses can be tested with the GLM approach. Thus, the researcher is freed from the shackles of posing only a limited variety of Research Hypotheses and can spend more effort testing the Research Hypotheses that are of real interest to a particular research study.

All parametric statistical procedures investigating a single-criterion variable are computational simplifications of the GLM and therefore can be tested with the GLM. In addition, many nonparametric procedures are subsets of the GLM. In the text where specialized statistical techniques are identified, the purpose is to assist those researchers familiar with these techniques, not to perpetrate the misguided impression that it is important to view each of these techniques as separate, necessary entities. The only compelling reason to recognize them is that they still appear in the literature. The statistical test should be in the background, at the service of statistically testing the Research Hypothesis. For instance, one should not say, "I did a *t* test," but rather one should say, "In order to statistically test the Research Hypothesis of interest, I conducted a *t* test."

Focus on the Criterion Variable Instead of on the Predictor Variables

Many researchers forget they are trying to "understand" their criterion variable and therefore place more emphasis on their predictor variables. Some researchers develop new treatments or procedures before they can account for a substantial amount of the variance in the criterion variable. If researchers do not know what accounts for the variance in the criterion, how can they develop intervention strategies?

The emphasis in this text is on accounting for the criterion variance by including in the model: (a) additional variables, (b) interactions of already included variables, and (c) functions of the variables other than linear ones. The amount of accounted-for variation in the criterion variable is measured through the R^2 value. We encourage researchers to transform the existing statistical tests of significance in the literature to R^2 values (see chapter 4). Indeed, many meta-type evaluations do just that (Rosenthal, 1984).

Multiple Predictor Variables

A single predictor variable usually will not account for all the variation in the criterion variable. Most social science researchers consider their constructs to be complexly determined yet employ very simple statistical models to test those constructs. On the other hand, carefully chosen predictor variables and well-thought-out relationships between those predictor variables and the criterion may require only a few variables to approach the ultimate goal of accounting for all the criterion variance. Chapter 12 even provides a real example of just one predictor variable accounting for all the criterion variance.

Predictor Variables Probably Needed from Various Areas

Many researchers believe that an inherent limit exists regarding the amount of variance in the criterion variable that can be accounted for by predictor variables, and they often stop their search for predictability much too soon. In research on humans, the measuring instruments usually do not meet very high levels of reliability and validity, and that foils our attempt to obtain an R^2 of 1.00. An R^2 of .60 or lower is typically reported, and that is too low. Some statistics authors suggest that two or three predictor variables will result in the maximum amount of variance accounted for in the criterion variable. The information obtained from models with low R^2 values is better than no information (Newman & Newman, 1999), but we argue that increasing the R^2 values of the models is a very desirable goal.

One should not be satisfied with achieving significant R^2 values if those values are low. As previously stated, it is our position that in an attempt to achieve the goal of obtaining an R^2 value of 1.00, which is theoretically attainable; researchers just may need to consider (a) additional variables—see chapter 6, (b) interactions of already included variables—see chapter 7, and (c) functions of the variables other than linear ones—see chapter 9.

Predictor Variables as "Pieces of Information"

It is valuable for the researchers to think of each variable as another "piece of information." There are two advantages to using this concept. First, the researchers may come to realize what other "information" is known about the subjects under study. Second, the concept leads to the calculation of degrees of freedom, which is usually difficult with statistical procedures other than the GLM.

When researchers consider predictor variables as pieces of information it will help them to answer questions such as: What information can we get and at what cost? Is this really new information? If we have already considered the two subtests, is the total test a new piece of information? If we have already decided who is in our sample, and we know that we are going to have two gender groups, do I need to consider both male and female as new pieces of information?

STATISTICAL ORIENTATION OF THE TEXT

The Generalizability of the General Linear Model

Almost all the statistical procedures encountered in an introductory statistics course, a correlational course, or an analysis of variance (ANOVA) course are computational simplifications of the GLM. The questions answered by all the "computationally simplified" procedures also can be answered by the GLM. Indeed, all parametric and most nonparametric Research Hypotheses investigating one criterion variable can be answered by the GLM. In addition, many Research Hypotheses that have not had computational simplifications developed for them can be tested by the GLM. Therefore, the readers of this text will be encountering a flexible, generally applicable statistical procedure that can replace most of the other procedures. The underlying similarity of all the statistical procedures should become evident.

Before the widespread availability of statistical computer software the GLM was computationally difficult to set up, and that is the reason why statisticians developed computational simplifications. Because the computer can do routine and complex operations easily and quickly, we can conveniently use the computer to do the necessary calculations. The researchers are then free from wondering if the "correct" statistical analysis is known or if the desired Research Hypothesis can be stated in such terms that one of the few statistical tests known by the researchers can be used to statistically test the Research Hypothesis. Thus, researchers can now let the desired Research Hypothesis "wag the tail" of the statistical procedure instead of the tails of the known statistical tests "wagging" the way the Research Hypothesis is stated.

Continuous Versus Categorical Representation of the Variable

Traditionally, two distinct statistical camps have existed—one concerned with continuous variables (the correlation camp), and one concerned with categorical variables (the ANOVA camp). One benefit of the GLM is that the statistical technique and the computer are indifferent to whether we want the variable to be treated as a continuous variable or a categorical variable. Whether to construct a continuous or categorical variable is a decision for the researchers to make. That decision depends on the Research Hypothesis, not on the statistical tools in the researchers' toolboxes.

We, the authors, are biased toward continuous variables for two reasons. First, most variables do occur naturally in a continuous fashion. For convenience, we often artificially dichotomize a continuous variable, producing such results as (a) a hire or no-hire decision, (b) a tall or short label, or (c) a like or dislike attitude. The value at which the researchers specify the cut point on the continuum to form the groups often becomes problematic and usually differs from situation to situation.

Second, we strongly believe that functional relationships occur in a continuous fashion. Behavior usually does not change abruptly as a result of being either below or above a certain cut point. If it does change abruptly, then it is highly unlikely that the a priori cut point is the "real" cut point. Initial consideration of continuous data would still allow the investigation of the existence of alternate cut points.

Hypothesis Testing Versus Data Snooping

Many researchers have used the GLM as a "data snooping" tool, specifically when they conduct a stepwise regression analysis (see chapter 6). The emphasis of this text is on the hypothesis-testing use of the GLM. It is important for the reader to know when the two uses are appropriate. Data snooping is valuable in the service of hypothesis generation, while the comparison of models is valuable in the service of hypothesis testing.

A Priori Comparisons Instead of Post Hoc Comparisons

A substantial amount of literature exists regarding which statistical technique to use after a significant ANOVA has been found. These techniques are referred to as *post hoc comparisons* because they test nondirectional hypotheses that were not originally stated. We suggest a single approach for adjusting the alpha level because of multiple non-independent tests of significance, that is, a Bonferroni-type adjustment (Newman, Groom, & Hodet, 1983; Newman, Fraas, & Laux, 2000), which we discuss further in chapter 4. The value of post hoc comparisons is found in the information they provide with respect to future data gathering and subsequent testing of a priori hypotheses.

In this text we make the case that there is little value in post hoc comparisons and much utility in a priori comparisons, comparisons that encourage the investigation of directional Research Hypotheses. The limited value of the post hoc comparisons should be placed within the framework of data snooping.

Interactions

When thinking about what pieces of information account for the criterion variance most researchers usually state "it depends." Whenever researchers use that language, they are referring to statistical interaction. Our position in this text is that interaction may well account for meaningful variance in many research areas. In traditional ANOVA, interaction is usually viewed as a source of variance that is not wanted, that complicates the situation, and that is not a meaningful question in its own right. We treat interaction as a possibly valuable piece of information that can be tested, just as any other piece of information can be tested (see chapter 7).

Covariate Adjustment

When groups are initially different, the technique of covariance adjustment can be used. The covariate must, though, have variance within groups, otherwise the criterion cannot be adjusted for that covariate, and the covariate remains as a competing explainer of the results. For instance, if a researcher is studying the criterion of the time it takes to run a mile, then pretreatment body fat would be a reasonable covariate, as those with less body fat can usually run faster. If a researcher conducts one treatment at one altitude and the other treatment at another altitude, then each subject at the one altitude would have the same covariate score on the altitude measure. Thus, altitude cannot be used as a covariate and remains as a competing explanation for whatever differences the researcher finds in the criterion. Similarly, if the two sites have, say, different levels of pollution, then the researcher cannot treat pollution as a covariate unless data were obtained on different days having different levels of pollution (see chapter 6).

Clarification of Possible Statistical Confusions

It has been our experience over the past 40 years that students who understand the GLM understand the big picture of statistics. Many concepts that were either confusing, skipped, placed on the back burner, or were incorrectly learned are clarified with the GLM. Be prepared for an "Aha" experience if you have some unresolved questions, such as the following:

1. How many subjects should I have?
2. How many variables should I have?
3. What does *interaction* really mean?
4. I think the relation between my predictor and my criterion is not linear; how can I reflect that relationship?
5. The GLM computer output provides the term *constant* or *intercept*; what does that mean?
6. What are *degrees of freedom* anyway?
7. What is the formula for the degrees of freedom for my Research Hypothesis?
8. My data do not meet the assumption of interval data; what do I do?
9. What does *significance* mean?
10. I found statistical significance, but do I have practical significance?

PURPOSES OF RESEARCH

Investigations into human behavior take many forms and are conducted for various reasons. The purpose of some is to describe, others to predict, and still others to improve (control) behavior. Underlying these behaviors are some expectations held by the researcher regarding a causal network behind observed behavior. Research designs and statistical procedures have been developed to help the researcher make decisions regarding the adequacy of the *description, prediction*, or *improvement* in relation to expected outcomes. These three functions are illustrated in the simplified example presented in the next section.

Description

Description refers to reporting data that have been obtained and applies only to the particular people (or other entities) from whom those data were obtained. In description, no inference is made to other data or to other people.

A group of social scientists may suspect that children from lower-income households seem to have less success in school than children from middle-income households. These researchers would expect children from lower-income households to have lower grade point averages (GPAs) when compared with the GPAs of their classmates from middle-income households. These expectations are based on some combination of intuition, past research, observations, and their theoretical framework.

To *describe* the expected relationship, the researchers will usually select a representative number of children from middle-income and lower-income households, with the formation of the income groups based upon some operational criteria, and then collect data on the criterion selected to represent school success (GPA in this case). Suppose a review of observed data reveal that 50 children from lower-income households have a mean GPA of 1.52 (on a five-point scale), and 50 children from middle-income households have an observed mean GPA of 3.68. The researchers confirm their expectation that children from lower-income households do have lower GPAs on the average than children from middle-income households. Further inspection of the individual GPAs may reveal there is some overlap between the two groups. Some children from lower-income households have higher GPAs than some children from middle-income households. In fact, some children from lower-income households have higher GPAs than the average GPA for children from middle-income households, and conversely, some children from middle-income households score lower than the average GPA for children from lower-income households. Low household income does *not always* "cause" poor school success; there is some overlap of observed scores among groups. Perhaps other predictor (or independent) variables also influence school success, a concern that will be discussed shortly.

Another way to *describe* these data is to report the strength of the relationship between household income and school success using a squared bivariate correlation coefficient (R^2). This value (R^2) represents the proportion of the total observed GPA variance explained by group membership. Suppose in this case that the R^2 value is equal to .45. Then the percentage of observed variance in the students' GPAs (i.e., the sum of the square of each subject's GPA from the overall mean GPA) that is explained by household income is 45%. The large difference in income group mean GPA values and the large percentage of variance in the students' GPAs accounted for by the household-income groups may be sufficient for the investigator to tentatively suspect *some* causal relationship between household income and GPA for this specific group of children. It is important to note that a more definite statement regarding causality would have to wait for manipulation.

Prediction

The process of using data from a smaller group (sample) of people (or measurable entities of any kind) to make estimates about a larger group (population) from which such data have not been gathered is called *prediction*. Our position in this text is that researchers use data to establish a prediction only *after descriptive data exist to support that prediction*. For example, a group of researchers may read the descriptive data on the relative school success of children from lower-income households and middle-income households that were reported in the previously described study. This second group of researchers may have noted that schoolroom activities approximate the activities in the middle-income households, and that lower-income households provide few of the classroom-like activities. The lack of preparation in the lower-income households could cause these children to be at a competitive disadvantage with the better prepared children from middle-income households.

In a situation such as the one just presented, the researchers usually desire to do more than describe a particular set of children; typically, they want to generalize from the sample of children they measured to a theoretical population of similar children. The researchers then choose children for the study that adequately represent the two theoretical subpopulations (i.e., a group of children from lower-income households and a group of children from middle-income households), and then the information from that sample of children can be used to generalize the results to all similar children from lower-income and middle-income households.

To know what generalization to make, one must determine how likely it is that the observed difference between means (or the R^2 value) would occur due to sampling variation. Suppose the researchers find with their sample in the descriptive phase of their study that the home preparation and school success has an R^2 value of .47. In addition, as depicted in Figure 1.1, the mean GPA values are 1.52 for the children from lower-income households (LIH) and 3.68 for the children from middle-income households (MIH).

The R^2 value of .47 can be subjected to a statistical test to determine how likely it is that the observed mean differences can be attributed to sampling variation. The technique that uses the GLM approach to statistically test the difference between two group means is discussed in chapter 4.

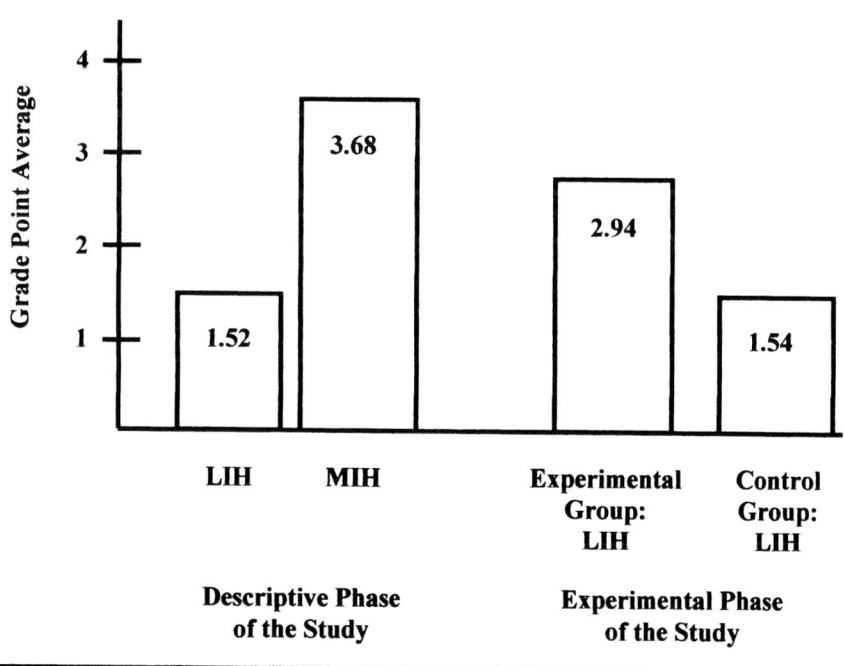

Figure 1.1.
Grade point average means for children from lower-income households (LIH) and middle-income households (MIH) obtained in the descriptive phase and the experimental phase of the study.

For a sample of 100 children, the observed R^2 and mean differences are highly unlikely to be due to sampling variation. On the basis of a statistical test of the difference between these two group means, the researchers would *tentatively* conclude that differences in income levels apparently "causes" differential school performance. Thus, if a new set of children from the two types of homes were selected, based on these researchers' predictive study, they would *predict* (or estimate), with some confidence, lower GPAs for children from lower-income households when compared with the GPAs of children from middle-income households. It is very important to note, however, that a definite inference of causality would have to wait for manipulation.

The causal interpretation just given is based on the theoretical network developed by the investigator. The theory bolstered by observation of differential household experiences in lower-income and middle-income households led to specific expectations. As with most research, other causal factors may be operating. The causal interpretation is always subject to modification based upon subsequent information.

Perhaps the difference between children from lower-income and middle-income households is due to nutrition or to expectations of teachers. These possible competing explanations have not been controlled in the present study. Some possible competing explanations can be eliminated by the design of the study, some can be statistically controlled, and most must be logically eliminated by the researcher.

Please note that determining causality is a process of logic and research design and not a process of statistics. Statistical verification is a necessary but never sufficient condition for causal interpretation. We present more on causality in chapter 10. Knowing the causes for GPA levels allows a meaningful change in the situation, which is the subject of the next section.

Improvement

Improvement involves making a change in a given situation that has been expected and verified to be a "cause." Many applied behavioral scientists want to improve the lot of humans and will attempt to manipulate the environment to bring about a controlled outcome. For example, the researchers previously discussed can *predict* that children from lower-income households will have lower GPAs than the children from middle-income households. But they may wish to "upset" the prediction and do something to change the situation so that the children for whom they would have predicted low GPAs will have higher GPAs.

They cannot, however, be very successful in improving the situation until they are certain of the cause of the low GPAs. Although the researchers found that income levels are related to GPA, they may discover that changing a child's household income does *not* raise the child's GPA (if household income is not the cause).

Improvement cannot be determined until predictor variables are identified as suspected causes of change in the criterion variable; and once those predictor variables are *manipulated* in a sample, they are found to be related to "improvement" in the criterion variable. Some variables, such as gender, cannot even be considered as ones that can be manipulated. Other variables will be difficult or too expensive to manipulate. Variables that can be manipulated may cause a change in other variables that are under consideration or were previously thought to be unrelated to the criterion. Additionally, some subjects may resist being "controlled." Finally, even if the researcher can change subjects on the predictor variable, the desired change in the criterion may not occur. It is an empirical world, and we will never know until we manipulate a predictor variable.

Manipulation to Search for Causal Factors

Suppose the researchers who investigated the performance of the children from lower-income and middle-income households in the descriptive phase of their study wanted to upset the predicted "poor" school success for the children from lower-income households (LIH). This goal would lead the researchers to manipulate (change) one of the factors suspected of causing poor school success in the experimental phase of their study. They may first select 100 children who represent the same theoretical population of children from LIH. The researchers would, on some random basis, assign 50 children from these 100 LIH children to an intervention program and 50 children to act as a natural control group. If home experience is the relevant causal variable, then a change in the home environment (e.g., introducing classroom-like experiences) theoretically should result in higher GPA performance than expected. There is nothing sacred about choosing 100 subjects; more subjects would have provided a more powerful test of the hypothesis (i.e., more likely to conclude the change is not due to random variation), and fewer than 100 would have provided a less powerful test of the hypothesis.

If these classroom-like experiences in the home were provided for the 50 experimental LIH children and not for the 50 control LIH children, a difference between the mean GPAs of these two groups favoring the experimental group should theoretically be found. When the data on the criterion are examined in the experimental phase of the study, it is found that the mean GPA for the experimental group (i.e., the enriched group) is 2.94, and for the control group the mean GPA is 1.54 (see Figure 1.1). The bivariate R^2 of .30 and its associated F test value are very unlikely to be due to sampling variation. Based on the results that the mean GPA of the experimental group (2.94) is higher than the mean GPA of the control group (1.54) in the experimental phase of the study and the lower-income-household groups in the descriptive phase of the study, one might conclude that the enriched home experiences apparently caused better school performance. However, note that the mean GPA of the experimental group (2.94) does not equal the mean GPA of the middle-income-household group (3.68) recorded in the descriptive phase of the study. The researchers have apparently identified one of the causal factors underlying lack of success, but some additional cause or set of causes is also operating. It could be that the experimental treatment was not as strong as the natural environment or that a related variable or set of variables was operating to influence the children's GPAs. Are children from MIH a bit healthier? Are they more motivated for success? The data to answer these questions are lacking in the present study.

The researchers also may note that the mean GPA of the control group (1.54), which was recorded in the experimental phase of the study, was close to the mean GPA of the lower-income-household group (1.52) observed in their initial study. The difference is so small that they would likely conclude that the two samples represent the same population. With these sample values (1.52 and 1.54), the researchers still do not know the population GPA mean, but they might assume that it is close to the average of the two sample means (1.53).

Causal conclusions are usually desired, and there are three requirements of such conclusions. First, there must be random assignment to the different treatment groups. Random assignment is different from random sampling, though the two often rely on the same operations. Random sampling from a population allows for generalization back to that population. Random assignment to treatment groups, which is often applied to a nonrandom sample, allows for isolation of the effectiveness of the manipulated independent variable—the treatment variable. Second, a manipulated variable is required for causal conclusion. Third, the researcher must have control over the predictor variable such that the threats to internal validity are eliminated or at least reduced. These research design issues are not discussed in detail in this statistics text, but can be explored further in Campbell and Stanley (1966) and Wilkinson and McNeil (1999).

COMPLEX DETERMINANTS OF BEHAVIOR

The simple example just presented was provided to illustrate the *descriptive, predictive*, and *improvement* aspects of research. At the present level of theoretical sophistication in behavioral research, such bivariate (single predictor and single criterion) research surely would be labeled naïve. It is "known" from past research that many other variables exist, such as test anxiety, measured intelligence, peer expectations, content of the material to be learned, and past special learning (e.g., as measured by standardized achievement tests) that enter into broad performance constructs such as GPA. Furthermore, these variables may not be additively related to performance. For example, Castaneda, Palermo, and McCandless (1956) showed that the way anxiety level affected a student's performance on a task depended upon task difficulty. Given a difficult task, low-anxious students tended to do better than other students; but given an easy task, high-anxious students tended to do best. This effect is called a *difficulty-by-anxiety interaction*. Statisticians use the term *interaction* when the effect of one variable *depends* upon another variable.

Besides the type of interaction discussed above, one might expect task difficulty to depend upon the relative brightness of the performing student. A difficult task for the average student may not be difficult for the exceptionally bright student. If one were to extend the Castaneda et al. (1956) study to include a task difficulty-by-anxiety-by-"intelligence" interaction, one would expect to explain more of the criterion variance and thus have less unexplained criterion variance.

Given this complex state of affairs, it seems that researchers who want to *describe, predict*, and *improve* need complex theoretical models and flexible statistical procedures that accurately reflect the complex models. The GLM, as presented in this book, is a flexible statistical procedure that meets the needs of a broad spectrum of behavioral researchers.

A Conceptual Model

The details and various applications of this flexible statistical procedure, the GLM, are presented in the rest of this text. But we first present a conceptual model that provides a way of considering variables that relate to the behavior the researchers are investigating. One of the major reasons for the grouping of variables discussed in this section is to stimulate the search for multiple variables, since consideration of multiple variables is usually necessary to account for variance in any criterion variable. The variable groupings presented here are applicable to behavioral research; other groupings could be developed for this or other areas of research.

A critical review of the behavioral literature reveals that the variation on the criterion is caused by a network of interrelated predictor variables. These predictor variables can be arbitrarily grouped into three categories: (a) person variables, (b) focal stimulus variables, and (c) context variables.

Person variables. Most areas of research into human behavior have substantiated that there are pervasive differences between individuals. In attempting to account for differing performances, one would want to specify the *person variables*, or the characteristics of the behaving individual that may influence that individual's degree of success.

Focal stimulus variables. There are many ways to measure any construct that a researcher wishes to consider a criterion variable. The ways in which those measures differ from one another are the *focal stimulus variables*; they are the characteristics of the instrument or task that is used to measure the criterion behavior. In attempting to account for differing performances, one would want to specify dimensions of the task to be completed that may influence an individual's degree of success.

Context variables. Many stimuli unrelated to the task can influence an individual's behavior. Such expectancies are neither part of the task nor within the person; they are part of the *context*, the situation, in which the criterion behavior occurs. In a sense, the context provides information regarding "payoff" or reinforcement associated with performance.

In an educational setting, context variables that might interact with student characteristics to influence response acquisition might be the physical condition of the classroom, the reward system for good grades used by the instructors, and peer expectations. For example, if a student's friends value school success, the student might work even while not valuing school achievement.

Symbolic Representation of the Conceptual Model

In view of the preceding discussion, one can say that *complex behavior* of a person is a function of (a) person characteristics, (b) focal stimulus characteristics, and (c) context characteristics. The preceding statement can be reduced to a quasi-mathematical model as a shorthand notation as follows:

$$Y = f(P, S, C) \hspace{3cm} \text{(Model 1.1)}$$

where:
  1. Y = the criterion;

2. P = person characteristics (variables that represent person characteristics);
3. S = focal stimulus characteristics;
4. C = context characteristics; and
5. f = the functional relationship of the variables in the three classes (P, S, C) as they relate to Y.

The equation can be read as follows: The criterion behavior of a number of individuals is a function of the characteristics of these individuals and their focal stimulus characteristics and context characteristics.

Please do not panic at the symbols. Symbol notation is used to simplify the expressions so they can be presented in the form of quasi-mathematical models, a practice that becomes extremely useful in later chapters of this text. There is no need to remember the particular symbols used here, or for that matter to understand fully the variables and examples used in this chapter. They are presented to show how one might go about grouping and exploring variables that may contribute to prediction. Indeed, some variables are difficult to classify. Accurate classification is not the goal, but increasing the amount of accounted-for variance in the criterion variable is the goal. The whole conceptual model for the study of complex behavior with examples of person, focal stimulus, and context variables is as follows:

$$Y = f(P, S, C) \hspace{3cm} \text{(Model 1.1)}$$

where:
1. Y = scores on a math achievement test for a set of individuals
2. P = person variables, and examples of person variables for this criterion are:
   a. convergent thinking ability (as measured by most standardized intelligence tests)
   b. divergent thinking ability (as measured by tests of creativity)
   c. symbol aptitude
   d. motivation
   e. sex role identification (masculinity-femininity)
   f. past learning relevant to success on the task
3. S = focal stimulus variables, and examples of focal stimulus variables (characteristics of the criterion measure) for this criterion are:
   a. length of the test
   b. ordering of items by difficulty
   c. numerical items versus "word problems"
   d. types of mathematics problems contained in the test
   e. number of operations required by each item
4. C = context variables and examples of context variables for this criterion are:
   a. peer expectancies

    b. adult expectancies
    c. physical plant (light, noise, heat)
    d. reward conditions

The functional relationships among the three groups of variables and the criterion (Y) can take many forms depending upon the research expectations of the researchers. This presentation of the quasi-mathematical model (Model 1.1) contains examples for each of the three groups that might explain observed behavior. The examples would, of course, vary from one criterion behavior to another. One task of research is to quantify the variables and then attempt to account for the observed criterion variation in terms of some weighted combination of the predictor variables. It is premature to delineate what functional relationships might exist between the predictor variables (i.e., P, S, and C) and criterion variable Y. In the next section we provide an example to give some ideas about how one might operationalize the three groups of predictor variables.

Accounting for Complex Behavior

Consider the task of predicting complex behavior, such as success in a job-training program. If the training program is costly, it would be worthwhile to predict the successful and unsuccessful trainees in terms of cost effectiveness (to avoid training many individuals who are unlikely to be successfully trained).

In view of the proposed conceptual model just presented, several aspects of the situation should be investigated. One might first examine the set of behaviors that are related to the criterion (Y), the observed terminal behavior. What are the task characteristics (represented by S) that might be relevant to the criterion? Such an examination might give the investigator a few notions regarding relevant human characteristics (represented by P) needed as prerequisite skills. Are specific abilities or special previous learning necessary for training success? What are the conditions (represented by C) surrounding the learning and testing setting? Is there peer pressure for success? Are the trainers placing pressure on the trainees? If so, will trainee anxiety be relevant? What are the intermediate payoff schedules? Must the trainee work through the program before some reinforcement is given? Will the need to achieve be relevant to the reinforcement schedule? If the trainee has failed often in the past, does the trainee expect to fail in this new task? If so, can the training context be manipulated to minimize these effects? Familiarity with the research literature should provide additional suggestions regarding the relevant variables that account for complex behavior. Indeed, the whole research endeavor focuses on selection of variables based upon theoretical expectations and past empirical findings regarding the relationships among sets of variables.

One might hypothesize that the desired terminal behavior (criterion) of the training program is related to spatial abilities (SA), anxiety level (Anx), ability to manipulate symbols (Sy), the amount of written work the job to be trained for will require (Wr), the length of the training program (Ln), and the number of trainees in the group (Nt). Note that the first three variables are person variables; the fourth variable is a focal stimulus variable; and the fifth and sixth variables are context variables. This set of variables can be cast as follows:

$$Y = f(\text{SA, Anx, Sy, Wr, Ln, Nt}) \qquad \text{(Model 1.2)}$$

where Y represents the observed terminal behavior of the trained individuals; and SA, Anx, etc., are defined as previously described.

The function sign (f) implies there are some mathematical functions that express the relationship of the predictors to the observed criterion. Types of commonly used functions are additive, multiplicative (allowing for interaction, discussed in detail in chapter 7), squared (second-degree polynomial, discussed in detail in chapter 9), square root, cubic, and trigonometric functions (e.g., sine and cosine).

One may assume that the scores on the training task (Y) are a sum of the six weighted predictor variables (a weighted additive function). Equation 1.1 expresses Model 1.2 as follows:

$$Y = a_1\text{SA} + a_2\text{Anx} + a_3\text{Sy} + a_4\text{Wr} + a_5\text{Ln} + a_6\text{Nt} \qquad \text{(Equation 1.1)}$$

The weights $a_1, a_2 \ldots, a_6$ might be chosen rationally or be empirically derived using some mathematical solution. In the domain of any science, researchers have seldom found a set of weights that satisfies the equality expressed in Equation 1.1. The reason for lack of equality is that theory, measurement, and expressed functions typically are not perfect. Errors of prediction are made due to incomplete or erroneous theory, inadequate measurement tools, and lack of ability to adjust relationships perfectly. What this means, then, is that the task of behavioral research is to develop more comprehensive theories, better measurement, and more appropriate quantitative procedures in the effort to minimize errors of prediction.

The equality expressed in Equation 1.1 is an ideal that is not observed in most research endeavors. There will usually be errors of prediction. That is, no matter how "good" the weights are, the sum of the weighted predictors will hardly ever be equal to the criterion score. Equation 1.1 can be adjusted to acknowledge errors of prediction and therefore satisfy the expressed equality. Equation 1.2 reflects the inclusion of the error component that satisfies the expressed equality is as follows:

$$Y = a_1\text{SA} + a_2\text{Anx} + a_3\text{Sy} + a_4\text{Wr} + a_5\text{Ln} + a_6\text{Nt} + E_1 \qquad \text{(Equation 1.2)}$$

Equation 1.2 can equivalently be expressed as follows:

$$Y = \hat{Y} + E_1 \qquad \text{(Equation 1.3)}$$

where E_1 represents the difference between the predicted score, which is the sum of the weighted scores ($\hat{Y}$), and the observed terminal performance (Y), that is, $E_1 = (Y - \hat{Y})$.

The GLM procedure derives the set of weights ($a_1, a_2 \ldots, a_6$) to minimize the sum of the squared differences between the observed criterion scores (Y) and the predicted criterion scores ($\hat{Y}$). The squared discrepancy between an observed score and a predicted score can be expressed symbolically as $(Y - \hat{Y})^2$ or as E_1^2. The sum of those squared errors across all subjects, or $\sum (E_1)^2$, has interesting properties that satisfy statistical distributions (e.g., the F statistic). Thus, with some manipulation, the least squares solution can be used inferentially for decision-making purposes. Chapters 2 and 4 treat statistical inference in greater detail.

Accounting for Complex Behavior with Alternate Functions

Suppose the weighted additive function expressed in Equation 1.2 yielded a large $\sum(E_1)^2$ for the group of N individuals. (The sum of the squared errors of prediction for the individuals is large.) Some of this error might be due to selection of inappropriate functions. Knowledge of past research might suggest to the investigators that the additive function is not adequate. They might suspect that Wr is geometrically related to Y. That is, the criterion scores are approximately the same for very low to middle Wr levels, but the criterion scores increase rapidly as Wr scores increase from medium to high. Furthermore, the investigators might expect anxiety (Anx) and length of training (Ln) to interact, such that anxiety might adversely influence terminal behavior, especially when the training period is long. These two circumstances can be reflected by adding two additional variables to Equation 1.2: (a) a squared function of Wr (Wr^2, reflecting the geometric expectation) and (b) a multiplicative function for anxiety and length of training (Anx * Ln, reflecting the interaction expectation). Note that the asterisk (*) will be used to indicate multiplication throughout this text.

The expanded equation would be as follows:

$$Y = a_1SA + a_2Anx + a_3Sy + a_4Wr + a_5Ln + a_6Nt + a_7Wr^2 + a_8(Anx * Ln) + E_2$$
$$\text{(Equation 1.4)}$$

where the value of the E_2 error term would be calculated in the same manner as the E_1 error term (i.e., $Y - \hat{Y}$). The error (E) is subscripted differently in Equation 1.4 because the error values for each subject will likely be different due to the two additional variables used for prediction in Equation 1.2.

If the eight variables in Equation 1.4 represent the functional relationship more adequately than the six variables in Equation 1.2, then the values for the predicted behavior ($\hat{Y}$) calculated with Equation 1.4 should be closer to their corresponding observed behavior (Y) values than the predicted values calculated using Equation 1.2. Therefore, Equation 1.4 should yield a smaller value for $\sum(Y - \hat{Y})^2$.

USING VENN DIAGRAMS TO REPRESENT ACCOUNTED FOR VARIANCE IN THE CRITERION VARIABLE

In this section we use Venn diagrams to depict the amount of variation in the criterion variable (say, math achievement) accounted for by the predictor variable(s). In these Venn diagrams each circle represents one unit of variance. The degree of predictability, which is measured by the model's R^2 value, is represented in a Venn diagram by the area of overlap between the predictor variable or variables and the criterion variable. Figures 1.2, 1.3, and 1.4 represent three possible scenarios.

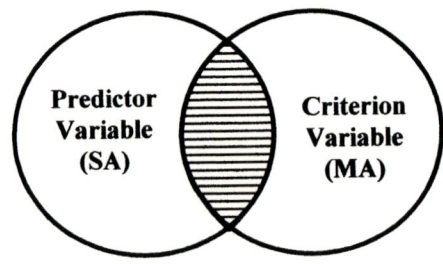

Figure 1.2.
A Venn diagram of a model in which the predictor variable (SA) accounts for 20% (area marked with horizontal lines) of the variance in the criterion variable (MA).

If only one predictor variable, say, spatial abilities (SA), is used to predict the math achievement criterion variable, a certain degree of predictability will be observed. The SA circle (predictor variable variance) overlaps with the MA circle (criterion variable variance). This overlap area, which is marked with horizontal lines, is assumed to be 20%, thus the R^2 value for the model that contained the criterion variable MA and the predictor variable SA is equal to .20. The unmarked area of the criterion variable represents 80% of the criterion variance, which is unaccounted for variance. Often this variance is labeled as *error variance* and considered to be due to measurement error. The unmarked area, though, may be due to omitting the relevant (and necessary) predictor variables. Consequently, this error is also often referred to as *model specification error*.

Adding another predictor variable to the prediction scheme (say, Sy, ability to manipulate symbols) might produce the situation depicted in Figure 1.3. Note that in Figure 1.3 the second predictor variable (Sy) also accounts for approx-

imately 20% of the criterion variance (areas marked with vertical lines and di-
amonds); but because the two predictor variables are correlated with each other,
the new predictor variable contributes only 10% of new overlap (diamond-
marked area), that is, unique overlap with the criterion variable.

Predictor variables will most likely be correlated in the real world. The concern is not how highly they are correlated; the concern is how much of the criterion variance they account for together. If the set of within-person predictor variables has not come close to the goal of accounting for 100% of the criterion
variance, then variables in addition to the within-person variables may account
for the criterion variance.

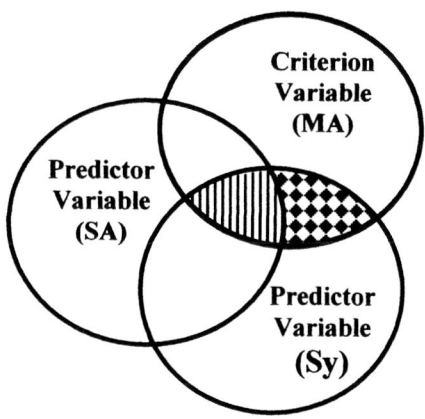

Figure 1.3.
A Venn diagram of a model in which the predictor variable Sy accounts for 10%
of unique variance (diamond-marked area) in the criterion variable (MA).

A criterion variable will most likely be a function of relevant focal stimulus
variables, context variables, and person variables. Figure 1.4 represents the ideal
situation, that is, 100% of the criterion variance is accounted for by a set of three
predictor variables (the area marked with horizontal lines). Note that the three
predictor variables in Figure 1.4 are represented by dashed circles because each
set that researchers could use would not always be expected to account for the
same amount of criterion variance, as is the case for the set depicted in Figure
1.4. The most important point for researchers to understand, however, is that
deciding which variables to use is based on theory or their knowledge base, and
the proportion of criterion variance accounted for by a given set is always an
empirical question to be answered by the data.

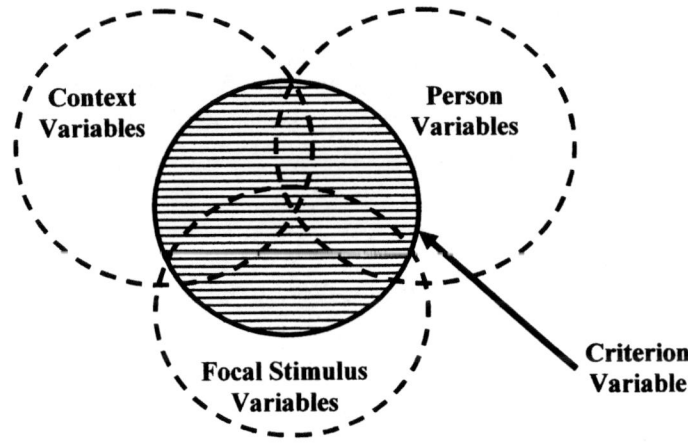

Figure 1.4.
A Venn diagram of a comprehensive model containing three sets of predictor variables (context, person, and focal stimulus) that account for 100% of the variation in the criterion variable (area marked with horizontal lines).

CHAPTER 2

HYPOTHESIS TESTING

INTODUCTION

In chapter 1 we suggested that research design and statistics are tools to help an investigator make decisions regarding the adequacy of *description, prediction,* and *improvement* of behavior. When testing expectations, one seeks to generalize from a small group of individuals to other individuals who are similar. We present a brief discussion of sampling theory for review purposes. Sampling theory is the basis for the steps in hypothesis testing which are as follows:

1. Statement of the Research Hypothesis: The researchers make a statement regarding the population that they are hoping to support. (The Research Hypothesis allows for a multitude of population values (e.g., $\mu > 5$ and $\mu_1 - \mu_2 \neq 0$.)
2. Statement of the Statistical Hypothesis: A statement is made about the population that is antithetical to the Research Hypothesis. (The Statistical Hypothesis specifies one population value (e.g., $\mu = 5$ and $\mu_1 - \mu_2 = 0$.)
3. Selection of the alpha level: The risk (probability) that the researchers are willing to take in rejecting a true Statistical Hypothesis is established. This risk level is referred to as the alpha level. (One minus alpha level is called the confidence level.)
4. The data are collected from a representative sample of the population.
5. The test statistic is calculated (e.g., t, F, chi-square, etc.).
6. Determination of significance is made. The following rules are used when determining significance based on a computer printout:
   a. If the reported probability is less than or equal to the stated alpha level, reject the Statistical Hypothesis, which indicates that the data support the Research Hypothesis.
   b. If the reported probability is larger than the stated alpha level, fail to reject the Statistical Hypothesis, which indicates that the data do not support the Research Hypothesis.

The following rules are used when determining significance based on statistical distributions:

  a. If the test statistic is greater than or equal to the distribution table value, reject the Statistical Hypothesis, which indicates that the data support the Research Hypothesis.
  b. If the test statistic is less than the distribution table value, fail to reject the Statistical Hypothesis, which indicates that the data do not support the Research Hypothesis.

In this text, we stress the interpretations obtained from statistical computer software. Specifically, printouts from SPSS for Windows and Microsoft Excel will be interpreted. Thus, in the application of these steps used to conduct hypothesis testing, emphasis will be placed on comparing the probability of the appropriate statistical test to the chosen alpha level.

THE POPULATION OF INTEREST

In a typical research situation, a *population* is a collection of individuals about whom one is seeking information. The population information to be estimated is assumed to have a *true* value. For example, theoretically there is a true mean value ($\sum X / N$) for the height of all men in the United States over the age of 21. This value is called a *parameter* (a population fact). There is a true value (another parameter) that represents the dispersion of the height of men from the true mean (population variance, $\sum(X - \overline{X})^2 / N$).

With most populations of interest, it is usually impossible to find and measure every individual, so it is impossible to arrive at the parameter value. A researcher therefore selects a subset of the population of interest and measures the individuals in the subset in order to make a statement about the population.

A *sample* is a collection of individuals who are a subset of individuals representing the research population. The sample information to be collected is assumed to be an *estimate* of the population parameter.

Whenever a random sample is drawn, the mean of that sample is the best estimate of the population mean, and is called an *unbiased estimate of the population mean*. On the other hand, the sample variance is called a *biased estimate of the population variance*. [An explanation of that statement can be found in other statistical texts, such as Stevens (2002) and Wackerly, Mendenhall and Scheaffer (2002).] While the sample variance, $\sum(X - \overline{X})^2 / N)$, is a biased estimate, the unbiased estimate of the population variance is $\sum(X - \overline{X})^2 / (N-1))$.

Both the sample mean and the sample variance are subject to sampling error. For example, assume that we are interested in a population of 100 men in a particular location. We could randomly select 20 men from this population, measure their heights, and calculate a mean value for these 20 men. This mean value is the best estimate of the average height of the population of 100 men that can be obtained from 20 men. Another sample of 20 men from this population can be randomly selected and measured, and a mean value of the measures can be calculated. The likelihood of these two means being exactly the same is very small. Yet one would expect them to be somewhat close in value. Furthermore, if one measured the total population and calculated the population mean, one would expect this parameter value and the two sample means to be very similar but not exactly the same. The difference between a population mean and a sample mean is assumed to be due to sampling error—of course this may not be exactly true because there may be systematic errors of measurement in the sample. If one assumes there are no systematic measurement biases, then the random errors of measurement should cancel out, and the sample mean would approximate the population mean. Errors of measurement are treated extensively elsewhere and are beyond the scope of this chapter.

In real-life situations, researchers are typically unable to obtain population parameters for the population of interest. For example, in a particular school system researchers may be interested in the relative merits of Method "e" when compared with Method "f" for 7th-grade students. The researchers could give Method e to half the 7th-grade students and give Method f to the other half. They could then use the mean scores (say, on some test designed to measure success in the subject matter being taught) of each of the two groups to *describe* the relative success of the two treatments for these 7th-grade students. Subsequently, the researchers would wish to use the data from these two groups to generalize to (or *predict* the success of these two methods for) a much broader population of students. In this example, they may wish to use the results of the methods for the two sample groups to generalize to 7th-grade students in this school for the next, say, two years. The researchers would be assuming that this year's 7th-grade class represents a random sample of the population of 7th-grade students in the school district for the three years of interest (this year plus the next two years). Of course, something may happen in the community or school to change the composition of the next two years' 7th-grade students, so that this year's students do not represent that changed population. The researchers' confidence in their ability to predict, generalize, or replicate the findings would then be reduced by an unknown amount. Their generalization might be totally invalid; without further data and analysis of that data, their original sample means are still the best information from which to predict the effects of the treatments. (Refer to Newman, McNeil, & Fraas, 2004, for a discussion on estimating the replicability of a study's findings.)

The researchers want to make a decision regarding which method to use for the next two years' 7th-grade students. Since those two groups of students are not yet seventh graders, they cannot expose them to the two methods and then decide. The researchers therefore want to use this year's 7th-grade students to *predict* for the next two years of 7th-grade students. Suppose there are 100 students in the present year's 7th-grade class. Each student can be assigned a number from 1 to 100; and using a table of random numbers, the researchers can assign 50 students to Method e and 50 students to Method f. This sampling procedure results in two treatment samples that are likely to be representative subsets of the total population being investigated. Random sampling does not guarantee that the samples are representative of the population, but it does minimize the chances that the samples are biased subsets of the population.

These students can be subjected to the treatment to which they were assigned, and then the criterion performance can be measured on some task the methods were designed to teach. Suppose the following information is obtained: (a) a mean value of 55 and variance of 25 for the sample receiving Method e, and (b) a mean value of 35 and variance of 25 for the sample receiving Method f. The question is: Do the two groups that originally represented the same population still represent the same population as a consequence of the treatment? Are the means of 55 for Method e and 35 for Method f different enough to consider Method e and Method f to be differentially effective? Or is the difference likely to be due to sampling variation? We now need to test whether the difference between the means is large enough to be considered "significant."

ANALYSIS OF VARIANCE

The analysis of variance (ANOVA) technique analyzes estimates of the population variance and yields a probability statement showing how likely it is that observed differences between means are due to sampling error. If the differences are highly unlikely to have been caused by sampling error (say 1 time in 1,000), one may be willing to conclude that, as a result of the two treatments, the two groups of students now represent different populations. Risk and cost must enter into the decision of what level of probability should be selected. These matters will be discussed later.

The F statistic is based upon the ratio between two estimates of the population variance. In research designs there are various ways to estimate population variance, some sensitive to group mean differences and others that are insensitive to group mean differences. In our present example, two estimates of interest are the within-group estimate of the population variance, and the among-group estimate of the population variance.

Within-Group Estimate of the Population Variance

The estimate of population variance using within-group information in analysis of variance usually is called *mean square within* (symbolized in many texts as MS_w). Our preference in this text is to refer to this variance estimate as the *within variance estimate* and to use the symbol $\hat{v}_w$, where (a) the small v is used to represent the population variance, (b) the hat (^) indicates an estimate, and (c) the subscript, w, indicates how the estimate is calculated (within group).

The $\hat{v}_w$ is an estimate of population variance based on how much the criterion scores of persons in each group differ from the mean score of their group and is calculated in several steps. First, the mean of one group is calculated, and that group mean is subtracted from each of the observed criterion scores (X) of the members of that group. Second, each of these discrepancy scores is then squared, and the squared scores are summed. This procedure is followed for each of the k groups. (In the above example, k is 2, the two groups being Method e and Method f.) The sums of squared discrepancy scores for all groups are then added (sum of squares within, or SS_w) and divided by the degrees of freedom within all the groups. The result is $\hat{v}_w$.

As typically presented, "degrees of freedom" is difficult to understand, though it is a simple concept. A more elaborate explanation of this matter is given later. A brief discussion is presented here because each estimate of the population variance has its own degrees of freedom.

A *variable* by definition can take on any of a range of numerical values. A variable is thus free to vary. Within Method e there are 50 students, and there are 50 observations on the criterion variable. Each of the 50 observations or scores is free to vary because each represents one of the 50 students and is therefore dependent on the performance of that student.

After obtaining the 50 scores for Method e on the criterion variable, one can calculate the mean of those 50 scores ($\sum X / N$). When the differences between the observed scores and the mean are calculated, are all 50 difference scores free to vary? (Note that it is a property of the mean that the sum of the difference scores between the mean and the observed scores is equal to zero.) Start with any one score, and it is free to be any value; that is, the first score is free to vary—it can be any value. Also, each subsequent score through observation 49 is free to vary. But the question is: What about the 50th score? That score is *not* free to vary because each of the preceding 49 scores is specified, and the group mean is known. Given the 49 scores and the mean, the 50th difference score can be only one value. Thus it is not free to vary.

The three scores of 4, 3, and 2 will help to explain the concept of degrees of freedom. The mean of these three scores is $\sum X / N = 9 / 3 = 3$. Now calculate the differences between each observed score and the mean. Given that the mean is 3, if one did not already know the actual scores, one might think that those three scores could take on an infinite number of values to get a mean of 3. If the first observed score is 4, then the difference score for that observed score is $4 - 3 = 1$. (That observed score was free to vary; it happened to be 4.) If the second observed score is 3, the difference score for it is $3 - 3 = 0$. That score was also free to vary. Now that it is known that two of the scores are 4 and 3, there is only one value the third score can be and still yield a mean of 3 for the three scores (and still have the difference scores sum to zero). That third score would, of course, be 2 because $2 - 3 = -1$. Of the three scores, only two are free to vary once the mean is known. To generalize: When the mean is known, one less than the total number of scores is free to vary (thus, when researching variability within a group, degrees of freedom is the total number of subjects less one, or $(N - 1)$.

When the problem of calculating $\hat{v}_w$ was presented, it was given that for Method e, the mean score was 55, and the sample variance was 25; for Method f, the mean score was 35, and the sample variance was 25. One could go back to the 100 observed scores (50 for each group) and calculate $\hat{v}_w$ from the procedure described above. It is easier, however, to use the knowledge of the group means and variances. The variance for Method e (25) was calculated by finding the difference between each score and the group mean, squaring and summing the difference scores, and dividing the sum by the number of scores (50), or $\Sigma(X - \overline{X}_e)^2 / N$. To obtain the unbiased estimate of the population variance ($\hat{v}_w$) one does not divide by N, but by degrees of freedom. The degrees of freedom here is $N - 2$, where N is the number of subjects, and 2 is the number of groups. The separate sums of squares for each group are summed together before dividing by degrees of freedom, so the sum of squares for each group is needed. If the Method e variance was 25 and $N = 50$, then from the above formula one can see that the sum of squares for Method e must equal 1,250 (i.e., 25 * 50). The variance and N are the same for Method f, so the sum of squares for Method f also must equal 1,250.

The $\hat{v}_w$ is calculated by adding the sum of squares of the k groups and then dividing by degrees of freedom. This example contains two groups, and the sum of squares for each is 1,250. The total degrees of freedom value is equal to the sum of the degrees of freedom for each group. In each group, a mean was calculated, so the degrees of freedom value for each group is 50 - 1 = 49, and the total degrees of freedom would be 98. One also could arrive at the total degrees of freedom by looking at the total number of observations in both groups (N) and subtracting the number of means calculated (k), or in this case 100 - 2 = 98.

The $\hat{v}_w$ is therefore calculated by using Equation 2.1:

$$\hat{v}_w = \frac{\Sigma(X - \overline{X}_e)^2 + \Sigma(X - \overline{X}_f)^2}{N - k} \qquad \text{(Equation 2.1)}$$

Substituting the appropriate values into Equation 2.1 produces the following:

$$\hat{v}_w = \frac{1{,}250 + 1{,}250}{100 - 2} = \frac{2{,}500}{98} = 25.5$$

The value of 25.5 is the best estimate of the population variance because it is not influenced by group mean differences. It is best in the sense that repeated calculations on random samples from the same population would result in variance estimates that deviate the least from the actual population variance. If the differences among the group means are small, then the estimate of the population variance using among-group information, which will be discussed next, should be close to the within-group estimate $\hat{v}_w$.

Among-Group Estimate of the Population Variance

The among-group estimate of the population variance is symbolized in this text as $\hat{v}_a$. The subscript "a" denotes that it is an estimate using among- or between-group information. The value for $\hat{v}_a$ is calculated by subtracting the mean of all the students regardless of group membership (this is often called *grand mean*, $\overline{X}_g$) from each of the k group means. For each group, this value is then squared and multiplied by the number of scores in the group. The product for each group is summed over groups (SS_a) and divided by degrees of freedom, which for this variance estimate is k - 1.

Degrees of freedom value for $\hat{v}_a$ is equal to the number of groups (k) minus one (1). Following the logic given for $\hat{v}_w$, each group mean is an observation free to take on any value, but once the grand mean is calculated using these means, only k - 1 means are free to vary. In this example, $\overline{X}_e = 55$, $\overline{X}_f = 35$, $\overline{X}_g = (55 + 35) / 2 = 45$. For this case, $\overline{X}_g$ can be derived by getting the average of the two means because the groups are of equal number. When groups differ in number, the above procedure is not applicable, but $\overline{X}_g$ can be obtained by summing all scores and dividing by *N*.

The $\hat{v}_a$ is calculated with the following equation:

$$\hat{v}_a = \frac{[(\overline{X}_e - \overline{X}_g)^2 * N_e] + [(\overline{X}_f - \overline{X}_g)^2 * N_f]}{k - 1} \qquad \text{(Equation 2.2)}$$

Substituting the appropriate values into Equation 2.2 produces the following value for this example:

$$\hat{v}_a = \frac{[(55 - 45)^2 * 50] + [(35 - 45)^2 * 50]}{2 - 1} = 10{,}000$$

The *F* Ratio

Two estimates of the population variance, $\hat{v}_w$ and $\hat{v}_a$ have just been discussed. Within-group estimates of the population variance are unaffected by group mean differences. Among-group estimates of the population variance are sensitive to group mean differences because they are based upon group mean variations from the grand mean. If there is no real difference between the two populations, then $\hat{v}_a$ will be a good estimate of the population variance and would be expected to be about equal to $\hat{v}_w$ (the best estimate of the population variance).

Consider the discussion of random sampling. When two samples are randomly drawn from a population, the sample means are expected to vary somewhat from each other. The difference between sample means is referred to as *sampling error*. The larger the number of subjects in a sample, the closer the sample means tend to be. Occasionally, however, large mean differences will be observed.

The *F* ratio represents the distribution of the ratios of two independent estimates of the population variance. It can be mathematically shown that the expected *F* ratio is 1.00 if the two samples are representative of a single population. The randomly selected groups will occasionally have large mean differences due to sampling variation. Therefore, an *F* ratio much greater than 1.00 would be obtained. Likewise, the randomly selected groups will occasionally have small mean differences due to sampling variation. In these instances, an *F* ratio less than 1.00 would be obtained. If the ratio of the two estimates is distributed as *F*, then for any set of degrees of freedom, the likelihood of an observed *F* ratio being due to sampling variation can be determined.

Figure 2.1 is an approximate distribution of *F* ratios with 1 (k - 1) and 98 (*N* - k) degrees of freedom. The area located to the right of 3.94 represents the 5% of the observed *F* ratios that will occur from the samples whose means are most discrepant. An *F* ratio of 3.94 or greater would be observed 5% of the time due to sampling variation. An *F* ratio of 6.90 or greater would be observed 1% of the time due to sampling variation. These values of 3.94 and 6.90 can be found in *F* distribution tables in many statistics books. In the example with Method e and Method f, the estimated population variance using within-group data ($\hat{v}_w$) was 25.5; and the among-group estimate ($\hat{v}_a$) was 10,000. The *F* ratio is calculated using Equation 2.3 as follows:

$$F = \frac{\hat{v}_a}{\hat{v}_w} \qquad \text{(Equation 2.3)}$$

Substituting the values for $\hat{v}_w$ (25.5) and $\hat{v}_a$ (10,000) into Equation 2.3 produces the following *F* ratio:

$$F = \frac{10,000}{25.5} = 392.15$$

The numerator degrees of freedom (df_n) value for this *F* test is equal to the denominator of Equation 2.2, which is 1 (i.e., 2 - 1), and the denominator degrees of freedom (df_d) value is equal to the denominator of Equation 2.1, which is 98 (i.e., 100 - 2).

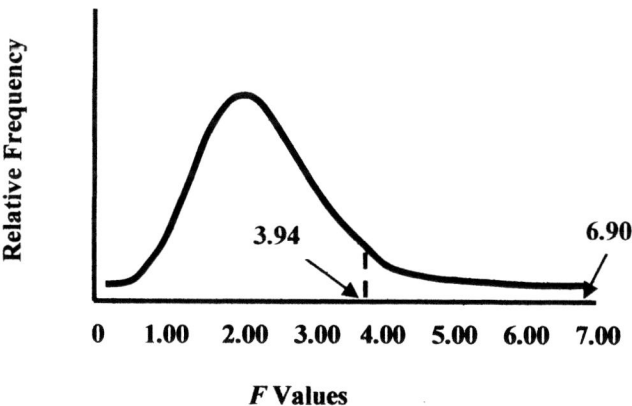

Figure 2.1.
An approximate distribution of F with 1 and 98 degrees of freedom.

Statistical tables indicate that an F of 392 with 1 and 98 degrees of freedom would be observed less than 1 time in a 100 due to sampling variation. Since $\hat{v}_a$ is so very large in relation to $\hat{v}_w$, one can conclude that the difference between the two observed means (Method e mean was 55 and Method f mean was 35) is unlikely to be due to sampling variation. Most likely, the two treatments have resulted in producing two populations: (a) one of 7th-grade students exposed to Method e and (b) another population of 7th-grade students exposed to Method f.

On the other hand, one must be aware that 1 time in 100 such a large mean difference is expected due to sampling variation. Suppose you are the decision maker regarding the selection of a method of instruction. What would you recommend? Your recommendation depends on whether you believe you have enough evidence to reject the following Statistical Hypothesis:

Statistical Hypothesis: The mean of the criterion measures is the same for each treatment population.

When should one reject the Statistical Hypothesis? This depends upon the amount of risk one wishes to take. In the case just presented, the researchers ask: If the observed sample mean difference is the 1-in-a-100 that will result from random sampling variation, how much *would* that mistake cost? On the other hand, if the sample results are indicative of the population, how much *would it not* cost to act as if they are? The probability that the present results came from a common population is very small. The researchers themselves must determine if this probability is tolerable. The probability of making a Type I error is completely under the control of the researchers, and is exactly what the researcher sets as the alpha level. (When multiple tests are performed on the same data, an inflated Type I error may occur, and corrections are usually called for. This issue will be discussed later.) Figures 2.2 and 2.3 represent studies in which two samples were drawn from a theoretical population.

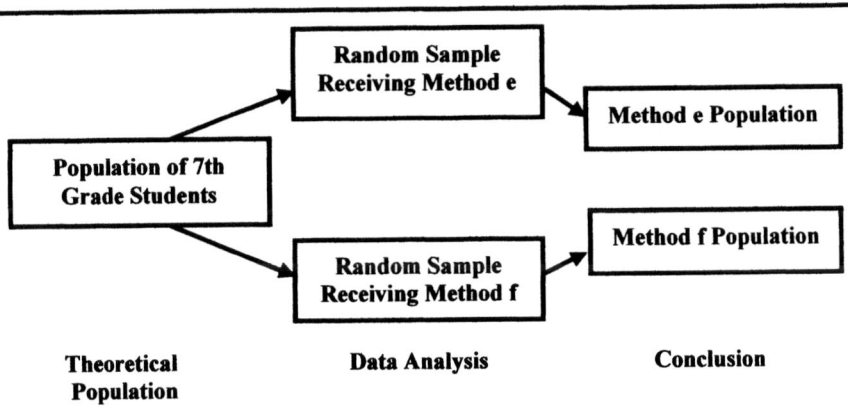

Figure 2.2.
Conclusion when the difference between group means is statistically significant.

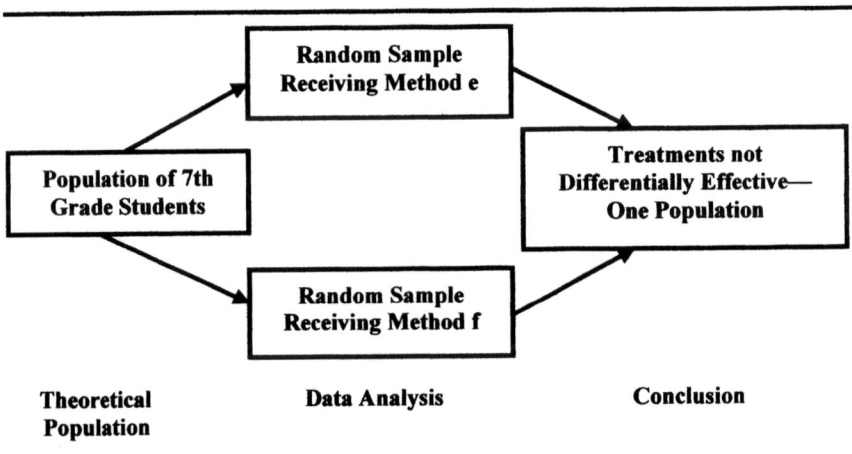

Figure 2.3.
Conclusion when the difference between group means is not statistically significant.

When statistical significance is obtained as a result of differential treatment effects, which is the case for the study presented in Figure 2.2, one can say with some degree of confidence that the two samples no longer represent the same population with respect to the criterion measure of concern. Those two samples now represent two populations, Method e population and Method f population.

The results obtained from the study depicted in Figure 2.3 represent the other state of affairs. Statistical significance has not been obtained, that is, the differential treatments have not produced a large enough difference to conclude that they have different effects on the criterion measure of concern. The two samples now represent the same population.

Type I Error

The decision to reject or fail to reject the Statistical Hypothesis is essentially one of risk taking. Usually, the probability level of either 5 times in 100 (alpha = .05) or 1 time in 100 (alpha = .01) is the decision point used by researchers in the behavioral sciences; that is, they are willing to risk making 5 errors (or 1) out of 100 rejections of a true Statistical Hypothesis (a Type I error). The choice of alpha is up to the decision makers and the degree of risk they wish to take. The level of alpha defines the probability of making a Type I error. Also, the decision makers must recognize that taking a conservative risk (e.g., 1 time in a 1000 instead of 1 time in a 100) regarding Type I error increases the probability of making a Type II error, which is discussed in the next section.

Researchers need to be cognizant of the value of alpha used by researchers in their fields of study. Some content areas have more reliable and valid measures. Hence a more conservative alpha (say, .001) may be reasonable. In less well-researched content areas, a more liberal alpha (say, .10) might be appropriate. When research is published, the decision made by the researcher will become public and undergo public scrutiny. The choice of alpha has to make sense to other researchers in order to be acceptable.

Type II Error

The decision makers have two possible conclusions they can make from the Statistical Hypothesis testing procedure. They can reject the Statistical Hypothesis (i.e., the rival or null hypothesis), or they can fail to reject the Statistical Hypothesis. One hopes that a correct decision is made, but because one never knows the population value(s), one can never be sure that the correct action has been taken. The Type I error, which was previously discussed, involves rejecting a Statistical Hypothesis when it should not be rejected. The researchers have complete control over the probability of error in the choice of the alpha level (when all assumptions are met). If alpha is .05, then 5% of the time the Statistical Hypothesis will be incorrectly rejected (if the Statistical Hypothesis is true).

On the other hand, the other decision—failing to reject the Statistical Hypothesis—also may be in error (Type II error). If the Statistical Hypothesis is not true of the populations, but the sampled data do not provide enough evidence to reject it, then an error is made. The probability of the Type II error decreases as sample size increases, and it also decreases as the alpha (probability of a Type I error) increases. The probability of Type II error also decreases as the discrepancy between the statistically hypothesized population value and the true population value increases. This difference between the statistically hypothesized population value and the true population value, which is expressed in various forms, is often referred to as the effect size.

The true population value is never known; therefore the probability of a Type II error being committed cannot be determined exactly. Because the consequences of making a Type II error are usually less devastating (e.g., failing to accept some new fact that is correct) than the consequences of making a Type I error (e.g., accepting a new fact that is incorrect), the alpha levels chosen by most researchers are low (e.g., .05 or .01, or .001). The content of the Research Hypothesis should always determine the cost of making either a Type I or Type II error. Though the probability of making a Type II error is difficult to calculate, the researchers need to be aware of the concept to realize that when they fail to reject the Statistical Hypothesis they *may* be making an error.

Related to the chance of making a Type II error is the *power* of a given statistical test. Power of a statistical test is equal to one minus the probability of committing a Type II error. In chapter 8 we discuss ways to increase the power of a statistical test, and thus decrease the chance of a Type II error.

Assumptions Related to the *F* Test

A few conditions must be met before the ratio between any two estimates of the population variance is distributed as *F*. There are four basic assumptions of concern.

1. All subjects in the treatment groups were originally drawn at random from the same parent *population*.
2. The variance (v) of the criterion measures is the same for each treatment *population*.
3. The criterion measure for each treatment *population* is normally distributed.
4. The means of the criterion measures are the same for each treatment *population*.

Figure 2.4 illustrates a situation in which Assumptions 2 and 3 are met for the case where two populations are under consideration. (Note that these assumptions refer to the populations.)

Since the populations are never available, the validity of the assumptions is never known for sure. Tests of significance are available to test Assumptions 2 and 3; but when using such tests, the researchers hope to fail to reject the Statistical Hypothesis. This situation is opposite to the usual hypothesis-testing procedure and should require different values of alpha.

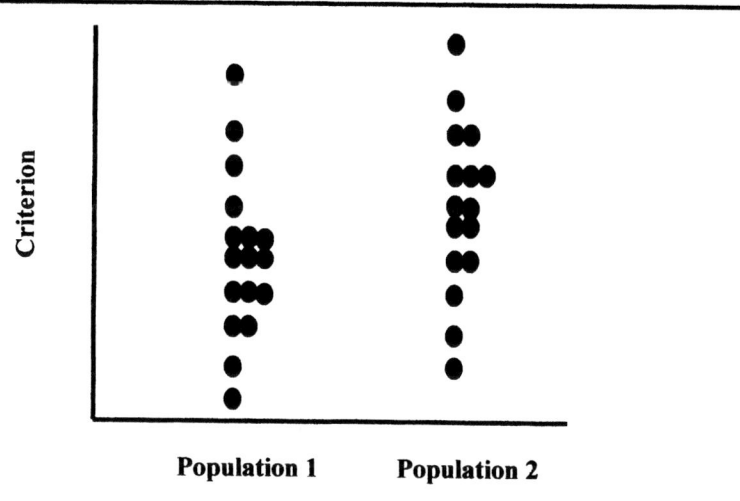

Figure 2.4.
Illustration of two populations that meet Assumptions 2 and 3 but not Assumption 4.

The desired results for the statistical tests of Assumptions 2 and 3 is to show that the *variances are not significantly different* and that the population distribution is not significantly different from a normal distribution, respectively. In both cases, the traditional hypothesis-testing process is reversed, in that an equivalent population state of affairs is desired. The Statistical Hypothesis, instead of the Research Hypothesis, is hoped to be true. If one wants to protect against falsely accepting this hypothesis, then alpha should be set at a much higher level, that is, .25 is the level that some literature suggest, but we take a less liberal stance and encourage use of a much higher alpha, say .60. Since a small number of subjects (i.e., study participants) leads to less likelihood of finding significance, this suggestion is particularly relevant when few subjects are available. If one really wants to establish "no difference," then one should obtain a large N size. Nonsignificance is particularly meaningful with a large N. (See the discussion of power analysis in chapter 8.)

Investigations into these assumptions have generally concluded that Assumptions 2 and 3 usually can be violated without seriously distorting the stated alpha level. The fourth assumption is really the study's Statistical Hypothesis being tested. The position we take in this text (and defend more fully in later chapters) is that attempting to obtain an R^2 close to 1.00, along with replicating findings, is a necessary and sufficient guard against any violation of assumptions.

The assumptions can be viewed as when a significant F is obtained, it could be due to any one of the four (or a combination) of the assumptions not being true. If the researcher is reasonably sure that Assumptions 1, 2, and 3 are true, then Assumption 4 can be rejected. For example, with degrees of freedom values of 1 and 100, an F of 6.90 or greater is observed 1 time in 100 due to sampling variation. If one specifies .01 as the amount of risk one will tolerate to reject a true Statistical Hypothesis, given an F larger than 6.90 and faith in the first three assumptions, one would reject the condition that the criterion means are equal for the two treatment populations. Woehlke, Elmore, and Spearing (1990) demonstrated how the assumptions underlying the General Liner Model can be investigated in the SPSS statistical computer software package, which is one of the two software packages utilized in this text.

Elaboration on Assumptions

The assumptions discussed in the previous section are applicable when one is dealing with treatment groups as in Figure 2.5, and they can be extended to cover a linear fit when the independent (predictor) variable is continuous as illustrated in Figure 2.6. Assumptions 2 and 3 regarding equal variance and normal distribution can applied to both situations.

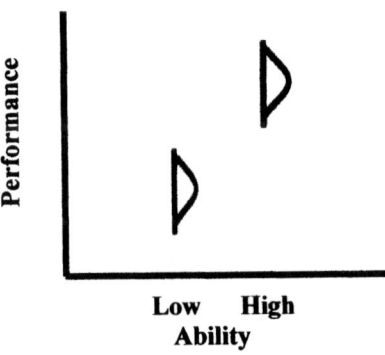

Figure 2.5.
A relationship between two groups and the criterion.

With respect to Figure 2.6 it is assumed that for each data point on the continuum of the predictor variable of ability, the criterion values have equal variance and are normally distributed in the population. Given a sufficiently large sample (say, greater than 100 subjects per variable), it is not too serious if there is a violation of the assumptions of equal variance and normal distribution. (One can investigate the residuals to see if some of the assumptions are tenable, as discussed in the section entitled "Inspection of Residual Plots" in chapter 4.)

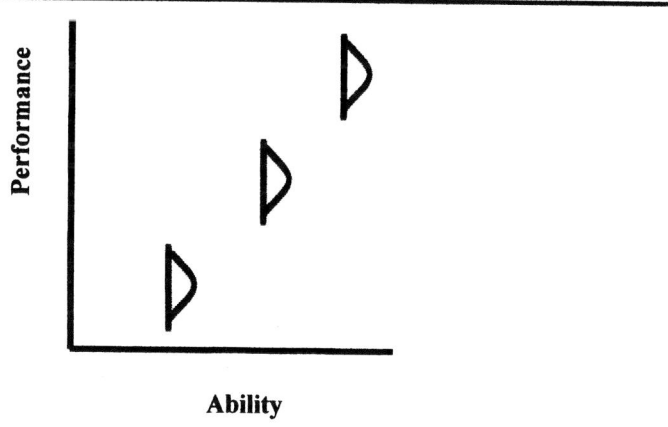

Ability

Figure 2.6.
A linear relationship between two continuous variables.

R^2—THE PROPORTIONAL ESTIMATE OF SUMS OF SQUARES

The F ratio also can be derived using R^2, which is a proportional estimate of the total observed sums of squares for each source of variance. In the two-group case of Method e and Method f, among-group sum of squares and within-group sum of squares values of 10,000 and 2,500 were calculated, respectively. A total observed sum of squares also could have been calculated by subtracting the grand mean from each of the 100 student's scores, squaring the difference scores, and then summing these 100 squared scores. This total sum of squares (SS_t) value would be 12,500, which is equal to the among-group sum of squares value of 10,000 plus the within sum of squares value of 2,500.

The range of R^2 is from .00 to 1.00 and is an expression of the proportion of total sample criterion variance accounted for by the particular set of predictor information. The total sample variance can be calculated by dividing the total sum of squares by the number of subjects (i.e., total variance = SS_t / N). The variance accounted for by knowing group means can be calculated by dividing the sum of squares among by the number of subjects (i.e., variance accounted for by group means = SS_a / N). The value for R_a^2, which uses among-group data, is calculated by using Equation 2.4, which is as follows:

$$R_a^2 = \frac{SS_a}{SS_t}$$

(Equation 2.4)

Substituting the values for SS_a (10,000) and SS_t (12,500) into Equation 2.4 produces the following the R_a^2 value:

$$R_a^2 = \frac{10,000}{12,500} = .80$$

In this example, SS_a is 80% of SS_t. The SS_a relies upon knowledge of which group (i.e., Method e or Method f) the subject is in; thus knowledge of group membership accounts for 80% of the total observed sum of squares (SS_t). The reader should note that such a high R^2 is usually not found when comparing treatment groups in the behavioral sciences.

R_w^2 (which uses within-group data) is calculated by dividing SS_w by SS_t, as indicated by the following equation:

$$R_w^2 = \frac{SS_w}{SS_t} \qquad \text{(Equation 2.5)}$$

For this example, the R_w^2 value is calculated as follows:

$$R_w^2 = \frac{2,500}{12,500} = .20$$

The sum of squares within reflects the variation within the groups that is not directly due to the two treatments. For this example, SS_w is 20% of the total sum of squares; thus 20% of the observed SSt is due to unknown sources.

What does introducing the notion of R^2 provide that was not known when using estimates of the population variance ($\hat{v}_a$ and $\hat{v}_w$)? The R^2 notion provides a different perspective regarding variation and research designed to explain variation. Indeed, this different perspective is one of the fundamental reasons for this text.

The R^2 Perspective

Given any one sample from a population, the best source from which to estimate the population variance is the observed variability of the sample. In the sample of 100 subjects given Methods e and f there was a total sum of squares of 12,500. The estimated population variance ($\hat{v}_t$), based on the sample, was found by dividing SS_t by degrees of freedom (df). The grand mean had to be calculated to get SS_t; therefore, degrees of freedom must be $N - 1$. Once the grand mean is known, only 99 scores are free to vary; the last score is fixed. Thus, $\hat{v}_t$ can be calculated by using Equation 2.6, which is as follows:

$$\hat{v}_t = \frac{SS_t}{N-1}$$ (Equation 2.6)

For this example, $\hat{v}_t$ is calculated as follows:

$$\hat{v}_t = \frac{12,500}{99} = 126.2$$

Using the R^2 perspective, given only a sample and a sample mean, 100% of the variance in the criterion scores is due to unknown sources. A researcher's goal is to find out what accounts for variance in the criterion—why some subjects score high, while other score low. Some of the variance might be due to different treatments, ability, motivation, and so on. The task of research is to seek out information that will account for all of the criterion variance, or for as much of it as is temporarily satisfactory.

In the case of the 100 7th-grade students, an estimated population variance ($\hat{v}_t$) of 126.2 was observed. The researcher then brings into the picture the information that these students were randomly assigned to two different methods of instruction. One might now wonder: Does this treatment information explain some of the criterion variance? As previously noted, R_a^2 was based upon knowledge of which students received which treatment. It was found that this treatment information explained 80% of the total variance ($R_a^2 = .80$).

Now that it is known how much criterion variance can be accounted for by using knowledge of treatment groups, can it be determined how likely it is that this large an R^2 value would be due to sampling error—as was done with the F ratio previously? As a matter of fact, it can; the F ratio can be calculated from R_a^2 and R_w^2 as well as from $\hat{v}_a$ and $\hat{v}_w$.

Using R_a^2 and R_w^2 (two proportional parts of the total sample variance); it can be determined whether the knowledge of group membership (Methods e and f) will explain a non-chance amount of the observed sample variance. The F test can be calculated using R_a^2 and R_w^2 with Equation 2.7:

$$F(df_n, df_d) = \frac{R_a^2 / (df_n)}{R_w^2 / (df_d)}$$ (Equation 2.7)

where:
1. R_a^2 = the proportion of unique observed variance due to knowledge of group membership;
2. R_w^2 = the proportion of unique observed variance due to unknown sources;
3. df_n = the number of group means free to vary once the grand mean has been calculated; and

4. df_d = the number of subjects whose criterion scores are free to vary once each group mean has been calculated.

The definitions for these four components are specific to the comparison of two groups. More general, and useful, definitions will be given in future chapters.

As previously determined, R_a^2 equals .80 and R_w^2 equals .20. The degrees of freedom numerator, (df_n) value is equal 1 (i.e., 2 - 1) because only one of the two group means was free to vary once the grand mean was calculated. The degrees of freedom denominator (df_d) value is equal to 98 (i.e., 100 - 2) because within each group the last (50th) score was not free to vary (i.e., the group mean plus the other 49 scores determined the last student's score).

Substituting these values into Equation 2.7, one obtains the following F value:

$$F(1,\ 98) = \frac{.80/1}{.20/98} = \frac{.80}{.002} = 392.15$$

This F value is exactly the same as obtained by substituting the values for $\hat{v}_a$ (10,000) and $\hat{v}_w$ (2,500) into Equation 2.4.

What has been gained by using the R^2 approach? Not only can one determine how likely it is that the R_a^2 is due to sampling error, but one also gets an idea of what proportion of the sample variance can be explained by the piece of information, in this case group membership. Furthermore, the R_w^2 shows how much ignorance still exists concerning the source of the observed criterion variance. For this example, $R_a^2 = 1 - R_w^2$, so the two points might seem redundant. This will not always be the case. The more general equation for F is developed in chapter 4.

Theoretical and Heuristic Benefits of R^2 in Analysis of Variance

The great amount of variance accounted for by the two methods in the example presented in this chapter is seldom observed in the behavioral sciences. Significant results often translate into extremely small R^2 values because large numbers of subjects can yield statistically significant results yet account for little variance. One study reported statistical significance with the R^2 value of .01. The authors likely thought that R^2 was similar to probability—the lower the better. One could paraphrase these authors' results as "knowing for sure about very, very little."

Heuristically, the use of R^2 keeps one's ignorance in full view because 1- R^2 is the proportion of variance yet to be accounted for in the criterion variable. If theoretical expectations of causal relationships are empirically found to account for less than 50% of the criterion variance, then it seems that additional "relevant" variables must enter into the theoretical model. The determination of variables that are relevant variables is empirically a function of the criterion variable and the other predictor variables that are already in the predictive system.

And, of course, such model building is in the spirit of chapter 1. Researchers should want as much as possible to reduce the unaccounted-for variance—the researcher's true enemy.

Many studies reported in journals give sufficient information to allow the reader to calculate the R^2 value. A preferable procedure would be one in which the researcher reported the R^2 in the published study, but this is not commonly done. R^2 is one way of representing effect size, and as such should be reported in all research. Increasing numbers of researchers are calling for the reporting of effect size, one of which is Thompson (1999). In chapter 4 (see the section entitled "Importance of Reporting Practical Significance") procedures are presented that enable the reader to calculate R^2 values from information commonly found in published work.

CHAPTER 3

VECTORS AND VECTOR OPERATIONS

INTRODUCTION TO VECTORS

The general linear model can be most easily understood within the framework of vector notation. If researchers are to develop a command over the use of GLM, they must become somewhat familiar with vector notation and vector operations. Once the researchers have grasped the basic components of vector algebra they can construct for themselves the models used to test a particular research question. Thus knowledge of vector algebra, coupled with mastery of the use of GLM, allows the researchers to ask the research question in the particular way they want and then to generate the statistical test. This approach may seem heretical to those researchers who were taught that they must frame their research question in a particular fashion or that only certain research questions are permissible.

Most people come into contact with vectors in their everyday lives; thus the mastery of vector algebra is not really a difficult chore. Many properties of vector algebra are simply generalizations from the algebra with which most readers are already familiar.

Why Use Vectors?

A vector is an ordered set of numbers that allows data to be represented in a very concise fashion. While there are other kinds of vectors, for the purposes of this text, a vector is simply an ordered set of numbers. Once the conventions are understood and the symbols are identified, a large amount of data can be represented in a small amount of space. A familiar example of vectors is shown in Table 3.1.

Table 3.1 is one that may be familiar. The same kind of measurement has been made for each entity represented (i.e., the noon temperature has been recorded for each city listed). The collection of temperatures is called a *vector* because each member of the collection is, in fact, a number; and the members (or *elements*) are in some particular order. That is, the temperature of 20 degrees was observed in Chicago, and the temperature of 30 degrees was observed in St. Louis.

Table 3.1
Example of a Single Vector: The Same Measurement
for Each of Several Entities

City	Noon Temperature (Degrees Fahrenheit)
Chicago	20
Miami	80
Pittsburgh	45
St. Louis	30

The data in Table 3.2 show another use of vectors, representing several measurements of a single entity. The information could have come from a biographical data sheet. If one were given the following vector (Y) and told that this vector represents the same type of information as presented in Table 3.2, then one would know that this individual is 72 inches tall, weighs 300 pounds, is 30 years old, has 12 years of education, is not married, and has no children:

$$Y = \begin{bmatrix} 72 \\ 300 \\ 30 \\ 12 \\ 0 \\ 0 \end{bmatrix} = \begin{matrix} \text{height of subject in inches} \\ \text{weight of subject in pounds} \\ \text{age of subject in years} \\ \text{years of education of subject} \\ 1 = \text{married; } 0 = \text{not married} \\ \text{total number of children} \end{matrix}$$

Table 3.2
Example of Several Measurements of a Single Entity

Height (inches)	68
Weight (pounds)	150
Age (years)	25
Schooling (years)	19
Married (yes = 1; no = 0)	1
Children (total number)	1

The data in Table 3.3 represent a combination of the kinds of vectors previously described. Here there are several observations of each of several entities. For the purposes of this text, the focus will be on these kinds of vectors, that is, vectors that represent the same kind of measurement of several subjects. Because there are four pieces of information about each entity or subject in Table 3.3, there are four vectors in Table 3.3. Someone has collected three pieces of information and then calculated a fourth piece of information, namely, batting average, from two of the prior pieces of information (i.e., "hits" divided by "at bats"). This fourth piece of information is considered to be a new piece of information, as discussed later in the chapter.

Table 3.3
Example of Several Vectors for Each Entity (Player)

Player	Home Runs	At Bats	Hits	Batting Average
James	3	120	60	.500
Sam	5	100	30	.300
Dick	15	110	22	.200
Jim	25	100	25	.250
Juan	20	120	48	.400

Definition of a Vector

A vector is an ordered set of numbers. The number of elements in a vector is called the *dimension of a vector.* With reference to Table 3.3 and given the order of the players, it is known that James hit three home runs because James is the subject corresponding to the first element of each vector, and the numerical value of the first element of the home run vector is 3. Also, since .400 is the fifth element of the batting average vector, and Juan is the subject corresponding to the fifth element, Juan's batting average is .400.

We should emphasize here, that is, any number can be an element of a vector. Positive numbers, zeroes, and decimal values have already been used. Negative numbers and common fractions are also valid candidates as elements of vectors.

The other property of a vector, that there is some order, should not be taken lightly. The order does not have to be inherent in the data; in fact, it usually is not. Referring again to Table 3.3 and the hits vector, there seems to be no order. The numbers are neither arranged from high to low nor from low to high. The reader should be well aware, though, that the order has already been defined by the first column of names.

Vectors as Pieces of Information

Some readers may more readily grasp the concept of vectors by thinking of them as pieces of information. One has a vector for each piece of information. How one decides to measure a variable, such as height, will determine how many pieces of information height will provide. If one measures height in inches as a continuous variable, then we know how tall a person is with just one piece of information. If, on the other hand, one measures height in terms of short, medium, and tall, then one would have three pieces of information.

Vector Notation

Vectors in this text will be represented with capital letters, sometimes with additional letters or numbers for further clarification. When a particular element of the vector is referred to, a subscript will be used. The subscript indicates the position of the element in the vector. The hits vector of Table 3.3 could be represented as H. For example, in vector H the particular element H_1 would be the number corresponding to James' hits (60), H_2 would be the number corresponding to Sam's hits (30), and so on. In general terms, the vector W with "t" elements can be represented by the symbol W, or by the bracketed symbols, or by the bracketed numbers; where the three dots mean "and so on continuing to." These three dots are necessary because the value of t is not known; there could be 50 elements of W, or 100, or just 4.

$$
W = \begin{pmatrix} W_1 \\ W_2 \\ W_3 \\ \cdot \\ \cdot \\ \cdot \\ W_t \end{pmatrix} = \begin{pmatrix} 6 \\ 10 \\ 3 \\ \cdot \\ \cdot \\ 17 \end{pmatrix}
$$

Example 3.1—Vector elements. The following vector contains five elements:

$$
X = \begin{pmatrix} 1 \\ 5 \\ 6 \\ 13 \\ -2 \end{pmatrix}
$$

Since this vector is represented by X, the element X_1 is equal to 1, the element X_2 is equal to 5, the element X_3 is 6, the element X_4 is 13, and the element X_5 is -2. Note that the elements X_6, X_7, etc., are not defined for this vector.

Continuous Vectors

When a vector contains more than two numbers, it is referred to as a *continuous vector*. The vector listed in the previous section is an example of a continuous vector. The numbers in the vector may be any one of the following measurement metrics: (a) *ordinal*, meaning that larger numbers have more of the construct; (b) *interval*, meaning that the difference between a "1" and a "2" is the same as the difference between a "2" and a "3"; and (c) *ratio*, meaning that besides the above characteristics, the scores in the vector have a true zero point. While most statistics texts make a point of requiring interval data for inferential statistics, most nevertheless provide many examples with ordinal data. As was discussed in chapter 1, we argue that the purpose of research is to find functional relationships, and therefore the measurement scale of the data is irrelevant. We present examples in chapter 9 and chapter 12 of well-known physical laws that take clearly ratio scales and make them ordinal in order to produce a high R^2.

Categorical Vectors

Many occasions will be found later to use what some call *categorical vectors*. These vectors are usually helpful in representing group membership. One number is assigned to entities that belong to the group under consideration and another number to those entities that do not belong to that group. Any two numbers could be used to designate the two groups, but it is more useful to use a value of 1 if an entity is a member of the group under consideration and a value of 0 if the entity is not a member of that group. *We cannot overemphasize that the judicious use of ones and zeros for representative group membership is a crucial aspect of the GLM.* Most of the statistical literature refers to these as dummy vectors because of the arbitrary assignment of the 1 and 0 values, but we prefer to think of these vectors as "smart vectors," because of their utility.

Example 3.2—Group membership vectors. Categorical vectors can be used to represent group membership not only when the group membership consists of two categories, but also when group membership involves more than two categories. To illustrate, suppose researchers are interested in studying the degree of political liberalism in the college student population, and they are interested in the effect on some behavior of three variables, namely, gender, college status, and marital status. With respect to the behavioral model introduced in chapter 1, the researchers think that the behavior they are interested in, which is the degree of political liberalism, depends on gender, college status, and marital status. The group membership vectors that need to be constructed for these three student characteristics (i.e., gender, college status, and marital status) are as follows:

1. The two gender vectors are:
   S1 = 1 if person is a male, 0 otherwise;
   S2 = 1 if person is a female, 0 otherwise;
2. The four college status vectors are:
   C1 = 1 if person is a freshman, 0 otherwise;
   C2 = 1 if person is a sophomore, 0 otherwise;
   C3 = 1 if person is a junior, 0 otherwise;
   C4 = 1 if person is a senior, 0 otherwise;
3. The two marital status vectors are:
   MYES = 1 if person is married, 0 otherwise; and
   MNO = 1 if person is not married, 0 otherwise.

Two vectors are formed to represent the two categories of gender, and two more vectors are constructed to represent the two categories of marital status, while it takes four vectors to represent the four categories of college status. Thus, three *sets* of vectors are constructed.

It is important for researchers to identify all possibly important categories for a given characteristic. For example, other researchers might not be satisfied with the set of group designations for marital status. They may believe than an additional marital status group could well be that of "divorced." This latter group might provide relevant information and thus increase the predictability of the behavior under consideration. It is the responsibility of the researcher to define the groups that are to be included. The GLM is flexible in that it will handle as many groups as the researcher is willing to define and expend the necessary time and energy on data collection. Even gender can be divided into more than two groups if one is willing to redefine the variable as "gender-role interests" and divide the continuum of responses to such a questionnaire into, say, three or four groups.

The three set of vectors for this illustration are as follows:

Person	S1	S2	C1	C2	C3	C4	MYES	MNO
1	1	0	1	0	0	0	0	1
2	0	1	0	0	0	1	1	0
3	1	0	1	0	0	0	1	0
4	1	0	0	0	0	1	1	0

Given these vectors, previously defined, one knows that the first person (the first element in each vector) is a male, a freshman, and is unmarried. Inspection of the second element in each vector reveals a female senior who is married.

ELEMENTARY VECTOR ALGEBRA OPERATIONS

Addition of Two Vectors

The addition of two vectors is defined as the addition of each element in one vector to the corresponding element in the other vector. An implicit requirement is that both vectors have the same number of elements in order for addition to be possible. The addition of two vectors, A and B, to produce a third vector, C, is written symbolically as:

$$C = A + B = \begin{bmatrix} A_1 \\ A_2 \\ A_3 \\ A_4 \\ A_5 \end{bmatrix} + \begin{bmatrix} B_1 \\ B_2 \\ B_3 \\ B_4 \\ B_5 \end{bmatrix} = \begin{bmatrix} A_1 + B_1 \\ A_2 + B_2 \\ A_3 + B_3 \\ A_4 + B_4 \\ A_5 + B_5 \end{bmatrix} = \begin{bmatrix} C_1 \\ C_2 \\ C_3 \\ C_4 \\ C_5 \end{bmatrix} = C$$

Example 3.3—Addition of two vectors. Consider the following vectors and the addition operation:

$$A = \begin{bmatrix} 1 \\ 0 \\ 3 \\ -4 \end{bmatrix} \qquad B = \begin{bmatrix} 2 \\ 1 \\ 5 \\ 7 \end{bmatrix} \qquad D = \begin{bmatrix} 3 \\ 0 \\ 4 \end{bmatrix}$$

$$C = A + B = \begin{bmatrix} 1 \\ 0 \\ 3 \\ -4 \end{bmatrix} + \begin{bmatrix} 2 \\ 1 \\ 5 \\ 7 \end{bmatrix} = \begin{bmatrix} 1+2 \\ 0+1 \\ 3+5 \\ -4+7 \end{bmatrix} = \begin{bmatrix} 3 \\ 1 \\ 8 \\ 3 \end{bmatrix} = C$$

Vectors A and B could be added because they have the same number of elements, that is, 4. The addition of the vectors B and D is not possible because these two vectors do not have the same number of elements.

Subtraction of Two Vectors

The subtraction of two vectors is as straightforward as the addition of two vectors. To subtract B from A to produce the vector C, one simply subtracts the corresponding elements of B from A.

Example 3.4—Subtraction of two vectors. Consider the following vectors and subtraction operation:

$$A = \begin{bmatrix} 2 \\ 0 \\ 7 \\ -4 \end{bmatrix} \qquad B = \begin{bmatrix} 1 \\ 1 \\ 5 \\ 7 \end{bmatrix} \qquad A - B = C$$

$$C = A - B = \begin{bmatrix} 2 \\ 0 \\ 7 \\ -4 \end{bmatrix} - \begin{bmatrix} 1 \\ 1 \\ 5 \\ 7 \end{bmatrix} = \begin{bmatrix} 2-1 \\ 0-1 \\ 7-5 \\ -4-7 \end{bmatrix} = \begin{bmatrix} 1 \\ -1 \\ 2 \\ -11 \end{bmatrix} = C$$

As was the case for addition, vector B could be subtracted from vector A because they have the same number of elements, that is, 4.

Multiplication of a Vector by a Number

Later in this text it is frequently necessary to multiply a vector by a number, so it is important to become familiar with this operation. The multiplication of a vector by a number is accomplished by multiplying each element of the vector by that number.

C = k * A (where k is a constant; and the "*" symbol indicates multiplication) is computed by multiplying each element of A by the constant k.

$$C = k * A = \begin{pmatrix} k * A_1 \\ k * A_2 \\ k * A_3 \\ . \\ . \\ . \\ k * A_T \end{pmatrix} = C$$

Example 3.5—Multiplication of a vector by a number. Suppose that several weight observations have been made on several entities, and the data need to be changed from the original unit of pounds to the new unit of ounces. Thus, because there are 16 ounces per pound, each pound observation must be multiplied by the constant 16. This operation can be represented in vector notation as:

Z = 16 * P

where the vector P represents the original observations in terms of pounds, and the vector Z represents the observations in the new units, ounces. Suppose that the original vector appeared as follows:

$$P = \begin{pmatrix} 2 \\ 4 \\ 6 \\ 3 \\ 0 \end{pmatrix}$$

Then the new vector Z would be:

$$Z = 16 * P = 16 * \begin{pmatrix} 2 \\ 4 \\ 6 \\ 3 \\ 0 \end{pmatrix} = \begin{pmatrix} 16 * 2 \\ 16 * 4 \\ 16 * 6 \\ 16 * 3 \\ 16 * 0 \end{pmatrix} = \begin{pmatrix} 32 \\ 64 \\ 96 \\ 48 \\ 0 \end{pmatrix} = Z$$

The reader should note that it makes sense to multiply every element of the pounds vector by 16 (the constant) because there are 16 ounces in each pound and each of the subject's pound information needs to be transformed to ounce information. The fourth element of vector Z, for instance, should represent (in ounces) the number of pounds corresponding to the fourth element of P. The fourth element of Z (48 ounces) is equivalent to the fourth element of P (3 pounds).

Example 3.6—The conciseness of multiplying a vector by a number. Another example is introduced here to illustrate the simplicity and conciseness of vector algebra. Suppose one wants to change vector of 100 observations reported in units of feet to units in terms of inches. Given that F is the vector of observations in feet, the vector I is the new vector of observations in inches, and k is the multiplication constant, which is 12, because there are 12 inches per foot. This operation can be represented with vectors in the following fashion:

$$I = k * F = 12 * F$$

Again, there is not just one multiplication implied by the above expression; every element of F is multiplied by the constant k to produce the corresponding element in I.

USEFUL PROPERTIES OF VECTORS

Combining the above knowledge of vectors with some knowledge about the properties of ordinary numbers, five useful properties can be expressed.

Property 1—A Vector Subtracted from Itself

The subtraction of a vector from itself yields the null vector, a vector of all zeros.

$$X + (-1) X = X - X = 0$$

The "0" symbol represents a vector in which every element is equal to 0. Such a vector is often called the *null vector*. The property becomes useful when one has an occasion to subtract a vector from itself. The result of subtracting a vector from itself is a vector in which every element is equal to zero, or the null vector 0, which can be illustrated as follows:

$$\left(\begin{bmatrix} X_1 \\ X_2 \\ X_3 \\ \cdot \\ \cdot \\ \cdot \\ X_t \end{bmatrix} + (-1) \begin{bmatrix} X_1 \\ X_2 \\ X_3 \\ \cdot \\ \cdot \\ \cdot \\ X_t \end{bmatrix}\right) = \left(\begin{bmatrix} X_1 \\ X_2 \\ X_3 \\ \cdot \\ \cdot \\ \cdot \\ X_t \end{bmatrix} - \begin{bmatrix} X_1 \\ X_2 \\ X_3 \\ \cdot \\ \cdot \\ \cdot \\ X_t \end{bmatrix}\right) = \begin{bmatrix} X_1 - X_1 \\ X_2 - X_2 \\ X_3 - X_3 \\ \cdot \\ \cdot \\ \cdot \\ X_t - X_t \end{bmatrix} = \begin{bmatrix} 0 \\ 0 \\ 0 \\ \cdot \\ \cdot \\ \cdot \\ 0 \end{bmatrix}$$

Property 2—Multiplication of the Sum of Multiple Vectors by a Constant

Multiplication of the sum of two or more vectors by a constant is equivalent to the multiplication of each vector by the constant and then the addition of the resulting products, as shown below (where a is the constant).

$$a * (X + Y) = (a * X) + (a * Y) = aX + aY$$

$$a * \left(\begin{pmatrix} X_1 \\ X_2 \\ X_3 \\ \cdot \\ \cdot \\ \cdot \\ X_t \end{pmatrix} + \begin{pmatrix} Y_1 \\ Y_2 \\ Y_3 \\ \cdot \\ \cdot \\ \cdot \\ Y_t \end{pmatrix} \right) = a * \begin{pmatrix} X_1 + Y_1 \\ X_2 + Y_2 \\ X_3 + Y_3 \\ \cdot \\ \cdot \\ \cdot \\ X_t + Y_t \end{pmatrix} = \begin{pmatrix} a*(X_1 + Y_1) \\ a*(X_2 + Y_2) \\ a*(X_3 + Y_3) \\ \cdot \\ \cdot \\ \cdot \\ a*(X_t + Y_t) \end{pmatrix}$$

$$= \begin{pmatrix} aX_1 + aY_1 \\ aX_2 + aY_2 \\ aX_3 + aY_3 \\ \cdot \\ \cdot \\ aX_t + aY_t \end{pmatrix} = \left(\begin{pmatrix} aX_1 \\ aX_2 \\ aX_3 \\ \cdot \\ \cdot \\ aX_t \end{pmatrix} + \begin{pmatrix} aY_1 \\ aY_2 \\ aY_3 \\ \cdot \\ \cdot \\ aY_t \end{pmatrix} \right) = \left(a \begin{pmatrix} X_1 \\ X_2 \\ X_3 \\ \cdot \\ \cdot \\ X_t \end{pmatrix} + a \begin{pmatrix} Y_1 \\ Y_2 \\ Y_3 \\ \cdot \\ \cdot \\ Y_t \end{pmatrix} \right) = aX + aY$$

Example 3.7—Simplification of two vectors multiplied by the same constant. Two of the Graduate Record Exam (GRE) sections are often added to represent the Total GRE score. Suppose that one wanted to multiply the total scores on the GRE of five subjects by a constant of 3. As illustrated below, one could multiply both the verbal (V) and quantitative (Q) sections by 3 and add these products; or one could first sum the verbal and quantitative sections and then multiply this sum (T) by the constant 3.

$$\text{Given: } V = \begin{pmatrix} 200 \\ 250 \\ 300 \\ 400 \\ 500 \end{pmatrix} \quad Q = \begin{pmatrix} 400 \\ 450 \\ 500 \\ 600 \\ 700 \end{pmatrix} \quad T = (V + Q) = \begin{pmatrix} 600 \\ 700 \\ 800 \\ 1000 \\ 1200 \end{pmatrix}$$

$$3 * T = \begin{pmatrix} 600 \\ 700 \\ 800 \\ 1000 \\ 1200 \end{pmatrix} = \begin{pmatrix} 1800 \\ 2100 \\ 2400 \\ 3000 \\ 3600 \end{pmatrix}$$

$$3*V+3*Q = \left(3* \begin{pmatrix} 200 \\ 250 \\ 300 \\ 400 \\ 500 \end{pmatrix} \quad 3* \begin{pmatrix} 400 \\ 450 \\ 500 \\ 600 \\ 700 \end{pmatrix} \right) = \left(\begin{pmatrix} 600 \\ 750 \\ 900 \\ 1200 \\ 1500 \end{pmatrix} + \begin{pmatrix} 1200 \\ 1350 \\ 1500 \\ 1800 \\ 2100 \end{pmatrix} \right) = \begin{pmatrix} 1800 \\ 2100 \\ 2400 \\ 3000 \\ 3600 \end{pmatrix}$$

The addition of the two vectors, before multiplying by the constant, is the most frequently used option. It involves one less mathematical operation than if one were to multiply both vectors by the constant and then add the results.

The primary reason for introducing these properties is that it will become necessary later in this chapter to reduce the number of vectors. Although the two sides of Property 2 are numerically the same, the left-hand side of Property 2 [a * (X + Y)], identifies only one vector, while the right-hand side identifies two vectors, [(a * X) + (a * Y)].

Property 3—Multiplication of a Vector by the Sum of Two Constants

Multiplication of a vector by the sum of two constants is equivalent to separately multiplying that vector by each constant, and then summing, as follows:

$$(a + b) * X = (a * X) + (b * X) = aX + bX$$

$$(a + b) * X = \begin{pmatrix} (a + b)X_1 \\ (a + b)X_2 \\ (a + b)X_3 \\ . \\ . \\ . \\ (a + b)X_t \end{pmatrix} = \begin{pmatrix} aX_1 + bX_1 \\ aX_2 + bX_2 \\ aX_3 + bX_3 \\ . \\ . \\ . \\ aX_t + bX_t \end{pmatrix} = \begin{pmatrix} aX_1 & + & bX_1 \\ aX_2 & + & bX_2 \\ aX_3 & + & bX_3 \\ & . & \\ & . & \\ & . & \\ aX_t & + & bX_t \end{pmatrix} = aX + bX$$

This illustration shows that when a vector is to be multiplied by the sum of two constants, one can either add the two constants together and then multiply the vector by the sum of the two constants or multiply the vector by the two separate constants and then add the resultant products. Adding the two constants first is much easier and quicker because fewer operations are involved.

Property 3 can be used to reduce the number of vectors. The left-hand side of Property 3 is simpler than the right-hand side, though they are numerically equal. The left-hand side clearly shows that there is only one vector; whereas the right-hand side contains two vectors that can be reduced, or simplified, to one vector. More on these notions will be presented in the section on linear dependencies.

Property 4—Multiplication of a Vector by Zero

Multiplication of a vector by zero yields a vector with all elements equal to zero, the null vector, as shown below.

$$(0) * X = 0$$

$$(0) * X = 0 * \begin{bmatrix} X_1 \\ X_2 \\ X_3 \\ \cdot \\ \cdot \\ X_t \end{bmatrix} = \begin{bmatrix} 0 * X_1 \\ 0 * X_2 \\ 0 * X_3 \\ \cdot \\ \cdot \\ 0 * X_t \end{bmatrix} = \begin{bmatrix} X_1 \\ X_2 \\ X_3 \\ \cdot \\ \cdot \\ X_t \end{bmatrix} = 0$$

No matter what elements the vector contains, if one multiplies the vector by zero, one will end with the null vector as the product.

Property 5—Multiplication of a Vector by One

Multiplication of a vector by one yields that same vector. Multiplication of any vector by the constant one (1) yields the same vector, as shown below.

$$1 * X = X$$

$$(1) * X = 1 * \begin{bmatrix} X_1 \\ X_2 \\ X_3 \\ \cdot \\ \cdot \\ X_t \end{bmatrix} = \begin{bmatrix} 1 * X_1 \\ 1 * X_2 \\ 1 * X_3 \\ \cdot \\ \cdot \\ 1 * X_t \end{bmatrix} = \begin{bmatrix} X_1 \\ X_2 \\ X_3 \\ \cdot \\ \cdot \\ X_t \end{bmatrix} = X$$

These last two properties may seem trivial, and indeed they are—but they are useful. The reader is encouraged to understand all the above five properties before proceeding. These properties of vectors, coupled with the idea of linear combinations to be discussed in the next section, form the structure of the GLM. The more adept one becomes with the ideas presented in the present chapter, the more adequately one can handle the building and simplification of linear regression models.

LINEAR COMBINATIONS OF VECTORS

Linear Combinations of Two Vectors

There will be many situations later where vectors will be combined and where it will be necessary to figure out if certain vectors are, in fact, linear combinations of other vectors. Therefore, the important idea of linear combinations needs to be defined.

Vector X is said to be a *linear combination* of vectors Y and Z if there exist two numbers (numerical constants called *weighting coefficients*), a and b (of which at least one is not zero), such that the following relationship holds:

$$X = (a * Y) + (b * Z)$$

This definition may become more understandable with the examples presented in this section and the following sections.

Example 3.8—Determining linear combinations. Consider the following vectors:

$$X = \begin{bmatrix} 3 \\ 4 \\ 5 \end{bmatrix} \qquad Y = \begin{bmatrix} 1 \\ 2 \\ 3 \end{bmatrix} \qquad Z = \begin{bmatrix} 1 \\ 0 \\ -1 \end{bmatrix}$$

Vector X is a linear combination of vectors Y and Z because:

$$X = (a * Y) + (b * Z), \text{ when } a = 2 \text{ and } b = 1, \text{ that is,}$$

$$(2 * Y) + (1 * Z) = 2 * \begin{bmatrix} 1 \\ 2 \\ 3 \end{bmatrix} + 1 * \begin{bmatrix} 1 \\ 0 \\ -1 \end{bmatrix} = \begin{bmatrix} 2+1 \\ 4+0 \\ 6-1 \end{bmatrix} = \begin{bmatrix} 3 \\ 4 \\ 5 \end{bmatrix} = X$$

Vector Y is a linear combination of vectors X and Z because:

$$Y = (c * X) + (d * Z), \text{ when } c = .5 \text{ and } d = -.5.$$

Thus,

$$(.5 * X) + (-.5 * Z) = .5 * \begin{bmatrix} 3 \\ 4 \\ 5 \end{bmatrix} + (-.5) * \begin{bmatrix} 1 \\ 0 \\ -1 \end{bmatrix} = \begin{pmatrix} 1.5 - 0.5 \\ 2.0 - 0.0 \\ 2.5 + 0.5 \end{pmatrix} = \begin{bmatrix} 1 \\ 2 \\ 3 \end{bmatrix} = Y$$

Therefore vector Y is a linear combination of vectors X and Z.

Some Special Linear Combinations of Vectors

A total test score vector that is computed by simply adding the two subtest scores is a linear combination of the two subtest vectors. In this instance, the weighting coefficients (i.e., a and b) are both equal to 1.

Example 3.9—Total score as a linear combination of subtest scores. Consider total GRE score, which is computed by adding the verbal and quantitative GRE subtest scores. Given the verbal vector (V) and the quantitative vector (Q):

$$V = \begin{bmatrix} 400 \\ 500 \\ 600 \\ 350 \end{bmatrix} \quad Q = \begin{bmatrix} 400 \\ 300 \\ 400 \\ 750 \end{bmatrix}$$

The total GRE vector (T) is formed as follows:

$$T = (1 * V) + (1 * Q) = V + Q$$

Therefore:

$$T = (V + Q) = \begin{bmatrix} 400 \\ 500 \\ 600 \\ 350 \end{bmatrix} + \begin{bmatrix} 400 \\ 300 \\ 400 \\ 750 \end{bmatrix} = \begin{bmatrix} 800 \\ 800 \\ 1000 \\ 1100 \end{bmatrix} = T$$

Another special case of a linear combination occurs when a vector is multiplied by a number. The weight of the "second vector" in this instance is zero; and because it is zero, the elements of the second vector are of no consequence. That is, it does not matter what the elements of the second vector are because it is already known that multiplication of any vector by zero will yield the null vector.

Example 3.10—Linear combination of two vectors. Consider the following vectors and operations:

$$A = \begin{bmatrix} 4 \\ 3 \\ 0 \\ 1 \end{bmatrix} \qquad B = \begin{bmatrix} B_1 \\ B_2 \\ B_3 \\ B_4 \end{bmatrix} \qquad C = \begin{bmatrix} 24 \\ 18 \\ 0 \\ 6 \end{bmatrix}$$

$$C = (6 * A) + (0 * B)$$

$$C = 6 * \begin{bmatrix} 4 \\ 3 \\ 0 \\ 1 \end{bmatrix} + 0 * \begin{bmatrix} B_1 \\ B_2 \\ B_3 \\ B_4 \end{bmatrix} = \begin{bmatrix} 6*4 & + & 0*B_1 \\ 6*3 & + & 0*B_2 \\ 6*0 & + & 0*B_3 \\ 6*1 & + & 0*B_4 \end{bmatrix}$$

$$C = \begin{bmatrix} 24 & + & 0 \\ 18 & + & 0 \\ 0 & + & 0 \\ 6 & + & 0 \end{bmatrix} = \begin{bmatrix} 24 \\ 18 \\ 0 \\ 6 \end{bmatrix}$$

The vector C is a linear combination of the vector A. The idea of a linear combination of vectors is not restricted to just two vectors. A vector may be a linear combination of more than two vectors. The following example will help to clarify this point.

Example 3.11—Linear combination of several vectors. Consider the following vectors:

$$A = \begin{bmatrix} 4 \\ 3 \\ 2 \\ 1 \end{bmatrix} \quad B = \begin{bmatrix} 1 \\ 0 \\ 0 \\ 0 \end{bmatrix} \quad C = \begin{bmatrix} 0 \\ 1 \\ 0 \\ 0 \end{bmatrix} \quad D = \begin{bmatrix} 0 \\ 0 \\ 1 \\ 0 \end{bmatrix} \quad E = \begin{bmatrix} 0 \\ 0 \\ 0 \\ 1 \end{bmatrix}$$

Vector A is a linear combination of vectors B, C, D, and E because:

$$A = (4 * B) + (3 * C) + (2 * D) + (1 * E)$$

The reader should verify the above statement by carrying out the implied multiplications. The vector B is not a linear combination of vectors C, D, and E because no weighting coefficients exist such that:

$$B = (a * C) + (b * D) + (c * E)$$

Mutually Exclusive Group Membership Vectors

Another special linear combination of vectors occurs when mutually exclusive group membership vectors are added.

Example 3.12—Representation of mutually exclusive vectors. Suppose that one had occasion to deal with the variables of gender and marital status. The group membership vectors may be defined as follows:

1. SF = 1 if subject is female, 0 otherwise;
2. SM = 1 if subject is male, 0 otherwise;
3. MY = 1 if subject is married, 0 otherwise; and
4. MN = 1 if subject is not married, 0 otherwise.

The values for these vectors are as follows:

	SF	SM	MY	MN
Sam	0	1	1	0
Sue	1	0	1	0
Sally	1	0	0	1
Jane	1	0	0	1
Jack	0	1	0	1
Joe	0	1	1	0

Vectors SM and SF are mutually exclusive group membership vectors, that is, all subjects belong to one or the other categories of male and female. Also, the married and not married categories exhaust all the possibilities of marital status (as far as the present researchers are concerned). Other categories of marital status could have been included, but evidently the researchers' question did not require any additional categories.

One way of checking to determine if the stated categories are, in fact, mutually exclusive is to compute the linear combination (using all weights equal to 1), of the vectors under consideration. If these vectors are in fact mutually exclusive, then they will consider each subject once and only once; that is, group membership vectors are represented by ones and zeros, and the resultant sum of the mutually exclusive group membership vectors will yield a vector with all elements equal to 1.

THE UNIT VECTOR

A vector with all its elements equal to 1 is called the unit vector and is symbolized as U. *Due to the frequent use of the unit vector, the symbol U is reserved for that vector.* Consider adding the two gender vectors presented in the previous section. Here a linear combination is being computed because the weights can be thought of as being equal to one (1) as in Example 3.9. To illustrate, consider the following:

$$(1 * SF) + (1 * SM) = \begin{bmatrix} 0 \\ 1 \\ 1 \\ 1 \\ 0 \\ 0 \end{bmatrix} + \begin{bmatrix} 1 \\ 0 \\ 0 \\ 0 \\ 1 \\ 1 \end{bmatrix} = \begin{pmatrix} 0+1 \\ 1+0 \\ 1+0 \\ 1+0 \\ 0+1 \\ 0+1 \end{pmatrix} = \begin{bmatrix} 1 \\ 1 \\ 1 \\ 1 \\ 1 \\ 1 \end{bmatrix} = U$$

Thus, vectors SF and SM are mutually exclusive because their sum is equal to the unit vector (U).

Vectors MY and MN also are mutually exclusive, which can be demonstrated as follows:

$$(1 * MY) + (1 * MN) = \begin{bmatrix} 1 \\ 1 \\ 0 \\ 0 \\ 0 \\ 1 \end{bmatrix} + \begin{bmatrix} 0 \\ 0 \\ 1 \\ 1 \\ 1 \\ 0 \end{bmatrix} = \begin{pmatrix} 1+0 \\ 1+0 \\ 0+1 \\ 0+1 \\ 0+1 \\ 1+0 \end{pmatrix} = \begin{bmatrix} 1 \\ 1 \\ 1 \\ 1 \\ 1 \\ 1 \end{bmatrix} = U$$

The reader should now have a good feeling for the fact that *the unit vector can be considered a linear combination of mutually exclusive group membership vectors*. In fact, the unit vector was relied upon to define mutually exclusive group membership vectors; so the above statement is simply a consequence of that definition—but a very important consequence, as will be shown in later chapters. The unit vector is assumed by computer programs to be in every regression model, so the relationship of the unit vector to the other vectors needs to be known.

PIECES OF INFORMATION

Linear Dependency

The idea of linear dependency is very important. It can be easily introduced at this point because it deals with linear combinations of vectors. A *linear dependency* occurs when one vector in a set of vectors can be expressed as a linear combination of the other vectors. Such a vector is said to be *linearly dependent* upon the other vectors. A linearly dependent vector, because it can be expressed in terms of other vectors, is redundant information—it is not a new piece of information. As such, it is not useful in terms of predicting behavior.

Example 3.13—Linearly dependent vectors. In Example 3.9, a total GRE score was expressed as the sum of the two subtest scores. The reader should verify that the verbal subtest (V) can be expressed as the total GRE (T) minus the quantitative subtest (Q), which can be expressed as $V = (1 * T) + (-1 * Q)$. It is also true that the quantitative subtest is a linear combination of the total GRE and the verbal subtest, which can be expressed as Q = (1 * T) + (-1 * V).

Any one of the three vectors in Example 3.9 is linearly dependent on the other two because it can be expressed as a linear combination of the other two. The total GRE can be expressed neither in terms of the verbal subtest alone or in terms of the quantitative subtest alone. That is, no weight (a) can be found such that: T = a * V, nor such that T = a * Q. Also, no weight (a) can be found such that V = a * Q.

Linear Independency

The discussion so far has centered on linearly dependent vectors. But one will generally want to figure out how many vectors in the set of vectors are linearly *independent*, which is the opposite of linearly dependent. If there is one linear dependency in a set of three vectors, then two vectors are linearly independent. In Example 3.14, the final result is a set of two vectors of information that is said to be *linearly independent*.

A vector is said to be *linearly independent* if that vector cannot be expressed as a linear combination of the other vectors in the set. If a vector can be expressed as a linear combination of the other vectors in the set, then it is redundant information and as such must be eliminated from the set of vectors when attempting to figure out the number of linearly independent vectors in a set of vectors. The following example is intended to clarify the determination of the number of linearly independent vectors.

Example 3.14—Determining the number of linearly independent vectors. Suppose that we have four pieces of information that come from a common instrument. Information that comes from the same place is often identified with the same capital letter and given a unique number after that letter as follows:

$$V1 = \begin{bmatrix} 1 \\ 2 \\ 3 \\ 4 \\ 5 \end{bmatrix} \quad V2 = \begin{bmatrix} 2 \\ 4 \\ 6 \\ 8 \\ 10 \end{bmatrix} \quad V3 = \begin{bmatrix} 4 \\ 8 \\ 12 \\ 16 \\ 20 \end{bmatrix} \quad V4 = \begin{bmatrix} 1 \\ 2 \\ 3 \\ 4 \\ 4 \end{bmatrix}$$

Vector V2 is a linear combination of the other vectors, because each element of V2 is twice the value of the corresponding element in vector V1. And therefore weighting coefficients can be found to form the following equality:

V2 = (2 * V1) + (0 * V3) + (0 * V4) or V2 = 2 * V1

Therefore, V2 is eliminated from the set of vectors for determining the number of linearly independent vectors in the set. There is now a potential set of three linearly independent vectors, V1, V3, and V4. But vector V3 is a linear combination of the remaining vectors as illustrated below:

$$V3 = (4 * V1) + (0 * V4) \text{ or } V3 = 4 * V1$$

Therefore, vector V3 is eliminated from the set of vectors for determining the number of linearly independent vectors in the set.

Two vectors, V1 and V4, remain. The problem is to find a weight (a) such that V1 = a * V4. Note that the first four elements of vector V4 must be multiplied by a weight of 1, whereas the fifth element must be multiplied by a weight of 1.25, in order to equal the elements of vector V1. Thus, there is no single weight that will suffice. Vectors V1 and V4 are thus linearly independent. Therefore, there are two *linearly independent* vectors in the set of vectors in Example 3.14.

One could have first eliminated vector V1 from the set of four vectors, because V1 = (.5 * V2) + (0 * V3) + (0 * V4). Then vector V3 could be eliminated because it is a linear combination of vectors V2 and V4, that is, V3 = (2 * V2) + (0 * V4). Examine whether vector V2 is linearly dependent upon vector V4. For the first four elements, the weighting coefficient would be 2, but for the last element, the weighting coefficient would have to be 2.5. Since vector V2 cannot be shown to be a linear combination of vector V4, there are two linearly independent vectors in the set of vectors in Example 3.14.

It also can be shown that V3 and V4 are two linearly independent vectors in the set of vectors in Example 3.14. It does not matter which two vectors remain in the set; the crucial point is that, in this set of four vectors, only two vectors contain new information. The other two vectors contain redundant information and thus would not increase predictability if used in a prediction equation.

Complexity

Complexity is operationally defined in the GLM as the number of linearly independent vectors. Thus, in Example 3.14, the level of complexity is two—two pieces of information exist—but in the following example, the level of complexity is five.

Example 3.15—Complexity and the number of linearly independent vectors. Consider the following vectors:

S1	S2	C1	C2	C3	C4	U
1	0	1	0	0	0	1
0	1	1	0	0	0	1
0	1	0	1	0	0	1
1	0	0	1	0	0	1
0	1	0	0	0	1	1
1	0	0	0	0	1	1

The first two vectors (i.e., S1 and S2) are the two gender vectors, and the next four vectors (i.e., C1, C2, C3, and C4) are the four class-rank vectors. Whenever dichotomous vectors, such as these vectors, are being considered with the unit vector (i.e., U), it is beneficial to leave the unit vector in the set. Of these seven vectors, two are linearly dependent. Vectors S1 and C1 are equal to the following:

$$S1 = (1 * U) + (-1 * S2)$$
$$C1 = (1 * U) + (-1 * C2) + (-1 * C3) + (-1 * C4)$$

Therefore, C1 and S1 are two linear dependencies in the set. No other vectors can be eliminated, and there are five pieces of information in the set of seven vectors. It is important to understand that one group membership vector in each mutually exclusive group can always be eliminated if the unit vector is in the initial set of vectors, which is the case when researchers use the SPSS for Windows and Microsoft Excel computer statistical software packages.

Most linear dependencies can be identified by knowing the variables and how they are defined. Indeed, one would seldom want to look at the actual data and attempt to find the weighting coefficients.

Suppose the following vectors were being considered:
1. U = 1 for all subjects;
2. M = 1 if male, 0 otherwise;
3. F = 1 if female, 0 otherwise;
4. X3 = 1 if freshman, 0 otherwise;
5. X4 = 1 if sophomore, 0 otherwise;
6. X5 = 1 if junior, 0 otherwise;
7. X6 = 1 if senior, 0 otherwise;
8. MA = math achievement Test A;
9. MB = math achievement Test B;
10. IQV = Verbal IQ;
11. IQNV = Nonverbal IQ; and
12. IQTOT = Total IQ (Verbal IQ plus Nonverbal IQ)

One of the gender vectors (M or F) can be eliminated; one of the college status vectors (X3, X4, X5, or X6) can be eliminated; and one of the IQ vectors can be eliminated. One probably would eliminate the total IQ, so that direct measures of both verbal and nonverbal IQ would remain in the analysis. Note that although two math achievement tests are being considered, it is highly unlikely that these two tests would be providing perfectly redundant information. To do so, they would have to be perfectly correlated, which is a very unlikely state of affairs. It could be that one of the math tests, though, is empirically linearly dependent upon the whole set of vectors. Again, this is not a *likely* state of affairs but a *possible* one. It is sometimes difficult to figure out linear dependencies by inspection of the model. Indeed, the data themselves determine whether a vector is linearly dependent. Fortunately, the computer solution will verify the actual number of linearly independent vectors, although one would want to rely on that procedure as a last resort. In the prediction of a given criterion, the number of nonzero weighting coefficients will be the number of linearly independent pieces of information.

In later chapters, the prediction of a criterion variable using a set of predictor variables will be discussed. The notion of linear dependencies will be used to eliminate redundancies from the predictor set. It is the case, though, that one of the major goals of research is to find weighting coefficients such that the criterion variable is linearly dependent upon the predictor set. It is therefore desirable to have linear dependency when considering the prediction of a criterion variable, whereas it is not desirable to have linear dependency when considering predictor variables.

CHAPTER 4

RESEARCH HYPOTHESES EMPLOYING

DICHOTOMOUS PREDICTOR VARIABLES

Two major concepts are presented in this chapter. First, Research and Statistical Hypotheses that involve dichotomous predictor variables and continuous criterion variables are examined along with the Full and Restricted Models used to statistically test those hypotheses. Second, the techniques used to statistically test Full Models that include multiple dichotomous predictor variables are presented. The techniques examined include: (a) the overall test of the differences among group means, (b) orthogonal contrasts, (c) a priori tests, and (d) post hoc comparisons. The use of dichotomous predictor variables in conjunction with a dichotomous criterion variable will be discussed in chapter 11.

THE STRUCTURE OF A DICHOTOMOUS VARIABLE

Researchers will encounter situations that require the use of one or more dichotomous variables. While a continuous variable contains numerous values from a measured continuum, a dichotomous variable is one that contains only two values. Although any two values could be used, the values 1 and 0 are usually used for various reasons, which will be discussed later in this chapter.

The 1 and 0 values in a dichotomous variable can represent a real dichotomy, such as 1 if alive, 0 if dead; or the dichotomy can be artificial, such as 1 if tall and 0 if short. Most dichotomies are artificial, the dividing line being the result of some arbitrary decision. One must keep in mind that most phenomena in the real world are of a continuous nature, and that imposing arbitrary boundaries probably will decrease predictability. More will be presented on this notion in later chapters.

RESEARCH HYPOTHESES INVOLVING TWO GROUPS

Research Hypotheses involving two treatments use dichotomous variables. Each of the treatments can be identified by a dichotomous "group membership vector." We first consider the simplest and most widely used situation involving dichotomous predictors—the two-group study. In such a study it is best if the researchers are able to randomly select and assign the study's participants to the two different treatments (groups) for reasons discussed in chapter 10. Researchers can use dichotomous variables, however, to analyze data to detect change even when the random sampling and random assignment have not been used. Researchers know, however, that in such studies the ability to attribute the change to a specific cause may be limited.

Derivation of the Full Model

The directional Research Hypothesis posed by the researchers for a study designed to access the effectiveness of two instructional methods regarding student math achievement may be stated as follows:

Directional Research Hypothesis: For a given population, the mean math achievement score for the experimental treatment is higher than the math achievement score for the comparison treatment.

To assess whether this Research Hypothesis is supported by the data collected by the researchers, a Full Model must be developed to allow each treatment (group) to have its own math achievement mean, which is accomplished through the use of two dichotomous predictor variables. The Full Model is constructed as follows:

$$Y = a_1 G1 + a_2 G2 + E_1 \qquad\qquad \text{(Model 4.1)}$$

where:
1. Y = the criterion variable of math achievement;
2. $G1 = 1$ if criterion score is from a student in experimental method, 0 otherwise;
3. $G2 = 1$ if criterion score is from a student in comparison method, 0 otherwise; and
4. a_1 and a_2 are least squares weighting coefficients calculated to minimize the sum of the squared values in the error vector E_1.

This model, which does not contain the unit vector, contains two linearly independent pieces of information in the form of the two dichotomous predictor variables. We will label the number of pieces of information in the Full Model as m1.

After formulating the Full Model, the task then becomes finding values for the a_1 and a_2 coefficients such that the sum of the squared values in E_1 is minimized. This task is generally accomplished by a computer solution, but several values for the coefficients are used in Table 4.1 to illustrate the task. Clearly, if all three students in G1 score 5 on the criterion test, the coefficient a_1 ought to be 5. Note that the value of 5 produces no error in prediction ($E_1 = 0$) for any of the students in G1. The next concern is to find the numerical value of the a_2 coefficient. One could use the most frequently occurring score, which is the value of 6. Using 6, one obtains perfect prediction for two of the three students, but over predicts by three units for the first student in G2. By using the mean score for students in G2 as the coefficient, one does not perfectly predict for any subject. But the objective for a coefficient in most statistical analyses, which is to achieve a minimum value for the squared error values, is satisfied when the group mean of 5 is used. Note that in Table 4.1 the sum of the squared error values is lowest when the group mean (i.e., the value of 5) is used as the coefficient. No other value for the coefficient will produce a lower sum of squared error values.

Table 4.1
Errors Produced by Various Values for Coefficients

Y	=	a_1G1	+	a_2G2	+	E_1	E_1	E_1
5		1		0		0	0	0
5		1		0		0	0	0
5		1		0		0	0	0
3	5 *	0	+ ? *	1		-3	-2	-1
6		0		1		0	1	2
6		0		1		0	1	2
						$\sum E_1^2 = 9$	$\sum E_1^2 = 6$	$\sum E_1^2 = 9$
						Using	Using	Using
						6 for a_2	5 for a_2	4 for a_2

Note that ones and zeros are used to reflect group membership information in the two dichotomous variable vectors. These values are used for two reasons. First, the mean of such a vector represents the proportion of subjects in the group represented by the vector. The means of a set of mutually exclusive vectors represent the proportion of subjects in each group and will therefore add to one; and if this is not so, then an error has been made. It has been our experiences that this error can often be attributed to a data entry error in the set of dichotomous variables. Second, and more important, the sample means for each group can be easily calculated when the values of 0 and 1 are used to form the vectors.

Most computer solutions to regression models will automatically insert a unit vector into the predictor set. The unit vector may or may not be linearly dependent, depending upon what other predictor vectors are in the system. For most Research Hypotheses, this automatic inclusion of the unit vector will cause no concern as the unit vector will be desired. A complete set of dichotomous vectors reflecting all group memberships, when used with the unit vector, will contain a linear dependency. For those readers using a matrix inversion solution, such as the solutions used in SPSS for Windows and Microsoft Excel, it would be best to omit linear dependencies from the model by omitting one of the group membership vectors. If a linear dependency is left in the model the SPSS for Windows program will analyze the model only after it automatically deletes one of the linearly dependent predictor variables, and Microsoft Excel program will not analyze the model—it will display an error message.

When the unit vector is included in the Full Model (Model 4.1) it is modified as follows:

$$Y = a_0 U + a_1 G1 + E_1 \qquad\qquad\qquad \text{(Model 4.1a)}$$

It should be noted that the researchers could have included variable G2 in the model rather than variable G1. Thus the Full Model could have been constructed as follows:

$$Y = a_0 U + a_2 G2 + E_1 \qquad\qquad\qquad \text{(Model 4.1b)}$$

The key point to understand is that Models 4.1, 4.1a, and 4.1b produce the same E_1 vector and the same R^2 value because they contain the same information.

With respect to Model 4.1a, the a_0 coefficient is equal to the mean math achievement score for the comparison method (G2), which was the method not explicitly represented by a variable in the model. The a_1 coefficient is equal to the mean math achievement score for the experimental method minus the mean math achievement score for the comparison method. Thus, the mean for the experimental method can be found by adding the G1 variable coefficient (a_1) to the unit vector coefficient (a_0).

Conceptually Deriving the Restricted Model

The Research Hypothesis has a corresponding Statistical Hypothesis that the condition stated in the Research Hypothesis is not true. Thus the directional Statistical Hypothesis for the study currently being analyzed is stated as follows:

> Directional Statistical Hypothesis: For a given population, the mean math achievement score for the experimental treatment is *not* higher than the math achievement score for the comparison treatment.

This Statistical Hypothesis implies that the two treatments could be considered to be a common treatment or that predictability is as good *without* knowledge as to which treatment a student received as it is *with* knowledge. The predictor variables in the Full Model reflect knowledge about which treatment a student received, but knowing only that the student was in the total sample would be reflected by the following model:

$$Y = a_0U + E_2 \qquad\qquad (Model\ 4.2)$$

Researchers should understand three points regarding this Restricted Model. First, we identify any model that reflects the condition stipulated in the Research Hypothesis as the Restricted Model. Second, this model contains only one piece of information, which is the unit vector. Thus, if we label the number of pieces of information contained in the Restricted Model as m2, then m2 = 1. Such a model is referred to as the Null Model. Third, since the value in the unit vector (U) for each student is 1, the coefficient for the unit vector (a_0) in this Null Model is equal to the sample mean math achievement score. Again, no other coefficient value will yield a smaller sum of squared error values for this model than will the sample mean of the criterion variable.

For the unique data in Table 4.1, the error vector in the Full Model will have the same values as will the error vector in the Restricted Model only because the overall mean is the same as each of the group means. For these data the additional knowledge about which treatment the student received did not help in predicting the criterion score. Another way of saying this is that the two treatments were not differentially effective.

Algebraically Deriving the Restricted Model

In the previous section the Full Model was constructed to directly reflect the Research Hypothesis. Another method that researchers can use to construct the Restricted Model places restrictions, which are implied by the Statistical Hypothesis, on the Full Model. Some researchers have found that making algebraic restrictions is easier than trying to conceptualize the Restricted Model directly.

If the two treatments are not differentially effective as stated by the Statistical Hypothesis, then the means of the two group populations will be equal. The a_1 and a_2 coefficients in the Full Model (Model 4.1) are estimates of the corresponding population parameters. Hence, restricting a_1 to equal a_2 would reflect in the sample what is being hypothesized about the population. Once the restriction is established, the researchers rewrite the Full Model with the restriction that $a_1 = a_2$. The steps used to construct the Restricted Model are as follows:

Step 1: Recall the Full Model

$$Y = a_1G1 + a_2G2 + E_1 \hspace{3cm} \text{(Model 4.1)}$$

Step 2: Impose the restriction $a_1 = a_2$ by replacing a_2 with a_1 in Model 4.1.

$$Y = a_1G1 + a_1G2 + E_2 \hspace{3cm} \text{(Model 4.3)}$$

It is important to realize that when restrictions are imposed on the Full Model, a different error vector is expected for the resulting Restricted Model. This is the reason that the error terms in the Full Model (Model 4.1) and Restricted Model (Model 4.3) have different subscript numbers.

Step 3: Simplify the terms in the Restricted Model (Model 4.3) either by collecting vectors being multiplied by like coefficients or by collecting coefficients that multiply like vectors (using operations described in chapter 3).

$$Y = a_1(G1 + G2) + E_2 \hspace{3cm} \text{(Model 4.3a)}$$

Step 4: Redefine vectors into their simplest forms. Since the sum of vectors G1 and G2 produces the unit vector, the Restricted Model constructed in Step 3 can be expressed as follows:

$$Y = a_1U + E_2 \hspace{3cm} \text{(Model 4.3b)}$$

Note that in the Full Model (Model 4.1) and the Restricted Model (Model 4.3b) the values of the a_1 coefficient will differ unless the mean math achievement scores for the two methods are equal, in which case the two method means will, of course, equal the mean of the total sample.

The researchers could have replaced a_1 with a_2 in Steps 2, 3, and 4, as follows:

Step 1a: Recall the Full Model: $Y = a_1G1 + a_2G2 + E_1$ \hspace{1cm} (Model 4.1)

Step 2a: Impose the restriction $a_1 = a_2$ by replacing a_1 with a_2 in Model 4.1.

$$Y = a_2G1 + a_2G2 + E_2 \qquad\qquad\qquad \text{(Model 4.4)}$$

Step 3a: $Y = a_2(G1 + G2) + E_2 \qquad\qquad\qquad \text{(Model 4.4a)}$

Step 4a: $Y = a_2U + E_2 \qquad\qquad\qquad\qquad\qquad \text{(Model 4.4b)}$

The important point for researchers to understand with respect to the derivation of the Restricted Model is that it does not matter in which way the restriction is imposed on the data; the Restricted Model results will essential be the same. The only difference in the two Restricted Models (Model 4.3b or Model 4.4b) is the symbol used for the unit vector coefficient, but both of these coefficients are equal to the sample mean of the criterion variable.

Remember that the Full Model (Model 4.1) did not include the unit vector. In this Full Model, the a_1 and a_2 coefficients equal the means of the comparison method and the treatment method, respectively. The restriction that a_1 is equal to a_2, which was placed on the Full Model, in effect requires the means of the comparison method and the experimental method to be equal. If the Full Model did contain a unit vector (Model 4.1a or Model 4.1b), which will be the case when researchers use SPSS for Windows and Microsoft Excel computer programs to estimate the Full Model, this restriction must be modified.

In the Full Model represented by Model 4.1a, the G2 variable (the comparison method variable) was not included in the model. Due to this fact the a_1 coefficient is equal to the mean math achievement score of the experimental method minus the mean math achievement score of the comparison method. Thus, a positive a_1 value indicates that the mean of the experimental method exceeds the mean of the comparison method, while a negative a_1 value reveals that the mean of the experimental method is less than the mean of the comparison method.

If the restriction requires the two group means to be equal, the researchers must set $a_1 = 0$ in the Full Model. Thus, the following three steps would be used to derive the Restricted Model from the Full Model that contained the unit vector (Model 4.1a):

Step 1: Recall the Full Model.

$$Y = a_0U + a_1G1 + E_1 \qquad\qquad\qquad \text{(Model 4.1a)}$$

Step 2: Impose the restriction $a_1 = 0$ on Model 4.1a.

$$Y = a_0U + 0*G1 + E_2 \qquad\qquad\qquad \text{(Model 4.5)}$$

Step 3: Multiply the G1 vector by 0.

$$Y = a_0U + E_2 \qquad\qquad \text{(Model 4.5a)}$$

Researchers should realize that the R^2 values for the previously discussed Full Model (Model 4.1) and the Restricted Model (Model 4.3b), which do not contain the unit vector, will match the R^2 values of the Full Model (Model 4.1a) and Restricted Model (Model 4.5a), which do contain the unit vector.

Regardless of whether unit vectors are or are not included in the models, if the R^2 for the Full Model is *not* significantly higher than the R^2 for the Restricted Model, the Research Hypothesis *cannot* be supported. This does not imply, however, that the Statistical Hypothesis is true for the population. Instead, such a finding simply means that *not* enough of a discrepancy existed from the placement of the restriction on the Full Model (i.e., $a_1 = a_2$ for the Model 4.1 or $a_1 = 0$ for the Model 4.5a) so that the researchers would be able to support the claim that in the population the two means are unequal.

The general F ratio, initially discussed in chapter 2, can be applied to the difference between the R^2 values of the Full and Restricted Models. If the R^2 for the Full Model is found to be significantly higher than the R^2 for the Restricted Model, then the Research Hypothesis can be accepted *if* the results show that the experimental method was *more* effective than the comparison method, as stated in the directional Research Hypothesis. This F test that can be used to statistically test the difference between the R^2 values of the Full and Restricted Models is discussed in the next section.

F TEST OF THE DIFFERENCE BETWEEN TWO R^2 VALUES

The general F formula for testing the difference between the R^2 values of the Full and Restricted Models is as follows:

$$F(df_n,\, df_d) = \frac{(R_F{}^2 - R_R{}^2)/(df_n)}{(1 - R_F{}^2)/(df_d)}$$

(Equation 4.1)

The $R_F{}^2$ is the proportion of observed criterion variance accounted for by the Full Model. The $R_R{}^2$ is the proportion of observed criterion variance that the Restricted Model explains. The difference between these two R^2 values $(R_F{}^2 - R_R{}^2)$, which is located in the numerator of the F test, is the proportion of unique variance in the criterion variable that the deleted predictor variable(s) explain. This value $(R_F{}^2 - R_R{}^2)$ is divided by the degrees of freedom of the numerator (df_n), which is the number of linearly independent vectors used to account for the proportion of variance difference between $R_F{}^2$ and $R_R{}^2$. That is, df_n is equal to the difference between the number of linearly independent vectors in the Full Model and the number of linearly independent vectors in the Restricted Model.

The denominator of the F test contains the value equal to 1 minus R_F^2 (i.e., $1 - R_F^2$), which is the proportion of variance unexplained by the Full Model (error variance). This value ($1 - R_F^2$) is divided by the degrees of freedom of the denominator (df_d). The df_d value is equal to the number of observations (N) minus the number of linearly independent vectors in the Full Model. In essence, df_d is the number of observations that are free to vary after coefficients for each of the linearly independent vectors in the Full Model have been calculated.

THE USE OF DIRECTIONAL HYPOTHESES

It is believed that most researchers would like to make some directional interpretations and should therefore state a directional Research Hypothesis. That is, before the data are collected, the researcher should state which treatment is expected to be more effective. There are often logical reasons one treatment may be expected (or at least hoped) to be more effective than the other treatment. Any experimental treatment should carry an expectation of whether it will yield a higher mean (or lower mean if that is the desired direction) than a comparison treatment on a particular criterion measure. The choice of either directional or nondirectional Research Hypotheses is not a *statistical* choice; it depends on theory, past research, and what the researcher wants to conclude from the analysis.

The Full and Restricted Models for directional Research Hypotheses are the same as for nondirectional ones. If the researchers state a nondirectional Research Hypothesis rather than a directional Research Hypothesis, they would lose statistical power unnecessarily. That is, the researchers would be less likely to determine that the new treatment was statistically significantly more effective than the old treatment. There is one exception, however, to the increased power level produced by the use of a directional Research Hypothesis. If the researchers state that the mean of the new treatment is higher than the mean of the old treatment, but the data analysis reveals that the opposite relationship exists between the treatment means, the test of the directional Research Hypothesis will have less power than the test of the nondirectional Research Hypothesis. If the researchers base their directional Research Hypothesis on a solid theoretical rationale, however, such a mistake should not occur.

When using directional Research Hypotheses, however, researchers must understand two points with respect to interpreting the statistical testing interpreting the statistical findings for such hypotheses. First, present statistical developments limit the directional Research Hypothesis to a single restriction on a Full Model. Thus, there will only be one degree of freedom for the numerator of the F ratio. The F ratio is equal to t^2 when there is only one degree of freedom in the numerator. Therefore, directional Research Hypotheses are sometimes referred to as *one-degree-of-freedom* hypotheses. If the degrees of freedom value in the numerator of the F ratio is greater than one, then a directional Research Hypothesis has not been tested.

Second, before interpreting "significant" results, one must be sure to determine whether the results were in the hypothesized direction. Assume the Research Hypothesis was stated as follows:

> Directional Research Hypothesis: For the population, the mean math achievement score for the experimental treatment is higher than the mean math achievement score for the comparison treatment.

For this Research Hypothesis (a directional one), the researchers would need to determine whether the mean of the experimental treatment was indeed higher than the mean of the comparison treatment, which can be assessed by examining the treatment coefficient in the Full Model (Model 4.1a). If the coefficient is in the hypothesized direction, then the (nondirectional) probability of the resultant F ratio should rightfully be divided by two. If the means turn out opposite to the situation that was hypothesized, then the probability of the resultant F ratio must be divided by two and then subtracted from one. In this latter case, significance would not be obtained. This point is discussed in more detail in the section of this chapter entitled "Applied Research Hypothesis 4.1."

GENERAL AND APPLIED RESEARCH HYPOTHESES

A number of General Research Hypotheses are discussed in the remainder of this text as a means of providing a "road map" for the testing process of various types of Research Hypotheses with the general linear model. In addition to these General Research Hypotheses, we provide many corresponding Applied Research Hypotheses. These Applied Research Hypotheses, which utilize SPSS for Windows and Microsoft Excel computer programs and data files that can be accessed from the internet site listed in Appendix A, "walk" researchers through an application of the corresponding General Research Hypotheses. The discussion of each Applied Research Hypothesis provides: (a) instructions on how to obtain the computer outputs for the Full and Restricted Models; (b) the interpretation of the key values related to the testing of the corresponding Statistical Hypothesis; and (c) the reporting of the findings as outlined by the Publication Manual of the American Psychological Association (2009), which will be referred to as the APA style of writing. The following two sections of this chapter present a discussion of the General Research Hypothesis 4.1 (GRH 4.1) and the Applied Research Hypothesis 4.1 (ARH 4.1), which address the issue of whether the criterion means of two groups differ.

GENRERAL RESEARCH HYPOTHESIS 4.1

The General Research Hypothesis 4.1 (GRH 4.1) addresses the type of analysis that involves a continuous criterion variable (Y) and a dichotomous predictor variable (Tr) that indicates the group to which each student is assigned. The dichotomous Tr variable contains a value of 1 for each student who is a member of the experimental group and a value of 0 for each student who is a member of the control group. If conducted as a nondirectional analysis, that is, a nondirectional Research Hypothesis was used, the purpose of the study is to determine whether the criterion means differ between the two groups. If the researchers are willing to suggest the criterion mean of the experimental group will exceed the criterion mean of the control group, a practice that we encourage, they would use a directional Research Hypothesis.

GRH 4.1 can be expressed as a directional Research Hypothesis or a nondirectional Research Hypothesis as follows:

Directional GRH 4.1: For the population of interest, the criterion mean of the experimental group is higher than the criterion mean of the control group.

Nondirectional GRH 4.1: For the population of interest, the criterion mean of the experimental group differs from the criterion mean of the control group.

As previously stated, we believe that researchers should strive to use directional Research Hypotheses whenever possible.

Regardless of whether the General Statistical Hypothesis (GSH 4.1) is expressed as a directional or a nondirectional hypothesis, it must restrict the means to be equal. Both forms of GSH 4.1 can be stated as follows:

Directional GSH 4.1: For the population of interest, the criterion mean of the experimental group is *not* higher than the criterion mean of the control group.

Nondirectional GSH 4.1: For the population of interest, the criterion mean of the experimental group *does not* differ from the criterion mean of the control group.

The Full Model that corresponds to either Research Hypothesis is as follows (Note that the unit vector is included in this model.):

$$Y = a_0 U + a_1 Tr + E_1 \hspace{3cm} \text{(Model 4.6)}$$

This Full Model has two pieces of information (m1 = 2), which are the unit vector (U) and the treatment group (Tr) vector. Remember that the phrase "pieces of information" is used interchangeably with linearly independent vectors.

As indicated by the directional Research Hypothesis, the researchers expect $a_1 > 0$. Thus, the corresponding restriction is $a_1 = 0$. For the nondirectional Research Hypothesis, the researchers expect $a_1 \neq 0$, and the corresponding restriction is $a_1 = 0$. Thus, the restriction is the same regardless of whether the researchers use the directional Research Hypothesis or the nondirectional Research Hypothesis.

Placing the restriction on the Full Model (Model 4.6) produces the following Restricted Model:

$$Y = a_0U + E_2 \tag{Model 4.7}$$

Note that this Restricted Model has only one piece of information (m2 = 1), which is the unit vector (U). The one restriction placed on the Full Model (i.e., $a_1 = 0$) caused the Restricted Model to have one less piece of information.

An F test of the difference between the R^2 values of the Full and Restricted Models, as discussed in chapter 2 and examined in greater detail in chapter 5, will determine whether the difference between the criterion means of the experimental group and the control group is statistically significant. The numerator degrees of freedom value for the F test would be equal to the m1 value minus the m2 value (i.e., 2 - 1 = 1), while the denominator degrees of freedom value would be equal to sample size (N) minus the m1 value (2).

If researchers used the directional General Research Hypothesis, it is important for them to understand one additional point regarding this statistical testing process. That is, since the researchers expected the mean of the experimental group to exceed the mean of the control group, the Statistical Hypothesis would not be rejected unless the value for a_1 was positive in addition to the probability of the F test being less than the established alpha level. A negative value for a_1 would lead the researchers to fail to reject GSH 4.1 regardless of the F test probability value level.

APPLIED RESEARCH HYPOTHESIS 4.1

In the previous section, the general concept of testing the difference between two group means through the analysis of Full and Restricted Models was presented. In this section, this type of analytic investigation is illustrated through the analysis of a data set.

Assume the researchers are interested in the following research question:

Is the mean mathematics score for students who are taught with a new method higher than the mean mathematics scores for students who are taught by the method currently used?

Note that this research question is directional. Thus, the researchers should construct a directional Research Hypothesis. If the researchers identified the group of students taught by the new method as the experimental group and the group of students taught by the currently used method as the control method, the Research Hypothesis, which is labeled Applied Research Hypothesis 4.1 (ARH 4.1), would be stated as follows:

Directional ARH 4.1: For a given population, the mean mathematics score for the experimental group is higher than the mean mathematics score for the control group.

The corresponding directional Statistical Hypothesis, which is labeled Applied Statistical Hypothesis 4.1 (ASH 4.1), is stated as follows:

Directional ASH 4.1: For a given population, the mean mathematics score for the experimental group is not higher than the mean mathematics score for the control group.

The variables used to statistically test ASH 4.1 are the criterion variable $X2$ and the predictor variable $X12$. The continuous criterion variable ($X2$) consists of a set of posttest scores recorded for the students who completed the study. The predictor variable ($X12$), which is a dichotomous variable, indicates which method was used to instruct each student. In the vector for variable $X12$ a value of 1 indicates that the student was instructed with the experimental method; while a value of 0 reveals that the student was instructed with the control method.

Full and Restricted Models

To determine whether the data support ARH 4.1, Full and Restricted Models are designed to reflect ARH 4.1 and ASH 4.1, respectively. When designing these models, it is important to note that the regression program contained in Microsoft Excel, which is the program used in this section, uses matrix inversion procedures to estimate the regression coefficients. (SPSS for Windows computer software also uses matrix inversion procedures.) Thus, group membership is represented by only one dichotomous group variable (i.e., $X12$).

It should be noted that if the value of 1 is assigned to every member of the experimental group in the dichotomous treatment group variable, the output obtained from either the Microsoft Excel program or the SPSS for Windows program is easier to interpret. To explain why, consider a Research Hypothesis such as ARH 4.1 in which it is hypothesized that the mean of the experimental group is higher than the mean of the control group. If the value of 1 is assigned to every student who is a member of the experimental group, the sign of the coefficient for the treatment variable will be positive *only* if the mean of the experimental group is indeed higher than the mean of the control group in the sample.

To depict the relationship between the experimental and control group means described in ARH 4.1, the Full Model must be constructed as follows:

$$X2 = a_0U + a_{12}X12 + E_1 \qquad\qquad \text{(Model 4.8)}$$

It is important to note that in order for this Full Model to accurately reflect ARH 4.1, which is a directional hypothesis, the value for a_{12} must be positive.

The restriction stated in ASH 4.1 requires the means of the experimental and control groups to be equal. Since the value for the a_{12} coefficient estimated by the Full Model (Model 4.8) is equal to the mean of the experimental group minus the mean of the control group, the restriction can be stated as follows:

$$a_{12} = 0 \qquad \text{(1 restriction)}$$

Placing this one restriction on the Full Model produces the following Restricted Model:

$$X2 = a_0U + E_2 \qquad\qquad \text{(Model 4.9)}$$

It is important to note two characteristics of this Restricted Model. First, it reflects the condition stated in ASH 4.1, that is, the two group means are equal. Second, this Restricted Model does not contain any predictor variables, just the unit vector, that is, it is a Null Model, and the R^2 value for a Null Model is equal to 0.

To determine whether the data support ARH 4.1, an *F* test of the difference between the R^2 value of the Full Model (Model 4.8) and the R^2 value of the Restricted Model (Model 4.9) is conducted using the Equation 4.1. When interpreting the results of this *F* test, it is important to note that ARH 4.1 is a directional hypothesis. Since ARH 4.1 and its corresponding Statistical Hypothesis (ASH 4.1) are directional, two conditions must exist before the researchers can reject ASH 4.1: (a) the value for the a_{12} coefficient in the Full Model must be positive and (b) the one-tailed *F*-test probability value, which is equal to one half of the two-tailed probability value, must be less than the established alpha level.

Analysis of ARH 4.1 with Microsoft Excel

The data used in connection with the Applied Research Hypothesis 4.1 have been stored in a Microsoft Excel file entitled "GLM DATA EXCEL FORMAT." This file can be accessed through the internet site listed in Appendix A. In addition, the data contained in this data file are also listed in Appendix B.

In this section the Microsoft Excel Regression menu is used to estimate the parameters of the Full and Restricted Models. Before the Regression menu can be displayed, however, the Data Analysis menu must be accessed. Click on the following to display this menu:

Data
    Data Analysis

Once these steps are completed, the Data Analysis menu will be displayed (see Exhibit 4.1).

Exhibit 4.1.
Microsoft Excel Data Analysis menu.

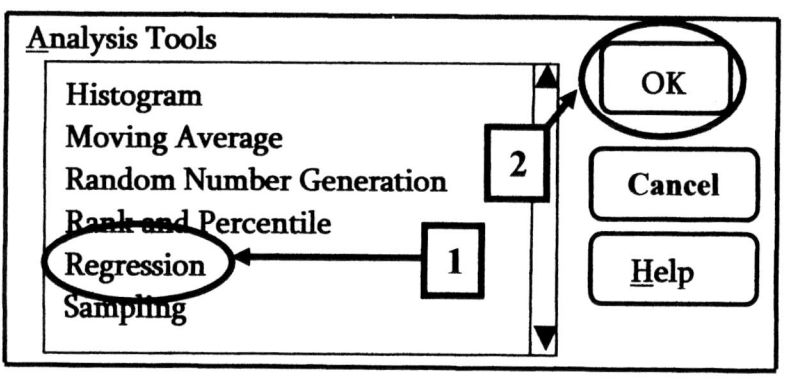

Complete the following steps to display the Regression menu:
1. Select the "Regression" menu (see Oval 1 in Exhibit 4.1).
2. Click on "OK" (see Oval 2 in Exhibit 4.1).
Upon completion of these two steps, the Excel Regression menu will appear as displayed in Exhibit 4.2).

To obtain the output for the Full Model (Model 4.8) used to statistically test ASH 4.1, which is the statistical hypothesis corresponding to ARH 4.1, the following four steps are completed in the Regression menu:

1. Click on the box next to "Input Y Range" (see Oval 1). Note that variable X2 is located in column B of the data file. Next, click and drag on the cells B1 to B61 in the data file to identify the X2 variable as the dependent variable.
2. Click on the box next to "Input X Range" (see Oval 2). Note that in the data file variable X12, which is the predictor variable for the Full Model, is located in column L. Next, click and drag on the cells L1 to L61 in the data file to identify this variable as the predictor variable.
3. Click on the box in front of "Labels" (see Oval 3).
4. Click on the "OK" button (see Oval 4).

After these four steps are completed, the output window for the Full Model (Model 4.8) will appear as displayed in Exhibit 4.3.

Exhibit 4.2.
Microsoft Excel Regression menu—Full Model (Model 4.8) for ARH 4.1.

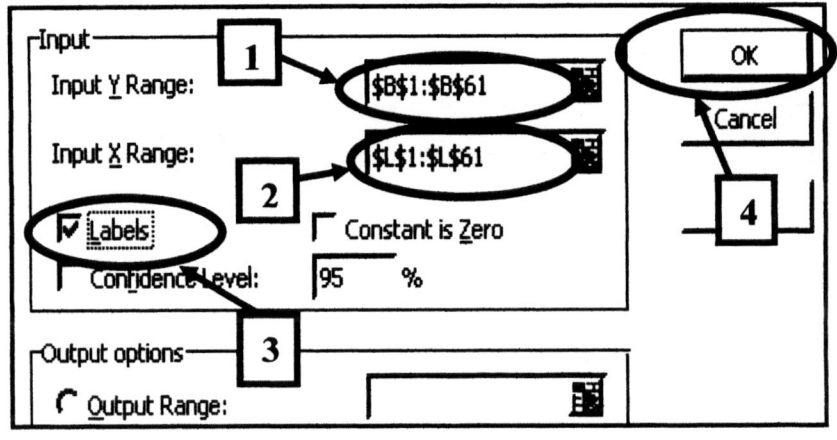

The key pieces of information on the Microsoft Excel output window (Exhibit 4.3) are as follows:

1. The R^2 value of the Full Model is .0003, which has been rounded to four decimal places (see Oval 1). Since the Restricted Model contains only the constant and error terms, the R^2 value for the Restricted Model is equal to .00. This value is not listed on the printout.
2. Since the Restricted Model used to reflect ASH 4.1 is the Null Model, the F test value listed in the ANOVA Table is the appropriate test of the difference between the R^2 values of the Full and Restricted Models. It is important to note that the F test listed in the ANOVA Table is the appropriate test only when the Restricted Model is the Null Model. The F

test value of the difference between the R^2 values of the Full and Restricted Models and its corresponding probability value are 0.015 and .903, respectively (see Oval 2). Since ARH 4.1 is a directional hypothesis, the probability value of .903 is divided by 2 to obtain the directional (one-tailed) probability value of .45.

3. The numerator degrees of freedom (*dfn*) and the denominator degrees of freedom (*dfd*) values used to calculate the F test of the difference between the R^2 values of the Full and Restricted Models are located in the rows entitled "Regression" and "Residual," respectively, under the box "*df*" (see Oval 3). Thus the *dfn* and *dfd* values are 1 and 58, respectively.

4. Recall that a value of 1 indicates that a given student is a member of the experimental group in the treatment variable. Due to this fact the unit vector coefficient in the Full Model is equal to the mean of the control group and the X12 variable coefficient is equal to the mean score of the experimental group minus the mean score of the control group. The unit vector value is located in the row entitled "Intercept" under the box "Coefficients," and the X12 variable coefficient is located in the row entitled "X12" under the same box (see Oval 4). Thus the mean of the control group is 23.81 and the mean of the experimental group exceeded the mean of the control group by 0.37 (see Oval 4)—the mean of the experimental group (24.18) is equal to the sum of those two values.

Exhibit 4.3.
Microsoft Excel regression output—Full Model (Model 4.8) for ARH 4.1.

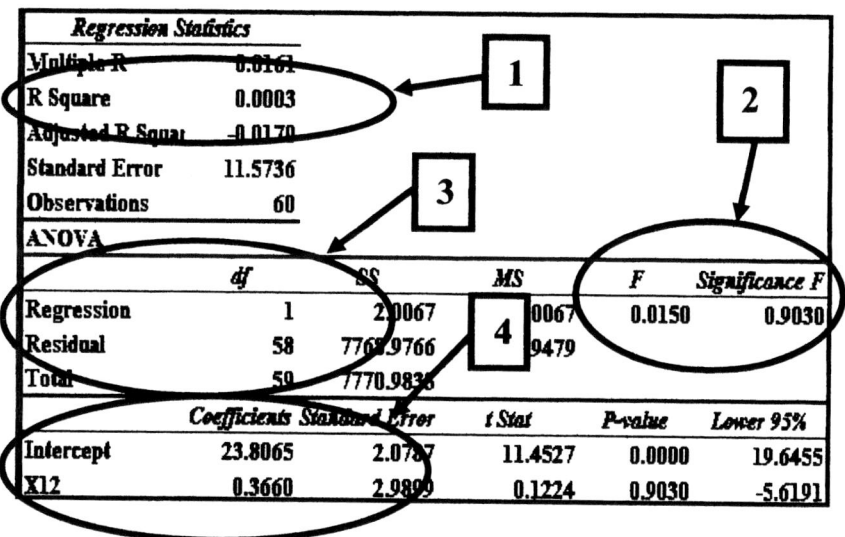

As previously stated, ARH 4.1 is a directional Research Hypothesis. That is the mean test score of the experimental group is expected to be higher than the mean test score of the control group. Thus, two conditions must exist before ASH 4.1 can be rejected (a) the regression coefficient for a_{12} in the Full Model must be positive and (b) the one-tailed F test probability value must be less than established alpha level, which is assumed to be set at .05. Although the a_{12} regression coefficient (0.37) is positive, the F test one-tailed probability value ($p / 2 = .45$) is greater than the established alpha level of .05, thus, ASH 4.1 is *not* rejected.

Further discussion of the concept of the directional (one-tailed) probability value is important. Unless a computer program indicates otherwise, the reported probability is for a nondirectional test. Whenever a directional Research Hypothesis is tested, the resulting probability should be divided by 2 and reported as such *if* the results are in the *correct direction*. If the results are in the *wrong direction*, then the reported probability value should be divided by 2 and that value subtracted from 1.

To illustrate the importance of adjusting the probability value for a directional test and determining whether the appropriate group had the higher mean, assume that the mean of the experimental group is quite a bit higher than the control group and the nondirectional probability is reported to be .06. Careless examination of the printout might lead one to say that the results were not significant at the .05 level. But the non-directional probability should be divided by 2, resulting in a p value of .03, which is less than the a priori alpha of .05 and hence significant. If the means were exactly opposite to the above results, the same nondirectional probability would result, but we would not want to conclude that significance has been obtained since the control group has a higher mean than the experimental group. Dividing the nondirectional probability of .06 by 2 and then subtracting that result from 1, that is, $1 - (.06 / 2)$, results in a directional p of .97 being reported, which is clearly not less than the a priori alpha of .05.

Writing the Results in APA Style

Once the statistical test results for ARH 4.1 are obtained, they must be communicated to interested people and decision makers. Often in the fields of education and the behavioral sciences such results are written using the guidelines presented in the Publication Manual of the American Psychological Association (2009). To illustrate how these results could be expressed in APA style, as previously stated, it is assumed that the X12 variable represented the experimental group and the criterion variable contained the participants' test scores. The results for ARH 4.1 could be written in APA style as follows:

Even though the mean test score for the experimental group (24.18) exceeded the mean test score of the control group (23.81) in the sample, as hypothesized, this difference was not statistically significant (F (1, 58) = 0.01, p / 2 = .45) at the one-tailed alpha level of .05. Therefore the two treatments must be considered equivalent. The proportion of unique variation in the test scores accounted for by whether a student was a member of the experimental or control group was less than .01 (R^2 = .0003).

A number of points regarding this results section should be noted. First, both the F and p letters are italicized. Second, the mean test values for the experimental and control groups, the F test value, and the p value are expressed to two decimal places. Generally in APA style, numbers will be expressed to two decimal places whenever possible. (Due to the very low R^2 value, it was expressed to four decimal places.) Third, in addition to the statistical test of the amount of variation in the test scores (X2) uniquely accounted for by the experimental group variable (X12), the effect size of the experimental group variable was noted. This effect size was expressed as the proportion of unique variation in the test scores accounted for by group membership (i.e., the difference between the R^2 values of the Full and Restricted Models).

Fourth, in APA style, a 0 is typed before the decimal point when the number is less than 1 *and* it is possible for the number to exceed 1. Since the probability value and the effect size, assuming the effect size is expressed as the unique proportion of variation in the test scores accounted for by the experimental group variable, cannot exceed 1, neither the probability value nor the effect size had a 0 typed before the decimal point. The F test value was expressed as 0.01, however, because an F value can exceed the value of 1. Fifth, the numerator degrees of freedom value (1) and denominator degrees of freedom value (58) used in the F test are listed along with the F value. Sixth, the fact that a directional hypothesis was tested is indicated by the use of p / 2 and the phrase "at the one-tailed alpha level of .05."

THE RELATIONSHIP OF THE R^2 VALUES AND THE F TEST TO OTHER STATISTICAL TECHNIQUES

The nondirectional Research Hypothesis in General Research Hypothesis 4.1 could have been phrased in correlational terminology as follows:

Nondirectional Research Hypothesis: For the population, treatment is correlated with the criterion.

Analogously, the directional hypothesis stated in correlational terminology is as follows:

Directional Research Hypothesis: For the population, the correlation between treatment variable and the criterion variable is positive.

To statistically test each of corresponding Statistical Hypotheses a dichotomous treatment variable (Tr) is correlated with a continuous criterion variable
(Y). The computational formula is referred to as a *point biserial correlation*. A
Pearson correlation between these two variables (i.e., Tr and Y) would yield the
same numerical value, and the square of either correlation would be equal to the
R^2 of the Full Model. The phrase "testing a correlation coefficient for significance" implies that the Statistical Hypothesis is being tested to ensure that the
population correlation is equal to 0 (therefore $R^2 = 0$). Notice that the statistically hypothesized population correlation, when squared, is the R^2 value of the Restricted Model. A specific computational formula exists for testing this hypothesis; and since it is a restricted case of the general F-test formula, the general F
test can be used as well.

The Research Hypotheses stated in the General Research Hypothesis 4.1
section of this chapter also could have been tested with a t test for the difference
between two means—an analysis of variance approach. Indeed, most statistical
advisors probably would have used the ANOVA approach rather than the correlational approach.

Most statistical texts separate correlational procedures from ANOVA procedures. The GLM approach does not make the distinction between these two
procedures. In fact, the GLM approach underscores the underlying isomorphism
of the two procedures and instead emphasizes the importance of the researcher
asking the question that is desired in the terminology that is desired. When one
is using the GLM, the emphasis is on the statistics answering the Research Hypothesis. In other statistical approaches, particularly the ANOVA approach, the
design of the study is often considered inseparable from the statistics.

Graphical Presentation

The isomorphism of the t test for the difference between two means and the
point biserial correlation coefficient also can be seen in a graphic presentation.
When one is considering the difference between two means, one has just that,
(i.e., two means), as illustrated in Figure 4.1. If there is no difference between
the two means, then the two means would be equally high on the Y axis. In this
case (Figure 4.1b), the straight line of best fit would be horizontal, and the Pearson correlation would be .00.

Now consider what the figure would look like in the correlational configuration. If one is correlating variable Tr with variable Y, then one would have
exactly the same figures as those in   Figure 4.1, the only difference would be
that a zero would represent each subject in one group and a one would represent
each subject in the other group. If there is a correlation between the two va-

riables, then there is a slope, and Figure 4.1a results. If there is no correlation between the two variables, then the line of best fit has a zero slope, or is parallel with the X axis, as in Figure 4.1b. The tests of significance of these two techniques look very different, but are not only conceptually the same but also mathematically the same (McNeil & Beggs, 1969).

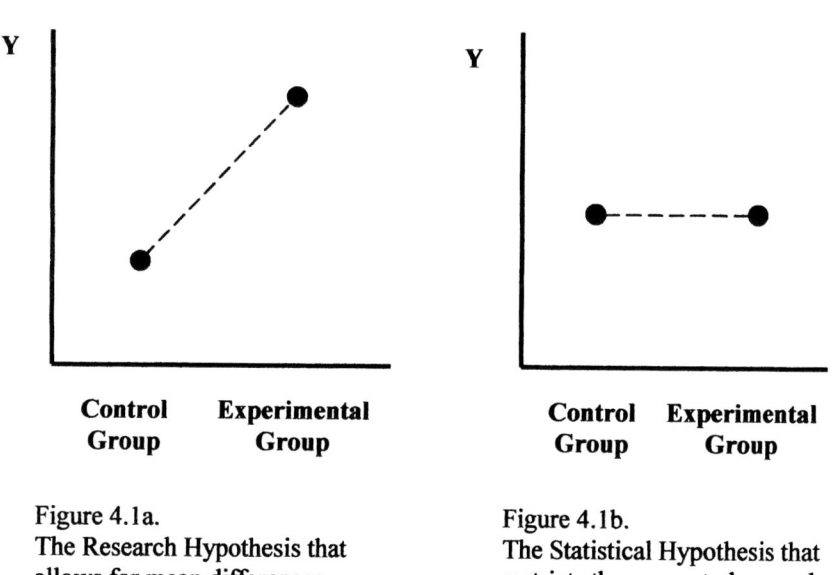

Figure 4.1a.
The Research Hypothesis that
allows for mean differences.

Figure 4.1b.
The Statistical Hypothesis that
restricts the means to be equal.

Figure 4.1.
Graphical presentations of the difference between two means.

Venn Diagrams

Venn diagrams, which were introduced at the end of chapter 1, provide assistance in understanding various Research Hypotheses. The Venn diagram for the present technique would consist of one predictor variable (e.g. a treatment variable) overlapping with the criterion variable. It is important to note that even though there are two groups, only one variable (identifying the treatment the subjects received) is being used to account for the variance in the criterion variable. The amount of overlap (diamond-marked area) in Figure 4.2 is the R^2 resulting from the Full Model.

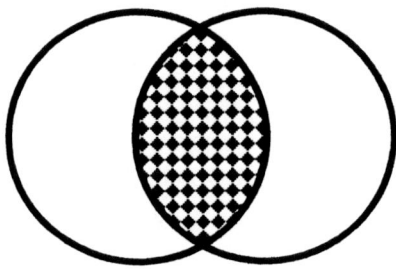

Figure 4.2.
Venn diagram of a *t* test for the difference between two means and the Point Biserial Correlation.

RESEARCH HYPOTHESES INVOLVING K GROUPS

Researchers often investigate multiple groups or treatments, instead of just two as discussed previously. (The analysis of variance analog would be the one-factor or one-way ANOVA.) With two groups, we need two dichotomous predictor vectors, while, with the number of groups being represented by k, a group membership vector needs to be constructed for each of the k groups. Thus, if there are four groups, there will be four group membership vectors. If the Full Model does not include the unit vector, the coefficients for each of these vectors would be the mean for that group. Under the usual Statistical Hypothesis, there is no difference between the group means, implying the restriction that all coefficients in the Full Model are equal. Again, both conceptually and mathematically, we arrive at the unit vector model as the Restricted Model.

If the unit vector is supplied by the computer program, as is the case for SPSS for Windows and Microsoft Excel, then one of the group membership vectors is linearly dependent and must be eliminated from the Full Model. In that case, the coefficient for the unit vector is the mean for the group whose vector has been eliminated, and each of the other coefficients is the difference between the mean for that group and the mean of the group whose vector has been omitted. The configuration of this design is presented in General Research Hypothesis 4.2 (GRH 4.2) and Applied Research Hypothesis 4.2 (ARH 4.2), which are located in a later section of this chapter. Whichever way the Full Model is represented, there will be (k -1) restrictions (as evidenced by the number of equal signs in the restrictions), and the Restricted Model will be the unit vector model.

It is important to understand that only a nondirectional Research Hypothesis can be tested due to the multiple restrictions being placed on the Full Model. If significance is found, all that can be said is that the k treatment criterion means are not equal, that is, the Research Hypothesis is supported by the data. The researchers could accept the Research Hypothesis and stop at that point, but in reality doing so does not say much. This point can best be understood through a graphical presentation of the group means.

Graphical Presentation

The k-group graph of group means (see Figure 4.3) is an extension of the two-group graph (see Figures 4.1a and 4.1b) in that there are k means. In Figure 4.3 Group 2 has the highest mean and Group 3 has the lowest mean. Let us assume that these results were significant, that not all four groups are equally effective on the criterion.

The four groups would yield the same degree of significance if Group 2 and Group 3 means were reversed. There are obviously many different possibilities that would satisfy the global Research Hypothesis. That is why we take the position that this type of hypothesis is not of great use to researchers. For practical purposes, one would like to know which treatment is most effective. Indeed, one may have some reason to believe that a given treatment will be more effective than another or a combination of several others. One approach to dealing with the lack of useful information is to conduct post hoc comparisons of the group means once a significant F test is revealed. The next section examines this approach.

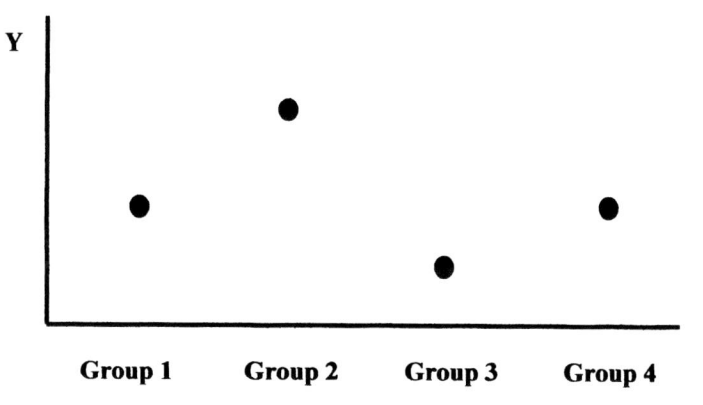

Figure 4.3.
Graphical presentation of the means for four groups.

ORTHOGONAL COMPARISONS—POST HOC

Most researchers are frustrated by the permissible conclusion with the omnibus one-way F test. The statement "There is a difference between group means," invariably leads one to ask, "Which specific means are different?" So, statisticians have developed intricate systems of post hoc comparisons that test more specific questions yet protect the initially stated alpha level stipulated by the researcher. Because each orthogonal contrast is a one-degree-of-freedom question, there can be as many orthogonal questions as there are degrees of freedom in the omnibus one-way F. The orthogonal contrasts can be specified either before the omnibus one-way F is calculated (referred to as *a priori comparisons*) or after the omnibus one-way F (referred to as *post hoc comparisons*). Since the post hoc questions were not specified before the data were collected and analyzed, no directional hypotheses can be tested with the post hoc comparisons. The use of orthogonal comparisons keeps the familywise alpha at the stated alpha for the omnibus F, while the effective alpha for each test is in reality the originally stated alpha divided by the number of tests. (See Newman, Fraas, and Laux (2000) for a discussion on the topic of adjusting the alpha level for multiple statistical tests.).

In the following discussion we use an example of four groups. The following restriction is placed on the hypotheses (comparisons) being tested: In all cases the hypotheses must be independent of one another (i.e., orthogonal), which simply means that what is found to be true with one hypothesis has no bearing at all on the other hypotheses. Whether a group of hypotheses (comparisons) does or does not meet this restriction can be determined mathematically by comparing the coefficients of the groups in the various hypotheses. Specifically, the coefficients must sum to zero, and the sum of the cross products must also sum to zero.

Even with just four groups, many sets of questions can be tested; the set chosen will be a function of how the four groups were configured, and, particularly, a function of the research interests of the investigator. Specific examples are provided in Exhibits 4.4 and 4.5. In Exhibits 4.4 and 4.5 the symbols mG1, mG2, mG3, and mG4 refer to the means of the four respective groups, while the symbols G1, G2, G3, and G4 refer to the group vectors. The values of 1, 0, 0, and -1 listed in Comparison 1 (see in Exhibit 4.4), are the weights by which the group vectors are multiplied to conduct the comparisons of interest. Note that the weights change for the other comparisons (see Exhibits 4.4, 4.5 and 4.6).

Combination of Groups

The three Research Hypotheses in Exhibit 4.4 indicate one orthogonal set of hypotheses. What this means is that if a researcher had stated an alpha of .05 for the omnibus one-way F test, then the same alpha would apply to the whole set of three hypotheses.

Exhibit 4.4.
A set of orthogonal comparisons.

Comparison 1—Research Hypothesis 1
  Nondirectional: mG1 ≠ mG4
  Directional: mG1 > mG4
  Statistical Hypothesis: mG1 = mG4
                    or
$$(1 * mG1) + (-1 * mG4) = 0$$
  Contrast Coefficients
  G1 G2 G3 G4
   1   0   0   -1

Comparison 2—Research Hypothesis 2
  Nondirectional: mG2 ≠ mG3
  Directional: mG2 > mG3
  Statistical Hypothesis: mG2 = mG3
                    or
$$(1 * mG2) + (-1 * mG3) = 0$$
  Contrast Coefficients
  G1 G2 G3 G4
   0   1   -1   0

Comparison 3—Research Hypothesis 3
  Nondirectional: (mG1 + mG4) ≠ (mG2 + mG3)
  Directional: (mG1 + mG4) > (mG2 + mG3)
  Statistical Hypothesis: (mG1 + mG4) = (mG2 + mG3)
                    or
$$(1 * mG1) + (1 * mG4) + (-1 * mG2) + (-1 * mG3) = 0$$
  Contrast Coefficients
  G1 G2 G3 G4
   1   -1   -1   1

Note: The symbols mG1, mG2, mG3, and mG4 represent the four group means, and symbols G1, G2, G3, and G4 represent the vectors for the four groups.

One can verify that these three hypotheses are indeed orthogonal by rearranging each Statistical Hypothesis so that zero appears by itself on one side of the equation. Then the contrast coefficients of the four groups (i.e., the values multiplied by the group means) are inserted under the box "Contrast Coefficients" in Exhibit 4.4.

Note the coefficients for Research Hypothesis 1 sum to zero, as do the coefficients for Research Hypothesis 2. Furthermore, the cross products of the coefficients for Research Hypothesis 1 and Research Hypothesis 2 (i.e., a given group coefficient in Comparison 1 is multiplied by the corresponding group coefficient in Comparison 2), also sum to zero $[(1 * 0) + (0 * 1) + (0 * -1) + (-1 * 0) = 0]$. Therefore Research Hypotheses 1 and 2 are orthogonal contrasts and the answer to Research Hypothesis 1 does not impinge on the answer to Research Hypothesis 2. Because Research Hypothesis 1 deals only with Groups 1 and 4, while Research Hypothesis 2 deals with Groups 2 and 3, this conclusion may seem like a trivial one (it is, and it was instructionally meant to be). On the other hand, Research Hypothesis 3 deals with all four groups, yet its coefficients sum to zero, and its cross products with those of Research Hypothesis 1 sum to zero. Thus Research Hypothesis 3 is orthogonal to Research Hypothesis 1. In addition, the sum of the cross products of Research Hypothesis 3 and 2 is equal to zero; thus they are orthogonal. One can conclude that the three research hypotheses are one set of orthogonal contrasts.

An example of when Research Hypotheses 1, 2, and 3 might be of interest is when a new treatment and an existing treatment are implemented in each of two schools. In this application, consider G1 (a vector) to be the new treatment in School 1, G2 to be the existing treatment in School 1, G3 to be the existing treatment in School 2, and G4 to be the new treatment in School 2. The primary hypothesis would be, "The mean for the new treatment is higher than the mean for the existing treatment," which would be tested with the contrast specified in Research Hypothesis 3. The hypothesis could be supported if there was a large difference in one school and little or no difference in the other school. If the researchers were concerned about this possibility, they could test Research Hypotheses 1 and 2. Research Hypothesis 1 compares the G1 and G4 groups, the new treatment in the two schools. Research Hypothesis 2 compares the existing treatment in the two schools.

In Exhibit 4.5, Research Hypotheses 4, 5, and 6 are another set of three orthogonal contrasts. While Research Hypothesis 5 and Research Hypothesis 2 are exactly the same, Research Hypothesis 4 and Research Hypothesis 6 are different from Research Hypotheses 1 and 3. The sum of the coefficients within each of the Research Hypotheses (i.e., Research Hypotheses 4, 5, and 6) is equal to zero, and their cross products sum to zero, thus these three research hypotheses constitute a different set of three orthogonal contrasts.

The set of orthogonal comparisons the researchers should use depends on the design of the study and the questions they have regarding the group means.

Indeed, there are many other sets of orthogonal contrasts. As in all research, the questions should guide the analysis. With post hoc comparisons, the researchers are limited to one less question than the number of groups.

Exhibit 4.5.
A second set of orthogonal comparisons.

Comparison 4—Research Hypothesis 4
  Nondirectional: $(mG1 + mG2 + mG3) / 3 \neq mG4$
  Directional: $(mG1 + mG2 + mG3)/3 > mG4$
  Statistical Hypothesis: $(mG1 + mG2 + mG3) / 3 = mG4$
                                or
            $(1 * mG1) + (1 * mG2) + (1 * mG3) + (-3 * mG4) = 0$
  Contrast Coefficients
    G1 G2 G3 G4
     1   1   1   -3

Comparison 5—Research Hypothesis 5
  Nondirectional: $mG2 \neq mG3$
  Directional: $mG2 > mG3$
  Statistical Hypothesis: $mG2 = mG3$
                            or
                $(1 * mG2) + (-1 * mG3) = 0$
  Contrast Coefficients
    G1 G2 G3 G4
     0   1 -1   0

Comparison 6—Research Hypothesis 6
Nondirectional: $mG1 \neq (mG2 + mG3) / 2$
Directional: $mG1 > (mG2 + mG3) /2$
Statistical Hypothesis: $mG1 = (mG2 + mG3) / 2$
                          or
            $(2 * mG1) + (-1 * mG2) + (-1 * mG3) = 0$
  Contrast Coefficients
    G1 G2 G3 G4
     2 -1 -1   0

Note: The symbols mG1, mG2, mG3, and mG4 represent the four group means, and symbols G1, G2, G3, and G4 represent the vectors for the four groups.

An example of when the Research Hypotheses 4, 5, and 6, which are listed in Exhibit 4.5, might be of interest is when the researchers are testing the effectiveness of three different new treatments (G1, G2, and G3) and one comparison treatment (G4). Since there are four groups, three orthogonal contrast questions can be asked, and if the questions are asked before inspection of the data, directional Research Hypotheses can be tested. Research Hypothesis 4 determines if the average of the three new treatment means is higher than the single comparison treatment mean. Research Hypothesis 5 tests if the second new treatment mean is higher than the third new treatment mean. Finally, Research Hypothesis 6 tests if the first new treatment mean is higher than the average of the other two new treatment means. As should now be clear, the design of the research and the desired conclusion(s) determine the choice of the hypotheses and whether the hypotheses are directional or nondirectional. No one choice is always correct; the choice will depend on the research questions.

Trend Analysis

When the treatments are ordered on some underlying continuum, one may want to investigate the trends in the data. Such a situation is as presented in Exhibit 4.6.

Exhibit 4.6.
A set of contrast coefficients with four groups—Trend analysis.

Comparison 7—Research Hypothesis 7 (linear)
  Nondirectional: $(-3mG1) + (-1mG2) + (1mG3) + (3mG4) \neq 0$
  Directional: $(-3mG1) + (-1mG2) + (1mG3) + (3mG4) > 0$
  Statistical Hypothesis: $(-3mG1) + (-1mG2) + (1mG3) + (3mG4) = 0$
                          or
            $(-3 * mG1) + (-1 * mG2) + (1 * mG3) + (3 * mG4) = 0$
  Contrast Coefficients
    G1 G2 G3 G4
    -3 -1   1   3

Comparison 8—Research Hypothesis 8 (quadratic)
  Nondirectional: $(1mG1) + (-1mG2) + (-1mG3) + (1mG4) \neq 0$
  Directional: $(1mG1) + (-1mG2) + (-1mG3) + (1mG4) > 0$
  Statistical Hypothesis: $(1mG1) + (-1mG2) + (-1mG3) + (1mG4) = 0$
                          or
            $(1 * mG1) + (-1 * mG2) + (-1 * mG3) + (1 * mG4) = 0$
  Contrast Coefficients
    G1 G2 G3 G4
     1 -1 -1   1

Exhibit 4.6 (continued).
A set of contrast coefficients with four groups—Trend analysis.

Comparison 9—Research Hypothesis 9 (cubic)
    Nondirectional: $(-mG1) + (3mG2) + (-3mG3) + (mG4) \neq 0$
    Directional: $(-mG1) + (3mG2) + (-3mG3) + (mG4) > 0$
    Statistical Hypothesis: $(-mG1) + (3mG2) + (-3mG3) + (mG4) = 0$
                                or
                $(-1 * mG1) + (3 * mG2) + (-3 * mG3) + (1 * mG4) = 0$
    Contrast Coefficients
    G1 G2 G3 G4
    -1   3   -3   1

Note: The symbols mG1, mG2, mG3, and mG4 represent the four group means, and symbols G1, G2, G3, and G4 represent the vectors for the four groups.

The question is: Does the criterion increase linearly with an increase in the underlying continuum (as in Research Hypothesis 7), or is there a *minimum* performance as in Research Hypothesis 8? (By reversing all the weights in Research Hypothesis 8, one could investigate *maximum* performance.)

Finally, with four groups a cubic trend may exist as stipulated in Research Hypothesis 9. Note that the coefficients for Research Hypothesis 7, 8, and 9 all sum to zero and that the cross products all sum to zero. Therefore, these three research hypotheses constitute another set of orthogonal contrasts for four groups.

Two Factors

Now suppose that the four groups differ not on just one underlying factor, as they did in the above examples, but on *two* underlying factors. Exhibit 4.7 posits the following example of two groups receiving the new treatment and two groups receiving the comparison treatment. Thus the first underlying factor is *treatment*, that is, new treatment versus comparison treatment.

One of the new treatment groups is scheduled in the morning (AM) and one is scheduled in the afternoon (PM). In the same manner, one of the comparison treatment groups is scheduled in the AM and one is scheduled in the PM. Thus the second factor is *time of treatment*, that is, AM versus PM.

What would be the Research Hypotheses of interest with this design? The researchers probably would want to compare the new treatments to the comparison treatments, and possibly the AM treatments to the PM treatments. These two hypotheses are developed first, and then we will turn our attention to the third orthogonal comparison.

The nondirectional Research Hypothesis for treatment (Research Hypothesis 10) would be as follows:

Exhibit 4.7.
A set of contrast coefficients—Two-Way analysis of variance.

Comparison 10—Research Hypothesis 10

Nondirectional: The two treatments, averaged across the two different time periods, are not equally effective.

$$(mG1 + mG2) / 2 \neq (mG3 + mG4) / 2$$

Directional. The new treatment, averaged across the two different time periods, is more effective than the comparison treatment.

$$(mG1 + mG2) / 2 > (mG3 + mG4) / 2$$

Statistical Hypothesis: $(mG1 + mG2)/2 = (mG3 + mG4) / 2$

or

$$(mG1 + mG2) = (mG3 + mG4) \text{ or } (mG1 + mG2) - (mG3 + mG4) = 0$$

or

$$(1 * mG1) + (1 * mG2) + (-1 * mG3) + (-1 * mG4) = 0$$

Contrast Coefficients

G1	G2	G3	G4
1	1	-1	-1

Comparison 11—Research Hypothesis 11

Nondirectional: The two time periods, averaged across the two treatments, are not equally effective.

$$(mG1 + mG3)/2 \neq (mG2 + mG4) / 2$$

Directional: The AM period, averaged across the two different treatments, is more effective than the PM period.

$$(mG1 + mG3)/2 > (mG2 + mG4) / 2$$

Statistical Hypothesis:

$$(mG1 + mG3)/2 = (mG2 + mG4) / 2$$

or

$$(mG1 + mG3) = (mG2 + mG4)$$

or

$$(mG1 + mG3) - (mG2 + mG4) = 0$$

or

$$(1 * mG1) + (-1 * mG2) + (1 * mG3) + (-1 * mG4) = 0$$

Contrast Coefficients

G1	G2	G3	G4
1	-1	1	-1

Exhibit 4.7 (continued).
A set of contrast coefficients—Two-Way analysis of variance.

Comparison 12—Research Hypothesis 12
    Nondirectional: The difference in effectiveness of the AM new treatment and the PM new treatment is different from the difference between the AM comparison treatment and the PM comparison treatment.
$$(mG1 - mG2) \neq (mG3 - mG4)$$
    Directional: The difference in effectiveness of the AM new treatment and the PM new treatment is greater than the difference between the AM comparison treatment and the PM comparison treatment.
$$(mG1 - mG2) > (mG3 - mG4)$$
    Statistical Hypothesis: The difference in effectiveness of the AM new treatment and the PM new treatment is the same as the difference between the AM comparison treatment and the PM comparison treatment.
$$(mG1 - mG2) = (mG3 - mG4)$$
or
$$(mG1 - mG2) - (mG3 - mG4) = 0$$
or
$$(1 * mG1) + (-1 * mG2) + (-1 * mG3) + (1 * mG4) = 0$$

Contrast Coefficients
G1 G2 G3 G4
 1 -1 -1   1

Note: G1 represents the new treatment AM group, G2 represents the new treatment PM group, G3 represents the comparison treatment AM group, and G4 represents the comparison treatment PM group.

Nondirectional Research Hypothesis 10: The means of the two treatments averaged across the two different time periods differ.

This Research Hypothesis would result in the orthogonal coefficients, which are listed for Research Hypothesis 10 in Exhibit 4.7. One could have stated this Research Hypothesis with a directional expectation, resulting in the same set of orthogonal coefficients. A directional Research Hypothesis for treatment (Research Hypothesis 10) may be stated as follows:

Directional Research Hypothesis 10: The two new treatment means averaged across the two different time periods exceeds the two comparison treatment means averaged across the two different time periods.

The nondirectional Research Hypothesis for time of treatment (Research Hypothesis 11) would be stated as follows:

Nondirectional Research Hypothesis 11: The means of the two time periods averaged across the two different treatments differ.

Again, the researchers could have stated this hypothesis with a directional expectation. Notice that the coefficients for Research Hypothesis 11 are orthogonal to those for Research Hypothesis 10. Research Hypotheses 10 and 11 are referred to as *main effects* hypotheses within the ANOVA framework. Unless stated directionally a priori, they are always tested in a nondirectional fashion.

Given these two orthogonal main effects contrasts, a third orthogonal contrast, which is presented in Research Hypothesis 12, would need to be specified. The nondirectional Research Hypothesis associated with the coefficients delineated in Research Hypothesis 12 is as follows:

Nondirectional Research Hypothesis 12: The difference between AM new treatment mean and PM new treatment mean is *different* from the difference between AM comparison treatment mean and PM comparison treatment mean.

Once again, the researchers could have stated this hypothesis with a directional expectation as follows:

Directional Research Hypothesis 12: The difference between AM new treatment mean and PM new treatment mean is *greater* than the difference between AM comparison treatment mean and PM comparison treatment mean.

Research Hypothesis 12, stated as a nondirectional hypothesis, is referred to in the ANOVA literature as the *test for interaction*, which is discussed in great detail in chapter 7.

The directional Research Hypothesis 10 could be tested with the following Full Model (Note that this model does not contain a unit vector.):

$$Y = a_1G1 + a_2G2 + a_3G3 + a_4G4 + E_1 \qquad \text{(Model 4.10)}$$

In terms of this Full Model the directional Research Hypothesis 10 can be expressed as follows:

$$(a_1 + a_2) / 2 > (a_3 + a_4) / 2 \text{ or } (a_1 + a_2) > (a_3 + a_4)$$

The restriction stipulated by the corresponding Statistical Hypothesis 10 would be as follows:

$$(a_1 + a_2) = (a_3 + a_4) \quad \text{or} \quad (a_1 + a_2 - a_4) = a_3 \quad \text{(1 restriction)}$$

Placing this restriction on the Full Model produces the following Restricted Model:

$$Y = a_1 G1 + a_2 G_2 + (a_1 + a_2 - a_4)G_3 + a_4 G_4 + E_2 \qquad \text{(Model 4.11)}$$

Multiplying $(a_1 + a_2 - a_4)$ by G_3 and collecting like coefficients produces the following equivalent Restricted Model:

$$Y = a_1(G1 + G3) + a_2(G2 + G3) + a_4(G4 - G3) + E_2 \qquad \text{(Model 4.11a)}$$

It is evident that the Full Model (Model 4.10) contains four pieces of information (i.e., G1, G2, G3, and G4) and it is also evident that the Restricted Model (Model 4.11a) contains three pieces of information (i.e., G1 + G3, G2 + G3, and G4 - G3). At first glance it is not evident that Model 4.11 also contains three pieces of information—it appears to have four pieces of information. A closer examination reveals, however, that Model 4.11 has only three coefficients—a_3 has been eliminated. In addition, a review of Model 4.11a, which is an equivalent model, clearly reveals that only three vectors (pieces of information) are present. Thus, researchers should realize that the number of pieces of information contained in a model is equal to the number of (non-zero) coefficients estimated by the model, which is also equal to the number of linerarly dependent variables in the model.

There are other ways to code group membership variables. One could use any two values, referred to as *nonsense coding* by Cohen and Cohen (1975) and by Williams (1987). One also could use the actual orthogonal weights (Williams, 1974a; 1974b). Each orthogonal comparison would be reflected by a vector. Thus the treatment main-effect vector would be: TREATMAIN = 1 if G1 or G2 and = -1 if G3 or G4. The time main-effect vector would be: TIMEMAIN = 1, if G1 or G3, and = -1, if G2 or G4. The third orthogonal contrast would be the interaction between time and treatment (discussed in Chapter 7): TIMETREAT = 1, if G1 or G4, and = -1, if G2 or G3. Using this coding scheme, Research Hypothesis 10 could be tested by using Models 4.12 and 4.13, which are "reparameterizations" of Models 4.10 and 4.11a:

$$Y = a_0 U + b_1 \text{TREATMAIN} + c_1 \text{TIMEMAIN} + d_1 \text{TIMETREAT} + E_1$$
$$\text{(Model 4.12)}$$

Research Hypothesis 10 stipulates the restriction $b_1 = 0$, which results in the following Restricted Model:

$$Y = a_0U + c_1TIMEMAIN + d_1TIMETREAT + E_2 \qquad \text{(Model 4.13)}$$

Note that Models 4.12 and 4.13 contain the unit vector.

Both of the Full Models (Models 4.10 and 4.12) contain four pieces of information, and they will produce the same R^2 value. In addition, both of the Restricted Models (Models 4.11a and 4.13) contain three pieces of information, and they also will produce the same R^2 value. Thus the F test of the difference between the R^2 values of Models 4.12 and 4.13 will yield the same value as the F test of the difference between the R^2 values of Models 4.10 and 4.11a. Researchers should understand that these are two equivalent ways to test the same hypothesis.

Research Hypothesis 11 could be tested by restricting $c_1 = 0$ in the Full Model (Model 4.12) resulting in the following Restricted Model:

$$Y = a_0U + b_1TREATMAIN + d_1TIMETREAT + E_3 \qquad \text{(Model 4.14)}$$

In a similar fashion, Research Hypothesis 12 could be tested by restricting $d_1 = 0$ in the Full Model (Model 4.12), resulting in the following Restricted Model:

$$Y = a_0U + b_1TREATMAIN + c_1TIMEMAIN + E_4 \qquad \text{(Model 4.15)}$$

Once again, the statistical tests produced by these models (i.e., the models that contain variables based on orthogonal weight coding) will match the statistical test results produced by corresponding models that contain variables with dummy coding (i.e., 0 and 1 coding).

ORTHOGONAL COMPARISONS—A PRIORI

Orthogonal comparisons were developed to attempt to identify where the significance was after the global, nondirectional question was found to be significant. Orthogonal comparisons also are appropriate when one has a priori expectations, particularly directional expectations. As indicated in Exhibits 4.4, 4.5, 4.6, and 4.7, each of those questions *could* have been directional if there was reason to believe a priori that the results would be in one direction. The only difference between post hoc and a priori is that in a priori one has a reason (based on theory or past research) to expect or want the results to be in a particular direction. Again, stating the Research Hypothesis in a directional manner allows one to make a directional conclusion.

Stating Questions of Interest

Another aspect of orthogonal comparisons needs to be clarified: The researcher should always ask the questions that are of interest. For instance, with four groups there are three degrees of freedom and therefore three orthogonal questions that can be asked. Research Hypotheses 1, 2, and 3 comprise one possible set. Research Hypotheses 4, 5, and 6 comprise another possible set. Research Hypotheses 7, 8, and 9 comprise another possible set. Research Hypotheses 10, 11, and 12 comprise another possible set. Which one is applicable depends upon the nature of the data and the question asked.

More complicated designs have been modeled with GLM. For instance, the Solomon four-group design was developed to assess the effect of pretest sensitization. Traditional ANOVA computational formulae have not been developed for this analysis, but Williams and Newman (1982) and Newman, Benz, and Williams (1990) illustrated the GLM approach to this design.

A researcher may be interested in only one or two of the questions in an orthogonal set. One does not have to test all the possible one-degree-of-freedom questions. If a researcher wishes to make a directional conclusion, then the contrasts must be stated before the data are collected. The present authors would argue that, in this situation, the omnibus one-way F does not need to be calculated, and the analysis can proceed directly to the stated questions (i.e., the omnibus one-way F was not a stated question of interest and therefore should not be tested).

A More Complicated Design

Perhaps the research design includes three slightly different comparison groups, and two slightly different discussion treatment groups, along with four other kinds of treatment groups. A Research Hypothesis of interest might be stated as follows:

Research Hypothesis: For a given population, the average of the two discussion treatment groups is higher than the average of the three comparison treatment groups, considering the variance of the four other kinds of treatment groups.

The antithesis of the Research Hypothesis (i.e., the Statistical Hypothesis) would be the following Statistical Hypothesis:

Statistical Hypothesis: For a given population, the average of the two discussion treatment groups is not higher than the average of the three comparison treatment groups, considering the variance of the four other kinds of treatment groups.

The Full Model reflecting the Research Hypothesis contains a dichotomous vector for each group, and is constructed as follows:

$$Y2 = a_0U + c_1C1 + c_2C2 + c_3C3 + t_1T1 + t_2T2 + t_3T3 + d_1D1 + d_2D2 + E_3$$

(Model 4.16)

where:
1. $Y2$ = the criterion variable;
2. U = the unit vector containing a 1 for each subject;
3. $C1 = 1$ if criterion from Comparison Group 1, 0 otherwise;
4. $C2 = 1$ if criterion from Comparison Group 2, 0 otherwise;
5. $C3 = 1$ if criterion from Comparison Group 3, 0 otherwise;
6. $T1 = 1$ if criterion from Treatment Group 1, 0 otherwise;
7. $T2 = 1$ if criterion from Treatment Group 2, 0 otherwise;
8. $T3 = 1$ if criterion from Treatment Group 3, 0 otherwise;
9. $T4 = 1$ if criterion from Treatment Group 4, 0 otherwise;
10. $D1 = 1$ if criterion from Discussion Group 1, 0 otherwise;
11. $D2 = 1$ if criterion from Discussion Group 2, 0 otherwise; and
12. $c_1, c_2, c_3, t_1, t_2, t_3, t_4, d_1,$ and d_2 are least squares weighting coefficients calculated so as to minimize the sum of the squared values in the error vector, E_3.

It is important to note that this Full Model contains a unit vector—thus one of the 9 variables used to represent the nine groups could not be entered into the model. Any one of the 9 variables could be eliminated; but due to the manipulations of the coefficients for the comparison and discussion groups required to form the Restricted Model it is best to eliminate one of the treatment group variables—we chose to eliminate the variable for the Treatment Group 4.

If the cells have the same number of subjects in each cell or are proportional, the restriction implied by the Research and Statistical Hypothesis is:

$$\frac{c_1 + c_2 + c_3}{3} = \frac{d_1 + d_2}{2}$$

If the number of subjects is not proportional, then the number of subjects in each cell needs to considered, as shown in a later chapter. The restrictions can be solved for any of the five weighting coefficients—we solve for c_1 here. Multiplying both sides by 3 yields:

$$c_1 + c_2 + c_3 = 3(d_1 + d_2) / 2$$

Subtracting c_2 and c_3 from both sides yields:

$c_1 = (3 / 2 * d_1) + (3 / 2 * d_2) - c_2 - c_3$

Now inserting the right-hand expression for c_1 into the Full Model yields:

$$Y2 = a_0U + [(3 / 2 * d_1) + (3 / 2 * d_2) - c_2 - c_3]C1 + c_2C2 + c_3C3 + t_1T1 + t_2T2 + t_3T3 + d_1D1 + d_2D2 + E_4$$

(Model 4.17)

Multiplying the C1 coefficients [i.e., $(3 / 2 * d_1)$, $(3 / 2 * d_2)$, $- c_2$, and $- c_3$] by C1 produces the following Restricted Model:

$$Y2 = a_0U + (3 / 2 * d_1)C1 + (3 / 2 * d_2)C1 - c_2C1 - c_3C1 + c_2C2 + c_3C3 + t_1T1 + t_2T2 + t_3T3 + d_1D1 + d_2D2 + E_4$$

(Model 4.17a)

Expanding and collecting terms results in the following Restricted Model:

$$Y2 = a_0U + c_2(C2 - C1) + c_3(C3 - C1) + t_1T1 + t_2T2 + t_3T3 + d_1[(D1 + (3 / 2 * C1)] + d_2[(D2 + 3 / 2 * C1)] + E_4$$

(Model 4.17b)

There are ($m1 = 9$) linearly independent pieces of information in the Full Model (Model 4.16) because one of the nine group vectors is linearly dependent upon the other eight group vectors plus the unit vector. There are ($m2 = 8$) good pieces of information in the Restricted Model (Model 4.17b). Therefore, $df_n = (9 - 8) = 1$ and $df_d = N - 9$. One could accept the Research Hypothesis, *if* the average of the two discussion coefficients [$(d_1 + d_2) / 2$] was greater than the average of the three comparison coefficients [$(c_1 + c_2 + c_3) / 3$], and if the *F* test of the difference between the R^2 values of the Full and Restricted Models produced a one-tailed probability value that was less than the established alpha.

Readers may consider some of the vectors contained in the Restricted Model as being rather strange and complex. They could easily be generated, however, by the SPSS for Windows and Microsoft Excel computer programs.

Advantage to Conceptualizing Post Hoc Comparisons in the GLM

The statistically sophisticated reader may see the similarities between what has just been presented and the myriad of techniques referred to as *multiple comparisons* or *post hoc comparisons*. The procedure presented in the previous section is similar to *planned comparison* techniques. We believe that three differences exist between what has just been presented and the body of traditional literature referred to as post hoc comparisons.

First, in the GLM approach as presented above, the researcher is forced to state the Research Hypothesis. The Research Hypothesis should be directional in nature allowing the researcher to make a conclusive statement. (The "planned comparison" literature does not usually consider the notion of directional hypothesis testing.)

Second, because the Research Hypothesis is stated by the researcher and is a well-thought-out analysis of the design, no adjustment to the resultant probability is necessary as is the case in *post hoc comparisons*. It should be noted that a priori (as contrasted to post hoc) comparisons do not differ from the regression presentation on this point, nor on the next point.

Third, the researcher does not initially have to compare the Full Model to the "unit vector model." (The post hoc comparisons require the one-way F to be significant before any specific comparisons can be made.) The "one-way F hypothesis" does not have to be tested for two reasons: (a) the one-way F question is not of interest to the researcher, and (b) global statistical significance might not be obtained, even though some specific questions of interest might produce significance. This could occur, for instance, when nine comparison groups and only one experimental group are examined. The numerous pair-wise tests of the 10 group means may not reveal even one significant difference, even though the mean of the one experimental group might well be higher than the average of the nine comparison group means.

GENERAL RESEARCH HYPOTHESIS 4.2

The General Hypothesis 4.2 (GRH 4.2) addresses the type of analysis that involves a continuous criterion variable (Y) and a set of multiple dichotomous predictor variables (G1, G2, . . ., G_k). The purpose of the analysis is to determine whether at least two criterion means differ among the k groups, where k represents the number of groups. Remember, in this type of analysis only a nondirectional Research Hypothesis can be used.

GRH 4.2 is expressed as a nondirectional Research Hypotheses as follows:

Nondirectional GRH 4.2: For a given population the k treatment criterion means are not all equal.

The corresponding General Statistical Hypothesis (GSH 4.2) is stated as follows:

Nondirectional GSH 4.2: For a given population the k treatment criterion means are equal.

The Full Model, which includes the unit vector, would contain only k - 1of the dichotomous group predictor variables:

Full Model: $Y = a_0U + a_1G1 + a_2G2 + \ldots + a_{k-1}G_{k-1} + E_1$ (Model 4.18)

The pieces of information (m1) contained in this Full Model is equal to $((k - 1) + 1)$, that is, $k - 1$ group vectors plus the unit vector. Thus, m1 is simply equal to k. For this analysis it does not matter which of the k group vectors is not included in the Full Model.

As indicated by the Research Hypothesis, the researchers expect $a_1 \neq 0$, $a_2 \neq 0, \ldots, a_{k-1} \neq 0$. Thus, the corresponding restrictions are $a_1 = 0$, $a_2 = 0, \ldots, a_{k-1} = 0$. Note that a total of $k - 1$ restrictions will be placed on this Full Model. The Restricted Model that results from the placement of these restrictions on the Full Model is as follows:

Restricted Model: $Y = a_0U + E_2$ (Model 4.19)

The number of pieces of information contained in this Restricted Model (m2) is 1, which is the unit vector. Thus $k - 1$ pieces of information have been deleted from the Full Model.

An F test of the difference between the R^2 values of the Full and Restricted Models will determine whether at least one of the differences among the criterion means of G1, G2, . . ., GK are statistically significant. The numerator degrees of freedom (df_n) value for the F test would be equal to m1 minus m2 (i.e., $df_n = k - 1$) and the denominator degrees of freedom (df_d) value would be equal to N (sample size) minus m1 (i.e., $df_d = N - k$.) With respect to this F test of the difference between the R^2 values of the Full and Restricted Models, it is important to remember that if significance is found, all that can be said is that not all of the k treatment criterion means are equal.

APPLIED RESEARCH HYPOTHESIS 4.2

Applied Research Hypothesis (ARH 4.2) illustrates the analysis of a global F test of four group means, which was the concept discussed in General Hypothesis 4.2. If this global F test was significant, this section illustrates the procedure used to conduct post hoc tests of the difference between pairs of means with Full Models. When analyzing ARH 4.2, the researchers addressed the research question: Do the criterion means of at least two groups in a set of four groups differ? If the data analysis provided a positive answer to this research question, the researchers addressed a second research question: Which group means differ?

The X2 variable in the GLM data set, which serves as the criterion variable for this analysis, is assumed to represent a set of test scores. The variables that represent these four groups are X16, X17, X18, and X19. In turn these groups are identified as follows:

1. The control group instructed in the morning (Control AM) is represented by the X16 variable.
2. The experimental group instructed in the morning (Experimental AM) is represented by the X17 variable.
3. The control group instructed in the afternoon (Control PM) is represented by the X18 variable.
4. The experimental group instructed in the afternoon (Experimental PM) is represented by the X19 variable.

To address the researchers' question of interest, the Research Hypothesis (ARH 4.2) is stated as follows:

ARH 4.2: For a given population, the test score means of the four groups are not all equal.

The corresponding statistical hypothesis (ASH 4.2) is as follows:

ASH 4.2: For a given population, the test score means of the four groups are equal.

To determine whether the data support ARH 4.2, Full and Restricted Models are designed to reflect ARH 4.2 and ASH 4.2, respectively. Since these models will be analyzed with the SPSS for Windows computer software, it is important to be aware of the fact the unit vector will be included in both models. Due to this characteristic of the computer program and the fact that the group variables (i.e., X16, X17, X18, and X19) are linearly dependent, only three of the four group variables are included in the model.

Full and Restricted Models

To depict the relationship among the group means described in ARH 4.2, the Full Model must be constructed as follows:

$$X2 = a_0U + a_{17}X17 + a_{18}X18 + a_{19}X19 + E_1 \qquad \text{(Model 4.20)}$$

This Full Model contains four pieces of information ($m1 = 4$), which are the unit vector and the vectors of the three groups explicitly included in the model.

The restrictions that must be placed on the Full Model in order for the Restricted Model to depict the relationship between the means of the groups described in ASH 4.2 are as follows:

$$a_{17} = 0, a_{18} = 0, \text{ and } a_{19} = 0 \quad \text{or} \quad a_{17} = a_{18} = a_{19} = 0 \quad \text{(3 restrictions)}$$

To understand these restrictions, it is important to note that the values for the regression coefficients a_{17}, a_{18}, and a_{19} will equal the differences between their respective groups and the mean of the group represented by the X16 variable, which was not entered into the Full Model (Model 4.20). The statistical hypothesis ASH 4.2 requires that these differences must be 0 (i.e., all four group means must be the same).

Placing the three restrictions on the Full Model (Model 4.20) produces the following Restricted Model:

$$X2 = a_0U + E_2 \qquad\qquad\text{(Model 4.21)}$$

This restricted model contains one piece of information ($m2 = 1$), which is the unit vector. Also note that this Restricted Model is the Null Model, yielding an R^2 value of .00. Once again, to determine whether the data support ARH 4.2, an F test of the difference between the R^2 values of the Full Model (Model 4.20) and the Restricted Model (Model 4.21) is conducted. For this F test, the numerator degrees of freedom value is equal to the number of restrictions or $m1 - m2$ (i.e., $4 - 1 = 3$); while the denominator degrees of freedom value is equal to the sample size (N) minus $m1$ (i.e., $60 - 4 = 56$). If the probability value for this F test was less than the alpha level of, say, .05, which was established by the researchers, the statistical hypothesis ASH 4.2 would be rejected, and the research hypothesis ARH 4.2 would be supported.

As discussed previously, if the statistical hypothesis ASH 4.2 is rejected, the researchers would have evidence that at least one of the many possible comparisons was significant. No information would be provided by the statistical test of ASH 4.2 regarding which one of those many possible comparisons was significant. The researcher may be interested in determining if a significant difference exists between the means of any two groups. Specifically, the researchers may want to conduct post hoc tests, such as the difference between each pair of group means. In such a case the researchers would be interested in six pairwise comparisons (the symbol μ represents the population mean): (a) μ_{17} vs. μ_{16}, (b) μ_{18} vs. μ_{16}, (c) μ_{19} vs. μ_{16}, (d) μ_{18} vs. μ_{17}, (e) μ_{19} vs. μ_{17}, and (f) μ_{19} vs. μ_{18}.

Note, that these six comparisons are not orthogonal or a priori. Thus, the alpha level used to test each comparison will be adjusted. The researchers may adjust the alpha level by dividing the alpha level of .05 by the number of comparisons being made, which is six for this example. Since the differences of six pairs of means are being statistically tested, the adjusted alpha level used in conjunction with each statistical test would be .008, which is the alpha level of .05 divided by 6. (See Newman, et al., (2000) for a discussion of a procedure for determining the level of adjustment needed in the alpha level for multiple statistical tests.)

The statistical tests of the differences among these pairs of means can be obtained by reviewing the results of the original Full Model (Model 4.21) and two variations of this Full Model. These variations of the original Full Model (Model 4.20) are constructed as follows:

$$X2 = a_0U + a_{16}X16 + a_{18}X18 + a_{19}X19 + E_3 \qquad \text{(Model 4.20a)}$$

$$X2 = a_0U + a_{16}X16 + a_{17}X17 + a_{19}X19 + E_3 \qquad \text{(Model 4.20b)}$$

These two additional Full Models are labeled Full Model 2 (Model 4.20a) and Full Model 3 (Model 4.20b).

Restrictions placed on the original Full Model (Model 4.20) will provide the statistical tests of the differences between the means of (a) μ_{17} vs. μ_{16}, (b) μ_{18} vs. μ_{16}, and (c) μ_{19} vs. μ_{16}. Restrictions placed on Full Model 2 (Model 4.20a) will provide the statistical tests of the differences between the means of (a) μ_{18} vs. μ_{17}, and (b) μ_{19} vs. μ_{17}. And restrictions placed on the Full Model 3 (Model 4.20b) will provide the statistical tests of the differences between the means of μ_{19} vs. μ_{18}.

Analysis of ARH 4.2 with SPSS for Windows

The GLM data used with the Applied Research Hypothesis 4.2 have been stored in a SPSS file entitled "GLM DATA SPSS FORMAT." This file can be accessed through the internet site listed in Appendix A. The data contained in this file are also listed in Appendix B. It should be noted that the SPSS for Windows menus and output sheets presented in this text were produced from Version 16.0. Certain "key" locations on the menus, such as the "OK" key, may vary from one version to another.

Once this file has been accessed with the SPSS for Windows computer software, the Linear Regression menu must be accessed to analyze the Full and Restricted Models used to reflect ARH4.2 and ASH 4.2, respectively. The Linear Regression menu is accessed by selecting the following sequence of menus:

Analysis
  Regression
    Linear

Once the Linear Regression menu is chosen, the menu presented in Exhibit 4.8 will be displayed.

The criterion variable (X2) and the predictor variables (X17, X18, and X19) contained in the original Full Model (Model 4.20) for ARH 4.2 must be entered into the Linear Regression menu. As illustrated in Exhibit 4.8, these variables are entered as follows:

1. Click on the X2 variable (see Oval 1) and click on the arrow key next to the "Dependent" box. This will identify X2 as the criterion variable.

2. While holding down the control key on the keyboard, click on the X17, X18, and X19 variables (see Oval 2). Release the control key and click on the arrow key next to the "Independent(s)" box. This will identify X17, X18, and X19 as the predictor variables.
3. Click on "OK" button (see Oval 3).

Exhibit 4.8.
SPSS Linear Regression menu—Original Full Model (Model 4.20) for ARH 4.2.

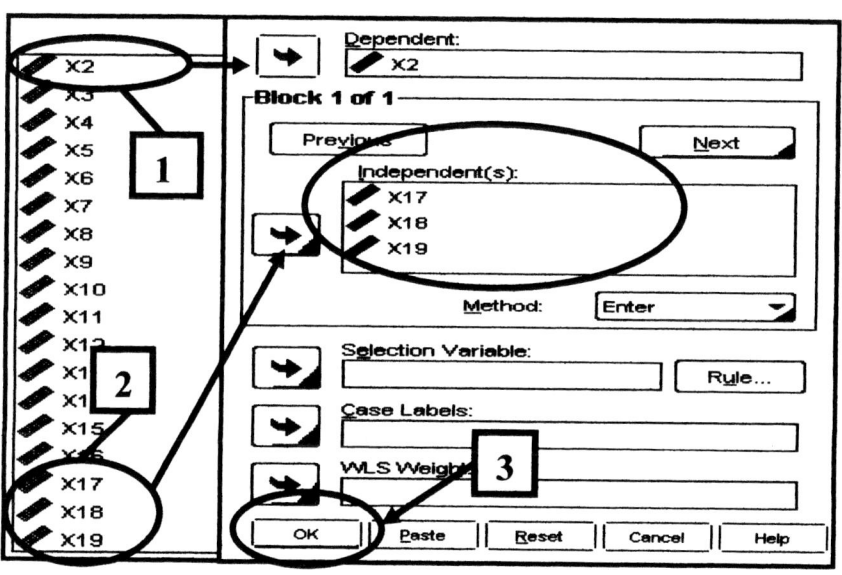

Once these three steps are completed, the regression analysis output contained in Exhibit 4.9 will be displayed. The key information on the output window is located in the following areas:

1. The original Full Model (Model 4.20) R^2 value is .729 (see Oval 1). Since the Restricted Model contains only the unit vector and error term, the R^2 value for the Restricted Model is equal to .00. This value is not listed on in the output window.
2. The numerator degrees of freedom (df_n) value and the denominator degrees of freedom (df_d) value are 3 and 56, respectively (see Oval 2). Since the number of restrictions placed on the original Full Model (Model 4.20) is 3, the df_n value is 3. The sample size of 60 minus the 4 estimated parameters (i.e., the three regression coefficients and the one

constant value) produces a df_d value of 56. These values are used in the
calculation of the F test.

3. The F test value listed in the ANOVA Table is the appropriate test of the
   difference between the R^2 values of the original Full Model (Model
   4.20) and the Restricted Model because the Restricted Model (Model
   4.21) is the Null Model. The F test value of the difference between the
   R^2 values of these Full and Restricted Models and its corresponding
   probability value are 50.23 and .00, respectively (see Oval 3). The probability value, which is listed under the "Sig." heading, can be expressed
   as $(p < .01)$.

Exhibit 4.9.
SPSS regression output—Original Full Model (Model 4.20) for ARH 4.2.

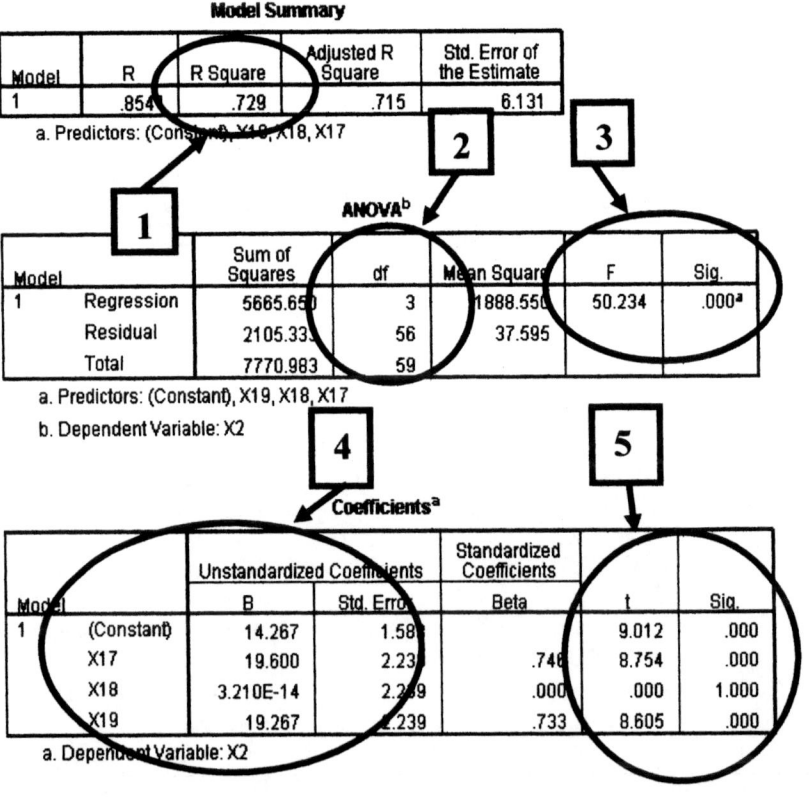

Based on the F test value (50.23) and its corresponding probability value ($p < .01$), the data provide sufficient evidence to reject ASH 4.2, which is the statistical hypothesis corresponding to ARH 4.2, at the .05 alpha level. Thus, the researchers are willing to state that for a given population, the four groups do not have equal test means.

As previously stated, once the researchers have concluded that for a given population, the four groups do not have equal test means, they may be interested in which pairs of group means differ. To determine which pairs of group means are significantly different, the researchers would examine the regression coefficients and their corresponding t tests produced by the analyses of three Full Models.

When reviewing the regression coefficients in the original Full Model (Model 4.20), which are presented in Exhibit 4.9, it is important to note that, the group variables X17, X18, and X19 were entered into the model, , but not variable X16. Due to this fact, the regression coefficients for X17, X18, and X19 indicate how much higher or lower their respective group means are compared to the mean for the group represented by the variable X16. Thus, the values contained in Exhibit 4.9 (see Oval 4) indicate the following regarding the various differences among the group means:

1. $\overline{X}_{17} - \overline{X}_{16} = 19.60$
2. $\overline{X}_{18} - \overline{X}_{16} = 0.00$ [Note: 3.210E-014 should be considered 0.00]
3. $\overline{X}_{19} - \overline{X}_{16} = 19.27$.

It should be noted that the value for the Constant (14.27) in Exhibit 4.9 (see Oval 4) is the mean of the predictor variable not entered into this Full Model, that is, the mean for the group represented by X16.

The t tests and their corresponding probability values for these differences among the pairs of means are as follows (see Oval 5):

1. $t_{17} = 8.75$, $p_{17} < .001$
2. $t_{18} = 0.00$, $p_{18} = 1.000$
3. $t_{19} = 8.61$, $p_{19} < .001$.

When using these probability values to determine which pairs of means are significantly different, the researchers would compare those values to the adjusted alpha level. As previously mentioned, this adjusted alpha level could be set at .008, which is the alpha level of .05 divided by the number of comparisons being made, which is six in this example. Since the probability values corresponding to the t tests for variables X17 and X19 are *less* than the adjusted alpha level of .008, the researchers would conclude that for a given population the means of groups X17 and X19 are higher than the mean of group X16 (i.e., $\mu_{17} > \mu_{16}$ and $\mu_{19} > \mu_{16}$). The probability value corresponding to the t test for variable X18, however, is *greater* than the adjusted alpha level. Thus, for a given population, the researchers are *not* willing to claim that the means of groups X18 and X16 are different (i.e., $\mu_{18} = \mu_{16}$).

Three additional statistical tests must be conducted to complete the testing of the differences between all possible pairs of means. The tests of the differences between the pairs of means of groups X18 and X17, and groups X19 and X17 are obtained from the output window for Full Model 2 (Model 4.20a), while the test of the difference between the means of groups X19 and X18 is obtained from the output window for the third Full Model (Model 4.20b). You may recall that in the original Full Model (Model 4.20), the variable X16 was not entered as a predictor variable, but variables X17, X18, and X19 were entered. This allowed the differences between the mean of the group represented by variable X16 and the means of the other three groups (i.e., for the groups represented by variables X17, X18, and X19) to be tested. In the Full Model 2 (Model 4.20a), variables X16, X18, and X19 were entered as predictor variables, but variable X17 was not. This allowed the differences between the mean of the group represented by variable X17 and the means of the other three groups (i.e., the groups represented by variables X16, X18, and X19) to be tested.

In Full Model 2 (Model 4.20a) the regression coefficients for X16, X18, and X19 indicate how much higher or lower their respective group means are compared to the mean for the group represented by the variable X17. Thus, the values contained in the "Coefficients" portion of the SPSS for Windows output for Full Model 2 (Model 4.20a) displayed in Exhibit 4.10 (see Oval 1) indicated the following regarding the various differences among the group means:

1. $\overline{X}_{18} - \overline{X}_{17} = -19.60$
2. $\overline{X}_{19} - \overline{X}_{17} = -0.33$.

Note that in this Full Model, the value for the Constant (33.87) contained in Oval 1 is the mean of the group represented by X17, which is the predictor variable not entered into the model.

The t tests and their corresponding probability values for these differences among the pairs of means, which are located in Oval 2 of Exhibit 4.10, are $t_{18} = -8.75$, $p_{18} < .001$; and $t_{19} = -0.15$, $p_{19} = .88$. Since the probability value corresponding to the t test for variable X18 is less than the adjusted alpha level of .008, but the probability value corresponding to the t test for variable X19 is greater than the adjusted alpha level, the researchers would conclude that for a given population the mean of group X18 is less than the mean of group X17, but they are not willing to claim that the mean of group X19 is different from the mean of group X17 (i.e., $\mu_{18} < \mu_{17}$ and $\mu_{19} = \mu_{17}$).

Full Model 3 (Model 4.20b) is needed to test the difference between the means of the final pair of means. The SPSS for Windows output for Full Model 3 is contained in Exhibit 4.10. In Full Model 3 (Model 4.20b), the regression coefficients for variables X16, X17, and X19 indicate how much higher or lower their respective group means are compared to the mean for the group represented by the variable X18. Once again, note that in the output for Full Model 3 (see Oval 3 of Exhibit 4.10) the value for the Constant (14.27) is the mean of the group represented by variable X18, which is the predictor variable not entered into the model.

Exhibit 4.10.
SPSS regression output for Full Models 2 and 3 (Model 4.20a and Model 4.20b)—Pairwise tests of means for ARH 4.2.

Full Model 2

Coefficients

Model		Unstandardized Coefficients		Standardized Coefficients		
		B	Std. Error	Beta	t	Sig.
1	(Constant)	33.867	1.583		21.392	.000
	X16	-19.600	2.239	-.746	-8.754	.000
	X18	-19.600	2.239	-.746	-8.754	.000
	X19	-.333	2.239	-.013	-.149	.882

Full Model 3

Coefficients

Model		Unstandardized Coefficients		Standardized Coefficients		
		B	Std. Error	Beta	t	Sig.
1	(Constant)	14.267	1.583		9.012	.000
	X16	4.402E-14	2.239	.000	.000	1.000
	X17	19.600	2.239	.746	8.754	.000
	X19	19.267	2.239	.733	8.605	.000

An examination of the regression coefficient for the X19 variable contained in Exhibit 4.10 (see Oval 3) indicates that $\bar{X}_{19} - \bar{X}_{18} = 19.27$. A review of the t test and the corresponding probability value for this difference between these pairs of means (see Oval 4), reveals that $t_{19} = 8.61$ and $p_{19} < .001$. Since this probability value is less than the adjusted alpha level of .008, the researchers would conclude that for a given population the mean of the group represented by variable X19 is significantly higher than the mean of the group represented by variable X18 (i.e., $\mu_{19} > \mu_{18}$).

To summarize the findings regarding the tests of the six pairs of means, which are presented in Table 4.2, it can be concluded that for a given population the means of the AM and PM Control groups do not significantly differ, and the means of the AM and PM Experimental groups also do not significantly differ.

However, the means of the AM and PM Control groups are less than the means of the AM and PM Experimental groups.

Table 4.2
Findings of the Multiple Comparisons Tests

Comparisons	Difference between Means	t Value	p
1. Experimental AM vs. Control AM	19.60	8.75	< .001
2. Control PM vs. Control AM	0.00	0.00	1.000
3. Experimental PM vs. Control AM	19.27	8.61	< .001
4. Control PM vs. Experimental AM	-19.60	-8.75	< .001
5. Experimental PM vs. Experimental AM	-0.33	-0.15	.882
6. Experimental PM vs. Control PM	19.27	8.61	< .001

Conclusions Drawn from Multiple Comparison Tests

In the previous section the researchers concluded that the means of groups X16 and X18 are less than the means of groups X17 and X19, that is, $\mu_{16} = \mu_{18} < \mu_{17} = \mu_{19}$, even though pair-wise multiple comparisons tests were conducted as nondirectional tests—a practice that is quite common. Some researchers would be hesitant, however, to state which group means are higher and which group means are lower when the nondirectional statistical tests of the differences in the group means were statistically significant; they would only be willing to state that those group means differed. Since the researchers in the previous section take a more liberal path regarding the conclusions drawn from the results of the nondirectional pair-wise tests of the group means, it is *extremely* important that the researchers call for the replication of those findings before too much importance is given to their findings.

It should be noted that researchers could use directional pair-wise tests based on theory or the researchers' experiences with the methods being investigated. If, in fact, the researchers do intend to use one-tailed pair-wise tests of the group means, it would not be necessary for them to conduct the overall F test of the group means. The point we believe is important to stress regarding this issue is that researchers need to spend time reflecting on their proposed hypotheses and the analytical methods they intend to use to test those hypotheses (see Newman, et. al., 2000 for a discussion of this point).

Writing the Results for ARH 4.2 in APA Style

The results for ARH 4.2 could be written in APA style as follows:

The differences among the mean test scores of the four groups accounted for 73% of the variation in the test scores, which was statistically significant at the .05 alpha level (F (3, 56) = 50.23, p < .01). This percentage of variation in the test scores accounted for by the groups would be classified as a large effect size according to Cohen (1988).

If the researchers continued the analysis by conducting the post hoc tests of the six pairs of group means, they would also include the following:

A review of the post hoc tests of the differences among the six pairs of means, which were conducted at the adjusted alpha level of .008, revealed no statistically significant difference between the means of the Control AM group (14.27) and the Control PM group (14.27) nor between the means of the Experimental AM group (33.87) and the Experimental PM group (33.54). The means of each of the two experimental groups, however, were significantly higher than the means of each of the two control groups.

Three points should be noted in the report writing by the researchers. First, the proportion of unique variation in the test scores accounted for by the group variables, which is equal to the difference between the original Full Model and the Restricted Model, should be used to gauge the effect size of the group variables. Second, researchers should consider classifying the effect size according to criteria suggested by Cohen (1988). Readers of the report or article may find the classification of the effect size as small, medium, or large informative. The third point relates to the importance of including in the report a Table similar to Table 4.2. Such a table would provide the readers of the report the critical statistical information regarding the multiple comparisons tests.

APPLIED RESEARCH HYPOTHESIS 4.3

Applied Research Hypothesis (ARH 4.3) illustrates the use of Full and Restricted Models to test the Comparison 11 in Exhibit 4.7. This comparison dealt with the difference between the average of two group means and the average of two other group means. For ARH 4.3, variables X16, X17, X18, and X19 represent the following:

1. The control group instructed in the morning (Control AM) is represented by the X16 variable.

2. The experimental group instructed in the morning (Experimental AM) is represented by the X17 variable.
3. The control group instructed in the afternoon (Control PM) is represented by the X18 variable.
4. The experimental group instructed in the afternoon (Experimental PM) is represented by the X19 variable.

In addition to these four group variables, the X2 variable is assumed to represent a set of test scores.

Applied Research Hypothesis (ARH) 4.3, which is directional, is as follows:

Directional ARH 4.3: For a given population, the new treatment, averaged across the two different time periods has a higher mean on the criterion test scores than the comparison treatment averaged across the two different time periods.

The corresponding statistical hypothesis (ASH 4.3) is as follows:

Directional ASH 4.3: For a given population, the new treatment, averaged across the two different time periods has the same mean on the criterion test scores as the comparison treatment averaged across the two different time periods.

Full and Restricted Models

To determine whether the data support ARH 4.3, the following Full Model is designed to reflect this applied research hypothesis:

$$X2 = a_0U + a_{17}X17 + a_{18}X18 + a_{19}X19 + E_1 \qquad \text{(Model 4.22)}$$

Before the Restricted Model used to reflect ASH 4.3 can be constructed, the restriction placed on the Full Model (Model 4.22) must be stated and manipulated. The restriction requires that the test scores (X2) of the treatment groups (X17 and X19), averaged across the two different time periods and the test scores (X2) of the control groups (X16 and X18), averaged across the two different time periods be equal. When using statistical computer programs that use a matrix inversion procedure to estimate the regression coefficients, such as SPSS for Windows or Microsoft Excel the restriction placed on the Full Model is as follows (if the number of participants in each group is the same):

$$(0 + a_{18}) / 2 = (a_{17} + a_{19}) / 2 \quad \text{(1 restriction)}$$

Once again, the values for the regression coefficients a_{17}, a_{18}, and a_{19} will equal the differences between their respective groups and the mean of the group represented by the X16 variable, which was not entered into the Full Model (Model 4.22). Due to this fact, the value of 0 is placed in the restriction for variable X16 (i.e., the mean of the X16 group does not differ from itself). Since each coefficient and the 0 value represents the difference between its respective group mean and the mean of the X16 variable, ASH 4.3 requires the mean of the average of 0 and a_{18} to equal the average of a_{17} and a_{19}.

This restriction can be reduced to the following $a_{18} = a_{17} + a_{19}$. Substituting this restriction into the Full Model results in the following Restricted Model:

$$X2 = a_0U + a_{17}X17 + (a_{17} + a_{19})X18 + a_{19}X19 + E_2 \qquad \text{(Model 4.23)}$$

Multiplying a_{17} by X18 and a_{19} by X18 produces the following Restricted Model:

$$X2 = a_0U + a_{17}X17 + a_{17}X18 + a_{19}X18 + a_{19}X19 + E_2 \qquad \text{(Model 4.23a)}$$

Collecting vectors that are multiplied by the same coefficient (i.e., vectors X17 and X18 which are each multiplied by a_{17}; vectors X18 and X19 which are each multiplied by a_{19}) results in the following Restricted Model:

$$X2 = a_0U + a_{17}(X17 + X18) + a_{19}(X18 + X19) + E_2 \qquad \text{(Model 4.23b)}$$

Variables labeled X20 and X21, which are equal to X17 + X18 and X18 + X19, respectively, are constructed and used as the predictor variables in the Restricted Model. Substituting these variables in the previous model and re-labeling the coefficients a_{17} and a_{19} as a_{20} and a_{21}, respectively, produces the following Restricted Model.

$$X2 = a_0U + a_{20}X20 + a_{21}X21 + E_2 \qquad \text{(Model 4.23c)}$$

An F test of the difference between the R^2 value of the Full Model (Model 4.22) and the R^2 value of the Restricted Model (Model 4.23c) is conducted to determine whether the data support ARH 4.3.

Analysis of ARH 4.3 with Microsoft Excel

The GLM data, which are listed in Appendix B, can be accessed from the internet site listed in Appendix A under the file name "GLM DATA EXCEL FORMAT." After this file has been accessed with the Excel computer software, the X20 and X21 variables must be generated to determine whether the data support ARH 4.3. As noted in the previous section, the variables X20 and X21 are formed by the addition of X17 and X18, and X18 and X19, respectively. Once these variables are constructed, the Microsoft Excel Regression menu will be used to estimate the Full and Restricted Models. Refer to the section in this chapter entitled "Analysis of ARH 4.1 with Microsoft Excel" for a discussion of the steps used to access the Regression menu.

Once the Regression menu has been accessed, the following steps are completed to obtain the output for the Full Model, Model 4.22 (see Exhibit 4.11):

1. Click on the box next to "Input $\underline{Y}$ Range" (see Oval 1). Note that variable X2 is located in column B in the data file. Next, click and drag on the cells B1 to B61 in the data file to identify the X2 variable as the dependent variable.
2. Click on the box next to "Input $\underline{X}$ Range" (see Oval 2). Note that in the data file the variables X17, X18, and X19, which are the predictor variables for the Full Model, are located in columns Q, R and S, respectively. Next, click and drag on the cells Q1 to S61 in the data file to identify these three variables as the predictor variables for the Full Model.
3. Click on the box in front of "Labels" (see Oval 3).
4. Click on the "OK" button (see Oval 4).

After these four steps are completed, the output window for the Full Model (Model 4.22) will appear as displayed in Exhibit 4.12.

Exhibit 4.11.
Microsoft Excel Regression menu—Full Model (Model 4.22) for ARH 4.3.

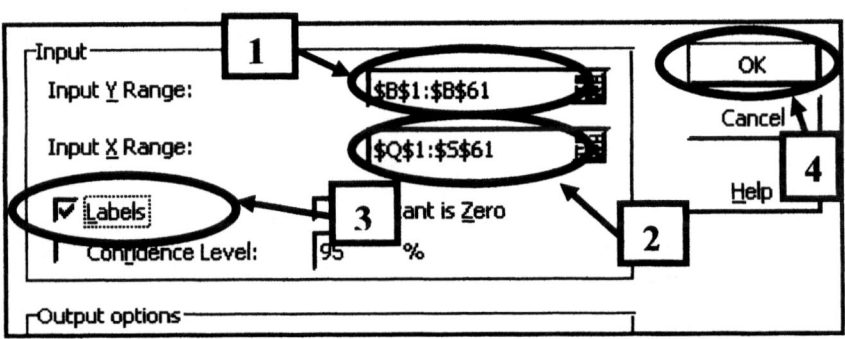

When reviewing the output for the Full Model it is important to remember that the Restricted Model used to test ASH 4.3 contains two predictor variables. Thus, the Restricted Model is not the Null Model. Since the Restricted Model is not the Null Model, the F test listed in the portion of the output window entitled "ANOVA Table" is not the F test needed to statistically test ASH 4.3. The appropriate F test must be conducted outside of the Regression menu.

Exhibit 4.12.
Microsoft Excel regression output—Full Model (Model 4.22) for ARH 4.3.

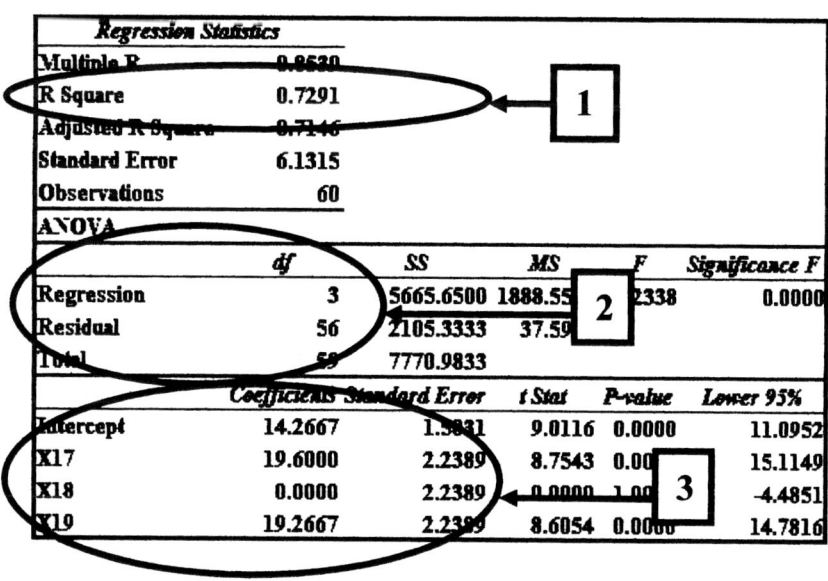

Five values are required to conduct the appropriate F test. A review of the formula used to calculate the F test for the difference between the R^2 values of a Full Model and a Restricted Model is helpful in identifying these five values. The F test formula used to statistically test ASH 4.3 is as follows:

$$F(df_n, df_d) = \frac{(R_F^2 - R_R^2) / (df_n)}{(1 - R_F^2) / (df_d)}$$

(Equation 4.1)

where:

1. The R_F^2 symbol represents the R^2 value of the Full Model.
2. The R_R^2 symbol represents the R^2 value of the Restricted Model.
3. The df_n symbol represents the numerator degrees of freedom value of the F test. The value for df_n is equal to the number of restrictions placed on the Full Model to obtain the Restricted Model, which in this case is 1. The value for df_n can also be found by subtracting the pieces of information in the Restricted Model from the number of pieces of information in the Full Model [(m1 - m2) = (4 - 3 = 1)]. This value also can be obtained from the output windows of the Full and Restricted Models. The df_n

value is equal to the regression degrees of freedom value for the Full Model minus the regression degrees of freedom value for the Restricted Model (3 - 2 = 1).

4. The df_d symbol represents the denominator degrees of freedom value of the F test. This value is 56 (N - m1, 60 participants minus the 4 pieces of information in the Full Model). This value also can be obtained from the output window of the Full Model. The df_d is equal to the residual degrees of value for the Full Model (56).

To conduct the appropriate F test of ASH 4.3, three of these pieces of information are obtained from the Full Model (Model 4.22) output listed in Exhibit 4.12. The first piece of information is the R^2 value of .729 (see Oval 1). The second and third key values are the Full Model regression degrees of freedom value of 3 and the residual degrees of freedom value of 56 (see Oval 2). The two other pieces of information, which are obtained from the Restricted Model printout, will be discussed presently.

In addition to the information required to conduct the F test of ASH 4.3, the means of the four groups, which will be used to calculate the mean test score of the experimental groups across the time periods and the mean test scores of the control groups across the time periods, are obtained from the Full Model (Model 4.22) output. The group means are derived from the values listed under the "Coefficients" box on the output window (see Oval 3 in Exhibit 4.12). Since the X16 variable was entered into the Full Model, the value listed on the "Intercept" row is the mean of the Control AM group (14.27). The values listed in the X17, X18, and X19 rows indicate how much higher or lower the respective group means are compared to the mean of the Control AM group (X16). Thus the mean of each of the other three groups is the sum of the mean of the Control AM group (i.e., the intercept value) and the value for a given variable listed under the "Coefficient" heading. Thus, the means of the three other groups are calculated as follows:

1. The Experimental AM group (X17) mean is equal to the sum of 14.27 and 19.60, which is equal to 33.87.
2. The Control PM group (X18) mean is equal to the sum of 14.27 and 0.00, which is equal to 14.27.
3. The Experimental PM group (X19) mean is equal to the sum of 14.27 and 19.27, which is equal to 33.54.

As previously mentioned, the calculation of the appropriate F test value for ASH 4.3 requires the use of five values, three of which are obtained from the Full Model (Model 4.22) output and two of which are listed on the Restricted Model (Model 4.23c) output. Thus, in addition to the Full Model output, the Restricted Model output is needed.

Before the output for the Restricted Model is obtained from the Excel computer software, the researchers must create the following variables (see Model 4.23b):

1. X20 = (X17 + X18).
2. X21 = (X18 + X19).

The X20 variable is created by typing "X20" in cell T1 and "= Q2 + R2" in cell T2. Copy the contents of the T2 cell and paste it into cells T3 to T61. To create the X21 variable, type "X21" in cell U1 and "= R2 + S2" in cell U2. The contents of the U2 cell is copied and pasted into cells U3 to U61.

Once these variables have been created and the Regression menu has been accessed, the output for the Restricted Model is obtained by completing the following steps (see Exhibit 4.13):

1. Click on the box next to "Input Y Range" (see Oval 1). Note that variable X2 is located in Column B in the data file. Next, click and drag on the cells B1 and B61 in the data file to identify the X2 variable as the dependent variable.
2. Click on the box next to "Input X Range" (see Oval 2). Note that in the data file the variables X20 and X21, which are the predictor variables for the Restricted Model, are located in columns T and U, respectively. Next, click and drag on the cells T1 to U61 in the data file to identify these three variables as the predictor variables for the Restricted Model.
3. Click on the box in front of "Labels" (see Oval 3).
4. Click on the "OK" button (see Oval 4).

After these four steps are completed, the output window for the Restricted Model (Model 4.23c) will be produced as displayed in Exhibit 4.14.

Exhibit 4.13.
Microsoft Excel Regression menu—Restricted Model (Model 4.23c) for ASH 4.3.

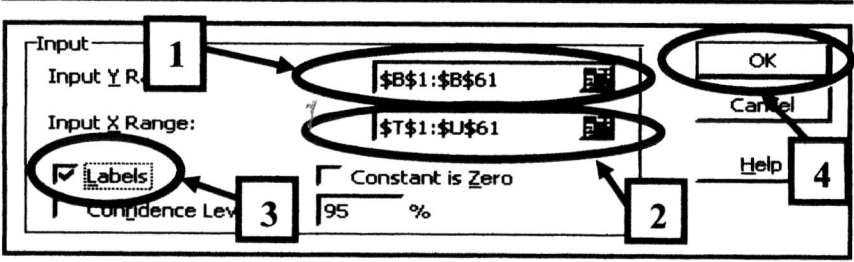

As previously discussed, five values are needed to conduct the appropriate F test for ASH 4.3. Three of these values were obtained from the Full Model (Model 4.22) output: (a) the R^2 value of .729, (b) the Full Model regression degrees of freedom value of 3, and (c) the residual degrees of freedom value of 56.

The other two values are obtained from the Restricted Model (Model 4.23c) output displayed in Exhibit 4.14. These Restricted Model values are (a) the R^2 value of .0001—see Oval 1 of Exhibit 4.14 and (b) the Restricted Model regression degrees of freedom value of 2—see Oval 2 of Exhibit 4.14.

Exhibit 4.14.
Microsoft Excel regression output—Restricted Model (Model 4.23b) for ASH 4.3.

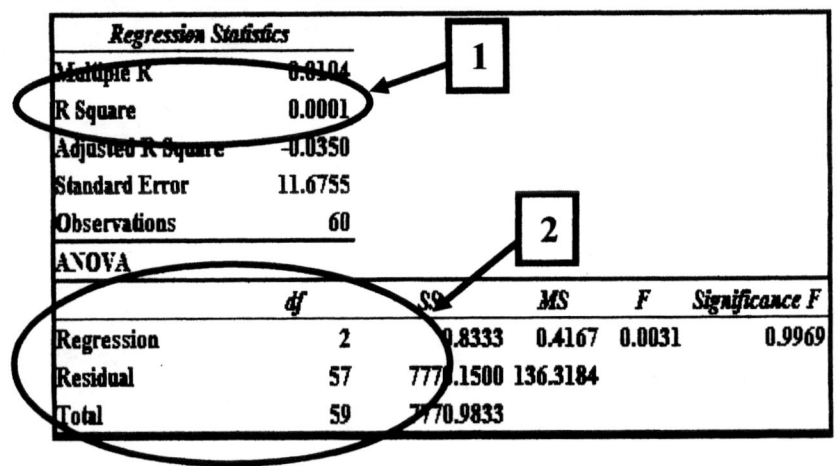

Regression Statistics					
Multiple R	0.0104				
R Square	0.0001				
Adjusted R Square	-0.0350				
Standard Error	11.6755				
Observations	60				

ANOVA

	df	SS	MS	F	Significance F
Regression	2	0.8333	0.4167	0.0031	0.9969
Residual	57	7770.1500	136.3184		
Total	59	7770.9833			

Remember, the Microsoft Excel Regression menu does not provide the appropriate F test for the difference between the R^2 values of the Full and Restricted Models designed for ARH 4.3. However, the appropriate F test value can be calculated with the use of the Microsoft Excel file entitled "F Test Calculation," which can be accessed from the internet site listed in Appendix A. A list of the commands needed to construct the Excel program is posted in Appendix C. Once this file is accessed, the following values are entered into the file (see Oval 1 in Exhibit 4.15):

1. The R^2 value of the Full Model (.729) is entered into row 1 of column B.
2. The R^2 value of the Restricted Model (.0001) is entered into row 2 of column B.
3. The regression degrees of freedom value for the Full Model (3) is entered into row 3 of column B.
4. The regression degrees of freedom value for the Restricted Model (2) is entered into row 4 of column B.
5. The residual degrees of freedom value for the Full Model (56) is entered into row 5 of column B.

Once these values are entered into the file, the F test value (150.62) and its corresponding probability value ($p < .01$) will be displayed (see Oval 2). Since ARH 4.3 is a directional hypothesis, the probability value would be divided by 2, which, of course, produces a one-tailed p value that is less than .01.

Exhibit 4.15.
Microsoft Excel F-test calculation for ARH 4.3.

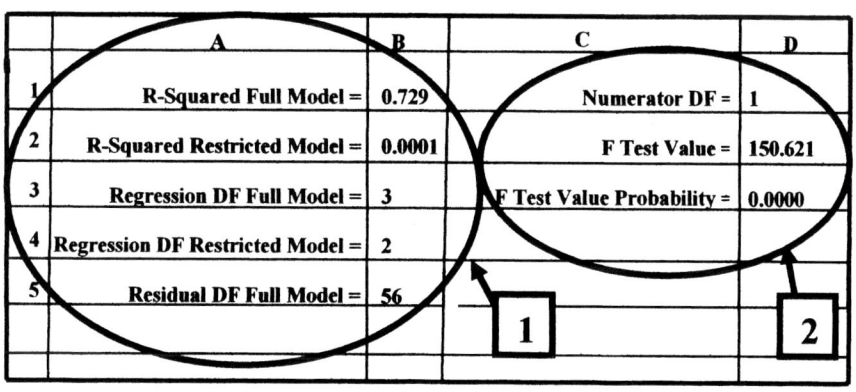

	A	B	C	D
1	R-Squared Full Model =	0.729	Numerator DF =	1
2	R-Squared Restricted Model =	0.0001	F Test Value =	150.621
3	Regression DF Full Model =	3	F Test Value Probability =	0.0000
4	Regression DF Restricted Model =	2		
5	Residual DF Full Model =	56		

Two conditions must be met in order for the statistical test results to support ARH 4.3, which is a directional hypothesis. First, the mean of the experimental groups across the time periods must exceed the mean of the control groups across the time periods. Using the four group means obtained from the Full Model (Model 4.22), the mean of the experimental groups across the time periods was calculated to be 33.70 (i.e., the sum of 33.87 and 33.53 divided by 2, assuming equal number of subjects in each group), while the mean of the control groups across the time periods was calculated to be 14.27 (i.e., the sum of 14.27 and 14.27 divided by 2, assuming equal number of subjects in each group). Since the mean of the experimental groups across the time periods exceed the mean of the control groups across the time periods, the first condition was met.

Second, the one-tailed probability of the F test value must be less than the established alpha level of .05. Since the one-tailed probability of the F test ($p / 2 < .01$) is less than the .05 alpha level (i.e., the alpha level established by the researchers), this second condition is met. Due to the fact that both conditions were met, we are willing to state that the data support the claim that for a given population, the experimental treatment, averaged across the two different time periods (X17 and X19) has a higher mean test score (X2) than does the comparison treatment averaged across the two different time periods (X16 and X18).

Writing the Results for ARH 4.3 in APA Style

The report of the researchers' findings regarding the statistical testing of ASH 4.3 could be written as follows:

> The mean of the experimental groups across the time periods was 33.70, while the mean of the control groups across the time periods was 14.27. The difference between these two means was significant at the .05 alpha level, F $(1, 56) = 150.62, p / 2 < .01$. The proportion of unique variation accounted for in the test scores by these two groups is .73, which is classified as a large effect size according to the criteria established by Cohen (1988).

Four points should be noted about the manner in which the results are reported. First, the mean test scores of the experimental and control groups across the time periods were reported. Second, the F test and its corresponding probability value were noted. It is important to stress that the probability value of the statistical test was reported. It is our position that this is consistent with the emerging notion of reporting the probability of the sample data occurring under the Statistical Hypothesis rather than imposing one's (arbitrarily chosen perhaps) alpha level on the readers. This position does not support the selection of alpha after the results are computed—instead it treats research readers as thinkers rather than as blind followers. A given journal author may adopt an alpha of .05 and find that the results are significant for the author. If the actual probability is not reported, a more conservative reader cannot determine whether the results are significant at .01 or not. However, if the author reports the actual probability of .007, the conservative reader knows that the Research Hypothesis is tenable for the reader as well as for the author. On the other hand, if the actual probability is .03 and is reported, then the Research Hypothesis is *not* tenable for the conservative reader, though the reader realizes that the Research Hypothesis is tenable for the author.

Third, the amount of unique variation in the test scores accounted for by the experimental and control groups across the time periods, which is equal to the difference between the R^2 values of the Full Model and the Restricted Model, was reported as the effect size. Fourth, this effect size was classified as a large effect size according to the criteria suggested by Cohen (1988). The editors of some journals will not require the effect size to be classified. However, many journal editors will require the effect size to be reported (Thompson, 1999).

IMPORTANCE OF REPORTING
PRACTICAL SIGNIFICANCE

It is important to note that the write up of the findings in every one of the Applied Research Hypotheses include an effect size value. Many researchers are aware that statistical significance can be obtained; yet practical significance may really not be obtained. Statistical significance is indeed a necessary but not sufficient condition for practical significance. Statistical significance simply indicates that something other than chance is operating, but statistical significance does not indicate to what degree that non-chance phenomenon is operating. Research studies using extremely large sample sizes often do find statistical significance, but little practical significance is obtained. The larger the sample size, the smaller the difference between two means needs to be before statistical significance is obtained. For example, one treatment mean could be 14.01 and the other 14.00, and if enough subjects were in the sample, statistical significance could be obtained. In most practical applications, a mean difference of .01 would not be of any importance.

When the Statistical Hypothesis is not true, increasing the sample size lowers the probability that a result of a given magnitude occurs by chance alone. The R^2 value, though, is not artificially inflated by the increase in sample size. Increasing sample size merely produces a more stable and closer estimate of the population R^2 value. This is one of the reasons that some researchers have been reporting R^2 values with their indexes of statistical significance. Persons who are writing a literature review should be encouraged to compute and include the R^2 value of each finding reviewed. When an F value is reported in the literature (when R^2 of the Restricted Model is .00), the R^2 value of the Full Model is:

$$R^2 = \frac{df_n * F}{df_d + (df_n * F)}$$

(Equation 4.2)

When a t is reported, the R^2 value is:

$$R^2 = \frac{t}{df + t^2}$$

(Equation 4.3)

For instance, a t of 2.4 with 116 subjects would be "highly significant" ($p < .01$); yet using Equation 4.2, the R^2 is found to be .05, showing that the researcher is accounting for only 5% of the phenomena. One might interpret this as "knowing something for sure about very little." In the past, probability values have been used to support theories. It is hoped that, in the future, R^2 values instead of probability values will be used to indicate the satisfaction one has with one's theoretical framework.

QUESTIONS FOR DR. GLM

After studying the concepts presented in this chapter, budding researchers may have various questions regarding those concepts. The questions posed in this section relate to: (a) sample size, (b) the traditional sequencing of hypotheses, and (c) inspection of residual plots.

Sample Size

Budding Researcher (Bud): How large a sample should I use?

Dr. General Linear Model (Dr. GLM): The desirability of large sample sizes has been an illusion for many years. As you have already learned, the number of people in a study enters into the calculation of the degrees of freedom denominator $(N - m_1)$. The F test of significance is adjusted for the number of people in the sense that, with a fixed R_F^2 and R_R^2, it is harder to get statistical significance with a smaller number of people. With reference to the F denominator degrees of freedom, the minimum sample size is one more subject than predictor variables in the Full Model. If there were only as many people as predictor variables in the Full Model, the denominator degrees of freedom would be zero (a researcher's bad dream). And if you were analyzing fewer people than predictor variables, the denominator degrees of freedom would be negative (a researcher's nightmare). If only as many people are used as predictor variables, the R^2 would always be 1.0, exemplifying the notion of producing a spurious result by overfitting the data.

I'm not saying you should *always* use a small number of people. What I'm trying to communicate is that you need only one more subject than the number of predictor variables in order to perform the F test of significance. Now, if a researcher has a good grasp of the functional relationship being studied, one needs only a few people. But I suspect that most relationships being studied in the behavioral sciences right now and in the near future will *not* yield extremely high R^2 values; thus larger sample sizes are needed to be fairly certain that one is getting a good estimate of the relationship in the population.

Unfortunately, there is no one recognized rule of thumb concerning the number of people to use. However, a researcher can use a number of guidelines that call for a larger recommended sample size: (a) the less tight a grasp the researcher has on the functional relationship; (b) the wider the range of values on the continuous predictor variables; and (c) the more reliance on data snooping rather than on theory and past research.

On the other hand, you must never let a small sample size derail an analysis. You must realize that small sample sizes reduce statistical power—the probability of finding significance if the Research Hypothesis is true in the population. Indeed, increasing the sample size will lower the probability value, while a change in the sample size has no systematic effect on the R^2 value (after the effects of overfitting have been eliminated—say, 20 people per predictor variable). Increasing the sample size merely results in the sample R^2 being a better estimate of the population R^2. That is to say, increasing the sample size tends to yield a sample R^2 that is *not* closer to one or to zero but to the population R^2, whatever it is. This seems to be a compelling reason to rely upon R^2 for decisions rather than solely on the probability value. (You might want to be sure to read the section on power in chapter 8.)

All findings require replication, and unexpected findings found in the analyses of small samples simply indicate a reduced chance of those findings being replicated. The proof is always in the prediction, and whether a finding from a small sample has merit rests on empirical replication of the finding.

Traditional Sequencing of Hypotheses

Bud: I have a number of hypotheses that I want to test. Is there a "correct order" in which to test them?

Dr. GLM: Whenever a researcher collects data, there are often reasons for several interrelated hypotheses to be tested. Sometimes the order of testing makes a difference, such as when making sure that certain assumptions are tenable. There are other times when the sequence depends on the design itself. At other times, the sequence depends on the desired conclusions. Therefore, there is no one "correct" way to sequence one's hypotheses. Indeed we have seen research studies that used many different sequences and used them very appropriately. The important point is that the Research Hypotheses should dictate each analysis, as well as the sequence.

Bud: The two-way ANOVA analogue discussed in this chapter seems to have a built-in sequence.

Dr. GLM: The two-way ANOVA usually does follow a specific sequence. The steps in this sequence are as follows:

Step 1: Test for interaction between the A factor and the B factor.

Step 2: If the interaction is significant, then stop.

Step 3: If the interaction is *not* significant, then test both the A main effect and the B main effect.

There may be times, though, when the interaction question is not of interest, say, when two curricula are being compared and the blocking variable of IQ is used. Although there may be an IQ-by-curriculum interaction, a given school district may not have the capacity for offering two different curricula—the district cannot afford to buy two sets of curricula and to train teachers in each of the two curricula. The purpose for including IQ as the blocking variable is to reduce the error variance in the analysis. Here, though, the researcher would not test for either an IQ-by-curriculum interaction or the IQ main effect. The IQ main effect is almost a given, and what could the researcher do, change some of the students' IQ levels? In such a case the researcher would directly test the A main effect and the B main effect.

Inspection of Residual Plots

Bud: Can you give me some advice regarding how I can determine if my data meet the assumptions?

Dr. GLM: It is always a good idea to look at a scatter plot of your data points to determine the shape of the data, the potential relationship between the predictor and criterion variables, to identify potential outliers, and to decide if some kind of data transformation is desirable. One of the most useful approaches for investigating the underlying assumptions of regression is to examine the residuals (the error vector—E), as it relates to the predicted values ($\hat{Y}$). If the underlying assumptions of regression are met, the residuals will be randomly and evenly scattered around a horizontal line—meeting the assumption of homogeneous variance (see Figure 4.4).

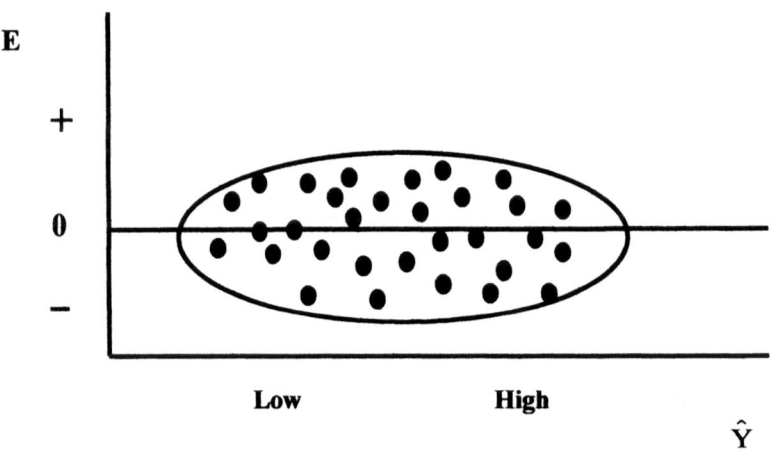

Figure 4.4.
A representative residual plot between the residuals (E) and the predicted values ($\hat{Y}$) when the assumption of homogeneous variance is met.

If the plot between the residuals and the predicted values is reflective of heterogeneous variance, one way the figure could look is as presented in Figure 4.5, like a megaphone. If there is a violation of the assumption of a linear relationship, the plot between the residuals and the predicted values may take on the shape illustrated in Figure 4.6, looking somewhat like an inverted U. It would not be too unusual to have more than one violation occur at the same time, and a plot of the residuals may take on a shape that is a combination of non-linearity and heterogeneity of variance. One can think of an inverted-U with greater variability at the low end and very little variability at the high end.

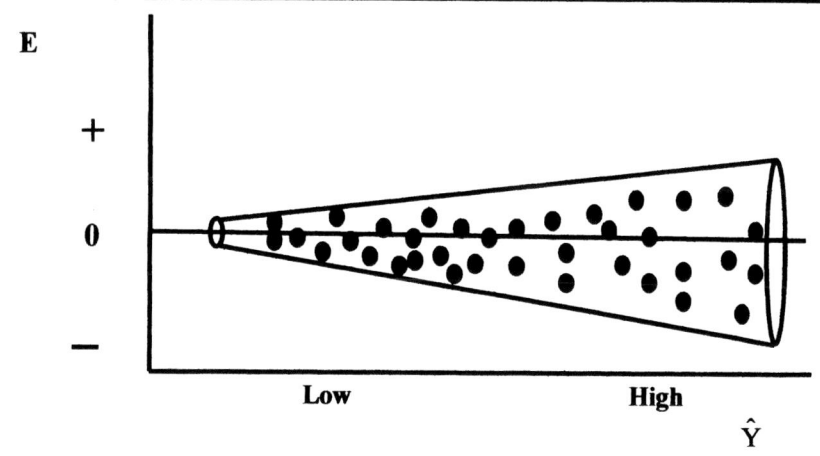

Figure 4.5.
A representative residual plot between the residuals (E) and the predicted values ($\hat{Y}$) when the assumption of homogeneous variance is not met.

Any systematic clustering of the residuals into a pattern other than a rectangular shape would be reflective of violations to the underlying assumption that in regression equations the residuals are normally distributed and independent. This is the basis of the statistical test of significance for regression (McNeil, Newman, & Kelly, 1996; Stevens, 2002). These violations can have drastic effects of Type I and Type II Error rates, and may require some conditioning of the data, such as data transformations or elimination of some extreme cases, so that the predicted scores will better fit the underlying assumptions of the statistical procedure.

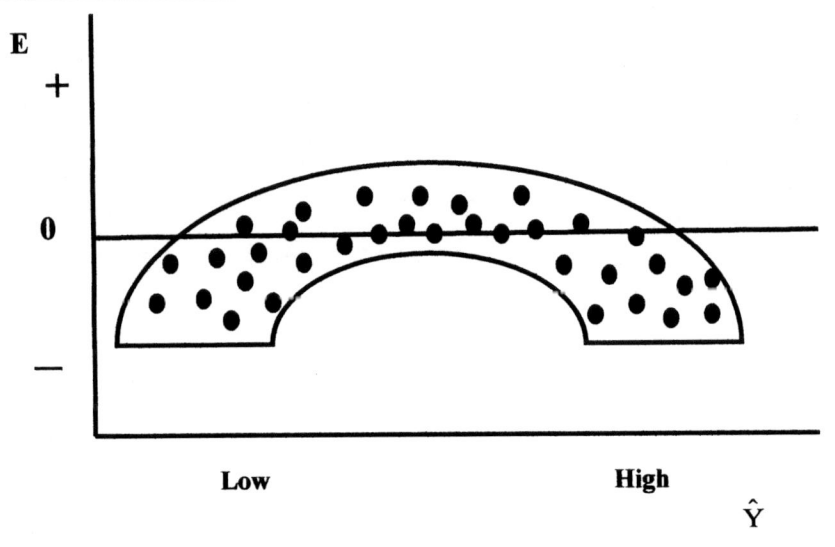

Figure 4.6.
A representative residual plot between the residuals (E) and the predicted values
($\hat{Y}$) when the assumption of linearity is not met.

CHAPTER 5

RESEARCH HYPOTHESES EMPLOYING
CONTINUOUS PREDICTOR VARIABLES

In the previous chapter, we discussed the use of the GLM with categorical predictor variables. The flexible GLM also can be used with continuous predictor variables. All the procedures are the same, but the way the Research Hypothesis is stated is often different because the data being analyzed are continuous instead of categorical.

The simple linear model and its components are presented. A simple data set is given to provide concrete meaning for these components. Since the calculation of linear components in complex models is an easy process with existing computer programs, procedures for calculating coefficients are discussed only briefly. The emphasis of this chapter is on stating the Research Hypothesis and constructing linear models designed to answer that Research Hypothesis. An actual problem (Applied Research Hypothesis 5.1), with data, computer output of the analysis, and an APA style write up are provided.

RESEARCH HYPOTHESES REQUIRING A
SINGLE STRAIGHT LINE OF BEST FIT

Given a criterion behavior (Y) exhibited by a group of individuals under study, one may wish to know if the variance in these criterion scores can be accounted for by one or more predictor variables. The information at hand to account for criterion behavior may take many forms (gender of the subject, previous test scores, knowledge of which treatment the subject received, etc.).

For illustrative purposes, suppose one is interested in establishing the following Research Hypothesis:

Research Hypothesis: There is a relationship between ability and performance.

Now suppose that scores on the performance behavior (Y), and the ability (X), which are hypothesized to be relevant to that performance behavior, are obtained for five subjects. Implied by the Research Hypothesis just given is a supposition that there *is* a systematic relationship between the X and Y variables. This relationship could take many forms, but since the particular form is not specified, the default assumption is that the relationship being investigated is a linear one. (Chapter 9 deals with nonlinear relationships, which we feel should more often be investigated).

Figure 5.1 is the graph of the observed performance and ability scores for the five individuals (A, B, C, D, and E). The description of this graph of scores involves using the formula for a straight line:

$$Y = a_0 + b_1X \qquad \text{(Equation 5.1)}$$

where:
1. $Y =$ the scores on the Y-axis variable, (performance scores);
2. $X =$ the scores on the X-axis variable (ability scores);
3. $a_0 =$ the point at which the straight line crosses the Y axis (called the *Y-intercept*); and
4. $b_1 =$ the increase in the Y-axis variable for every one-unit increase in the X-axis scores (*called the slope of the line*).

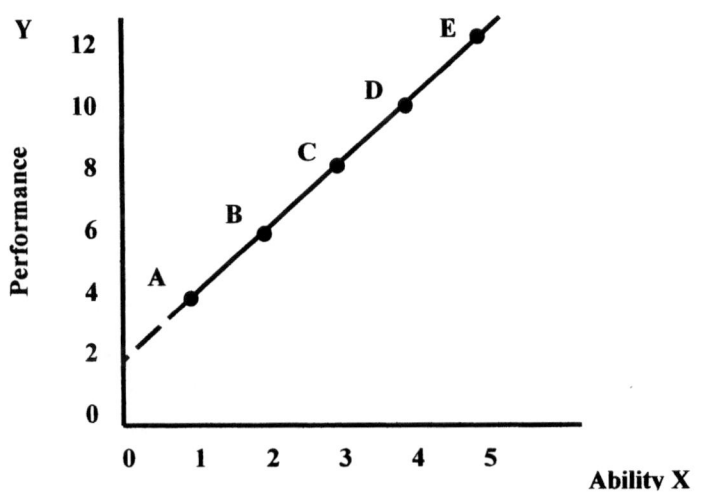

Figure 5.1.
The observed X and Y scores for five individuals.

In Figure 5.1, notice the line intersects each of the five points. Calculating values for a_0 and b_1 in Equation 5.1 can specify this line. What the value of the slope (b_1) reveals about this line can best be understood by noting that Person A scored 1 on ability and 4 on performance, while Person B scored 2 on ability and 6 on performance. Thus, for a one-unit increase in ability (i.e., from 1 to 2), performance increased two units (i.e., from 4 to 6). For every one-unit increase in X, there is a two-unit increase in Y. Therefore, b_1 (in Equation 5.1) must equal 2. The value for a_0 identifies the point where the specified line crosses the Y axis. Thus, the value of a_0 is equal to the Y value when the X value is set equal to 0.

Vectors for the Straight Line

So far, $\hat{Y} = a_0 + 2X$ has been obtained, where $\hat{Y}$ represents the predicted Y, and $b_1 = 2$. To complete the equation, assume for the moment that $a_0 = 0$. In vector form the equation would be as follows:

$$\hat{Y} \quad = \quad 0 * U \quad + \quad 2 * X$$

$$
\begin{array}{l}
\text{Person A} \\
\text{Person B} \\
\text{Person C} \\
\text{Person D} \\
\text{Person E}
\end{array}
\begin{bmatrix} 2 \\ 4 \\ 6 \\ 8 \\ 10 \end{bmatrix}
= 0 *
\begin{bmatrix} 1 \\ 1 \\ 1 \\ 1 \\ 1 \end{bmatrix}
+ \; 2 *
\begin{bmatrix} 1 \\ 2 \\ 3 \\ 4 \\ 5 \end{bmatrix}
$$

Note that a_0 can be represented as $a_0 * U$. Indeed, the value of a_0 is added to each subject. Also note that the first element in vector X is 1 and must be the score on X for Person A. Also, the X score of 5 must represent the score on X for Person E. The vector U provides a one for every person and, when multiplied by a_0, adjusts each score by a constant amount. This is the reason that the coefficient for the unit vector, a_0, is often referred to as the *regression constant*.

When each element in X is multiplied by 2 (the value of b_1) and added to the value of "0 times the unit vector," the result is the $\hat{Y}$ vector. In finding values for a_0 and b_1, one is trying to make $\hat{Y}$ equal to Y. If the value of a_0 is really 0, as has been assumed above, then $\hat{Y}$ should equal Y. Note that with $a_0 = 0$, however, $\hat{Y}$ and Y are not equal, as demonstrated as follows:

	Y		$\hat{Y}$		$Y - \hat{Y}$
Person A	4		2		2
Person B	6		2		2
Person C	8		2		2
Person D	10		2		2
Person E	12		2		2

To make $\hat{Y}$ equal to Y, a value of 2 must be added to every score in $\hat{Y}$. This means that $a_0 = 2$ (not 0). Now with $a_0 = 2$ and $b_1 = 2$, a line of perfect fit is obtained, as illustrated below and depicted in Figure 5.2.

	$\hat{Y}$	=	2 * U	+	2 * X		$Y - \hat{Y}$
Person A	4		1		1		0
Person B	6	= 2 *	1	+ 2 *	2		0
Person C	8		1		3		0
Person D	10		1		4		0
Person E	12		1		5		0

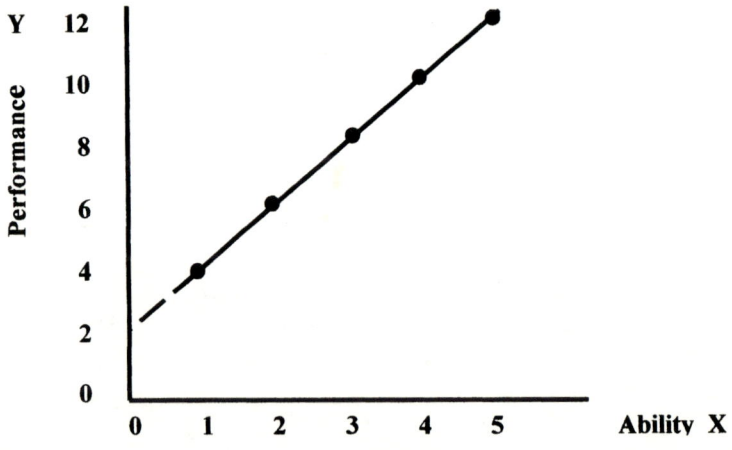

Figure 5.2.
The line of best fit from the Model: $\hat{Y} = 2 + 2X$.

None of the observed ability scores was zero, but if there were an X score of zero, the value of Y would be 2, and the line of best fit would cross the Y axis at the Y value of 2, which is therefore the intercept for these data. The original

Research Hypothesis for which these data were obtained was: There is a (linear) relationship between ability and performance. Figure 5.2 shows that there is a linear relationship between X and Y; without exception, as X increases, so does Y. And since the increase has a constant value (i.e., 2), Y is a linear function of X.

Statistical Test of the Straight Line

The observed relationship presented in the preceding section may be due to sampling error, as discussed in chapter 2. If one wished to generalize beyond these five subjects, one would want to ask: How likely is it that the observed linear relationship between X and Y in the sample is due to sampling variation? This can be determined by the *F* statistic. In the data that were collected, there was only one person at each of the five X scores. If the study included more subjects and therefore more observations at each point on X, it would be unlikely that all subjects with a particular score on X would have had the same criterion (Y) score. Each point on the X variable has a population of people with that score. In any particular study, the individuals at the scale point X = 2 are a sample of the population of people who have an X score of 2—a statement that holds true for all scale points on X. When a line is fit to the data, then all observed squared deviations on Y from that line for all scale points on X can be viewed as a within-group sum of squares. Knowledge of X cannot explain these deviations. The difference between this within-group sum of squares and the total sum of squares yields the variation explained by knowledge of the estimated linear relationship between X and Y.

For the data in Figure 5.2, every data point is located on the line. Thus, there is perfect prediction for these data, a very unlikely situation. Figure 5.3 more clearly shows the notion of variability about the line of best fit for each value of the predictor variable X. Note that in Figure 5.3 two subjects have an ability score of 6, but one is at a performance level of 5, whereas the other is at a performance level of only 3. Figure 5.3 also portrays an unlikely situation in that the criterion means for each ability level fall directly on the line of best fit.

We will begin our discussion of the statistical test of a straight-line relationship by assuming that the data displayed in Figure 5.2 were generated by the study. The researchers would begin the study by posing the following research question:

Does a linear relationship exist between the criterion variable Y and the predictor variable X?

The Research Hypothesis that addresses this research questions is as follows:

Research Hypothesis: For the population of interest, a linear relationship does exist between the criterion variable Y and the predictor variable X.

The corresponding Statistical Hypothesis is as follows:

Statistical Hypothesis: For the population of interest, a linear relationship does not exist between the criterion variable (Y) and the predictor variable (X).

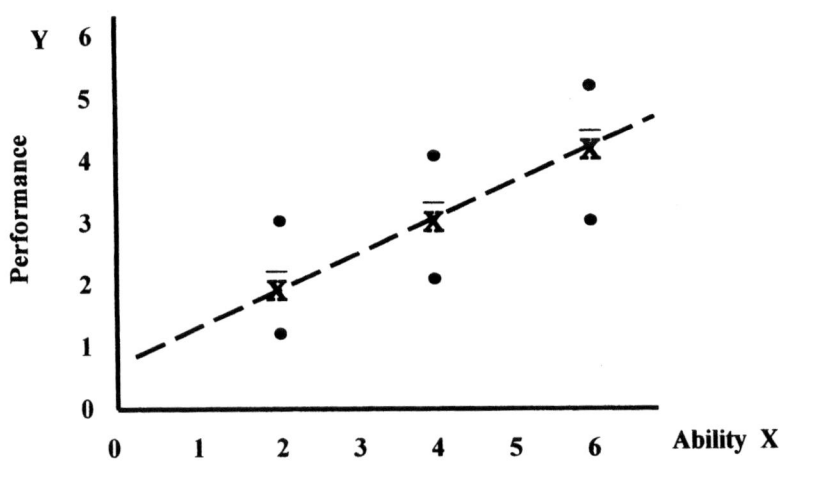

Figure 5.3.
Criterion variability for persons having the same X scores.

To test a Research Hypothesis, one must compare the accuracy of two linear models used to fit the same data. One model reflects the state of affairs dictated by the Research Hypothesis and the other model reflects the state of affairs required by the Statistical Hypothesis. The two models provide two estimates of the population criterion variance accounted for by the predictor variable. An F test can then be calculated from these two variance estimates.

Since the Research Hypothesis specifies a systematic straight-line relationship, the Full Model used to depict this relationship is as follows:

$$Y = a_0U + b_1X + E_1 \qquad \text{(Model 5.1)}$$

The Statistical Hypothesis states that a linear relationship between ability and performance does not exist. Thus this hypothesis implies that the best prediction is the mean of all the criterion scores, or, in other words, that the line of best fit is horizontal—implying that the slope is equal to zero. Thus the restriction placed on the Full Model by the Statistical Hypothesis is as follows:

$b_1 = 0$ (1 restriction)

When this restriction is placed on the Full Model (Model 5.1) it effectively eliminates variable X from the model. Thus, the Restricted Model is as follows:

$Y = a_0U + E_2$ (Model 5.2)

Two points should be noted regarding these two models. First, the error vector is identified differently in the Full Model (Model 5.1) than in the Restricted Model (Model 5.2). This is done to imply that the Full Model is allowed to have a different degree of predictability (and therefore error) than is the Restricted Model. The criterion vector is the same in both models since the two models are reflections of the same Research Hypothesis. That is, the same data are being "modeled" in two different ways. The Full Model simply uses more predictor information than does the Restricted Model in accounting for the variance in a given criterion. Second, the coefficient for the unit vector will likely be different in the two models. Indeed, if the coefficient for X is other than zero in the Full Model, the unit vector coefficient will be different in the two models.

Before we continue our discussion of the statistical test of the Statistical Hypothesis with the use of the Full and Restricted Models, a comment regarding the coefficients generated for these models is in order. It is important to note that the values contained in the Y and X variable vectors were "raw values," that is, the students' actual performance and ability scores. Thus, the coefficients generated for the models are referred to as "raw score coefficients", or "non-standardized coefficients." If the values contained in the vectors of all of the variables are standardized so that they have mean values of 0.0 and standard deviation values of 1.0, then the coefficients are referred to as "standardized coefficients," or "standard partial regression coefficients," or" beta coefficients." Seldom is there any value in standardizing the data, so non-standardized coefficients will be used throughout the text. Now we return to the discussion of the statistical test of the Statistical Hypothesis.

Testing the straight line with an F test using variance estimates. To begin the testing of the Statistical Hypothesis, the error sum of squares value for the Full Model (ESS_F) is calculated. To illustrate the calculation of this value, consider the following vectors for the Full Model (Model 5.1):

$$Y \quad = \quad a_0 * U \quad + \quad b_1 * X \quad + \quad E_1 \quad (E_1)^2$$

$$\begin{bmatrix} 4 \\ 6 \\ 8 \\ 10 \\ 12 \end{bmatrix} = 2 * \begin{bmatrix} 1 \\ 1 \\ 1 \\ 1 \\ 1 \end{bmatrix} + 2 * \begin{bmatrix} 1 \\ 2 \\ 3 \\ 4 \\ 5 \end{bmatrix} + \begin{bmatrix} 0 \\ 0 \\ 0 \\ 0 \\ 0 \end{bmatrix} \quad \begin{bmatrix} 0 \\ 0 \\ 0 \\ 0 \\ 0 \end{bmatrix}$$

$$\sum (E_1)^2 = 0 = ESS_F$$

If each element in the error term (E_1) is squared and then summed $[\sum(E_1)^2$ or $ESS_F]$, the resulting value is equal to 0.

Dividing the error sum of squares value of this Full Model by the number of data points free to vary (df_w), as in equation 5.2, yields a within estimate of the criterion variance $(\hat{v}_w)$ value:

$$\hat{v}_w = \frac{ESS_F}{df_w} \qquad \qquad \text{(Equation 5.2)}$$

The df_w value is equal to the sample size (N) minus the number of coefficients used to estimate the line of best fit. With the sample size of 5, the df_w value for the single straight-line model is 3 (i.e., $5 - 2$) because the 2 coefficients in the Full Model (i.e., a_0 and b_1) that had to be calculated to estimate the straight line of best fit were subtracted from the sample size. Thus the $\hat{v}_w$ value is calculated as follows:

$$\hat{v}_w = \frac{0}{3} = 0$$

Next, the error sum of squares value for the Restricted Model (ESS_R) is calculated. To illustrate the calculation of this value, consider the vectors presented at the end of this paragraph for the Restricted Model (Model 5.2). To minimize the sum of the squared elements in E_2, the coefficient a_1 will, in this case, equal the mean of the scores in the criterion vector Y. The mean of Y is 8; therefore, $a_0 = 8$. Each element of E_2 is the difference of the individual score from the criterion mean. The first element in the unit vector is multiplied by the coefficient of 8, and the resultant value is 8. The first element in Y is 4; thus the first element in E_2 is $(4 - 8) = -4$. Vector E_2 contains the deviation of each individual's score from the criterion mean. When these elements are squared and summed, the error sum of squares for the Restricted Model (ESS_R) is obtained. No value other than 8 (the criterion mean) for a_0 will give a smaller error sum of squares value for the Restricted Model. The reader may wish to try other values for a_0 to see what happens to E_2, and, more importantly, what happens to $\sum(E_2)^2$.

$$
Y \quad = \quad a_0 * U \quad + \quad E_2 \qquad (E_2)^2
$$

$$
\begin{bmatrix} 4 \\ 6 \\ 8 \\ 10 \\ 12 \end{bmatrix} = 8 * \begin{bmatrix} 1 \\ 1 \\ 1 \\ 1 \\ 1 \end{bmatrix} + \begin{bmatrix} -4 \\ -2 \\ 0 \\ +2 \\ +4 \end{bmatrix} \qquad \begin{bmatrix} 16 \\ 4 \\ 0 \\ 4 \\ 16 \end{bmatrix}
$$

$$
\sum (E_2)^2 = 40 = ESS_R
$$

After each error term is calculated, each element in the error term (E_2) is squared and then summed, which produces an error sum of squares value for the Restricted Model (ESS_R) of 40. The error in prediction will generally be lower in the Full Model than in the Restricted Model. This is definitely the case for Model 5.1 ($EES_F = 0$) and Model 5.2 ($EES_R = 40$).

Once again the question is: How likely is it that the decrease in error in prediction (or increase in predictability) is due to randomness? The increase in predictability would be the difference between the two sums of squared errors ($ESS_R - ESS_F$). Dividing this difference by the number of pieces of information that account for the difference (df_p), which is the number of pieces of information in the Full Model minus the number of pieces of information in the Restricted Model, yields another variance estimate labeled $\hat{v}_p$. Thus, $\hat{v}_p$ is calculated with the following equation:

$$
\hat{v}_p = \frac{ESS_R - ESS_F}{df_p} \qquad \qquad \text{(Equation 5.3)}
$$

Since the Full Model (Model 5.1) contains two pieces of information (i.e., the unit vector and variable X) and the Restricted Model (Model 5.2) has only one piece of information (i.e., the unit vector), the value for $\hat{v}_p$ is 1 (i.e., 2 - 1). Recall that ESS_F and ESS_R are 0 and 40, respectively. Thus the value for $\hat{v}_p$ is calculated as follows:

$$
\hat{v}_p = \frac{40 - 0}{1} = 40
$$

Due to the fact that $\hat{v}_w$ is based on all the predictor information in the Full Model, it is considered the best estimate of the criterion variance. The variance estimate $\hat{v}_p$ is the variance estimate influenced by the "degree of worth" or "gain in accuracy" achieved by the predictor (in this case X) included in the Full Model that is absent from the Restricted Model. The F test value is calculated using Equation 5.4, which is as follows:

$$F(df_p, df_w) = \frac{\hat{v}_p}{\hat{v}_w} \qquad \text{(Equation 5.4)}$$

Substituting the values for $\hat{v}_w$ (0), $\hat{v}_p$ (40), df_p (1) and df_w (3) into Equation 5.4 produces the following:

$$F(1,\ 3) = \frac{40}{0} = \text{infinity!!!}$$

Of course, an F value of infinity, which was the result of each point being located on the straight line, would allow the researchers to reject the Statistical Hypothesis. Thus, the data support the claim that a linear relationship does exist between performance and ability scores.

Testing the straight line with the F *test using* R^2 *values.* At the end of chapter 2 and in chapter 4 the F test formula was expressed in terms of R^2 values. Since that formula is more easily generalized to other Research Hypotheses, the F test stated in Equation 5.4 will be translated to R^2 terms in Equation 5.5, which is as follows:

$$F(df_p, df_w) = \frac{(R_F^2 - R_R^2)/(df_p)}{(1 - R_F^2)/(df_w)} \qquad \text{(Equation 5.5)}$$

It is the case that $(R_F^2 - R_R^2)$ yields the gain in accuracy due to the additional information in the Full Model. This gain is compared to the error variation shown by the denominator in Equation 5.4 as ESS_F and in Equation 5.5 as $(1 - R_F^2)$. Thus, the amount of gain is compared against the measure of random error variation. If this gain is large in comparison to the measure of random variation, as determined by the magnitude of F and its associated chance probability distribution, then the Statistical Hypothesis may be rejected.

In general, the degrees of freedom for the numerator and denominator will be referred to as df_n and df_d, respectively. Therefore, the F test used to test the difference between the R^2 values of the Full Model (Model 5.1) and the Restricted Model (Model 5.2) can be calculated with Equation 4.1, which was introduced in Chapter 4. This equation is as follows:

$$F(df_n, df_d) = \frac{(R_F^2 - R_R^2)/(df_n)}{(1 - R_F^2)/(df_d)}$$ (Equation 4.1)

When using Equation 4.1 to calculate the F test for the performance and ability scores graphed in Figure 5.2, the R^2 values for the Full Model (Model 5.1) and the Restricted Model (5.2) need to be calculated.

The total sum of squares (SS_T) value can be calculated by subtracting the criterion mean from each criterion score, squaring this deviation, and then summing all the squared deviations. The total sum of squares value for the data under consideration is 40. Once the total sum of squares (SS_T) and the error sum of squares (ESS_F) values are known, the R^2 value for the Full Model (R_F^2) can be calculated with Equation 5.6.

$$R_F^2 = \frac{(SS_T - ESS_F)}{(SS_T)}$$ (Equation 5.6)

Since the ESS_F for the Full Model (Model 5.1) is equal to 0 and the SS_T value for the criterion scores is 40, the R_F^2 value is calculated using Equation 5.6 as follows:

$$R_F^2 = \frac{(40 - 0)}{(40)} = 1.00$$

Once the total sum of squares (SS_T) and the error sum of squares (ESS_R) values are known, the R^2 value for the Restricted Model (R_R^2) can be calculated with Equation 5.7, which is as follows.

$$R_R^2 = \frac{(SS_T - ESS_R)}{(SS_T)}$$ (Equation 5.7)

Since the ESS_R for Model 5.2 is equal to 40 and the SS_T value for the criterion scores is 40, the R_R^2 value is calculated using Equation 5.7 as follows:

$$R_R^2 = \frac{(40 - 40)}{(40)} = .00$$

Note that the R_R^2 value for this Restricted Model is .00. This will be case when the Restricted Models only contains the unit vector, that is, it is a Null Model. This will not always be the case, however. Often, several predictors are under consideration, and the researcher wishes to restrict one of the predictors from the Full Model to test the proportion of unique variance that that one piece of information adds to the others when used with them. The Restricted Model will then contain information in addition to the unit vector. Such Restricted Models will be discussed in the next chapter.

In the previous section we discussed the calculation of degrees of freedom in terms of how many pieces of information are contained in the models. Degrees of freedom also can be conceptualized in terms of the number of linearly independent predictor vectors in the Full and Restricted Models and the number of subjects. Recall that the notion of linearly independent vectors was introduced at the end of chapter 3.

The degrees of freedom in the numerator (df_n) value in Equation 4.1 is equal to the difference between the number of linearly independent vectors in the Full Model (symbolized as m1) and the number of linearly independent vectors in the Restricted Model (symbolized as m2). One way to conceptualize the degrees of freedom for the numerator of the F test is to ask: How many linearly independent predictors were deleted from the predictors in the Full Model to form the Restricted Model?

Before we discuss the number of linearly dependent variables in the Full Model (Model 5.1) and the Restricted Model (Model 5.2) it is important to understand why the criterion variable and the error vector are not considered to be linearly independent vectors. The number of linearly independent vectors in a model would include neither the criterion vector nor the error vector, only the predictor information. The coefficients for the predictor vectors are mathematically determined to make the criterion vector linearly dependent upon the weighted combination of the predictor vectors; therefore the criterion vector is not a linearly independent vector. A perfect fit will usually not be possible; therefore, an error vector is included. But note that the values in the error vector are determined only after the other coefficients are determined; therefore the error vector is not a predictor vector. Now we return to the discussion of determining the number of linearly dependent vectors in the Full and Restricted Models.

The Full Model (Model 5.1) has two linearly independent vectors (i.e., the unit vector and the X predictor variable), thus m1 is equal to 2. The df_d value is equal to the sample size minus m1. Since the sample size is 5, the df_d value is 3 (i.e., $5 - 2$). The Restricted Model (Model 5.2) has only one linearly independent vector (i.e., the unit vector), thus m2 = 1. Since df_n is equal to m1 minus m2, the value for df_n is 1 (i.e., 2 − 1).

Substituting the values for R_F^2 (1.00), R_R^2 (.00), df_d (3), and df_n (1) into Equation 4.1 produces the following:

$$F = \frac{(R_F^2 - R_R^2)/(df_n)}{(1 - R_F^2)/(df_d)} = \frac{(1.00 - .00)/(2 - 1)}{(1 - 1.00)/(5 - 2)} = \frac{(1/2)}{(0/3)} = \text{infinity!!!}$$

This F value of infinity, of course matches the F test value calculated using the variance estimates. Since computer statistical programs provide the R^2 values of the Full and Restricted Models, we will use Equation 4.1 rather than Equation 5.4 to calculate the F test in the remaining chapters of this text.

A SECOND ILLUSTRATIVE EXAMPLE OF A SINGLE STRAIGHT LINE OF BEST FIT

The second example contains data for six subjects that are not as systematic as those in the first example and are closer, therefore, to the real-world state of affairs. The Research Hypothesis under consideration is the same as for the first example and can be stated in several ways, for example—For the population of interest:

1. A relationship exists between ability and performance.
2. A linear relationship exists between ability and performance.
3. For every unit of increase in ability, there is a constant rate of change in performance.
4. The correlation between ability and performance is other than zero.
5. The linear correlation between ability and performance is other than zero.
6. Ability is linearly predictive of performance.

The respective rival or Statistical Hypotheses for these Research Hypotheses can be stated as—For the population of interest:

1. A relationship does not exist between ability and performance.
2. A linear relationship does not exist between ability and performance.
3. For every unit of increase in ability, there is no change in performance.
4. The correlation between ability and performance is zero.
5. The linear correlation between ability and performance is zero.
6. Ability is not linearly predictive of performance.

The careful reader will note that the term "linear" is not in statements 1 and 4. We explore in detail in chapter 9 lines of best fit that are other than linear. In fact, a straight line is only one of an infinite number of possibilities. One should be careful to insert the term "linear" whenever the investigation is limited to just a straight line. Research Hypotheses 1 and 4 are found in the literature quite often, but represent the default statements in 2 and 5. We encourage the researchers to be precise in their language by saying exactly what they are testing, and replacing statements 1 and 4 with 2 and 5.

The observed data points for each of the six subject's ability and performance are given in Figure 5.4. The six data points reflected in the scattergram seem to follow a trend; however, a single straight line cannot be cast that would go through all the data points. Two people had an ability score of 2, but their performance scores differed, that is, one had a performance score of 3 and the other had a performance score of 1. A similar type of discrepancy is noted for ability levels 4 and 6. A perfect linear relationship does not exist because a single straight line does not fit all the data points. A line of best fit, though, can be cast that will minimize the sum of the squared distances from that line (the sum of the squared elements in the error vector).

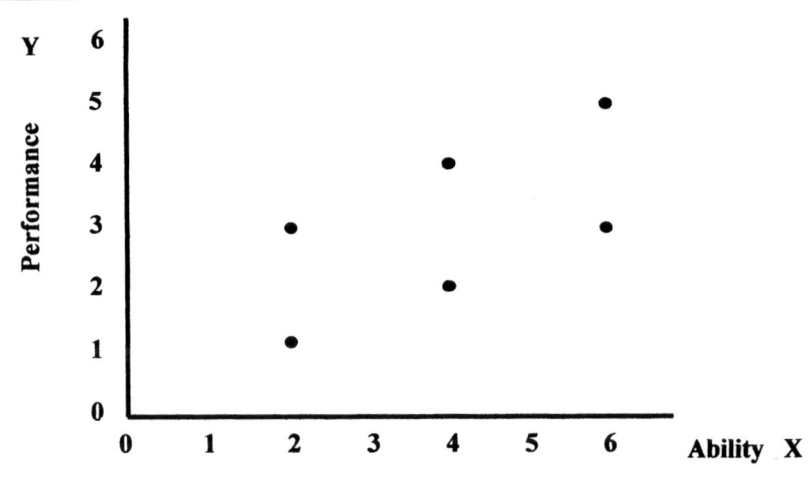

Figure 5.4.
Observed performance scores (Y) and ability scores (X) for six individuals.

Full and Restricted Models

Assume the researchers who conducted this study posed the following research question:

Does a positive linear relationship exist between ability scores and performance scores?

Since this is a directional question (i.e. a *positive* relationship is expected), the researchers would use the following directional Research Hypothesis:

> Research Hypothesis: For the population of interest, a positive linear relationship exists between the ability scores and performance scores.

The corresponding directional Statistical Hypothesis is as follows:

> Statistical Hypothesis: For the population of interest, a positive linear relationship does not exist between ability scores and performance scores.

The Full Model required by the Research Hypothesis is the same single straight-line model as the one constructed for the first example. We know that this is the model by answering three questions:

1. What are we trying to understand? (Answer: The subjects' performance scores, which formed the criterion variable Y, are used.)
2. What information about the subjects are we going to use to predict the criterion? (Answer: The subjects' ability scores, which formed the predictor variable X, are used.)
3. What functional relationship are we going to investigate? (Answer: A linear one is being investigated because that functional relationship was specified in the Research Hypothesis.)

These answers lead to the formation of the following Full Model:

$$Y = a_0 U + b_1 X + E_3 \qquad \text{(Model 5.3)}$$

The task is to find values for a_0 and b_1 so as to minimize the sum of the squared elements in E_3, which we will do without the aid of a computer program for instructional reasons.

First, a_0 and b_1 will be solved intuitively, and then the coefficients will be derived formally. The vector representation of the Full Model is as follows:

$$Y \quad = \quad a_0 * U \quad + \quad b_1 * X \quad + \quad E_3$$

$$\begin{pmatrix} 1 \\ 3 \\ 2 \\ 4 \\ 3 \\ 5 \end{pmatrix} = a_0 * \begin{pmatrix} 1 \\ 1 \\ 1 \\ 1 \\ 1 \\ 1 \end{pmatrix} + b_1 * \begin{pmatrix} 2 \\ 2 \\ 4 \\ 4 \\ 6 \\ 6 \end{pmatrix} + \begin{pmatrix} ? \\ ? \\ ? \\ ? \\ ? \\ ? \end{pmatrix}$$

Refer to the scattergram presented in Figure 5.4 and place a point halfway between the two scores for the individuals who scored 2 on the ability measure.

The corresponding performance score (i.e., the Y value) is 2, which is the mean criterion score for these two individuals. The mean criterion scores can be obtained in the same manner for ability scores 4 and 6 were 3 and 4, respectively. As indicated by the graph contained in Figure 5.5, one can easily plot a straight line for these three calculated performance scores (i.e., criterion scores of 2, 3, and 4). This plotted line is the line of best fit; that is, the sum of the squared elements in the error vector (E_3) is minimized.

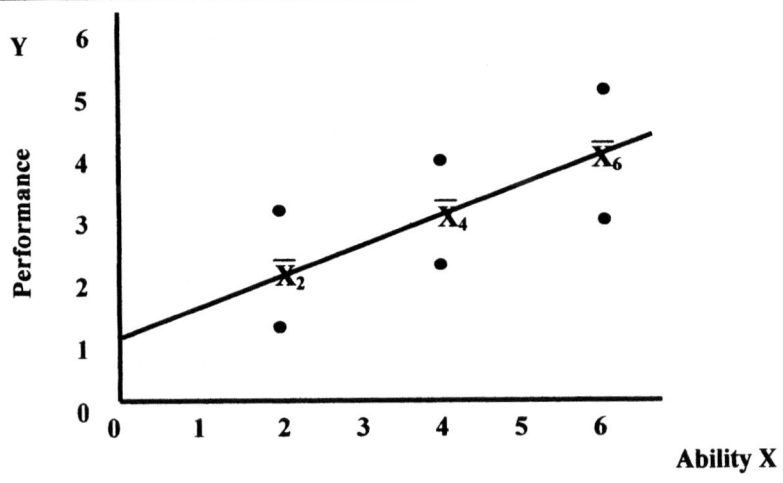

Figure 5.5.
An intuitive selection of the line of best fit.

The line in Figure 5.5 is straight and passes through the mean data points; thus the line of best fit has been obtained. What is the value for a_0? The line intersects the Y axis at 1; therefore, $a_0 = 1$. Also, a two-unit increase on X from 0 to 2 gives a one-unit increase on Y; thus a one-unit increase in X results in a .5 unit increase in Y. One can check this value by noting that the value on the line at X = 2 is 2, and at X = 4 is 3. So for each unit increase on X, a .5 increase on Y is observed. The next step is to calculate E_3 with these coefficients, and then square each of the elements of E_3, as illustrated on the following page.

The first element in vector E_3 is -1. This value was obtained by multiplying a_0 times U (i.e., 1 * 1) and adding that to b_1 times the first element of X (i.e., .5 * 2), which gives a predicted score of 2. Solving for E_3, the observed score (1) minus the predicted score (2) equals -1. This procedure is followed to obtain each element in vector E_3. Squaring each element in the error vector E_3, and summing yields the error sum of squares found with the Full Model. No other values for a_0 and b_1 than 1 and .5, respectively, yield as small an error sum of squares.

$$Y \quad = \quad 1 * U \quad + \quad .5 * \quad X \quad + \quad E_3 \quad E_3^2$$

$$\begin{bmatrix} 1 \\ 3 \\ 2 \\ 4 \\ 3 \\ 5 \end{bmatrix} = 1 * \begin{bmatrix} 1 \\ 1 \\ 1 \\ 1 \\ 1 \\ 1 \end{bmatrix} + .5 * \begin{bmatrix} 2 \\ 2 \\ 4 \\ 4 \\ 6 \\ 6 \end{bmatrix} + \begin{bmatrix} -1 \\ +1 \\ -1 \\ +1 \\ -1 \\ +1 \end{bmatrix} \begin{bmatrix} 1 \\ 1 \\ 1 \\ 1 \\ 1 \\ 1 \end{bmatrix}$$

$$\sum (E_3)^2 = 6$$

Statistical Test of the Full and Restricted Models

The Statistical Hypothesis sets $b_1 = 0$, reflecting the notion that there is no linear relationship between ability and performance. The sample data indicate that b_1 is not 0, but that $b_1 = .5$. However, is this value of .5 merely a chance sampling discrepancy from a true population value of 0? To answer this question, the F test between a Full Model and Restricted Model can be calculated and evaluated according to a predetermined alpha level, say, .05.

To test the directional Statistical Hypothesis the restriction specified in that hypothesis is placed on the Full Model (Model 5.3). This restriction, which does not allow the ability scores to be a positive predictor of the performance scores, can be expressed as $b_1 = 0$. Placing this one restriction into the Full Model deletes the X variable from the Full Model, resulting in the following Restricted Model:

$$Y = a_0U + E_4 \tag{Model 5.4}$$

Recall that the value for the unit vector coefficient estimated for the Full Model will generally not be the same as the unit vector coefficient value estimated for the Restricted Model, which is indeed the case for these data.

The value for a_0 in the Restricted Model that will produce the minimal sum of squared elements in E_4 is the criterion mean of Y, $\sum Y / N$, or $18 / 6 = 3$. Therefore, the a_0 value in the Restricted Model is 3. The vector representation of the Restricted Model, including E_4 and $(E_4)^2$, is as presented on the following page.

As in the case of the first example, when only the criterion mean is used to predict the criterion, the sum of the squared elements in the error vector is numerically equal to the total sum of squares (SS_T). For the data under consideration, $SS_T = 10$. To obtain the R_F^2, one must place the sum of squares explained by that model over the total sum of squares.

$$
Y \quad = \quad 3 * U \quad + \quad E_4 \qquad (E_4)^2
$$

$$
\begin{pmatrix} 1 \\ 3 \\ 2 \\ 4 \\ 3 \\ 5 \end{pmatrix} = 3 * \begin{pmatrix} 1 \\ 1 \\ 1 \\ 1 \\ 1 \\ 1 \end{pmatrix} + \begin{pmatrix} -2 \\ 0 \\ -1 \\ +1 \\ 0 \\ +2 \end{pmatrix} \qquad \begin{pmatrix} 4 \\ 0 \\ 1 \\ 1 \\ 0 \\ 4 \end{pmatrix}
$$

$$
\sum (E_4)^2 = 10
$$

The sum of squares in vector E_3 is 6, which is the sum of squares *not* accounted for by knowledge of ability. The total sum of squares is 10. The proportion of variance accounted for in the Full Model (R_F^2) is the difference between the total sum of squares (SS_T) and $\sum (E_3)^2$ (or ESS_F), divided by the total sum of squares. Substituting the values for SS_T (10) and ESS_F (6) into Equation 5.7 produces the following R^2 value for the Full Model (R_F^2):

$$
R_F^2 = \frac{(SS_T - ESS_F)}{(SS_T)} = \frac{(10 - 6)}{(10)} = \frac{4}{10} = .40
$$

The proportion of variance accounted for in the criterion variable (Y) by the Restricted Model is equal to the total sum of squares (SS_T) minus the sum of the squared elements in E_4 (or ESS_R) over the total sum of squares. In this case, the sum of the squared elements in E_4 is equal to the total sum of squares value of 10. When the Restricted Model contains only the unit vector, the ESS_R will be equal to the ESS_T, since there is no differential predictability with just the unit vector. Therefore, the R^2 for the Restricted Model is calculated as follows with Equation 5.7:

$$
R_R^2 = \frac{(SS_T - ESS_R)}{(SS_T)} = \frac{(10 - 10)}{(10)} = \frac{0}{10} = 0
$$

It is *not* important for one to memorize these formulae. They are presented here only to help those who may have relied too heavily on ANOVA.

Before the F test value for testing the difference between the R^2 values of the Full Model (Model 5.3) and the Restricted Model (Model 5.4) can be calculated using Equation 4.1, the numerator and denominator degrees of freedom values must be determined. The numerator degrees of freedom (df_n) value is equal to the difference between the number of linearly independent vectors in the Full Model, which is 2 (i.e., the unit vector and the X variable) and the Restricted Model, which is 1 (i.e., the unit vector). The denominator degrees of freedom (df_d) value is equal to the sample size (N) minus the number of linearly independent vectors in the Full Model. Since this study used the data from 6 subjects, df_d is equal to 4 (i.e., $6 - 2$).

Substituting the values for R_F^2 (.40), R_R^2 (.00), df_n (1), and df_d (4) into Equation 4.1 produces the following F test value:

$$F(1, 4) = \frac{(.40 - .00)/(2 - 1)}{(1 - .40)/(6 - 2)} = \frac{(.40/1)}{(.60/4)} = \frac{.40}{.15} = 2.67$$

When determining whether this F value is sufficient large to reject the Statistical Hypothesis it is important to recall that this hypothesis is directional. Thus, the critical value to which this calculated F value is compared must reflect this fact, which is done by locating the F value in an F-distribution table, which is known as the critical F value, that corresponds to the df_n and df_d values of 1 and 4, respectively, and the alpha value of .10. Note that even though the alpha level is doubled from the established .05 level to .10 only to allow a one-tailed F value to be located, the test is still conducted at an alpha level of .05. Since the calculated F value of 2.67 is less than the critical one-tailed F value of 4.54, which can be located in an F-distribution table, the Statistical Hypothesis is not rejected. Thus, the data do not support the claim that a positive linear relationship exists between the performance scores and ability scores in the population.

Two points should be noted regarding this directional test of a linear relationship. First, with the use of a computer statistical program, such as SPSS for Windows and Microsoft Excel, the testing procedure used in this example and the preceding one will produce a probability value for the F test of the difference between the R^2 values of the Full and Restricted Models. When a nondirectional Statistical Hypothesis is being tested, it will be rejected if this F test probability value is less than the alpha value. When a directional Statistical Hypothesis is being tested the two-tailed F test probability value, which will be listed on the computer output window, must be divided by 2 before it is compared to the alpha level. Even if this one-tailed F test probability value is less than the alpha level, another value must be considered before one can determine whether the hypothesis should be rejected, which brings us to the next point.

In a set of directional hypotheses a specific relationship between the criterion and predictor variables is specified (i.e., either a positive or negative linear relationship). Thus, before a directional Statistical Hypothesis can be rejected not only does the one-tailed F-test probability value need to be less than the alpha level, the sign of the predictor variable must match the sign of the relationship specified in the Research Hypothesis.

After reading the material regarding the two examples of straight-line relationships, one should not assume that researchers commonly encounter the relationships depicted in those examples. That is, seldom will researchers discover that all of the data points fit a straight line perfectly, which was the case for the first example; nor will the line often go through all the mean data points, which was the case for the second example. Therefore, mathematical formulae are usually required to find the coefficients for the line of best fit, as discussed in the following section.

MATHEMATICAL CALCULATION OF THE SINGLE STRAIGHT LINE OF BEST FIT

The data used up to this point were constructed to provide intuitive solutions to obtaining the line of best fit. In a bivariate case where the line of best fit is not intuitively obvious, the following equations can be used to obtain the slope coefficient (b_1) and the Y-intercept coefficient (a_0) that will minimize the sum of the squared elements in the error vector:

$$b_1 = \frac{\sum XY - \dfrac{(\sum X)(\sum Y)}{N}}{\sum X^2 - \dfrac{[(\sum X)^2]}{N}}$$

(Equation 5.9)

$$a_0 = \overline{Y} - (b_1 * \overline{X})$$

(Equation 5.10)

To obtain the values for a_0 and b_1 for the data set listed for the second example, the following calculations are completed:

Y	X	X^2	XY
1	2	4	2
3	2	4	6
2	4	4	8
4	4	16	16
3	6	36	18
5	6	36	30
$\sum Y = 18$	$\sum X = 24$	$\sum X^2 = 112$	$\sum XY = 80$

$$\overline{Y} = (\textstyle\sum Y / N) = (18 / 6) = 3$$

$$\overline{X} = (\textstyle\sum X / N) = (24 / 6) = 4$$

Substituting these values into the previous two formulae (Equations 5.9 and 5.10) produces the following values:

$$b_1 = \frac{80 - \dfrac{(24)(18)}{6}}{112 - \dfrac{(24)^2}{6}} = \frac{(80 - 72)}{(112 - 96)} = \frac{8}{16} = .5$$

$$a_0 = [3 - (.5)(4)] = (3 - 2) = 1$$

The coefficients ($b_1 = .5$ and $a_0 = 1$) are the same values previously obtained using the intuitive approach.

It should be apparent that the real world seldom yields data that can be solved intuitively, and indeed real-world problems extend beyond the bivariate case such that many predictors are usually needed to account for the criterion variance. However, as complex models are investigated, the reader should be aware that the complex models break down into subsets of the basic linear model that has the form of Full Models 5.1 and 5.3. The mathematical solution of coefficients for the multiple predictor variable set becomes involved and has been treated extensively elsewhere (Maxwell & DeLaney, 1990). The intent of this chapter, as well as the entire text, is to explicate conceptual research problems and the linear models required to answer the resulting hypotheses. We let the computer do the calculating.

POINTS TO NOTE REGARDING THE INVESTIGATION OF LINEAR RELATIONSHIPS

A number of points should be understood when examining and understanding the methods discussed in this chapter used to test for such relationships. First, when one continuous predictor is used, one is attempting to account for the criterion variance with one predictor variable; hence the Venn diagram is the same as in the previous chapter. That is, the proportion of overlap of the two circles is equal to the R^2 of the Full Model.

Second, the single straight-line model and the Pearson Product Moment Correlation (commonly referred to as "correlation coefficient") are isomorphic.

Moment Correlation. The fact that the Pearson Product Moment Correlation only measures the linear relationship between two variables is thus made clear in the regression formulation. Readers can verify this fact by analyzing the 6 pairs of data points listed in the previous section. If one uses a computer statistical program to calculate the correlation coefficient for these data, the resulting value will be .632. The square of this value, which is .40, is equal to the R^2 value obtained for Model 5.3 (the Full Model used to estimate the straight line relationship between the 6 pairs of values). Thus, the single straight-line model provides the same information as does the correlation coefficient.

Third, a single continuous predictor allows a single straight line to fit the data. If the data do not conform to the single line, or to a straight line, then the single straight-line model will not fit it well and the R^2 will be substantially lower than 1.00. If it appears that there is a systematic relationship in the data that goes beyond a single straight line, then another model of the data needs to be tested. The types of models used to test for multiple predictors (chapter 6), interaction (chapter 7), and nonlinear relationships (chapter 9) are extensions of the material in this chapter.

Fourth, both the Full and Restricted Models used in conjunction with the statistical testing of a single straight-line hypothesis contain the unit vector. Most computer regression programs provide the unit vector automatically because of its widespread use in various types of models. Since the unit vector is included automatically, researchers do not have to "manually" enter it into the computer program. The researchers must remember that the unit vector has been automatically included, however, when determining the number of linearly independent vectors in the model. In addition, researchers should be aware that even though the unit vector coefficient is often referred to as the "regression constant" (i.e., the value does not change from one person to the another) it is labeled differently by various computer statistical programs—SPSS uses the term "Constant" and Microsoft Excel use the term "Intercept."

GENERAL RESEARCH HYPOTHESIS 5.1

The General Research Hypothesis 5.1 (GRH 5.1) provides a template for an analysis of a continuous criterion variable and a continuous predictor variable. Researchers could state the Research Hypothesis as either a directional hypothesis or a nondirectional one as follows:

Directional GRH 5.1: For some population, X is positively related with Y1 in a linear manner.

Nondirectional GRH 5.1: For some population, X is related with Y1 in a linear manner.

Regardless of whether researchers decide to use a directional or nondirectional Research Hypothesis (remember, we strongly urge researchers to consider the use of directional hypotheses), the corresponding General Statistical Hypotheses (GSH 5.1) are as follows:

Directional GSH 5.1: For some population, X is *not* positively related with Y1 in a linear manner.

Nondirectional GSH 5.1: For some population, X is *not* related with Y1 in a linear manner.

A Full Model is designed to include the variables contained in the General Research Hypothesis. For GRH 5.1, the Full Model is as follows:

$$Y1 = a_0U + b_1X + E_1 \qquad \text{(Model 5.5)}$$

where:
1. $Y1$ = criterion;
2. $U = 1$ for all subjects;
3. X = predictor score for subject; and
4. a_0 and b_1 are least squares coefficients calculated so as to minimize the sum of the squared values in the error vector.

The number of pieces of information contained in this Full Model (Model 5.5) is 2 (i.e., $m1 = 2$). Note that the unit vector allows for a nonzero coefficient, an indicator of a piece of information, thus the unit vector is considered as a piece of information in the model.

The directional General Research Hypothesis requires $b_1 > 0$, while the nondirectional hypothesis requires $b_1 \neq 0$. Despite the different requirements of the directional and nondirectional General Research Hypotheses, the restriction placed on the Full Model to reflect the condition stated in the General Statistical Hypothesis is as follows:

$b_1 = 0$ (1 restriction)

Placing this restriction on the Full Model results in the following restricted Model:

$$Y1 = a_0U + E_2 \qquad \text{(Model 5.6)}$$

The number of pieces of information contained in this Restricted Model (Model 5.6) is 1, thus $m2 = 1$.

An F test of the difference between the R^2 values of the Full and Restricted Models will determine whether the linear relationship between the variables X and Y, which is estimated by the Full Model, is statistically significant. The numerator degrees of freedom (df_n) value for the F test is equal to m1 (2) minus m2 (1), that is, 1; while the denominator degrees of freedom (df_d) value is equal to N (sample size) minus m1, that is, $df_d = N - 2$. If this F test reveals that the difference between the R^2 values of the Full and Restricted Models is indeed statistically significant, the researchers would reject the General Statistical Hypothesis 5.1. However, if the researchers posed a directional Research Hypothesis, a significant F test would lead to the rejection of the General Statistical Hypothesis only if the value for the b_1 coefficient in the Full Model was positive. If this coefficient was negative and they used a directional Research Hypothesis, the General Statistical Hypothesis would not be rejected regardless of the F test value. (Refer to the discussion regarding this point as presented in the section entitled Applied Research Hypothesis 4.1.)

APPLIED RESEARCH HYPOTHESIS 5.1

Applied Research Hypothesis 5.1 (ARH 5.1) illustrates the procedures researchers would use to investigate whether a continuous predictor variable is a positive linear predictor of a continuous criterion variable. For ARH 5.1 the variable X3, which is the predictor variable, represents a set of scores recorded for students on Test A at the end of their third-grade year, while the variable X2, which is the criterion variable, consists of a set of test scores recorded for those same students on Test B at the end of their fourth-grade year. The researchers are interested in the following research question:

For a given population, is Test A a positive linear predictor of Test B?

Since the investigation undertaken by the researchers involves a directional test, ARH 5.1 is expressed as a directional Research Hypothesis as follows:

Directional ARH 5.1: For a given population of fourth graders, X3 is a positive linear predictive of X2.

The corresponding Statistical Hypothesis 5.1 (ASH 5.1) is as follows:

Directional ASH 5.1: For a given population of fourth graders, X3 is not a positive linear predictive of X2.

Full and Restricted Models

To depict the relationship stated in ARH 5.1, the Full Model is constructed as follows:

$$X2 = a_0U + a_3X3 + E_1 \qquad\qquad \text{(Model 5.7)}$$

As stated in ARH 5.1, the researchers expect the value for a_3 to be positive.

Since ASH 5.1 states that X3 is not a positive linear predictive of X2, the following restriction must be placed on this Full Model to reflect this statement:

$a_3 = 0$ (1 restriction)

Placing this restriction on the Full Model produces the following restricted Model:

$$X2 = a_0U + E_2 \qquad\qquad \text{(Model 5.8)}$$

Note that this Restricted Model is the Null Model.

Analysis of ARH 5.1 with Microsoft Excel

The data file entitled "GLM DATA EXCEL FORMAT" can be accessed from the internet site (see Appendix A). Once this file has been read into Microsoft Excel and the Regression menu has been displayed (refer to the section in chapter 4 entitled "Analysis of ARH 4.1 with Microsoft Excel"), the following steps are completed to specify the Full Model for ARH 5.1 (see Exhibit 5.1):

1. Click on the box next to "Input $\underline{Y}$ Range" (see Oval 1). Note that variable X2 is located in column B in the data file. Next, click and drag on the cells B1 to B61 in the data file to identify the X2 variable as the dependent variable.
2. Click on the box next to "Input $\underline{X}$ Range" (see Oval 2). Note that variable X3 is located in column C. Next, click and drag on the cells C1 to C61 in the data file to identify X3 as the predictor variable for the Full Model.
3. Click on the box in front of "Labels" (see Oval 3).
4. Click on the "OK" button (see Oval 4).

After these four steps are completed, the output window for the Full Model (Model 5.5) will appear as displayed in Exhibit 5.2. The key pieces of information listed in this output window are the following:

1. The R^2 value of the Full Model is .928 (see Oval 1). Remember, the Restricted Model is a Null Model. Thus, the R^2 value for the Restricted Model is .00, which is not listed on the output.
2. Since the Restricted Model is the Null Model, the F test listed in the "ANOVA" table of the output window is the appropriate F test for ASH 5.1. The F test of the difference between the R^2 values of the Full Model (.927) and the Restricted Model (.00) listed in the "ANOVA" ta-

ble is 747.02 (see Oval 2). The probability value of this *F* value is .0000, which can be expressed as *p* < .01.

Exhibit 5.1.
Microsoft Excel Regression menu—Full Model (Model 5.5) for ARH 5.1.

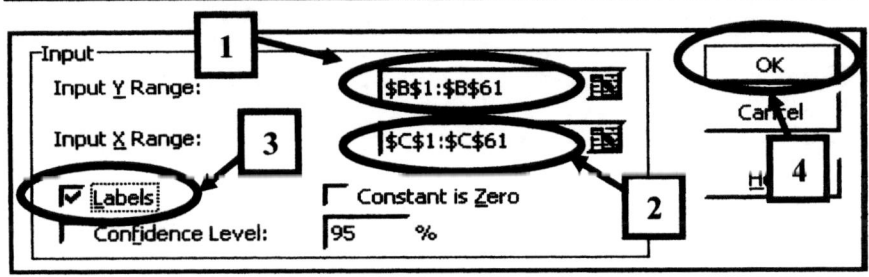

Exhibit 5.2.
Microsoft Excel regression output—Full Model (Model 5.5) for ARH 5.1.

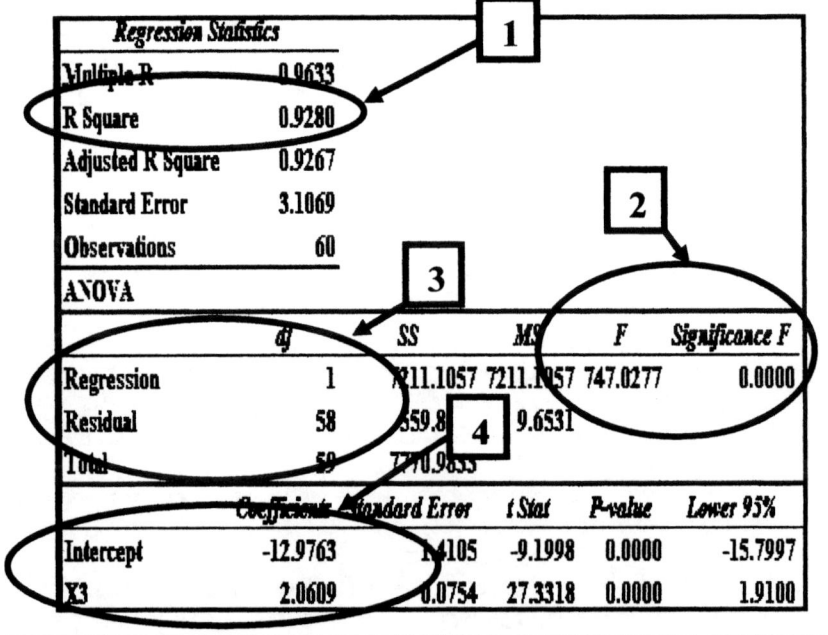

3. The numerator degrees of freedom value (1) and the denominator degrees of freedom value (58) are located in Oval 3. These values were

used in the calculation of the F test of the difference between the R^2 value of the Full Model (.928) and the R^2 of the Restricted Model (.00).

4. The regression coefficient for the X3 predictor variable is 2.06 (see Oval 4).

Since the one-tailed F test probability value ($p / 2 < .01$) is less than the established alpha level of .05, and the coefficient a_3 is positive, ASH 5.1 is rejected. Thus, the data support ARH 5.1. The positive sign of the regression coefficient for variable X3 ($a_3 = 2.06$) indicates that positive increases in the X3 values are associated with positive increases in the X2 values. Thus, the data support that X3 is a positive linear predictor of X2.

Writing the Results for ARH 5.1 in APA Style

The analysis of ARH 5.1 may be expressed in APA style as follows:

The difference between the R^2 values of the Full and Restricted Models, which measured the proportion of variation in the set of fourth-grade test scores accounted for by the linear component of the relationship with the third-grade scores, was .93. In addition, the coefficient for the Test A scores variable was positive ($a_3 = 2.06$). Since this coefficient was positive and the difference between the R^2 values of the Full and Restricted models was statistically significant at the alpha level of .05, $F(1, 58) = 747.03$, $p / 2 < .01$, the data support the claim that Test A is a positive linear predictor of Test B. In addition to being statistically significant, the proportion of unique variation accounted for in the Test B scores by the linear component of the relationship with the Test A scores is classified as a large effect size according to the criteria established by Cohen (1988).

Three points should be noted regarding this write-up of the study's results. First, the proportion of unique variation accounted for in the fourth-grade test scores by the linear component of the relationship, which was the difference between the R^2 values of the Full and Restricted Models, was reported. Second, the F test of the difference between the R^2 values of the Full and Restricted Models, and its corresponding probability value were also reported. Third, the difference between the R^2 values of the Full and Restricted Models was identified as the effect size and classified as large.

QUESTIONS FOR DR. GLM

After studying the concepts presented in this chapter a budding researcher (Bud) had a number of questions for Dr. General Linear Model (Dr. GLM). The questions posed in this section relate to: (a) miscellaneous questions about the type of research hypotheses that regression models help test—non-nil research hypotheses and (b) reporting the p value for a statistical test in addition to whether the results were significant or not significant.

Miscellaneous Questions about Research Hypotheses that Regression Models Help Test—Non-Nil Null Hypotheses

Bud: I notice that most of the restrictions so far involve the value of 0. Why is that?

Dr. GLM: The restrictions follow from the Research Hypothesis. All of the Research Hypotheses so far (except some of the multiple comparison ones) have called for the restriction of 0. But each of these Research Hypotheses could have been stated so that some other value was restricted. For example, one might be interested in the following research question:

For a given population, is the criterion mean for the new treatment (Treatment A) more than 2 units higher than the criterion mean for the existing treatment (Treatment B)?

To address this research question, the researchers would create a modified original criterion variable. A value of 2 is subtracted from the scores in the original criterion variable (Y) but only for the scores recorded for the students in Treatment A; the scores for the students in Treatment B remain unchanged. Once this modified criterion variable (Y_{Mod}) is generated, the following Full Model is constructed and analyzed (the dichotomous variable T is used to identify which treatment each student received—Treatment A = 1 and Treatment B = 0):

$$Y_{Mod} = a_0 + a_1 T + E_1 \qquad\qquad \text{(Model 5.9)}$$

When the modified criterion variable is included in the Full Model the value for the a_1 coefficient is equal to the amount by which the criterion mean of Treatment A exceeds the criterion mean of Treatment B by more than 2 points. Thus the restriction placed on the Full Model is as follows:

$$a_1 = 0 \quad \text{(1 restriction)}$$

The Restricted Model produced by placing this restriction on the Full Model is as follows:

$$Y_{Mod} = a_2 U + E_2 \qquad\qquad \text{(Model 5.10)}$$

Note that this Restricted Model is the Null Model.

The results of the F test of the difference between the R^2 values of the Full and Restricted Model would indicate whether the data support the Research Hypothesis. The term used in the literature for this kind of statistical test is "nonnil" because the difference between the group means was not set equal to 0. Note that the researchers could have used any meaningful value, that is, the difference between the two group means need not be hypothesized to exceed 2.

This type statistical procedure can be applied to a research question that involves a continuous predictor variable and a continuous criterion variable. To illustrate assume the following research question is posed:

Is an increase of one unit in the predictor variable (X) associated with an increase in the criterion variable (Y) that is greater than 2 units?

Before the Full Model is constructed the criterion variable is modified to reflect that the value of the slope of the regression line must exceed 2 units. Since the slope of the regression line must exceed 2, the coefficient for the predictor variable X is set equal to 2, which is expressed as (2 * X). Each value in the modified criterion variable (Y_{Mod}) is constructed by subtracting each (2 * X) value from its corresponding original criterion variable value. Once this modified criterion variable is formed, the Full Model is constructed as follows:

$$Y_{Mod} = a_0U + a_1X + E_1 \qquad \text{(Model 5.11)}$$

When the modified criterion variable is used, the a_1 coefficient value will equal the amount by which the slope of the regression line differs from 2. Thus the restriction placed on the Full Model is as follows:

$$a_1 = 0 \quad \text{(1 restriction)}$$

Placing this restriction into the Full Model results in the following Restricted Model:

$$Y_{Mod} = a_0U + E_2 \qquad \text{(Model 5.12)}$$

Note that this Restricted Model is the Null Model. As was the case for the previous example, the F test of the difference between the R^2 values of the Full Model and the Restricted Model, which is calculated with Equation 4.1, can be used to statistically address the research question.

BUD: What value should be used?

Dr. GLM: The researcher's desired conclusion should guide the choice. If the researcher is content with stating "The mean of Treatment A is higher than the mean of Treatment B," then the value would be 0. The value 0 is by far the most often used value—so much so that the term "Null Hypothesis" is used in most statistics texts. The reason we use the term "Statistical Hypothesis" instead of the term "Null Hypothesis" is to acknowledge that values other than zero can be used and to encourage researchers to use values other than 0. Students, researchers, and even some statistics instructors often ask us: Why some particular value such as 2? We counter by asking: Why 0?

The value of 0 has been used so often that many forget where it comes from, what it means, and that the researcher has a choice of what value to use. Researchers should be able to provide the rationale for selecting 0 just as people expect them to explain the use of any other value. The statement, "the mean of Treatment A is higher than the mean of Treatment B" is less compelling than the statement, "the mean of Treatment A is more than 2 points higher than the mean of Treatment B" just as this later statement is less compelling than the statement, "the mean of Treatment A is more than 6.16 points higher than the mean of Treatment B." The question is: Which statement would more likely compel you to adopt Treatment A? Assuming that adopting the new educational treatment has cost implications for new textbooks, workbooks, teacher training, and possibly parental communication, the requirement that the new treatment be just "better" may not be enough. All of these costs, as well as others, should be taken into account in the determination of how much better Treatment A should be before switching from the existing treatment to the new treatment.

Your questions on this topic are very important. You may want to refer to a paper written by Newman, Fraas, and Herbert (2002) that examines the testing of non-nil null hypotheses. In this paper the authors suggest incorporating practical significance in the statistical testing procedure of the difference between group means.

Reporting the *p* Value for a Statistical Test in Addition to Whether the Results were Significant or Not Significant

BUD: I see that some research articles report the *p* value listed on the computer printout.

Dr. GLM: In this text we take the traditional stance that the researcher should choose an alpha level before analysis of the data. Which particular value of alpha to use is difficult to determine. The costs of rejecting a true Statistical Hypothesis, as well as accepting a false Research Hypothesis must be taken into account. These dollar costs are so difficult to determine that most researchers use the default alpha value of .05 or .01—values that are almost exclusively used in the social science literature.

I agree, though, with the growing trend in the number of studies that report *p* values. The value of reporting the actual *p* value is that readers can make a decision for themselves regarding the significance (or lack thereof) of the results. But I also believe that it is important for the researcher to "take a stand" with respect to setting their alpha levels prior to the data analysis stage of the study. But each researcher may have a different view on an appropriate alpha level. By reporting the actual probability of a given statistical test, the readers of the study can make their own decisions regarding whether the data do or do not support the Research Hypothesis.

CHAPTER 6

MULTIPLE CONTINUOUS PREDICTOR VARIABLES

In the previous chapter we presented a limited case of researching behavior. Although a single continuous variable is definitely a parsimonious way of researching behavior, most behavior is probably a function of more than one variable. Therefore, there is a need for statistical models that can include more than one continuous variable. In this chapter, we present the more general case of multiple continuous predictor variables.

We begin with the simplest case of more than one continuous predictor—the case of two continuous predictors. General issues related to multiple continuous predictors are discussed. First, the overall "value" of the variables in the model is tested. Second, the value of a particular variable in the model is tested. Third, the full intent of chapter 1 is fulfilled by using Full and Restricted Models to compare two competing models of behavior.

TWO CONTINUOUS PREDICTOR VARIABLES

Rationale

In chapter 5 we presented research questions that required the models used to address those questions to contain only one predictor variable. It is important for readers to be aware of the limiting aspects of such research questions and their corresponding models. Therefore this section allows us to reflect more closely the multivariable approach to research we presented in chapter 1.

Graphical Representation

While the model that contains a single continuous predictor can be represented by a single straight line in a two-dimensional space, the model that contains two continuous predictor variables requires another dimension. Even with this additional dimension, however, the model can be represented by a single flat plane in a three-dimensional space. The two-dimensional ellipse of data points (see Figure 6.1a) becomes a three-dimensional football of data points, with a plane of best fit (see Figure 6.6). The models constructed to reflect the relationships depicted in Figure 6.1a, Figure 6.1b, Figure 6.1c, and Figure 6.1d are as follows:

$$Y = a_0U + a_1X + E_1 \qquad \text{(Model 6.1—Figure 6.1a)}$$

$$Y = a_0U + a_1X + a_2Z + E_2 \qquad \text{(Model 6.2—Figure 6.1b, Figure 6.1c, and Figure 6.1d)}$$

The coefficients for these models are determined so that the plane fits the data as well as possible, minimizing the squared distance of the actual data points from the predicted data points that are on the *line* of best fit for Model 6.1 and the *plane* of best fit for Model 6.2.

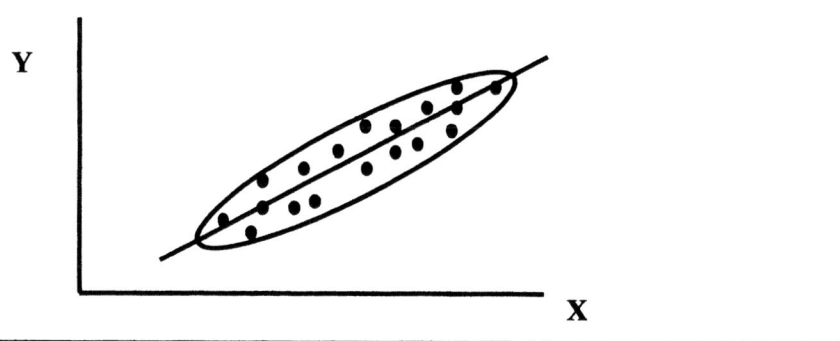

Figure 6.1a.
Graphical representation of a model with one predictor variable.

The two predictor variables in Model 6.2 could have relationships with the criterion variable that form different planes of best fit. In Figure 6.1b, increases in the values of the predictor variable X are associated with *positive* changes in the values of the criterion variable Y, while increases in the values of the predictor variable Z are associated with *negative* changes in the values of the criterion variable Y. Figure 6.1c depicts the opposite situation. That is, increases in the values of the predictor variable X are associated with *decreases* in the values of the criterion variable Y, while increases in the values of the predictor variable Z are associated with *positive* changes in the values of the criterion variable Y.

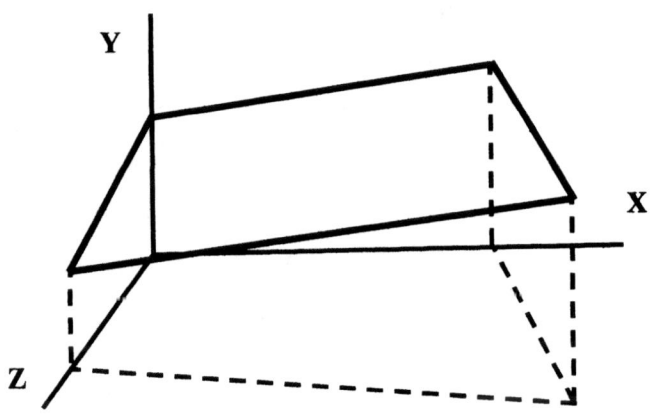

Figure 6.1b.
Graphical representation of a model with two predictor variables.

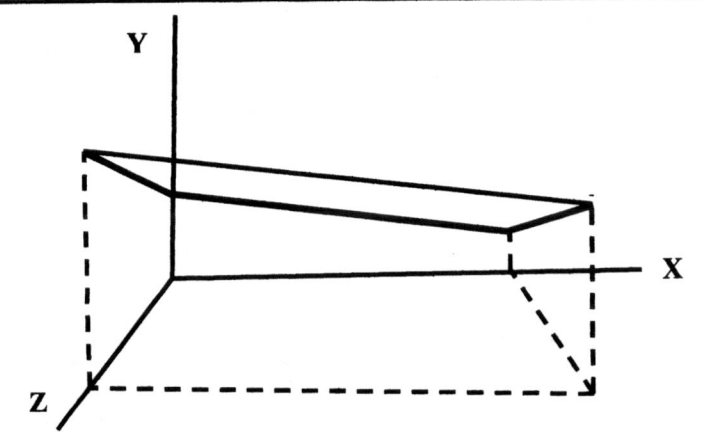

Figure 6.1c.
A plane in which X is negatively related to Y and Z is positively related to Y.

In Figure 6.1d increases in the values of both predictor variable X and Z are associated with *increases* in the values of the criterion variable Y. We leave for the reader the task of graphically depicting the fourth possibility, that is, one in which increases in the values of both predictor variables X and Z are associated with *decreases* in the values of the Y criterion variable.

Figures 6.2a and 6.2b provide two depictions of the relationship between variables X and Y when variable Z accounts for no additional variance in the Y variable. If variable Z accounts for no additional variation in the criterion Y variable, variable Z is in actuality not needed in the model; then the plane of best fit is perpendicular to that axis, as depicted in Figure 6.2a. This situation reduces to a two-dimensional ellipse of data, requiring only a single line of best fit as in Figure 6.2b.

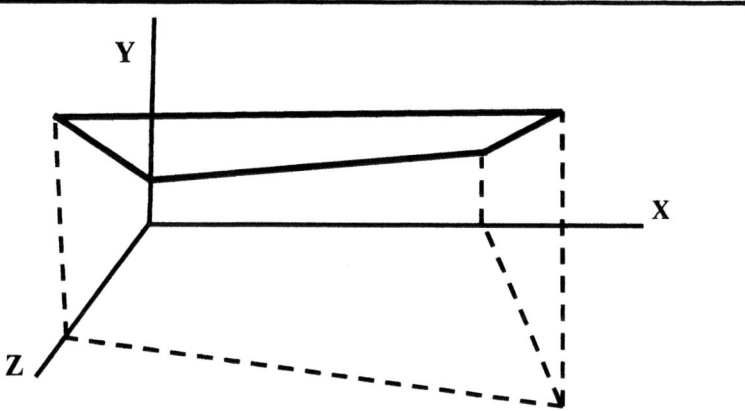

Figure 6.1d.
A plane in which both X and Z are positively related to Y.

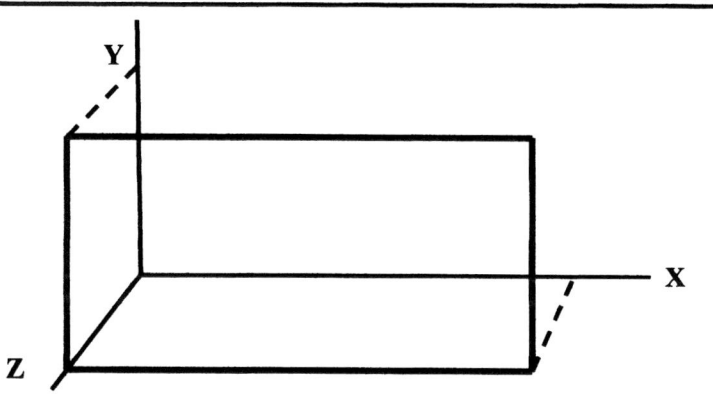

Figure 6.2a.
The plane of best fit when variable Z is not related to Y and it is included in the model.

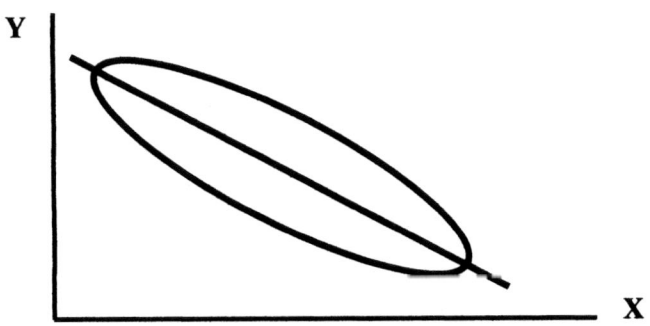

Figure 6.2b.
The line of best fit when variable Z is not related to Y and it is not included in the model.

Research Hypothesis

When two continuous predictor variables are used to predict a criterion, the Research Hypothesis posed by the researchers could take the following form:

Research Hypothesis (Version A): The two predictor variables (X and Z) account for variance in the criterion variable (Y).

An alternate way of stating this Research Hypothesis would be as follows:

Research Hypothesis (Version B): The two predictor variables (X and Z) predict the criterion variable (Y).

A Statistical Hypothesis can be stated for each version of the Research Hypothesis as follows:

Statistical Hypothesis (Version A): The two predictor variables (X and Z) do *not* account for variance in the criterion variable (Y).

Statistical Hypothesis (Version B): The two predictor variables (X and Z) do *not* predict the criterion variable (Y).

Regardless of which versions of the Research and Statistical Hypotheses are used, the researchers need to construct Full and Restricted Models to reflect those hypotheses.

Full and Restricted Models

The following Full Model is constructed to reflect the relationship stipulated by the Research Hypothesis, regardless of which version is used:

$$Y = a_0 U + a_1 X + a_2 Z + E_1 \qquad \text{(Model 6.3)}$$

The Research Hypothesis *allows* both the a_1 and a_2 coefficient values to differ from 0. It is important to note that the Full Model allows for three possibilities with respect to the values of a_1 and a_2 coefficients: (a) both the a_1 and a_2 coefficient values differ from 0; (b) the a_1 coefficient value is equal to 0, while the a_2 coefficient value differs from 0; and (c) the a_1 coefficient value differs from 0, while the a_2 coefficient value is equal to 0.

The Statistical Hypothesis, regardless of which version is used, requires that *both* the a_1 and a_2 coefficient values must equal 0. Thus the restrictions are as follows:

$$a_1 = 0 \text{ and } a_2 = 0 \quad \text{or} \quad a_1 = a_2 = 0 \qquad \text{(2 restrictions)}$$

Placing these restrictions on the Full Model (Model 6.3) produces the following Restricted Model:

$$Y = a_0 U + E_2 \qquad \text{(Model 6.4)}$$

Two points should be noted with respect to this Restricted Model. First, although the unit vector is not specified overtly in the Research Hypothesis, to include the unit vector in each regression model has become an automatic default. This is because the unit vector acts like a scale adjuster, making the mean of the predicted criterion equal to the mean of the actual criterion. The rationale for using a scale adjuster relies on the nature of how variables are scaled in the behavioral sciences. In most situations, one would not be able to assume that the value of zero on X will predict the value of zero on Y, thus, the need for the scale adjuster. If one did not use both the U and X vectors, the line estimated by the model would be forced to go through the origin identified as (0, 0); or with a second predictor variable, say, vector Z, the plane estimated by the model would be forced to go through the origin identified as (0, 0, 0). Second, this Restricted Model contains only the unit vector, that is, it is the Null Model, which has an R^2 value of .00.

Testing of the Statistical Hypothesis

The difference between the R^2 values of the Full Model (Model 6.3) and Restricted Model (Model 6.4) is statistically tested with Equation 4.1, which is as follows:

$$F(df_n, df_d) = \frac{(R_F^2 - R_R^2)/(df_n)}{(1 - R_F^2)/(df_d)}$$

(Equation 4.1)

where:
1. R_F^2 = the R^2 value of the Full Model;
2. R_R^2 = the R^2 value of the Restricted Model;
3. df_n = the number of linearly independent variables in the Full Model minus the number of linearly independent variables in the Restricted Model; and
4. df_d = the sample size minus the number of linearly independent variables in the Full Model.

Since more than one restriction was placed on the Full Model, this F test must be conducted as a two-tailed test. Thus, the Research Hypothesis would be rejected if the two-tailed F test probability is less than the established alpha level.

Conclusion based on the statistical test results. It is important for the researchers to know exactly what can and cannot be stated if the Research Hypothesis is rejected. The discussion of this point is facilitated by an examination of the Venn diagrams presented in Figures 6.3, which are designed to represent the Full and Restricted Models. It should be noted that in the Venn diagrams that depict the two models, the unit vector has the same value for each subject, and thus it provides no differential predictability. Due to this fact, the unit vector does not appear in the Venn diagram.

An examination of the Venn diagram in Figure 6.3a, which is designed to reflect the Full Model (Model 6.3), reveals that the two predictor variables contained in the Full Model (labeled variables X and Z), are correlated (i.e., their circles overlap) and *both* of these predictor variables account for some of the variation in the Y criterion variable (i.e., their circles overlap the circle for the Y criterion variable—the diamond-marked area). It is important to understand that the relationships among the variables could be different than the one predicted here—the following section discusses some of the possible relationships.

A review of the Venn diagram contained in Figures 6.3b indicates that the Restricted Model contains neither of the predictor variables. When the difference between the R^2 values of these two models is statistically tested, the results will indicate if the variation in the criterion variable (Y) accounted for by *both* predictor variables (X and Z) is significant. The key point illustrated by the two Venn diagrams is that this test only indicates if the amount of variation in the criterion variable (Y) accounted for by the *set* of two predictor variables is significant. That is, these models are not designed to test whether each of the predictor variables account for a statistically significant amount of variation in the criterion variable.

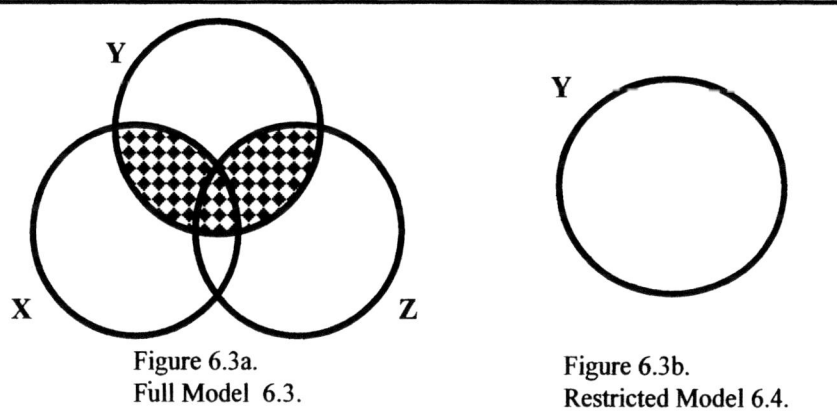

Figure 6.3a.
Full Model 6.3.

Figure 6.3b.
Restricted Model 6.4.

Figure 6.3.
Venn diagram for the Full Model (Model 6.3) and Restricted Model (Model 6.4.)

The thoughtful reader might suggest that the Full Model should be compared to two different Restricted Models, one that contains only the X predictor variable and the other Restricted Model contains only the Z predictor variable. Although the same Full Model would be used in conjunction with the two Restricted Models, the use of the two different Restricted Models implies that *two different questions* are being tested. It is important to note that neither of those two questions will match the question being tested when the original Restricted Model, the one that contains none of the predictor variables, is used. (The statistical testing of the amount of variation in the criterion variable accounted for by each variable is discussed later in this chapter.)

Venn Diagrams for Various Relationships among the Variables

Regardless of which type of research question is posed, when the analysis involves more than one predictor variable, it is important for the researchers to understand the various relationships that may exist among the variables. In the next section we examine through the use of Venn diagrams the various relationships that may exist among the predictor variables and among the predictor and criterion variables.

Our discussion of the various types of relationship among the variables contained in a Full Model begins with predictor variables that are unrelated to each other but each of which is correlated with the criterion variable. Figures 6.4a and 6.4b contain two Venn diagrams that represent two Full Models. When reviewing a Venn diagram for a given model, it is important to recall from the previous section, that the unit vector contained in the model is not part of its Venn diagram.

In Figure 6.4a the Full Model is assumed to contain the criterion variable (Y), a unit vector (U) and predictor variable (A). The circles A and Y represent the variance in the predictor variable and the criterion variable, respectively. The overlap area of these two circles (diamond-marked area) represents the amount of variance in the criterion variable (Y) accounted for by the variance in the predictor variable (A), and the remaining portion of the Y circle represents the error variance in the criterion variable.

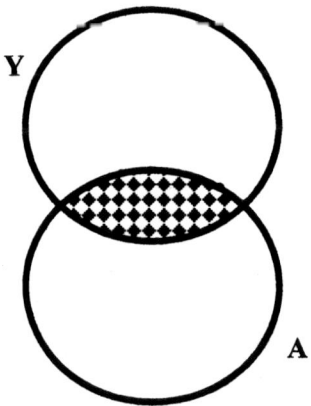

Figure 6.4a.
One-Way ANOVA.

In Figure 6.4b the variables labeled variable B and variable A*B are now placed in the Full Model along with the original predictor variable (A). One can think of these variables as the A-main effect (A), the B-main effect (B), and the interaction effect (A*B) in a two-way ANOVA. Since the circles for the three predictor variables do not overlap, the predictor variables are uncorrelated. (This type of relationship among the predictor variables is the same as the one that exists for the orthogonal comparisons listed in Exhibit 4.7 of chapter 4.) Even though the predictor variables are uncorrelated with each other, the overlap of each of the predictor variable circles (i.e., the A, B, and A*B circles) with the criterion variable circle (Y) reveals that each predictor variable does account for some of the variation in the criterion variable, which is represented by the diamond-marked areas.

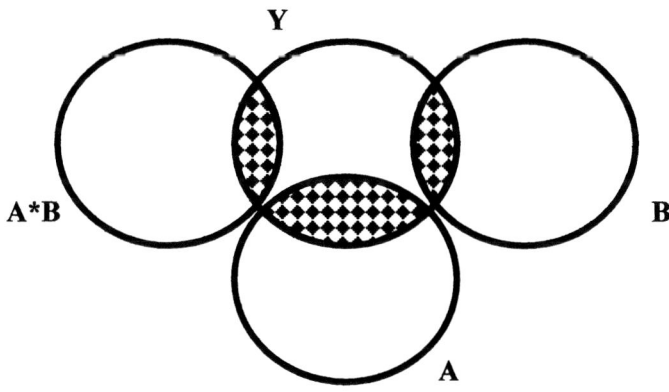

Figure 6.4b.
Orthogonal contrasts with four groups in a two-way ANOVA.

It is important to note two results of using a Full Model that contains variables A, B, and A*B when the relationships among the variables are those illustrated in the Venn diagram contained in Figure 6.4b. First, when the B and the A*B predictor variables were included in the analysis, the error variance in the Y criterion variable, which is the area in circle Y not overlapped by any predictor variable, is further reduced. The amount by which the error variance in the Y criterion variable is reduced is represented by the areas in the Y circle that overlapped with the B and A*B circles. Thus, this reduction in the error variance of the criterion variable, which was caused by the additions of variables B and A*B to the Full Model, will increase the power of the statistical tests of each predictor variable when conducted with this Full Model. This is one advantage afforded researchers when they include additional predictor variables that account for some of the variation in the criterion variable.

Second, when the B and the A*B predictor variables, which are uncorrelated with each other and with the A predictor variable, were included in the analysis, the amount of variance in the criterion variable accounted for by variable A remains the same. That is, the overlap area of the circles Y and A is the same in Figures 6.4a and 6.4b. The reason for this is that the B and A*B variables account for totally separate amounts of variance in the Y criterion variable. When predictor variables are correlated, which is a more realistic representation of the type of data encountered by researchers, the amount of variation accounted for by Venn diagram becomes a little more complicated, as illustrated next.

Before a Venn diagram that represents a Full Models containing correlated predictor variables is discussed, we will illustrate through correlation values why the amount of variation accounted for in the criterion variable is not just simply the sum of the amount of variation in the criterion variable accounted for by each predictor variable.

Suppose we have two predictor variables X and Z, and the criterion variable is represented by Y. Assume the relationships among these three variables are as follows: (a) $r_{xy} = .50$, (b) $r_{zx} = .50$, and (c) $r_{zy} = .50$. Based on these correlation values, we know that a 25% overlap exists between variables X and Y (i.e., $r_{xy}^2 = .50^2 = .25$), and a 25% overlap exists between variables Z and Y (i.e., $r_{zy}^2 = .50^2 = .25$). These two percentages of variance cannot simply be added, however, because variables X and Z are themselves correlated (i.e., $r_{zx} = .50$).

If we do not know the degree of correlation between the predictor variables X and Z, any number of correlational situations could be occurring among the three variables, including the three types of relationships depicted in the Venn diagrams presented in Figures 6.5a and 6.5b. In the Venn diagram contained in Figure 6.5a, predictor variables X and Z are uncorrelated, and the amount of accounted for variance in the criterion variable Y (25 units) was totally accounted for by variable X. In Figure 6.5b the Venn diagram, once again, depicts a situation in which the predictor variables X and Z are also uncorrelated. In this case, however, the total amount of accounted for variance in the criterion variable Y was 50 units, with equal amounts accounted for by the two predictor variables (i.e., 25 by X and 25 by Z).

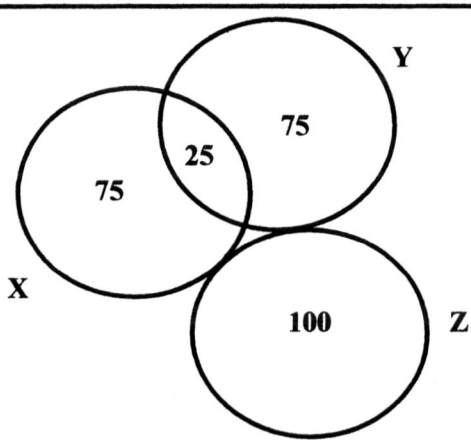

Figure 6.5a.
Variables X and Z are not correlated, variables X and Y are correlated, and variables Z and Y are not correlated.

In the Venn diagram contained in Figure 6.5c, the predictor variables are correlated (i.e., the circles representing variables X and Z overlap), a situation that is quite common in many fields of study, including the fields of social science, education, business, and economics. In this Venn diagram each variable accounts for 15 units of *unique* variance (i.e., variance only accounted for by a given variable) in the criterion variable (Y). However, if predictor variable Z is removed from the analysis, variable X accounts for 25 units of unique variance in the criterion variable Y (i.e., the 15 units uniquely accounted for by X and the 10 units that previously was jointly accounted for by X and Z).

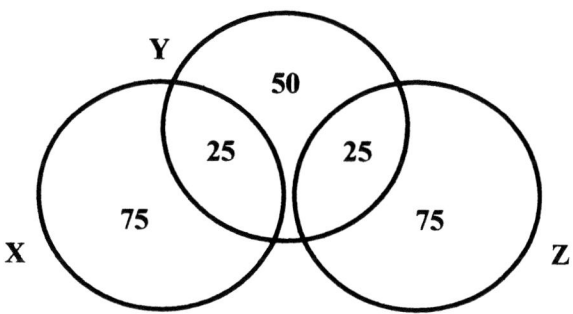

Figure 6.5b.
Variables X and Z are not correlated and variables X and Z are correlated with variable Y.

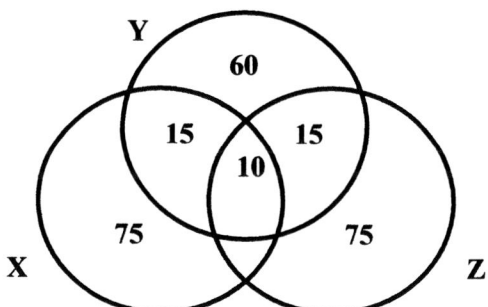

Figure 6.5c.
Variables X and Z are correlated and variables X and Z are correlated with variable Y.

In like manner, if predictor variable X is removed from the analysis, variable Z accounts for 25 units of unique variance in the criterion variable Y. When correlated predictor variables are used to account for variation in the criterion variable, it is important for researchers to understand the concept of "*unique variation in the criterion variable accounted for by a given variable*" as compared to "variation accounted for in the criterion variable by a given variable." In later chapters of this text this concept will form the basis of the discussion of the difference between the concepts of "group adjusted means" and "group means."

COMPARING THE STATISTICAL TESTS OF R^2 CHANGE VALUE AND THE MULTIPLE CORRELATION VALUE

In a previous section a Full Model (Model 6.3) and a Restricted Model (Model 6.4) were designed to determine if two variables (variables X and Z) were predictor variables of a criterion variable (Y). For every model that includes at least one predictor variable, which is the case for this Full Model, a multiple correlation value ($R_{Y.XZ}$) is calculated. The square of this multiple correlation value is equal to the R^2 value of the Full Model (R_F^2). Since the Full Model (Model 6.3) contained two predictor variables X and Z, the square of its multiple correlation value is calculated from the bivariate Pearson moment correlation coefficients for the three variables (i.e, r_{YX}, r_{YZ}, and r_{ZX}) as follows:

$$R_{Y.XZ}^2 = \frac{r_{XY}^2 + r_{YZ}^2 - 2(r_{YZ} * r_{YX} * r_{ZX})}{1 - r_{ZX}^2}$$

(Equation 6.1)

The question addressed in this section is:

Does one obtain the same results for the statistical test of the difference between the R^2 values of the Full and Restricted Models and the statistical test of the square of the multiple correlation value?

An F test of the difference between the two R^2 values of the Full and Restricted Model is conducted by using Equation 4.1, which is as follows:

$$F(df_n, df_d) = \frac{(R_F^2 - R_R^2)/(df_n)}{(1 - R_F^2)/(df_d)}$$

(Equation 4.1)

The multiple correlation value can also be tested with Equation 4.1 even though Equation 4.1 does not *visually* match the information contained in Equation 6.1. Certain terms can be substituted in Equation 4.1, however, that will produce a better visual match.

Since the Restricted Model labeled Model 6.4 contains only the unit vector, its R^2 value (R_R^2) is equal to .00. Thus, the difference between the R^2 values of these two models ($R_F^2 - R_R^2$) is simply equal to the R^2 value for the Full Model (i.e., $R_F^2 - 0 = R_F^2$). In addition, as previously stated, the square of the multiple correlation value ($R^2_{Y.XZ}$) is equal to the R^2 value of the Full Model (R_F^2); thus, $R^2_{Y.XZ}$ can be substituted for ($R_F^2 - R_R^2$) and (R_F^2) in Equation 4.1.

One can also substitute the terms k and $N - k - 1$ for the terms df_n and df_d, respectively. The letters k and N represent the number of predictor variables (not counting the unit vector) and the sample size, respectively. It should be noted that the 1 in the $N - k - 1$ term represents the unit vector. Placing the appropriate substitutions in Equation 4.1 produces the following F-test formula:

$$F = \frac{R^2_{Y.XZ} \, / \, k}{(1 - R^2_{Y.XZ}) \, / \, (N - k - 1)}$$

(Equation 6.2)

Although Equations 4.1 and 6.2 will produce the same F-test value for $R^2_{Y.XZ}$ and ($R_F^2 - R_R^2$), Equation 6.2 is not as versatile. It will provide a test of $R^2_{Y.XZ}$ that is equivalent to the test of ($R_F^2 - R_R^2$) calculated with Equation 4.1 only when the Restricted Model is the Null Model, that is, it contains only the unit vector.

EXPANDING THE NUMBER
OF PREDICTOR VARIABLES

Rationale

The discussion in chapter 1 implied that a researcher probably would want to include multiple predictors when attempting to understand a particular phenomenon. The construction and analysis of models that contain multiple predictors offers the researcher the chance to "model" the phenomenon as complexly as it really is determined. We have never understood why researchers would want to reflect what they acknowledge to be a complexly determined phenomenon with an overly simplified one- or two-variable model.

Graphical Representation

One may think of a two-predictor variable situation as previously presented in Figure 6.1a as a football with an X and Z axes for the predictor variables and a Y axis for the criterion variable (see Figure 6.6). Now assume the football is spinning through a fourth dimension—time. That is, in addition to the criterion variable Y and the predictor variables X and Z, Time is now included in the model (see Figure 6.7). The reader's mind (as well as each author's) is obviously being challenged by the four-dimension case. Since we live a three-dimensional world, a geometric representation of more than three predictor variables, which requires a four-dimensional depiction, is difficult to visualize.

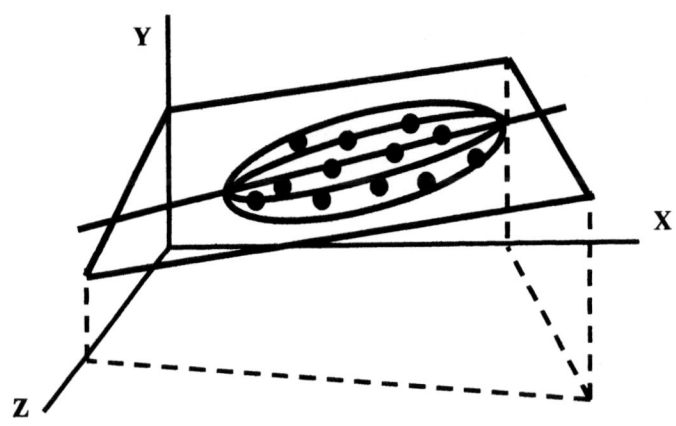

Figure 6.6.
A graphical representation of a model in which the criterion variable is Y and the two predictor variables are X and Z.

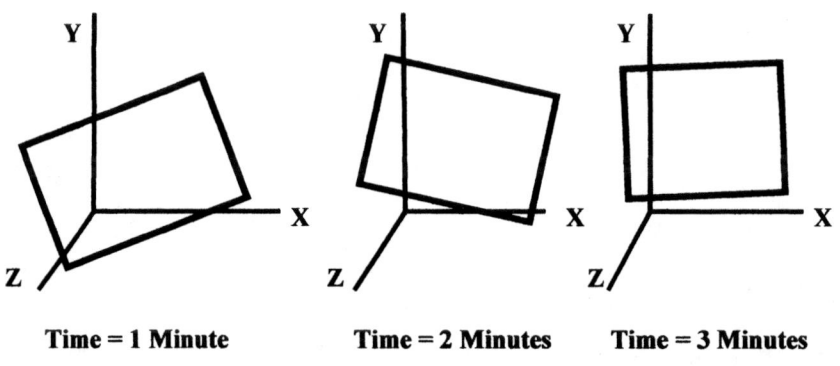

Figure 6.7.
A graphical representation of four dimensions with time being the fourth dimension.

One can have a better sense of a four-dimensional model by visualizing three snapshots of the plane of best fit, one for each of three separate time periods, as depicted in Figure 6.7. Notice that for given values of the fourth dimension of time (such as 1 minute, 2 minutes, and 3 minutes) a plane of best fit is formed. Although somewhat difficult to visualize in one's mind, such complex relationships can be modeled by researchers and estimated, especially with the aid of computer programs.

Venn Diagrams

As the graphical representation becomes more involved with multiple variables, so does the Venn diagram representation. But it is instructive to develop a hypothetical Venn diagram representation when variables are added one at a time. To illustrate, Models containing two, three, and four predictor variables are represented in Figure 6.8.

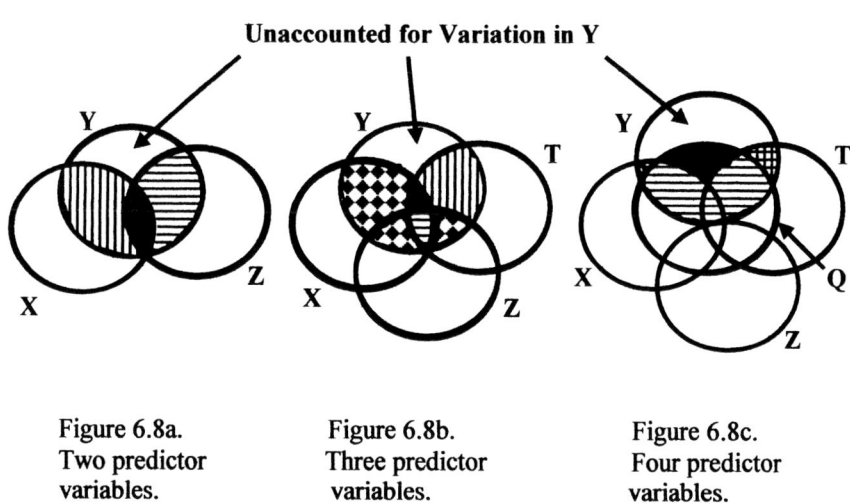

Figure 6.8a.
Two predictor
variables.

Figure 6.8b.
Three predictor
variables.

Figure 6.8c.
Four predictor
variables.

Figure 6.8.
Venn diagrams for complex models.

In the model with the two predictor variables X and Z, each of these variables accounted for unique amount of variance in the Y-criterion variable. As indicated in Figure 6.8a, the amount of unique variation accounted for by variables X and Z are the areas marked with vertical lines and horizontal lines, respectively. The amount of variation in the Y-criterion variable they both shared is shaded in black. Note that the unmarked area in the Y-variable circle represents the amount of variation in the Y-criterion variable that is not accounted for by the predictor variables. When the third variable T is added to the model, the amount of variance in Y uniquely accounted for by T is marked with vertical lines (see Figure 6.8b). Once again, note that the unmarked area in the Y-variable circle represents the amount of variation in the Y-criterion variable that is not accounted for by the predictor variables.

In Figure 6.8c, predictor variable Q is added to the model. Although highly correlated with the criterion (determined by the large amount of overlap between Q and Y, which is marked with the area shaded in black and horizontal lines), its circle also overlaps with the circles of the other predictor variables. When the Q variable is added to the model containing variables T, X, and Z, the increase in R^2 value produced by this model will equal only the unique variance accounted for by Q, is only the area shaded in black. Even with these four predictor variables, notice that not all of the variation in the Y-criterion variable is accounted for by the predictor variables, as indicated by the unmarked area in the circle representing the Y variable.

As indicated in chapter 1, inclusion of several variables from a particular area (such as within person) may lead the researcher to a point of diminishing returns. Some authors suggest that five or six variables will be the upper limit of useful variables (Hinkle, Wiersma, & Jurs, 1994). We strongly feel that all behavior is predictable and that researchers simply have to find the right variables (or the right functional relationships—interactions and nonlinear relationships, which will be discussed in later chapters).

It should be noted, however, that when researchers increase the number of predictor variables in their Full Model, the size of the sample must be a consideration. If the ratio of the number of cases (i.e., sample size) to the number of predictor variables becomes "too small," the R^2 value and weighting coefficients can be unstable as measured for one sample versus another sample. Stevens (2002) suggested that researchers strive to have at least 5 cases to 1 predictor-variable ratio. This is not to suggest that researchers should consider placing a limit on the number of predictor variables used, but rather they must strive to have sufficient sample sizes to support the estimation and testing of their models. The emphasis should always be placed on the Research Question—if the question calls for more variables, then the researcher is obligated to collect data on more participants.

GENERAL RESEARCH HYPOTHESIS 6.1

The discussion of General Research Hypothesis 6.1 (GRH 6.1) provides a template for constructing Full and Restricted Models used to test whether a set of predictor variables can predict a criterion variable. As previously discussed in this chapter, the Research Hypothesis could be stated a number of ways for such study. GRH 6.1 is stated as follows:

GRH 6.1: For a given population, the set of predictor variables containing X1, X2, . . ., Xk predict the criterion Y.

It should be noted that when researchers consider a specific Research Hypothesis, the number of predictor variables should not be constrained by their statistical concerns (e.g., sample size), rather it is the responsibility of the researchers to pose a *theoretically sound* Research Hypothesis and then concern themselves with the possible statistical issues. For example, if a theoretically sound Research Hypothesis requires the inclusion of a substantial number of predictor variables, it is the responsibility of the researchers to ensure that the study has a sample size sufficient to allow for the statistical testing of the corresponding Statistical Hypothesis.

The General Statistical Hypothesis 6.1 (GSH 6.1), which corresponds to GRH 6.1, is as follows:

GSH 6.1: For a given population, the set of predictor variables containing X1, X2, . . ., Xk does not predict the criterion Y.

Implied by GRH 6.1 is the following Full Model:

$$Y = a_0U + a_1X1 + a_2X2 + \ldots + a_kXk + E_1 \tag{Model 6.5}$$

Since the antithesis of the GRH 6.1 is that the set of predictor variables does not predict the criterion Y, the following restrictions are placed on this Full Model:

$$a_1 = 0; a_2 = 0; \ldots; a_k = 0 \quad \text{or} \quad a_1 = a_2 = \ldots; a_k = 0 \quad \text{(k restrictions)}$$

Placing these restrictions on the Full Model (Model 6.5) results in the following Restricted Model:

$$Y = a_0U + E_2 \tag{Model 6.6}$$

The F test (Equation 4.1) of the difference between the R^2 values of the Full and Restricted Models is applicable to the testing of this Research Hypothesis because (a) the same criterion is in both models, (b) the unit vector is in both models, and (c) the Restricted Model is a restricted subset of the Full Model (for these restrictions the Restricted Model is the Null Model). It should be noted that since more than one restriction is being made (i.e., more than one predictor variable is deleted from the Full Model), a directional Research Hypothesis cannot be tested, thus the researchers must use a two-tailed test.

As previously discussed, it is important to understand what this F test is in fact testing. Since all of the predictor variables were deleted from the Full Model to form the Restricted Model, the statistical test of the Statistical Hypothesis GSH 6.1 only deals with the *combined* predictive power of the predictor variables. This does not mean that each one of the predictor variables is in fact able to predict the criterion variable. Thus, researchers need to understand this lack of information regarding which specific variables do in fact serve as predictor variables when posing and testing such a Research Hypothesis. If researchers want to know more about which variables do in fact predict the criterion, they should pose a Research Hypothesis which when tested will reveal such information (see the sections in this chapter that contain Applied Research Hypotheses 6.2, 6.3, and 6.4).

APPLIED RESEARCH HYPOTHESIS 6.1

For Applied Research Hypothesis 6.1 (ARH 6.1), the predictor variables X3, X4, and X5 contain test scores from Test A, Test B, and Test C, respectively. The criterion variable for ARH 6.1 is X2, which is assumed to represent a set of test scores for Test D.

The researchers posed Applied Research Hypothesis (ARH) 6.1 as follows:

ARH 6.1: The set of test scores consisting of Test A (X3), Test B (X4), and Test C (X5) predict the Test D (X2) scores.

The corresponding statistical hypothesis (ASH 6.1) is as follows:

ASH 6.1: The set of test scores consisting of Test A (X3), Test B (X4), and Test C (X5) does not predict the Test D (X2) scores.

Full and Restricted Models

To determine whether the data support ARH 6.1, Full and Restricted Models are designed to reflect ARH 6.1 and ASH 6.1, respectively. The Full Model is constructed as follows:

$$X2 = a_0U + a_3X3 + a_4X4 + a_5X5 + E_1 \qquad \text{(Model 6.7)}$$

The restrictions placed on the Full Model that are needed to produce the Restricted Model are as follows:

$a_3 = 0$, $a_4 = 0$, and $a_5 = 0$ or $a_3 = a_4 = a_5 = 0$ (3 restrictions)

Placing these restrictions on the Full Model produces the following Restricted Model:

$$X2 = a_0U + E_2 \qquad \text{(Model 6.8)}$$

Note that this Restricted Model is the Null Model.

Analysis of ARH 6.1 with SPSS for Windows

Once the SPSS data file entitled "GLM DATA SPSS FORMAT" has been accessed from the internet site (see Appendix A) and the commands needed to access the Linear Regression menu (see chapter 4) have been completed, the Linear Regression menu presented in Exhibit 6.1 will be displayed.

The criterion variable (X2) and the predictor variables (X3, X4, and X5) contained in the Full Model (Model 6.7) for ARH 6.1 must be entered into the regression menu. As illustrated in Exhibit 6.1, these variables are entered as follows:

1. Click on the X2 variable (see Oval 1) and click on the arrow key next to the "Dependent" box. This will identify X2 as the criterion variable.
2. While holding down the control key on the keyboard, click on the X3, X4, and X5 variables (see Oval 2). Release the control key and click on the arrow key next to the "Independent(s)" box. This will identify X3, X4, and X5 as the predictor variables.
3. Click on the "OK" button (see Oval 3).

Exhibit 6.1.
SPSS Linear Regression menu for ARH 6.1.

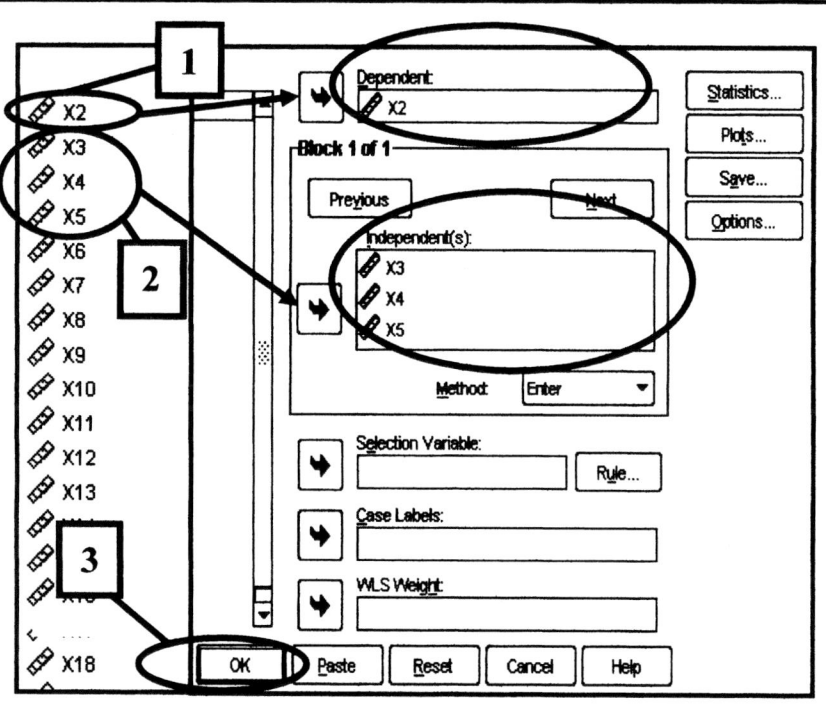

Once these three steps are completed, the regression analysis output contained in Exhibit 6.2 will be displayed. The key information on the output window for the Full Model (Model 6.7) is located in the following areas:

1. The Full Model R^2 value is .934 (see Oval 1). Since the Restricted Model contains only the constant and error terms, the R^2 value for the Restricted Model is equal to .00. This value is not listed on the output window.

2. The numerator degrees of freedom (df_n) value and the denominator degrees of freedom (df_d) value are 3 and 56, respectively (see Oval 2). Since the number of restrictions placed on the Full Model is 3, the df_n value is 3. The sample size of 60 minus the 4 estimated parameters (i.e., the three regression coefficients and the one constant value) produces a df_d value of 56. These values are used in the calculation of the F test.

3. Since the Restricted Model is the Null Model, the F test listed in the "ANOVA" table of the SPSS output window is the appropriate test for statistical hypothesis ASH 6.1. The F test value of the difference between the R^2 values of the Full and Restricted Models and its corresponding probability value are 265.773 and .000, respectively (see Oval 3).

Exhibit 6.2
SPSS regression output—Full Model (Model 6.7) for ARH 6.1

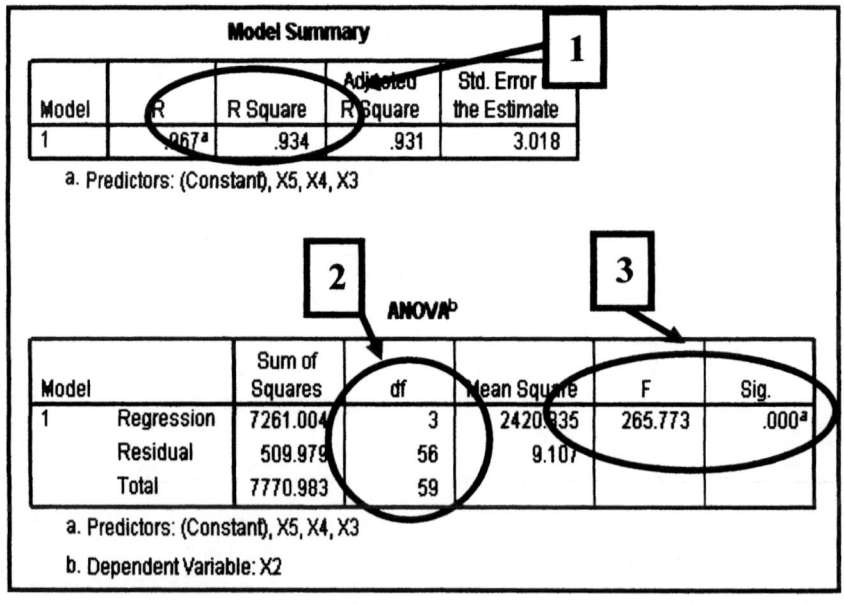

Model Summary

Model	R	R Square	Adjusted R Square	Std. Error the Estimate
1	.967ª	.934	.931	3.018

a. Predictors: (Constant), X5, X4, X3

ANOVAᵇ

Model		Sum of Squares	df	Mean Square	F	Sig.
1	Regression	7261.004	3	2420.335	265.773	.000ª
	Residual	509.979	56	9.107		
	Total	7770.983	59			

a. Predictors: (Constant), X5, X4, X3
b. Dependent Variable: X2

Since the F test probability value ($p < .01$) is less than the established alpha level of .05, the proportion of variance accounted for in the criterion variable X2, which was .93 (i.e., the difference between the R^2 values of the Full and Restricted Models), was statistically significant. Thus, the Statistical Hypothesis (ASH 6.1) is rejected, and the data support the Research Hypotheses (ARH 6.1), which states that the set of variables X3, X4, and X5 predict the criterion variable X2.

It is important to understand that even though the data support ARH 6.1, this conclusion does not address the issue of which variables are or are not "valuable" in this prediction. This analysis also does not allow for a directional interpretation, because df_n is not equal to 1, and the Research Hypothesis (ARH 6.1) was not directional. Thus, the information gained from such a Research Hypothesis may be of limited value. (This issue will be addressed in the sections entitled "General Research Hypothesis 6.2" and "Applied Research Hypotheses 6.2, 6.3, and 6.4.")

Writing the Results for ARH 6.1 in APA Style

The results for ARH 6.1 could be written in APA style as follows:

The proportion of variance in the Test D scores accounted for by the variance in the scores of Test A, Test B, and Test C was .93, which was statistically significant at the .05 alpha level, $F (3, 56) = 265.77, p < .01$. This proportion of accounted for variance in the Test D scores is a large effect size according to criteria established by Cohen (1988).

It is important to understand that this discussion of the statistical test results for ASH 6.1 does not include any statement about the amount of unique variation accounted for by each of the predictor variables, that is, whether each variable serves as a predictor of the criterion variable. In the next section we will discuss how one might ask a definitive question regarding the contribution of each variable in a set of variables, and how to construct and test a model designed to address that question.

TESTING ONE VARIABLE'S CONTRIBUTION

Rationale

In response to the limitation identified in the previous sections regarding the testing of a *set* of predictor variables, one might be interested in focusing on the "value added benefit" of a particular variable. If a predictor variable does not add a significant amount of variance accounted for in the criterion variable to the amount accounted for by the variables already in the model, then that variable can be omitted from further consideration. On the other hand, it may be valuable to know how a predictor variable affects the criterion variable, within a set of other predictors. We often use the terminology "over and above" or "unique variation accounted for" when investigating the amount of value added by the inclusion of a variable in a set of predictor variables.

Graphical Representation and a Venn Diagram

How a predictor variable affects the criterion variable within a set of other predictor variables can be phrased in geometric terms as: Is a k-dimensional space needed, or can one less dimension be just as effectively used to reflect the data? If the F test is significant, then the larger space is needed. If the F test is *not* significant, then the smaller space is all that is needed.

When each Research Hypothesis in a set of Research Hypotheses focuses on the unique variance in the criterion variable (Y) accounted for by the targeted predictor variable, say, X, the R^2 value of the corresponding Full Model is not the value of major interest (the areas in Figure 6.9 marked with vertical lines, horizontal lines and shaded in black)—it is, however, used in the denominator of the F test employed to test the given Research Hypotheses. The focus of the researchers is on the unique variation accounted for in the criterion variable accounted for by the targeted predictor variable (variable X), that is, the amount of variation in variable Y accounted for by variable X over and above the variation accounted for by variable Z. Thus the area of focus in the Venn diagram contained in Figure 6.9 is the unique overlap between the Y and X variable circles, which is the area marked with vertical lines.

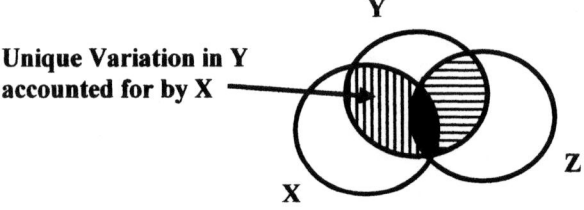

Figure 6.9.
Unique variance accounted for by the X predictor variable—The value added model.

GENERAL RESEARCH HYPOTHESIS 6.2

The purpose of General Research Hypothesis 6.2 (GRH 6.2) is to provide a guide for the procedure used to test the amount of unique variance accounted for in the criterion variable by a given predictor variable in a set of predictor variables.

GRH 6.2 is stated as a nondirectional hypothesis as follows:

GRH 6.2 Version 1: For a given population, variable Xk accounts for additional variance in the criterion variable, over and above the amount of variance accounted for by variables X1, X2, . . ., Xk – 1.

An alternative version of GRH 6.2 is as follows:

GRH 6.2 Version 2: For a given population, variable Xk is a predictor of the criterion variable over and above variables X1, X2, . . ., Xk – 1.

Later in this section, it will be shown that researchers could appropriately pose a directional Research Hypothesis.

Corresponding to GRH 6.2 is the General Statistical Hypothesis 6.2 (GSH 6.2), which can be stated as to correspond to either version of GRH 6.2 as follows:

GSH 6.2 Version 1: For a given population, variable Xk *does not* account for additional variance in the criterion variable, over and above the amount of variance accounted for by variables X1, X2, . . ., Xk-1.

GSH 6.2 Version 2: For a given population, variable Xk is *not* a predictor of the criterion variable over and above variables X1, X2, . . ., Xk-1.

The Full Model that reflects GRH 6.2, which contains all of the predictor variables mentioned in GRH 6.2, is as follows:

$$Y = a_0U + a_1X1 + a_2X2 + \ldots + a_kXk + E_1 \quad \text{(Model 6.9)}$$

The restriction implied by the GSH 6.2 is as follows:

$$a_k = 0 \quad \text{(1 restriction)}$$

Note that only one restriction will be placed on the Full Model (Model 6.9). Thus, it would be appropriate for researchers to pose a directional Research Hypothesis.

If researchers chose to state a directional Research Hypothesis, it is best to revise Nondirectional Version 2 of the GRH 6.2 as follows:

GRH 6.2 Version 3: For a given population, variable Xk is a *positive* predictor of the criterion variable over and above variables X1, X2, . . ., Xk - 1.

Placing the one restriction on the Full Model (Model 6.9) results in the following Restricted Model:

$$Y = a_0U + a_1X1 + a_2X2 + \ldots + a_{k-1}Xk - 1 + E_2 \quad \text{(Model 6.10)}$$

Note that variable Xk has been deleted from this Restricted Model, that is, the last predictor variable in the model is Xk − 1.

As always, the F test can be used to test the difference between the R^2 values of the Full Model (Model 6.9) and the Restricted Model (Model 6.10). The F test value is calculated using Equation 4.1, which was previously presented in this chapter. There are k continuous predictors in the Full Model (Model 6.9), along with the unit vector, therefore m1 = k + 1 (i.e., the Full Model has k + 1 pieces of information). With one restriction, m2 = (k + 1) - 1, which indicates that the Restricted Model (Model 6.10) has (k +1) - 1 pieces of information. Since the difference between m1 and m2 is 1, the value for df_n in Equation 4.1 is equal to 1. Thus, when testing GSH 6.2, Equation 6.3, which is listed below, can be substituted for Equation 4.1.

$$F(df_n, df_d) = \frac{(R_F^2 - R_R^2)/(1)}{(1 - R_F^2)/(df_d)} \quad \text{(Equation 6.3)}$$

The reduction in the R^2 from the Full Model to the Restricted Model is a function of the one variable that is restricted from the Full Model. Therefore, any difference in variance accounted for is directly attributable to that one variable.

Since the F test of the unique variance accounted for in the criterion variable by variable Xk has a numerator degrees of freedom value of 1, researchers could also use the t of the coefficient for variable Xk in the Full Model (Model 6.9). The use of both the F test and the t test will be illustrated in the next section.

APPLIED RESEARCH HYPOTHESES 6.2, 6.3, AND 6.4

In a previous section of this chapter we tested the Applied Statistical Hypothesis 6.1 (ASH 6.1), which corresponded to Applied Research Hypothesis 6.1 (ARH 6.1). Recall that ARH 6.1 dealt with the amount of variation in X2 accounted for by a *set* of three predictor variables (i.e., X3, X4, and X5). Instead of being interested in ARH 6.1, researchers may want to obtain an estimate of the amount of unique variation accounted for in the criterion variable by each of these three predictor variables, and determine whether each amount is statistically significant. Thus, this section presents an illustration of how researchers could apply the testing procedures discussed in connection with the General Research Hypothesis 6.2.

In this section we will assume that X3, X4, and X5 contain test scores recorded for Test A, Test B, and Test C, respectively. In addition, we will assume the criterion variable is X2, which represents a set of test scores for Test D. In keeping with our view that researchers should pose directional research questions whenever possible, we also are assuming that directional research questions are used in this section. These research questions are stated as:

Research Question 6.2: Are the Test A scores positive predictors of the Test D scores when the scores from Test B and Test C are also used as predictors?

Research Question 6.3: Are the Test B scores positive predictors of the Test D scores when the scores from Test A and Test C are also used as predictors?

Research Question 6.4: Are the Test C scores positive predictors of the Test D scores when the scores from Test A and Test B are also used as predictors?

The three directional Applied Research Hypotheses (ARH 6.2, ARH 6.3, and ARH 6.4) and the corresponding Applied Statistical Hypotheses (ASH 6.2, ASH 6.3, and ASH 6.4) corresponding to these three research questions are as follows:

ARH 6.2: For the population, variable X3 is a positive predictor of the criterion variable X2 when X4 and X5 are also used as predictor variables.

ASH 6.2: For the population variable, X3 is not a positive predictor of the criterion variable X2 when X4 and X5 are also used as predictor variables.

ARH 6.3: For the population variable, X4 is a positive predictor of the criterion variable X2 when X3 and X5 are also used as predictor variables.

ASH 6.3: For the population variable, X4 is not a positive predictor of the criterion variable X2 when X3 and X5 are also used as predictor variables.

ARH 6.4: For the population variable, X5 is a positive predictor of the criterion variable X2 when X3 and X4 are also used as predictor variables.

ASH 6.4: For the population, variable X5 is not a positive predictor of the criterion variable X2 when X3 and X4 are also used as predictor variables.

The researchers could use either of two methods to statistically test these three research hypotheses. One method, which we labeled "The Full and Restricted Models Approach," involves the construction of a Full Model and three Restricted Models. Once these models are constructed and estimated, the difference in the R^2 value of the Full Model and each Restricted Model is statistically tested with an F test. The other method, which we labeled, "The Coefficient t Test Approach," utilizes the t tests of the regression coefficients estimated for the three predictor variables in the Full Model.

The Full and Restricted Models Approach

In the application of The Full and Restricted Models Approach only one Full Model is required to reflect ARH 6.2, ARH 6.3, and ARH 6.4. This Full Model is constructed as follows:

$$X2 = a_0U + a_3X3 + a_4X4 + a_5X5 + E_1 \qquad \text{(Model 6.11)}$$

Based on the Statistical Hypotheses, three restrictions will be placed on this Full Model. It is very important to note, however, that each restriction is separately placed on the model. These restrictions are as follows:

$$a_3 = 0 \quad \text{(Restriction 1)} \quad a_4 = 0 \quad \text{(Restriction 2)} \quad a_5 = 0 \quad \text{(Restriction 3)}$$

Restricted Models 1, 2, and 3 are generated when Restrictions 1, 2, and 3, respectively, are separately placed on the Full Model. These Restricted Models are as follows:

$$X2 = a_0U + a_4X4 + a_5X5 + E_2 \qquad \text{(Restricted Model 1—Model 6.12)}$$

$$X2 = a_0U + a_3X3 + a_5X5 + E_3 \qquad \text{(Restricted Model 2—Model 6.13)}$$

$$X2 = a_0U + a_3X3 + a_4X4 + E_4 \qquad \text{(Restricted Model 3—Model 6.14)}$$

These three Restricted Models reflect the statements contained in the three Applied Statistical Hypotheses.

Analysis of ARH 6.2, 6.3, and 6.4 with SPSS for Windows

Once the SPSS data file entitled "GLM DATA SPSS FORMAT" has been accessed from the internet site (see Appendix A) and the commands needed to access the regression menu (see chapter 4) have been completed, the regression menu presented in Exhibit 6.3 will be displayed.

Exhibit 6.3.
SPSS Linear Regression menu—Restricted Model 1 (Model 6.12) for ASH 6.2.

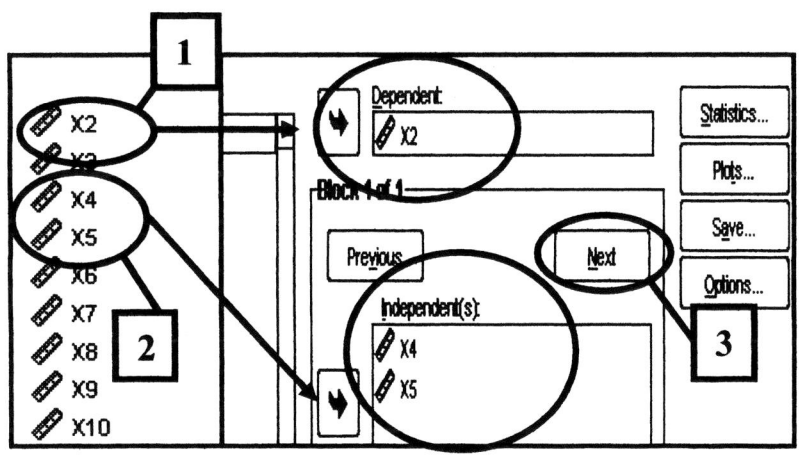

It is important to note that when using SPSS for Windows to statistically test the difference between the R^2 values of a Full Model and each Restricted Model, the variables contained in the Restricted Model are entered first. Thus, to statistically test ASH 6.2 with SPSS for Windows, the criterion variable X2 and the predictor variables X4 and X5 contained in Restricted Model 1 (Model 6.12) are entered into the regression menu first as follows (see Exhibit 6.3):

1. Click on the X2 variable (see Oval 1) and click on the arrow key next to the "Dependent" box. This will identify X2 as the criterion variable.
2. While holding down the control key on the keyboard, click on the X4 and X5 variables (see Oval 2). Release the control key and click on the arrow key next to the "Independent(s)"box. This will identify X4 and X5 as the predictor variables.
3. Click on the "Next" button (see Oval 3).

After these three steps are completed, the menu contained in Exhibit 6.4 will be displayed.

To obtain the Full Model (Model 6.11) used to reflect ARH 6.2, the following steps are completed in this menu:

1. Click on the X3 variable (see Oval 1) and click on the arrow key next to the "Independent(s)" box. This will include X3 along with X4 and X5 as the predictor variables for the Full Model.
2. Click on the "Statistics" button (see Oval 2).

Upon completion of these two steps, the Statistics menu will be displayed (see in Exhibit 6.5).

Exhibit 6.4.
SPSS Linear Regression menu—Full Model (Model 6.11) for ARH 6.2.

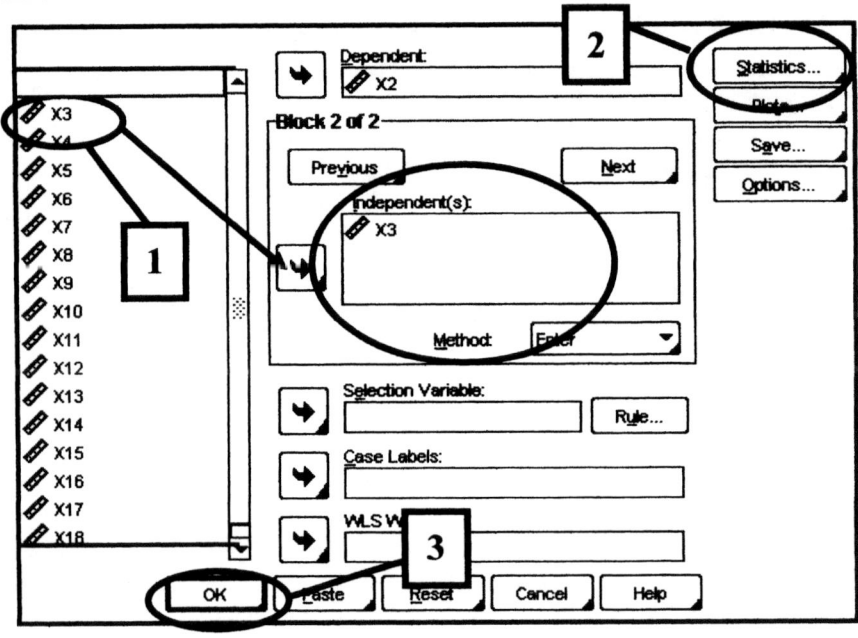

Exhibit 6.5.
SPSS Statistics menu for ARH 6.2.

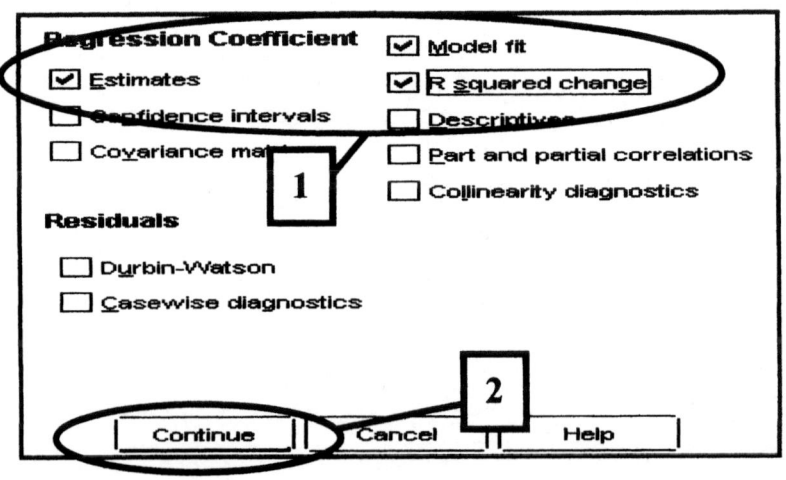

The following steps are completed in the Statistics menu:

1. Click on the boxes in front of "R squared change," "Model fit," and "Estimates" (see Oval 1).
2. Click on the "Continue" button (see Oval 2).

Once these steps are completed, the regression menu will be displayed again, as shown in Exhibit 6.4. Click on the "OK" button in this menu (see Oval 3 in Exhibit 6.4). Once this task is completed the output window for the Full Model and Restricted Model 1 will be displayed (see Exhibit 6.6). Note that only the "Model Summary" and "Coefficients" tables of the output window are displayed in Exhibit 6.6.

Exhibit 6.6.
SPSS regression output for ASH 6.2 and ARH 6.2.

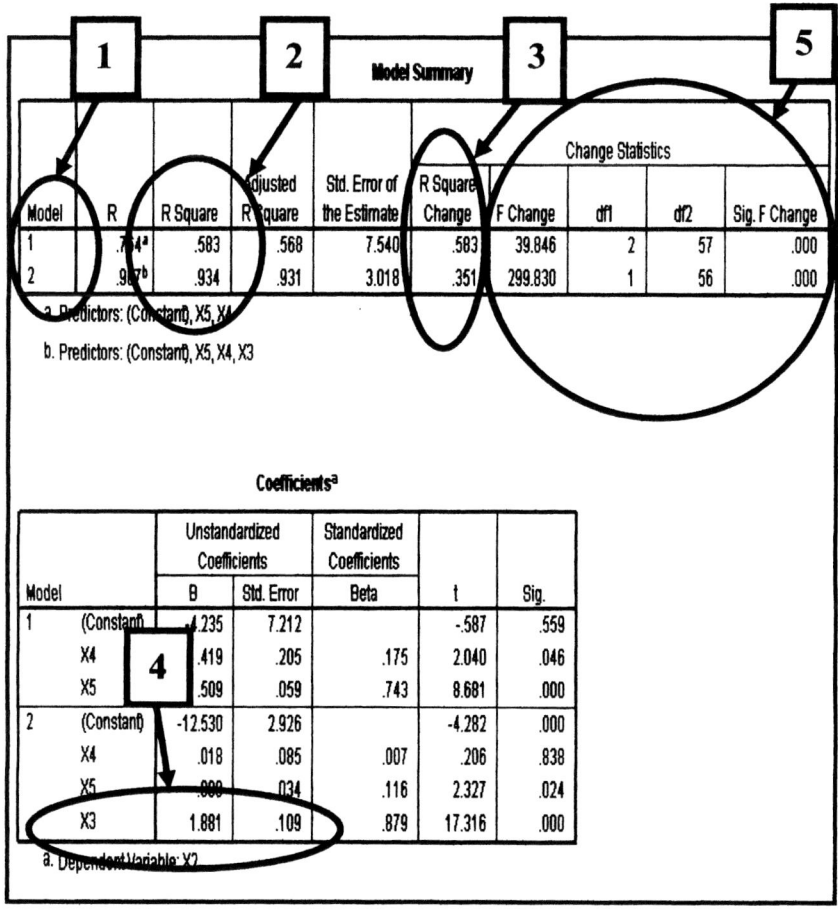

Model Summary

Model	R	R Square	Adjusted R Square	Std. Error of the Estimate	R Square Change	F Change	df1	df2	Sig. F Change
					Change Statistics				
1	.764ᵃ	.583	.568	7.540	.583	39.846	2	57	.000
2	.967ᵇ	.934	.931	3.018	.351	299.830	1	56	.000

a. Predictors: (Constant), X5, X4
b. Predictors: (Constant), X5, X4, X3

Coefficientsᵃ

Model		Unstandardized Coefficients B	Std. Error	Standardized Coefficients Beta	t	Sig.
1	(Constant)	-4.235	7.212		-.587	.559
	X4	.419	.205	.175	2.040	.046
	X5	.509	.059	.743	8.681	.000
2	(Constant)	-12.530	2.926		-4.282	.000
	X4	.018	.085	.007	.206	.838
	X5	.080	.034	.116	2.327	.024
	X3	1.881	.109	.879	17.316	.000

a. Dependent Variable: X2

The key pieces of information contained in Exhibit 6.6 are as follows:

1. The output identifies Restricted Model 1 and the Full Model as Model 1 and Model 2, respectively (see Oval 1).
2. The R^2 values for Restricted Model 1 and the Full Model are .583 and .934, respectively (see Oval 2).
3. The difference between the R^2 values of Restricted Model 1 and the Full Model is .351 (see Oval 3).
4. The regression coefficient for variable X3 is 1.881 (see Oval 4).
5. The F test of the difference in the R^2 values of the two models and its corresponding probability value are 299.83 and .000 ($p < .01$), respectively (see Oval 5). Since the Research Hypothesis is directional and the coefficient for variable X3 was positive as hypothesized, the two-tailed probability value is divided by 2 to obtain the one-tailed probability value, which is also less than .01.

Since the regression coefficient for X3 is positive and the one-tailed probability value ($p / 2 < .01$) of the F test is less than the alpha level of .05, the statistical hypothesis ASH 6.2 is rejected. Thus the data support the statement that variable X3 is a positive predictor of the criterion variable X2 when X4 and X5 are also used as predictor variables.

When testing the statistical hypotheses ASH 6.3 and ASH 6.4, the steps followed in the SPSS for Windows menus are the same as the ones employed in the testing of 6.2 (of course, the variables entered for the Restricted Models and the Full Model change). For ASH 6.3, the variables of X3 and X5, which are the predictor variables in Restricted Model 2 (Model 6.13), are entered first in the "Independent(s)" box of the regression menu. After the "Next" key is clicked, the predictor variable X4 is entered in the "Independent(s)" box, which is the additional predictor variable needed to form the Full Model (Model 6.11). The output for the two models used to test ASH 6.3, which are Model 6.11 (Full Model) and Model 6.13 (Restricted Model 2), is contained in Exhibit 6.7.

To test ASH 6.4, the predictor variables of X3 and X4, which are the predictor variables in Restricted Model 3 (Model 6.14), are entered first in the "Independent(s)" box of the regression menu. After the "Next" key is clicked, the predictor variable X5 is entered in the "Independent(s)" box, which is the additional predictor variable needed to form the Full Model (Model 6.11). The output for the two models used to test ASH 6.4, which are Model 6.11 (Full Model) and Model 6.14 (Restricted Model 3), is also contained in Exhibit 6.7. (Note that only the "Model Summary" and "Coefficients" tables of the output windows are displayed in Exhibit 6.7.)

The information contained in Exhibit 6.7, which is used to test the statistical hypothesis corresponding to ASH 6.3 is as follows:

1. The output identifies Restricted Model 2 and the Full Model as Model 1 and Model 2, respectively (see Oval 1).
2. The R^2 values for Restricted Model 2 and the Full Model are .934 and .934, respectively (see Oval 2). It should be noted that the R^2 values of

these two models are the same when expressed to only three decimal places. By clicking on the "Model Summary" table and double clicking on each R^2 value, one will find that the R^2 values of Restricted Model 2 and the Full Model are .93432 and .93437, respectively.

3. The difference between the R^2 values of Restricted Model 2 and the Full Model is .000 (see Oval 3). Again, note that the difference between the two R^2 values is 0 when displayed to three decimal places, but it is equal to .00005 when displayed to five decimal places.

4. The regression coefficient for variable X4 is .018 (see Oval 4).

Exhibit 6.7.
SPSS regression output for ASH 6.3, ARH 6.3, ASH 6.4, and ARH 6.4.

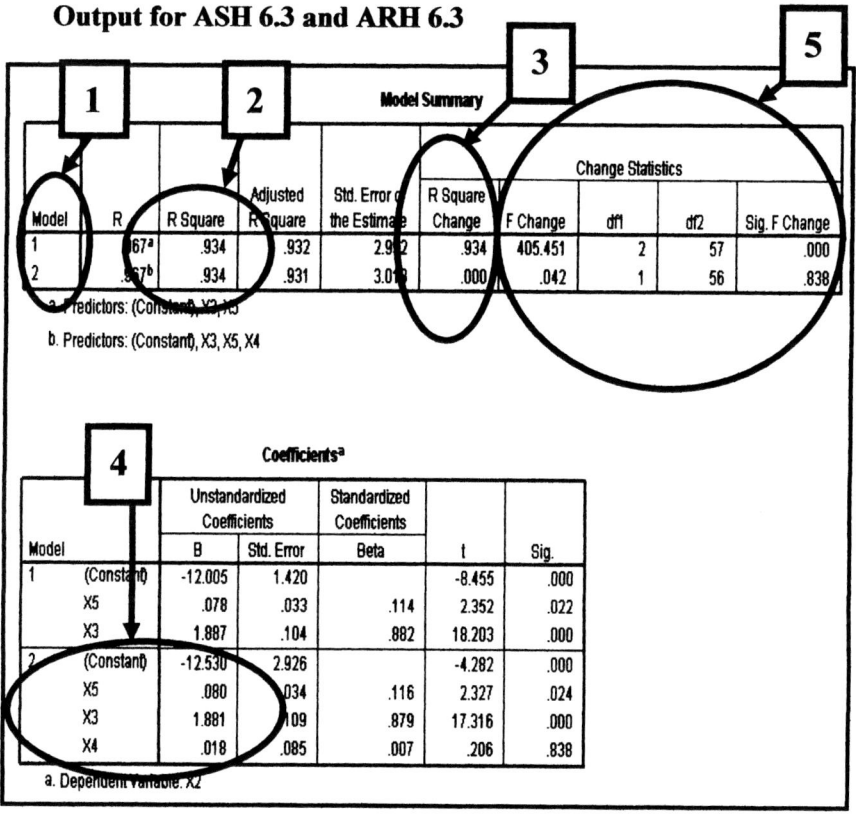

Output for ASH 6.3 and ARH 6.3

Model Summary

Model	R	R Square	Adjusted R Square	Std. Error of the Estimate	R Square Change	F Change	df1	df2	Sig. F Change
					Change Statistics				
1	.967ª	.934	.932	2.9?2	.934	405.451	2	57	.000
2	.?7b	.934	.931	3.0?8	.000	.042	1	56	.838

a. Predictors: (Constant), X3, X5
b. Predictors: (Constant), X3, X5, X4

Coefficientsa

Model		Unstandardized Coefficients B	Unstandardized Coefficients Std. Error	Standardized Coefficients Beta	t	Sig.
1	(Constant)	-12.005	1.420		-8.455	.000
	X5	.078	.033	.114	2.352	.022
	X3	1.887	.104	.882	18.203	.000
2	(Constant)	-12.530	2.926		-4.282	.000
	X5	.080	.034	.116	2.327	.024
	X3	1.881	.109	.879	17.316	.000
	X4	.018	.085	.007	.206	.838

a. Dependent Variable: X2

5. The F test of the difference in the R^2 values of the two models and its corresponding probability value are .042 and .838, respectively (see Oval 5). Since the Research Hypothesis is directional and the coefficient

for variable X5 was positive as hypothesized, the two-tailed probability of .838 is divided by 2 to obtain the one-tailed probability of .419.

Exhibit 6.7 (continued).
SPSS regression output for ASH 6.3, ARH 6.3, ASH 6.4, and ARH 6.4.

Output for ASH 6.4 and ARH 6.4

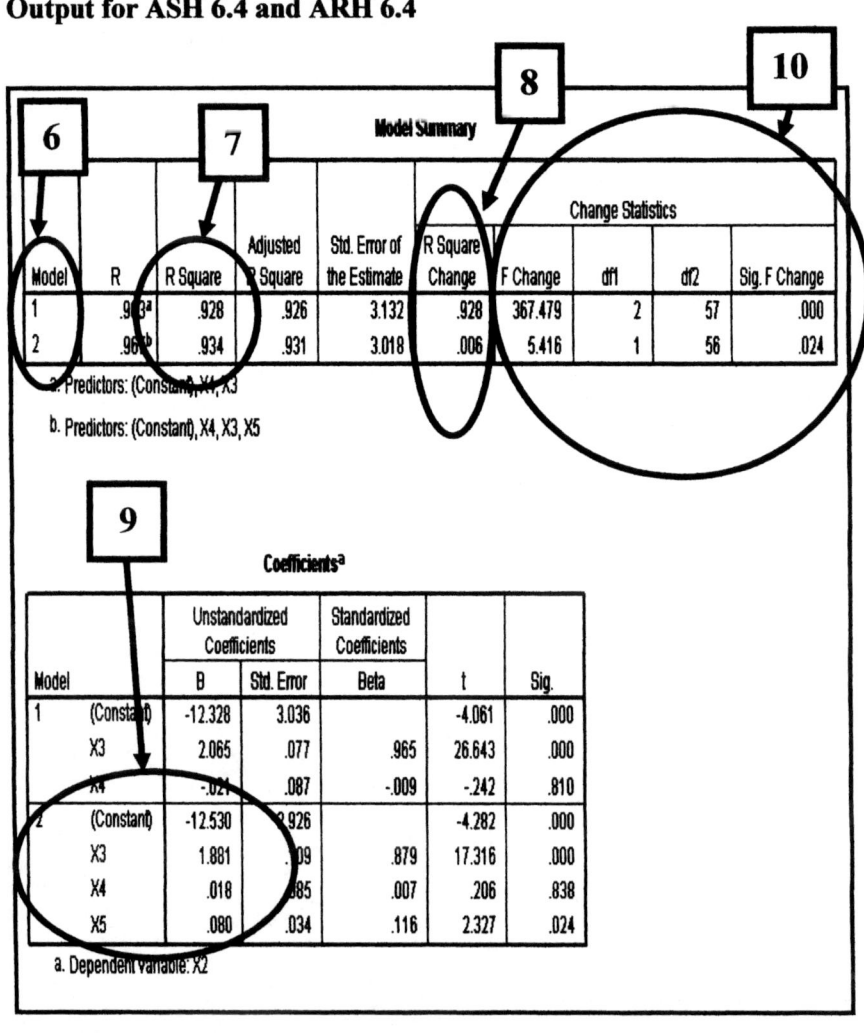

Model Summary

							Change Statistics			
Model	R	R Square	Adjusted R Square	Std. Error of the Estimate	R Square Change	F Change	df1	df2	Sig. F Change
1	.963ᵃ	.928	.926	3.132	.928	367.479	2	57	.000
2	.966ᵇ	.934	.931	3.018	.006	5.416	1	56	.024

a. Predictors: (Constant), X4, X3
b. Predictors: (Constant), X4, X3, X5

Coefficientsᵃ

Model		Unstandardized Coefficients		Standardized Coefficients		
		B	Std. Error	Beta	t	Sig.
1	(Constant)	-12.328	3.036		-4.061	.000
	X3	2.065	.077	.965	26.643	.000
	X4	-.021	.087	-.009	-.242	.810
2	(Constant)	-12.530	2.926		-4.282	.000
	X3	1.881	.109	.879	17.316	.000
	X4	.018	.085	.007	.206	.838
	X5	.080	.034	.116	2.327	.024

a. Dependent Variable: X2

It should be noted that the *F* test value differs from 0 and its corresponding probability value is less than 1.0 due to the fact that the difference between the R^2 values is slightly higher than 0. If the difference between the two R^2 values is exactly 0, the F test value would equal 0.0 and its probability value would equal 1.0.

The information contained in Exhibit 6.7, which is used to test the statistical hypothesis ASH 6.4, is as follows:

1. Restricted Model 3 and the Full Model are identified as Model 1 and Model 2, respectively (see Oval 6).
2. The R^2 values for Restricted Model 3 and the Full Model are .928 and .934, respectively (see Oval 7).
3. The difference between the R^2 values of Restricted Model 3 and the Full Model is .006 (see Oval 8).
4. The regression coefficient for variable X5 is 0.080 (see Oval 9).
5. The F of the difference in the R^2 values of the two models and its corresponding probability value are 5.416 and .024, respectively (see Oval 10). Since the Research Hypothesis is directional and the coefficient for variable X5 was positive as hypothesized, the two-tailed probability of .024 is divided by 2 to obtain the one-tailed probability of .012.

A review of the output for the statistical hypothesis ASH 6.3 indicates that the regression coefficient for variable X4 (.018) was positive. However, the difference between the R^2 values of the Full Model and Restricted Model 2 was less than .001. The one-tailed probability value ($p / 2 = .42$) of the F test used to test this difference was not statistically significant at the .05 alpha level. Thus, the ASH 6.3 was not rejected, which indicates that the researchers would conclude that variable X4 is *not* a positive predictor of the criterion variable X2 when X3 and X5 are also used as predictor variables.

A similar review of the output for the statistical hypothesis ASH 6.4 reveals that the difference between the R^2 values of the Full Model and Restricted Model 3 is equal to .006. Examination of the one-tailed probability value ($p / 2 = .01$) of the F test, which was used to test this difference, reveals that this difference was statistically significant at the .05 alpha level. Thus, ASH 6.4 was rejected. Based on this finding, the researchers would conclude that variable X5 is a positive predictor of the criterion variable X2 when X3 and X4 are also used as predictor variables.

The Coefficient t Test Approach

It is important to note that in the Full and Restricted Model Approach one coefficient was restricted to equal 0 in the testing of each of the three statistical hypotheses (i.e., ASH 6.2, ASH 6.3, and ASH 6.4). When one restriction is placed on a Full Model, the df_n value is 1 for the F test used to test the difference between the R^2 values of the Full Model and the resulting Restricted Model. In such a case, the F value calculated by The Full and Restricted Models Approach will equal the square of the t value used to statistically test whether the regression coefficient of a given predictor variable differs from 0. Since the F and t^2 values will be equal in such tests, their probability values also will be equal.

Thus, researchers can test statistical hypotheses such as ASH 6.2, ASH 6.3, and ASH 6.4 by interpreting the t-test values and the probability values of the regression coefficients contained in the Full Model. As previously stated, this approach is referred to as the "Coefficient t Test Approach."

When using The Coefficient t Test Approach, in conjunction with SPSS for Windows, to statistically test ASH 6.2, ASH 6.3, and ASH 6.4, the researchers would construct the following Full Model:

$$X2 = a_0U + a_3X3 + a_4X4 + a_5X5 + E_1 \qquad \text{(Model 6.11)}$$

Since this Full Model is the same as the Full Model used in the testing of ASH 6.1, its output would be obtained in the same manner as previously presented in the section of the text that dealt with the testing of ASH 6.1. The key portion of the SPSS for Windows output for the Full Model is the portion entitled "Coefficients," which is displayed in Exhibit 6.8.

The key information contained in the output window displayed in Exhibit 6.8 that is used to test the statistical hypotheses corresponding to ARH 6.2, ARH 6.3, and ARH 6.4 is as follows:

1. The regression coefficient for the predictor variable X3 is 1.881 (see Oval 1). The t test value for this coefficient is 17.316, and the corresponding probability value is printed as .000, thus, $p < .01$ (see Oval 2). Since the Research Hypothesis is directional and the coefficient for variables X3 was positive as hypothesized, this probability value, which is a two-tailed value, would be divided by 2 to obtain the one-tailed probability value. Of course this one-tailed probability value is also less than .01.

2. The regression coefficient for the predictor variable X4 is 0.018 (see Oval 1), which is positive as hypothesized. The t test value for this coefficient is 0.206 and the corresponding probability value is .838 (see Oval 2). The one-tailed probability value is .419.

3. The regression coefficient for the predictor variable X5 is 0.080 (see Oval 1), which is positive as hypothesized. The t test value for this coefficient is 2.327 and the corresponding probability value is .024 (see Oval 2). The one-tailed probability value is .012.

It should be noted that the one-tailed probability values obtained from the t tests of the three regression coefficients were the same as one-tailed probability values of the corresponding F tests of the differences between the R^2 values of the Full Model (Model 6.11) and the Restricted Models (Models 6.12, 6.13, and 6.14). Thus, the conclusions the researchers would draw from the results of The Coefficient t Test Approach would match the conclusions drawn from the results of The Full and Restricted Models Approach. That is, statistical hypotheses ASH 6.2 and ASH 6.4 are rejected, but ASH 6.3 is not rejected. The researchers concluded the following: (a) variable X3 is a positive predictor of the criterion variable X2 when X4 and X5 are also used as predictor variables, (b) variable

X4 is not a positive predictor of the criterion variable X2 when X3 and X5 are also used as predictor variables, and (c) variable X5 is a positive predictor of the criterion variable X2 when X3 and X4 are also used as predictor variables.

Exhibit 6.8.
SPSS regression output for ARH 6.2, ARH 6.3, and ARH 6.4—Coefficients t Test Approach.

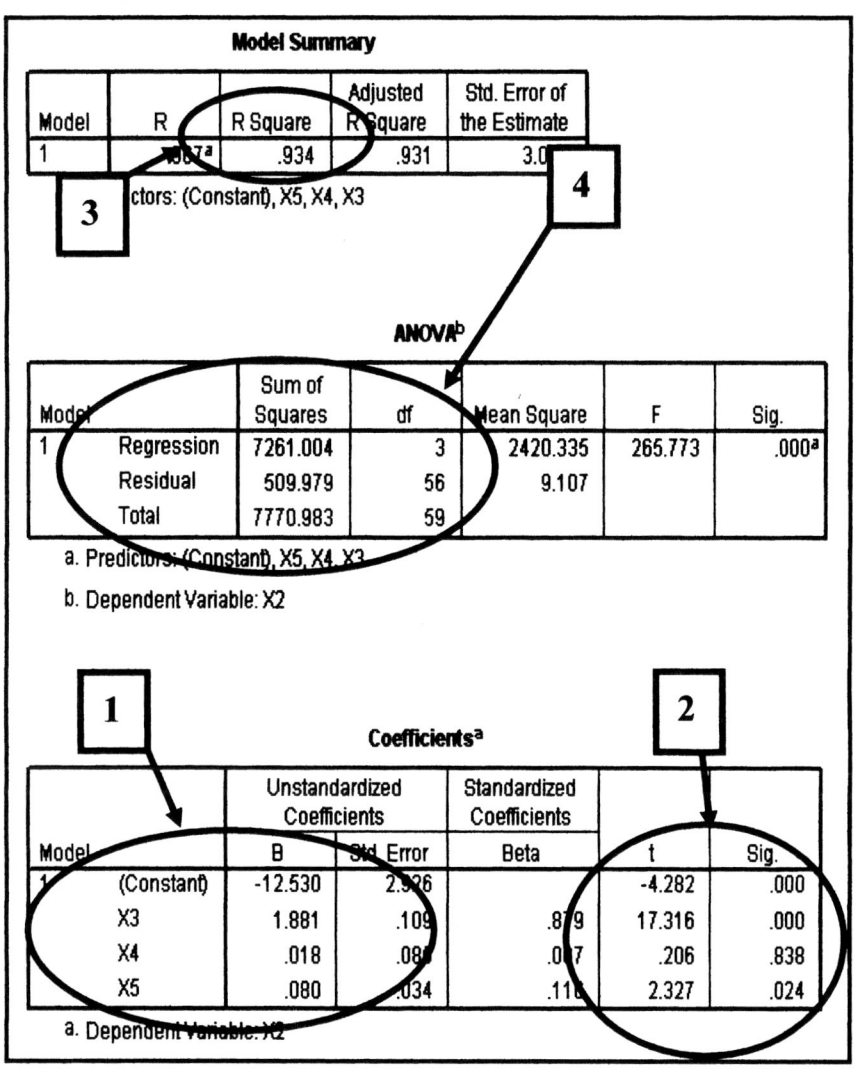

Model Summary

Model	R	R Square	Adjusted R Square	Std. Error of the Estimate
1	.967ᵃ	.934	.931	3.0

ctors: (Constant), X5, X4, X3

ANOVAᵇ

Model		Sum of Squares	df	Mean Square	F	Sig.
1	Regression	7261.004	3	2420.335	265.773	.000ᵃ
	Residual	509.979	56	9.107		
	Total	7770.983	59			

a. Predictors: (Constant), X5, X4, X3
b. Dependent Variable: X2

Coefficientsᵃ

Model		Unstandardized Coefficients		Standardized Coefficients		
		B	Std. Error	Beta	t	Sig.
1	(Constant)	-12.530	2.926		-4.282	.000
	X3	1.881	.109	.879	17.316	.000
	X4	.018	.08	.007	.206	.838
	X5	.080	.034	.11	2.327	.024

a. Dependent Variable: X2

Although The Coefficient t Test Approach provides the same results regarding the statistical testing of the three statistical hypotheses, unlike The Full and Restricted Models Approach, it does not directly provide the amount of unique variation accounted for by each of the predictor variables. These values can be obtained, however, by employing Equation 6.4:

$$\Delta R^2 = \frac{(t)^2 (1 - R^2)}{df_d}$$

(Equation 6.4)

where:
1. The ΔR^2 symbol represents the change in the R^2 value associated with the inclusion of a given predictor variable in the model.
2. The t^2 symbol represents the square of the t-test value for the predictor variable.
3. The quantity $(1 - R^2)$ is the proportion of the variation in the criterion variable not accounted for by the predictor variables in the model.
4. The df_d symbol is the degrees of freedom of the denominator for the model, which is equal to the sample size minus a value equal to the number of predictor variables in the model plus 1. Note that the SPSS for Windows output labels df_d as "Residual."

When using Equation 6.4 to calculate the unique amount of variation in variable X2 accounted for by a given predictor variable, three values must be obtained from the output window displayed in Exhibit 6.8. These values for predictor variable X3 are (a) the coefficient t-test value of 17.316 (see Oval 2), (b) the R^2 value for the model of .934 (see Oval 3), and (c) the df_d value—Residual df— of 56 (see Oval 4). Substituting these values into the formula produces the following value for the unique amount of variation in X2 accounted for by X3:

$$\Delta R^2 = \frac{(17.316)^2 (1 - .934)}{56} = .35$$

Thus, the proportion of unique variation in the X2 variable accounted for by the predictor variable X3 when variables X4 and X5 are also used as predictor variables is .35. (This is the same value as the difference between the R^2 in Models 2 and 1 in Oval 3 in Exhibit 6.6.)

This value could also be obtained by using the Microsoft Excel file entitled "Change in R-Squared," which is located in the internet site listed in Appendix A. In addition, the commands needed to construct this program are listed in Appendix C. Once this program is accessed, the window contained in Exhibit 6.9 will be displayed. To calculate the change in the R^2 value for variable X3 the R^2 value of the model (.934), the t-test value of the regression coefficient (17.316), and the df_d value (56) are placed in cells B1, B2, and B3 of the file (see Oval 1), respectively. Once these three values are entered, the R^2 change value will be listed in cell D1 (see Oval 2).

The same procedure is used to calculate the corresponding change in R^2 values for variables X4 and X5. The change in R^2 value for variable X4 is less than .0001, while the change in R^2 value for variable X5 is equal to .006. Note that since these three values match the ones obtained from The Full and Restricted Models Approach, researchers can report the values as effect sizes for the predictor variables even when using the "Coefficient t Test Approach."

Exhibit 6.9.
Calculation of the change in the R^2 value for variable X3 with Microsoft Excel.

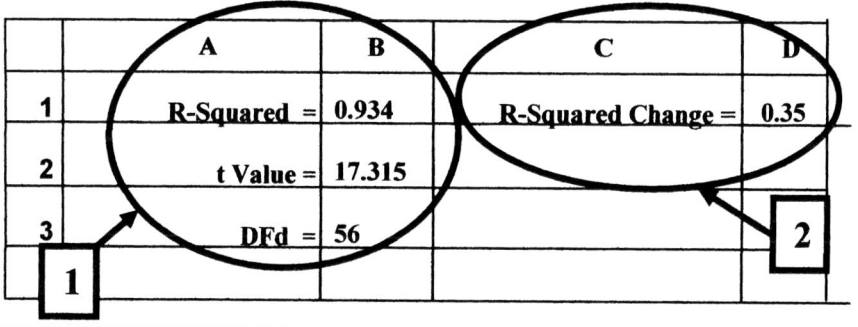

	A	B	C	D
1	R-Squared =	0.934	R-Squared Change =	0.35
2	t Value =	17.315		
3	DFd =	56		2
1				

Writing the Results for ARH 6.2, ARH 6.3, and ARH 6.4 in APA Style

The results obtained for ARH 6.2, ARH 6.3, and ARH 6.4 may be written in APA style as follows:

> The regression coefficients for Test A (1.88), Test B (0.02), and Test C (0.08) were positive as hypothesized. As indicated by the t-test values of the coefficients, only the positive coefficients for Test A ($t = 17.32$, $p / 2 < .01$) and Test C ($t = 2.33$, $p / 2 = .01$) were significant at the one-tailed alpha level of .05. The amounts of unique variation in Test D accounted for by Test A (.35) and Test C (.006), which are used as effect sizes for these predictor variables, were classified as large and small, respectively, according to criterion established by Cohen (1988).

It should be noted that this conclusion, which was written for the testing of the statistical hypotheses ASR 6.2, ASR 6.3, and ASR 6.4 is considerably more informative than the one written for the testing of ASR 6.1. Thus, it behooves researchers to develop research hypotheses and corresponding statistical hypotheses that will produce results that will be the most informative.

MULTIPLE CONTINUOUS PREDICTORS:
COMPARING TWO MODELS OF BEHAVIOR

Rationale

In the Applied Research 6.1 section of this chapter, the Full Model, which consisted of two or more predictor variables, was compared to a Restricted Model that contained none of the predictor variables. Researchers will often need to deal with research questions that will require only some of the predictor variables to be deleted from the Full Model, such as in the previous section, thus the Restricted Model will no longer be the Null Model. The researchers can think of such Full and Restricted Models as two competing ways of modeling, or explaining behavior. One model has fewer variables and is thus more parsimonious. The other model, because it has more predictor variables, will yield a higher R^2.

The cost of the additional pieces of information, as compared to the benefit of a higher R^2 can be accessed through the use of the F test calculated with Equation 4.1. The F test simply assures the researchers that the significant variables do better than what random variables would do. Alternatively, the researchers could require that each piece of information result in, say, an R^2 increase of 10%. If such an approach is used, it is essential for the researchers to provide the rationale used to establish the comparison value (i.e., the 10% value used in the example).

Research Hypothesis, Full Model, and Restricted Model

The comparison of two models of behavior requires that the same behavior be considered and that one model is a restricted case of the other model. Reviewing the situation the other way, one model is an extension of the other model. In either case, one model is likely to be the "accepted" or "traditional" model of behavior. The researcher either has identified a simpler way to explain the behavior *or* has identified additional mediating or necessary variables. Since most R^2 in the behavioral sciences are pathetically low, the researcher has probably identified some additional variables that were not specified in the original theory, but could:

1. have recently become feasible to measure,
2. have surfaced as necessary or mediating variables,
3. interact with variables already in the model, and/or
4. specify an untapped nonlinear relationship.

The last three possibilities will be dealt with extensively in later chapters. The first two cases, that is, the inclusion of new variables into the analysis, will now be discussed.

A Research Hypothesis that calls for the inclusion of two additional variables would be stated as follows:

Research Hypothesis: Dr. Smith's six-variable model of behavior can be improved with the addition of variables X7 and X8.

The corresponding Statistical Hypothesis is as follows:

Statistical Hypothesis: The six variables in Dr. Smith's model cannot be improved with the addition of variables X7 and X8.

The Full Model required to reflect the relationships contained in the Research Hypothesis is:

$$Y = a_0U + a_1X1 + a_2X2 + \ldots + a_6X6 + a_7X7 + a_8X8 + E_1 \qquad \text{(Model 6.15)}$$

The restrictions required by the Statistical Hypothesis, which indicates that variables X7 and X8 will not account for additional variation in the criterion variables, are as follows:

$$a_7 = 0 \text{ and } a_8 = 0 \quad \text{or} \quad a_7 = a_8 = 0 \quad \text{(2 restrictions)}$$

Placing these restrictions into the Full Model (Model 6.15) results in the following Restricted Model:

$$Y = a_0U + a_1X1 + a_2X2 + \ldots + a_6X6 + E_2 \qquad \text{(Model 6.16)}$$

One should note that the Restricted Model is Dr. Smith's model, and that the Full Model is the proposed improved model.

The F test as calculated with Equation 4.1 can again be used to determine whether the difference between the R^2 values of the two models is statistically significant. Since the Full Model (Model 6.15) contains 9 pieces of information and the Restricted Model (Model 6.16) has 7 pieces of information, the F value would be calculated as follows:

$$F(2, N-9) = \frac{(R_F^2 - R_R^2)/(2)}{(1 - R_F^2)/(N-9)} \qquad \text{(Equation 6.5)}$$

Whatever increase there is in the R^2 value from the Restricted Model to the Full Model can be attributed to the two variables that have been added to the Full Model. If one variable is "valuable," the value of the other variable (the increase in R^2) will be averaged over both variables. (This is shown in the numerator of the F test.) In such a case, significance may not be obtained, whereas if only the one "valuable" variable had been tested, significance might have been obtained. Whenever a researcher tests more than one variable simultaneously such a risk is taken. Only when a one-degree-of-freedom question is tested do we know exactly what piece of information is responsible for the decrease in R^2

within the context of all the variables remaining in the Restricted Model. Refer to the section in this chapter entitled "Applied Research Hypotheses 6.2, 6.3, and 6.4" for a discussion of the techniques researchers can use to statistically test the amount of variation in the criterion variable accounted for by one predictor variable when the Full Model contains other predictor variables.

TESTING THE VALUE ADDED WITH STEPWISE REGRESSION PROCEDURES

Testing the "value added" of a particular variable has often been conducted by researchers through the use of one or more of a family of analytic procedures known as *stepwise regression analysis*. One type of the stepwise regression procedures is known as *forward* stepwise regression analysis. In this procedure, the computer software completely takes over and takes the researcher on a gigantic fishing expedition. In Step 1, a model is identified that consists of the variable most highly related to the criterion—the one that accounts for the most variance in the criterion.

Step 2 consists of testing the value added of *each* of the other variables. Thus, with, say, nine predictor variables, once the first step is concluded and the one "best" variable is identified, then each of the other eight are tested with the format discussed in this section. Eight tests of significance are performed in Step 2, and Step 2 is finished when the computer identifies the best two-variable model. Step 3 starts with this two-variable model and performs the statistical test discussed in this section seven times, once for each of the remaining seven variables, with the result being the best three-variable model. Step 4 starts with this three-variable model and performs the statistical test discussed in this section six times, once for each of the remaining six variables, with the result being the best four-variable model.

This process continues until one of several criteria imposed by the researcher is met: (a) a minimum R^2 value is obtained; (b) a pre-specified p value is not met; or (c) all nine variables are entered into the model. If the four-variable model turns out to be the "best" model, a total of 30 (9 in Step 1, 8 in Step 2, 7 in Step 3, and 6 in Step 4) F tests have been calculated. Many users of stepwise regression analysis either forget this or were never aware of it in the first place. This procedure is one of the most successful fishing expeditions a researcher can go on because a "best" model will always be identified. But the technique should be viewed for what it is—a fishing expedition for *identifying* Research Hypotheses for future verification; it is *not* for verifying hypotheses (since none were asked) on the data at hand.

Researchers can use another version of the stepwise regression procedure called *backwards* stepwise regression procedure. In particular, the backwards stepwise procedure works in just the opposite direction, starting with the best nine-variable model and then identifying the variable that can be eliminated from the nine variable set in order to result in the best eight-variable model. If the researchers have their minds set on applying a stepwise regression procedure to their data, which is a practice we do not encourage, then we recommend that one use the backward stepwise method. The reason for this recommendation is that models produced by this procedure will never have smaller R^2 values than the R^2 values calculated for models derived from the forward stepwise method— in the multivariable world variables often work together and their contribution may well exceed the sum of the parts.

There are several problems inherent in the stepwise regression procedure. First and foremost, it is not a theory driven procedure. Second, a multitude of tests of significance are conducted. Thus, the issue is: Should one adjust the alpha levels used for the statistical tests for the multiple tests, and if so by how much? Third, researchers often obtain different answers from forward and backward stepwise computer programs. Fourth, researchers can acquire different answers depending upon the probability values chosen for inclusion and for exclusion of the predictor variables. Fifth, it is possible to obtain different sets of statistically significant predictor variables from a forward stepwise analysis than the backward stepwise analysis.

Although we discuss stepwise regression in more detail at the end of this chapter, it is due to the aforementioned problems inherent in the stepwise regression procedure that we are not advocates of its use. Because of the various problems inherent in the stepwise regression procedure, we maintain that stepwise regression should only be used to develop hypotheses, not to test hypotheses. Furthermore, researchers should always be sure to determine if the same results obtained from the analysis of one sample can be obtained from another sample, that is, be sure to cross validate their findings.

QUESTIONS FOR DR. GLM

The budding researcher (Bud) had a number of questions for Dr. General Linear Model (Dr. GLM) after reading this chapter. The questions posed relate to: (a) identifying criterion and predictor variables, (b) hypothesis and data snooping, and (c) sequencing of hypotheses.

Identifying Criterion and Predictor Variables

Bud: What variable do I use as my criterion?

Dr. GLM: Strangely enough, that question is often asked of us statisticians, but it can never be specifically answered by anyone but the researcher. The criterion variable is the variable that measures the construct of interest to the researcher. The researcher is intrigued that people vary on this variable and sets out to account for the variability in this variable. Often the construct is measurable by more than one variable, although the GLM approach as discussed in this text can handle only one criterion variable at a time. You could use the same predictor model to predict two different criterion variables separately. A friend of mine named Spaner played around with that idea and published his thoughts on it in 1970.

Another notion about variables is that a particular variable may be a criterion variable for one researcher and a predictor variable for another researcher. This is often the case in developmental studies and studies on academic success.

Bud: How do I know what variables to use as predictors?

Dr. GLM: Most important, the nature of the predictor variables you use should emanate from a combination of past research and your theoretical structure. Given this as a premise, what specifically do you want to know about predictors?

Bud: How should I measure the constructs I've decided to use as predictors?

Dr. GLM: Two researchers may agree on using a given construct as a predictor, but they might use two quite different measures of that construct. You must remember that the R^2 values of the models used in the study are produced by the analysis of actual data rather than the construct you hope those data measure or at least provide a good approximation of that construct. But, for instance, you must always keep in mind that the measure of IQ often used in statistical analyses is most likely not synonymous with the construct of IQ. Until someone is able to "see" the IQ construct, we must act as if the operational measure is at best a "good" approximation. The goodness of that approximation is, in measurement terms, referred to as *validity*.

You also should be aware that the measures need not be of an interval scale, as demonstrated by my friends McNeil and Kelly in 1970. Many authors of statistics texts state that only interval data can be analyzed by the GLM. But when developing models, the researcher's goal is to account for as much of the variance in the criterion variable as possible. McNeil and Kelly presented several examples in which non-interval data resulted in an R^2 close to 1.0

I would like to stress that the researcher does not need to follow any of the "guidelines" about using "recognized" tests or measurements. The predictor variables can be measured in any way the researcher sees fit. If a high R^2 value is obtained, more credit to the researcher. There is nothing sacred about any of the prevalent measures in any of the behavioral sciences. Indeed, the many years of attention paid to certain well-known tests may have been a constraining factor on the growth of knowledge about relationships in that field. On the other hand, a low R^2 value may not indicate that the criterion variable is unrelated to the constructs represented by the set of predictor variables. A low R^2 value may be the result of the set of predictor variables being poor measures of the constructs. In such a case, the researcher is not in need of additional variables but rather better measures of the constructs.

Bud: Should all my chosen predictor variables be uncorrelated?

Dr. GLM: When predictor variables are highly correlated, the variables are said to exhibit a condition known as multicollinearity. That is another of those guidelines I would rather you ignored. Most multiple correlation literature encourages the use of uncorrelated predictor variables. ANOVA designs are set up such that the predictor variables are uncorrelated. But the real world is never fashioned that way. More importantly, the ultimate measure of the value of a predictor variable is its effect on the R^2. If the R^2 is significantly increased by the inclusion of a correlated predictor variable, then by definition that variable is a good variable (in that set of predictor variables).

Lewis-Beck (1980) provided four recommendations if you want to avoid highly correlated predictors (referred to as *multicollinearity*). First, one can obtain a larger sample to see if the correlation is an artifact of the particular sample. Second, one could combine the correlated predictors into a more encompassing predictor through the use of factor analysis. Third, one could restrict the use of the regression equation to that of prediction, and not try to extend the model to that of interpretation. Fourth, one could discard one of the "offending variables," but that would be a "willful commission of specification error"—that is, omitting a predictor that you originally considered relevant.

There is a special predictor variable—one that is uncorrelated with the criterion but highly correlated with other predictor variables—and it has been given the name *suppressor variable*. This kind of variable (whatever name you want to give it) is a predictor variable that was of no value in the bivariate way of looking at the criterion but is indeed valuable in the multivariate model. Because most past research is of a bivariate nature, you might want to have less faith that past research will give you many insights into productive multivariate hypotheses. Another of my friends stated the results of multivariate studies quite succinctly: "Multivariate statistical procedures have the darnedest habit of doing what they are designed to do (considering context in finding optimal linear combinations), rather than simply confirming what we think we know from examining univariate statistics taken out of context, one variable at a time" (Harris, 1992, p. 11).

Some suppressor-variable situations will merely be rescaling problems. Others will depict intricate and inseparable relationships between distinct constructs. These will be the variables that "complexly" measure the "complex behavior" so often discussed in the behavioral sciences.

Bud: How many predictor variables should I use in my analysis?

Dr. GLM: You've got to use as many predictor variables as it takes to result in an R^2 that satisfies you. And I hope you will not be satisfied until you get an R^2 close to 1.00. You have an infinite number of predictor variables at your disposal. You choose to measure a limited number of variables from an infinite number of possibilities. And once those limited number have been measured, the possibilities of polynomials and interactions are unlimited. Because you are the researcher, it is up to you to choose from the unlimited set of transformed variables those that are to be used in the analysis. This decision is a very difficult one, for the value of a variable is in its ability to increase R^2, not in whether it is a linear interaction term, or a third-degree term, and so forth. But if you choose to defend the inclusion of predictor variables on some grounds other than practical utility, remember that the originally measured variable (the linear term) must be defended just as much as the second-degree term, or as much as any other term for that matter. And, as humans are probably too frail to be able to posit reasons for including one variable out of an infinite set, the final decision to include a predictor variable thus should be deferred to pragmatic reasoning—whether or not the variable increases the R^2 value.

Hypothesis and Data Snooping

Bud: What is the relationship between data snooping and hypothesis testing?

Dr. GLM: How one sequences one's analyses depends upon one's theory, design, and sequence of questions. Most of the questions will, one hopes, be posited before the data are collected, while some data-snooping hypotheses may be entertained. The answers to the a priori hypotheses are definitive for the population from which the people were sampled. The answers to the data-snooping investigations are not definitive but are bases for future research.

The story of Archimedes is appropriate here. When he discovered the relationship between volume and displacement while taking his bath, he was so excited that he went running through the streets naked shouting "Eureka! Eureka!" Since he had discerned a law, which by definition has an R^2 of 1.00, a law that was extremely parsimonious and quite replicable, he had every right to run naked. Present-day researchers should have as much excitement about their research and as much investment in the results that they would be willing to run through the streets of their hometown crying "Eureka!"

Researchers would want to be sure to be correct in their pronouncements, however, before they go "running naked through the streets shouting eureka," that is, before publishing their results in a professional journal. If the researchers are wrong—possibly causing a retraction of the article in a future issue of the journal—that would be embarrassing. Contemplating such a situation should lead researchers to give more serious thought about the choice of variables measuring the construct, the design of treatments or choice of intact groups, the sequencing of hypotheses, and a priori specification of these Research Hypotheses. The researchers would definitely *not* want to "run naked" if something were discovered from a data-snooping analysis that had not been expected. Modesty would require a replication of the unexpected findings before such a public display.

Sequencing of Hypotheses

Bud: If I don't need to follow the multiple-comparison route of sequencing of hypotheses, what is the procedure developed from the computer point of view?

Dr. GLM: I do not recommend this procedure for hypothesis-testing purposes, but you should be aware of it for hypothesis-generating purposes. Most computer statistical packages have a *stepwise regression program*. A priori hypotheses are not tested with such a program. The hypothesis-testing sequence is controlled only very slightly by the researcher; although some stepwise programs allow for an "importance" indicator for each variable. But specific models generally cannot be constructed and tested with stepwise programs.

The researcher places into the stepwise program the criterion variable and all the predictor variables of interest. The idea is to find which variables form the "best" predictor model for that criterion. Forward selection procedures find the one variable in the predictor set that is most highly correlated with the criterion. This variable, with the unit vector, is referred to as the best model at Step 1. Then the program searches the variables to find which predictor, in combination with the two already in the system, will yield the highest R^2. Once found, these three variables become the best model at Step 2. The process continues until all variables are in the system or until no variable can be found such that the increase in R^2 is significant at a level specified either by the program or by the user.

The stepwise procedure is almost always applied to continuous variables. Few applications have involved interactions or nonlinear terms. Few have involved a set of dichotomous variables designed to represent group membership. The major reason my researchers do not use a set of dichotomous predictor variables is predicated on the problem created when only one of the dichotomous predictor variables is selected for inclusion in the model. Recall that for a set of three dichotomous variables at least two of those variables need to be placed in the model to represent group membership. What interpretation would the researchers give for such a result?

Two friends of mine, Williams and Lindem, developed in 1979 a procedure, referred to as *setwise regression*. Instead of using single variables, the researcher is allowed to define *sets* of variables. While sets can be defined on a logical basis, the use of a setwise procedure seems mandatory when binary-coded variables are included and there are more than two categories involved. The setwise procedure includes one set at a time in a stepwise fashion.

Some problems arise with the stepwise approach. There are many hypotheses being tested, none of which has been specified by the researcher. The resulting "best" model will most likely be quite drastically overfit, and therefore replication of that model is unlikely. Also, the stepwise procedure may stop too soon, in that two variables considered simultaneously might significantly increase the R^2, whereas neither one of them may separately significantly increase the R^2. Some stepwise versions allow variables already in the system to be deleted if they are not making a contribution in a much larger model. Versions that do not allow for this additional flexibility may end with an inferior best model.

Some of these problems with forward stepwise regression are resolved with backward stepwise regression. In this approach, all variables are considered as the best model at Step 1. Then the variables (or sets, as in setwise regression) are evaluated one at a time to find which one will, when omitted from the system, reduce the R^2 the least. That variable is then omitted, and the model using all the variables except that one is the best model at Step 2. The process is repeated for Step 3 and so on, until all of the variables have been omitted or until omitting any of the variables would result in a significant loss in predictability. The backward procedure is preferable, though many hypotheses are tested (resulting in a large probability of making at least one Type I error). Thus, the resulting model may not replicate because of overfitting. At any rate, the particular hypothesis you want to test very likely will not appear—and if it does appear, you may have difficulty recognizing it.

In essence, stepwise regression programs perform a hypothesis formulation function, whereas the regression procedures discussed in this text are concerned with testing those formed hypotheses. Thayer (1990) reviewed all of the stepwise options and provided recommendations for obtaining the best model for *explanatory* purposes. Thayer describes in excellent detail the use of various stepwise procedures and admonishes:

> When variables are selected for a regression model, the stepwise method can be helpful if the initial choice of variables is chosen as much as possible using theory, the defaults of the computer program used are not used automatically, more than one computer run is done using different variable selection methods, and the final model is chosen through an intelligent process, not automatically using the final model generated by the computer program. When the model is described, all subjective decisions made in the model selection process should be reported. (p. 67)

Bud: What method for sequencing hypothesis testing can you recommend from the researcher's point of view?

Dr. GLM: Build your hypotheses upon *past research* (yours or someone else's) *and your theoretical views*. If you have enough information and enough confidence in your theory to state directional Research Hypotheses, then state them in the order of your interest and expectations, and test them in that order. You can expect statistical support for such hypotheses, and you should be able to progress toward causal interpretation.

Bud: What about those areas of my theory where past research has given me no clues about what functional relationships to expect? I can't state directional Research Hypotheses about those relationships, can I?

Dr. GLM: You make a good point. I am emphatic about the researcher stating the Research Hypothesis. The Research Hypothesis should be a statement of expected outcome based upon past relevant research, theoretical relationships, and synthesis of relevant constructs. But most past research has been bivariate in nature; that is, only one predictor variable was used to account for criterion variance, and the amount of criterion variance accounted for has usually been quite small. Therefore, theory development has been held back, and so there is little knowledge from which one can synthesize. Data snooping is one way of finding functional relationships. As long as one constantly remembers to replicate the findings on a new sample of data, I see nothing wrong with "seeing what works." Using a stepwise regression program is one way to discover predictor variables; accidental (serendipitous) findings are another way. Guidelines about how to obtain serendipitous findings are nonexistent, but you should be willing to cross-validate a variable that is valuable in a pilot study. Many important variables in the sciences have been discovered accidentally, although their value was repeatedly tested (cross-validated) before they were widely accepted. Mosteller and Tukey (1968) aptly summarized how the researcher should react to results found through data-snooping when they said, "Here we must stop our calculations with indications and be careful to think of our results only as hints about what to study next, rather than as established results."

To give you an idea of what data-snooping is all about, I give you a "data-snooping analogy," in which the bread is analogous to a meaningful discovery, and the telling of friends about the news is analogous to publishing one's results for the benefit of the profession.

Suppose that you have been asked by your significant other to go down to the milk discount store to buy some milk. Now, if you do exactly as you are asked, you will go directly to the milk counter, pick up the milk, pay for the milk, and return home.

Let us suppose that you snoop around instead, and you accidentally notice that this store has bread at an exceptionally low price. In fact, at the price listed, you might label this "the best bread buy in town." If you are the least bit economically minded, you will take note of this price and purchase a loaf or two, especially if you are aware of procedures by which bread can be stored for a long period.

You are not "sure" that this low price will prevail for long, but you would consider informing some of your friends about the "best bread buy in town." You probably would indicate your reservations by saying something like, "Yesterday I accidentally discovered that the milk discount store had a good price on bread. You might want to go down there and see if they still have this good bargain; I really do not know if they still have that good buy." That is, you are somewhat reluctant to yell too loudly about the bargain until the bargain can be verified. The longer the store retains this low price the louder and more frequently you will announce your discovery to more of your friends. If the bargain remains long enough and enough people are able to verify your finding, then the "best bread buy in town" becomes a lawful fact.

That lawful fact might not have been discovered if you had not snooped around while at the milk discount store looking for milk. An auxiliary bargain discovered while looking for something completely different must be verified on subsequent trips before much faith can be placed in the stability of that bargain.

As a researcher, please take the analogy to heart. Snooping around in your data is not antithetical to good research. If you find a meaningful relationship while snooping around, then you are obligated to replicate your results before you say too much about them. Your Research Hypotheses should be based upon theory, which is often loosely based on past research. Often there is little past research upon which to base your hypotheses. Furthermore, the past research is often of poor quality and bivariate in nature and therefore possibly uninformative or even misleading. When one snoops around in data, that study must be considered to be a *pilot* (past research) upon which hypotheses are developed for the replication sample.

It could be argued that researchers have drastically held back scientific advancement by not snooping around in their data. It seems very unfortunate to collect large amounts of data and then to look at those data with blinders. We should take those blinders off—but simultaneously remember to leave our "replicators" on!

Bud: Can I use multiple regression procedures to help in my data snooping?

Dr. GLM: Data-snooping is easily done within the GLM approach. All your measured variables (about which you have hypotheses, or hunches, or you just want to see what's going on) can be included in a Full Model. If many predictor variables are being used, many people should be used, so that the resulting functional relationship is not due entirely to the idiosyncrasies of the sample at hand.

The R^2 for the Full Model using all measured predictor variables could be compared against the unit vector model to determine if any significant criterion variance is being accounted for by the set of predictor variables. If you want to limit your predictor set to, say, one half of the number of predictors you started out with, and you want to find the best ones, then one of the stepwise regression approaches could be used.

Bud: In some articles I notice that the researcher identifies "the most important predictor."

Dr. GLM: I am totally against doing so for five reasons.

First, when one uses multiple predictors, one is acknowledging that the behavior under consideration is multiply determined—that various variables work together to generate the level of the criterion for each individual. These variables are allowed to be correlated in GLM, as they probably are in the real world. It is then antithetical to both the researcher's stance and the real world data to identify "the most important."

Second, researchers usually base their decision of the most important as the variable that is brought into the model on the first step. But that variable is simply the one that has the highest bivariate correlation with the criterion. You wouldn't need to consider all the predictors in one GLM to identify that variable!

Third, that "most important variable" may not be subject to manipulation, or it may cost a lot to manipulate it.

Fourth, the analysis does not tell the researcher that the participants will respond to a change on that variable, unless the variable has been manipulated, which is usually not the case.

Fifth, if you change humans on one predictor, then there will likely be a change on the other predictors since the predictor variables are likely correlated.

That is, when you change one predictor, you cannot hold humans constant on all the other predictors. And following the logic above, there is really no way to predict how humans will respond to being changed on any one variable, let alone all the variables, unless you have already conducted the manipulation.

CHAPTER 7

INTERACTION

Behavior is often a result of more than the effects of one or more variables alone; behavior is usually a result of variables in combination. Previous chapters have illustrated how researchers can consider dichotomous predictor variables as "main effects," and continuous predictor variables as "linear effects."

In this chapter we illustrate how one can use the GLM to model behavior when the effect of one predictor "depends upon" another predictor. Statisticians call this phenomenon interaction and represent it mathematically with multiplication of the variables. See also Lewis-Beck (1980) for a discussion of interaction as indicated by "it depends."

When you think about it, most behavior is not a function of just one variable by itself but depends on the presence or absence (dichotomous variable) or numerical value (continuous variable) of another variable. Interaction has traditionally been viewed as a difficult concept and as one that researchers would rather not have around because its presence clouds the interpretation of the main effects. We take a strong stance in treating interaction as a phenomenon to be expected, as another predictor variable, and as a clarifying phenomenon. If interaction does exist, then we need to include it in our model. We need to be able to test for it rather than casually assume that it does not exist. Most importantly, we need to be able to interpret interaction if it does exist in our data. Based on our experience, interaction will likely exist. Few causal statements are simply "X causes Y," but rather "X causes Y, depending on the value of Z."

We discuss interaction from three vantage points, following the structure provided in the previous three chapters. The first section illustrates how GLM can reflect interaction between two dichotomous predictor variables. The second section illustrates how GLM can reflect interaction between a dichotomous predictor variable and a continuous predictor variable. The third section illustrates how GLM can reflect interaction between two continuous predictor variables. In the final section, we discuss the relative advantages in employing dichotomous and continuous predictor variables.

INTERACTION BETWEEN DICHOTOMOUS
PREDICTOR VARIABLES WITH TWO LEVELS

We first discuss interaction between dichotomous predictor variables when each predictor has only two levels. An example with more than two levels is presented at the end of the section.

Rationale

We repeat Exhibit 4.7 from chapter 4 as Exhibit 7.1 for the purpose of discussing Research Hypothesis 12, the interaction hypothesis. In chapter 4 we discussed the two main-effect hypotheses, Research Hypothesis 10 and Research Hypothesis 11. Remember that with four groups these three hypotheses (Research Hypotheses 10, 11, and 12) comprised one set of orthogonal hypotheses.

In this section we will relate these Research Hypotheses to the concept of interaction. The nondirectional Research Hypothesis listed under Research Hypothesis 11 states that the two time periods, averaged across the two different treatments, are not equally effective. The nondirectional Research Hypothesis listed under Research Hypothesis 12 qualifies that difference by not averaging across the two different treatments. Instead, Research Hypothesis 12 actually specifies that the difference between the two time levels will be different in the two treatments. Research Hypothesis 12 states that whatever difference exists at one level, that difference will not be the same at the other level. Thus, as indicated in Exhibit 7.1, Research Hypothesis 12 stipulates the following relationship between group means:

$$(mG1 - mG2) \neq (mG3 - mG4).$$

The corresponding contrast coefficients are as follows: +1, -1, -1, and +1.

If the difference between means of Groups 1 and 2 differs from the difference between the means of Groups 3 and 4 (i.e., the data support Research Hypothesis 12), the main-effect statement of Research Hypothesis 11, even if statistically significant, is not equally true at each level. It may even be the case, as demonstrated in the next section, that the overall main effect may be a result of a large difference at one level and no difference at the other level (or even a small opposite difference).

Research Hypothesis 12 could have been stated as:

The difference in effectiveness of the AM new treatment and the AM comparison treatment is different from the difference between the PM new treatment and the PM comparison treatment.

This Research Hypothesis results in the following relationship between the four group means: $(mG1 - mG3) \neq (mG2 - mG4)$.

The corresponding contrast coefficients are as follows:

+1, -1, -1, and +1.

Exhibit 7.1.
A set of contrast coefficients—Two-Way Analysis of Variance.

Comparison 10—Research Hypothesis 10
    Nondirectional: The two treatments, averaged across the two different time periods, are not equally effective.
$$(mG1 + mG2) / 2 \neq (mG3 + mG4) / 2$$
    Directional: The new treatment, averaged across the two different time periods, is more effective than the comparison treatment.
$$(mG1 + mG2) / 2 > (mG3 + mG4) / 2$$
    Statistical Hypothesis: $(mG1 + mG2) / 2 = (mG3 + mG4) / 2$
$$\text{or}$$
$$(mG1 + mG2) = (mG3 + mG4) \text{ or } (mG1+ mG2) - (mG3+ mG4) = 0$$
$$\text{or}$$
$$(1 * mG1) + (1 * mG2) + (-1 * mG3) + (-1* mG4) = 0$$
    Contrast Coefficients
    G1 G2 G3 G4
    1   1 -1 -1

Comparison 11—Research Hypothesis 11
    Nondirectional: The two time periods, averaged across the two treatments, are not equally effective.
$$(mG1 + mG3) / 2 \neq (mG2 + mG4) / 2$$
    Directional: The AM period, averaged across the two different treatments, is more effective than the PM period.
$$(mG1 + mG3) / 2 > (mG2 + mG4) / 2$$
    Statistical Hypothesis:
$$(mG1 + mG3) / 2 = (mG2 + mG4) / 2$$
$$\text{or}$$
$$(mG1 + mG3) = (mG2 + mG4)$$
$$\text{or}$$
$$(mG1 + mG3) - (mG2 + mG4) = 0$$
$$\text{or}$$
$$(1 * mG1) + (-1 * mG2) + (1 * mG3) + (-1 * mG4) = 0$$
    Contrast Coefficients
    G1 G2 G3 G4
    1 -1   1 -1

Exhibit 7.1 (continued).
Set of contrast coefficients—Two-Way Analysis of Variance.

Comparison 12—Research Hypothesis 12
Nondirectional: The difference in effectiveness of the AM new treatment and the PM new treatment is different from the difference between the AM comparison treatment and the PM comparison treatment.
$$(mG1 - mG2) \neq (mG3 - mG4)$$
Directional: The difference in effectiveness of the AM new treatment and the PM new treatment is greater than the difference between the AM comparison treatment and the PM comparison treatment.
$$(mG1 - mG2) > (mG3 - mG4)$$
Statistical Hypothesis: The difference in effectiveness of the AM new treatment and the PM new treatment is the same as the difference between the AM comparison treatment and the PM comparison treatment.
$$(mG1 - mG2) = (mG3 - mG4)$$
or
$$(mG1 - mG2) - (mG3 - mG4) = 0$$
or
$$(1 * mG1) + (-1 * mG2) + (-1 * mG3) + (1 * mG4) = 0$$

Contrast Coefficients
G1 G2 G3 G4
 1 -1 -1   1

Note: G1 represents the new treatment AM group, G2 represents the new treatment PM group, G3 represents the comparison treatment AM group, and G4 represents the comparison treatment PM group.

Since these contrast coefficients are the same as the contrast coefficients listed for Research Hypothesis 12, this alternative way of stating Research Hypothesis 12 does actually test the same question. This formulation, however, emphasizes the differences between treatments rather than the differences between the time periods. Thus the two different comparisons involve the same interaction effect, and, if found to be significant, it qualifies not only Research Hypothesis 11, the time main effect, but also Research Hypothesis 10, the treatment main effect. That is, finding a significant interaction between A and B qualifies the interpretation of both the A main effect and the B main effect.

Geometric Interpretation

Figure 7.1 illustrates three possible situations of interaction in a two-by-two ANOVA design. Each graph in Figure 7.1 represents the same amount of interaction, although the impact on the main effects interpretation for treatment is very different.

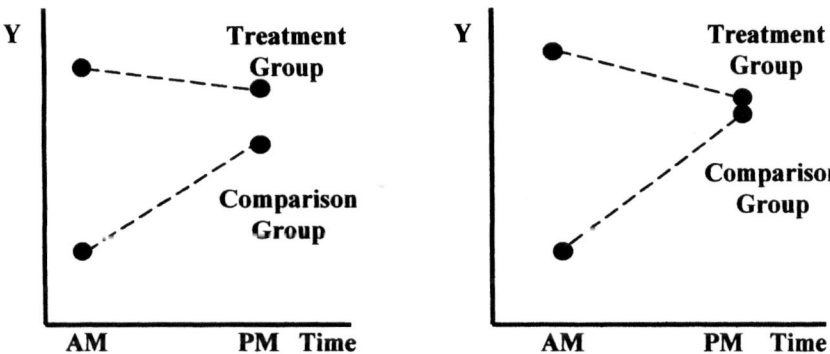

Figure 7.1a.
AM treatment group mean > AM comparison group mean and PM treatment group mean > PM comparison group mean.

Figure 7.1b.
AM treatment group mean > AM comparison group mean and PM treatment group mean ≈ PM comparison group mean.

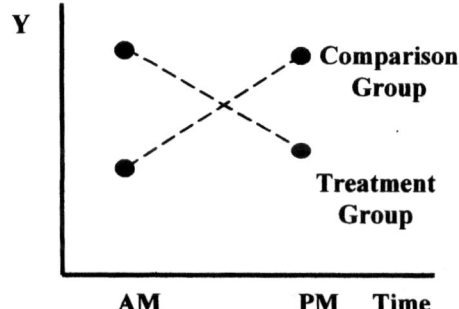

Figure 7.1c.
AM treatment group mean > AM comparison group mean and PM comparison group mean > PM treatment group mean.

Figure 7.1.
Possible interaction results.

In Figure 7.1a the criterion means for both the AM and PM treatment groups are higher than the corresponding means for the comparison groups, particularly for the AM groups. In Figure 7.1b the criterion group means are once again higher for the treatment groups than the corresponding criterion means for the comparison groups, but the difference is essentially nil for the PM groups, that is, it appears that it really does not matter which treatment is implemented in the PM. If only one treatment can be implemented in both the AM and the PM, then the treatment ought to be the one for both of the interactions depicted in Figures 7.1a and 7.1b.

A very different state of affairs is presented in Figure 7.1c, however. In this case the criterion mean is higher for the AM treatment group than it is for the AM comparison group. The opposite situation exists, however, for the PM groups. That is, the criterion mean is lower for the PM treatment group than it is for the PM comparison group. In this situation the decision to implement or not implement the treatment is crucial. The researchers would hope that the decision makers had the capability of implementing or not implementing the treatment, depending upon time of day. That is, to maximize the outcome variable, the decision makers would implement the treatment in the AM and not implement it during the PM.

Venn Diagrams

When all four groups have an equal number of subjects (or the number of subjects in the four groups are proportional) the two-way ANOVA effects (i.e., the A-main effect, the B-main effect, and the A * B interaction effect) will be orthogonal. In such a case the amount of variance in the criterion variable accounted for by a given effect is not impinged on by the other effects.

Figure 7.2 contains three possible states of affairs through Venn diagrams when the three effects are orthogonal. Notice that since these effects are orthogonal, the circles that represent them do not overlap. In Figure 7.2a each of the two main effects account for some of the variance in the criterion variable, whereas the interaction effect does not. All three sources account for some of the variance in the criterion variable in Figure 7.2b. Finally, in Figure 7.2c neither main effect accounts for some of the variance in the criterion variable but the A * B interaction effect does account for some of the variance.

Full Model, Restricted Model, and F test for Research Hypothesis 12

Suppose that one wanted to test the directional hypothesis in Research Hypothesis 12. For each group to have its own mean, the Full Model must allow for all four group means:

$$Y = a_1M1 + a_2M2 + a_3M3 + a_4M4 + E_1 \qquad \text{(Model 7.1)}$$

where:
1. Y = the criterion;
2. $M1 = 1$ if subject in new treatment AM, 0 otherwise;
3. $M2 = 1$ if subject in new treatment PM, 0 otherwise;

4. M3 = 1 if subject in comparison treatment AM, 0 otherwise; and
5. M4 = 1 if subject in comparison treatment PM, 0 otherwise.
It is important to note that Model 7.1 does not contain the unit vector.

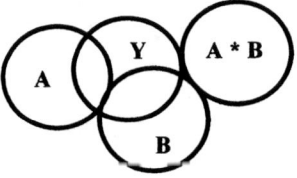

Figure 7.2a.
A and B Main Effects,
are significant, but not
the A*B Interaction.

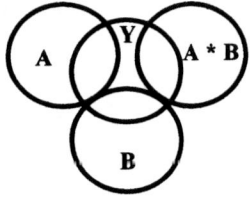

Figure 7.2b.
A and B Main Effects,
and the A*B Interac-
tion are significant.

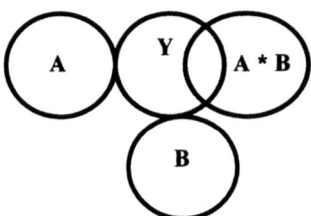

Figure 7.2a.
Interaction is significant,
but the A and B Main Ef-
fects are not.

Figure 7.2.
Possible two-way ANOVA results represented in Venn diagrams.

The restriction implied by Research Hypothesis 12 is:

$a_1 - a_2 = a_3 - a_4$ (1 restriction)

Solving for a_1 produces the following restriction:

$a_1 = a_2 + a_3 - a_4$ (1 restriction)

Forcing this restriction into the Full Model (Model 7.1) yields:

$$Y = (a_3 - a_4 + a_2)M1 + a_2M2 + a_3M3 + a_4M4 + E_2 \qquad \text{(Model 7.2)}$$

Multiplying the coefficients in the parentheses by M1 results in the following model:

$$Y = a_3M1 + (-a_4M1) + a_2M1 + a_2M2 + a_3M3 + a_4M4 + E_2 \qquad \text{(Model 7.2a)}$$

Collecting variables with the same coefficients results in the following model:

$$Y = a_2(M1 + M2) + a_3(M1 + M3) + a_4(M4 - M1) + E_2 \qquad \text{(Model 7.2b)}$$

The three vectors in this Restricted Model (Model 7.2b) have a rather strange appearance. The presence of three vectors in this Restricted Model is what one would expect when one restriction was placed on four pieces of information in the Full Model (Model 7.1). Model 7.2b can be transformed, however, so that the variables contained in it are more recognizable.

To understand this transformation one must first consider that the sum of the M2 and M1 variables is equivalent to a vector that identifies members of the new treatment groups. Thus, (M1 + M2) can be replaced by variable labeled New, with the variable New identifying participants who are members of the M1 or M2 treatment groups. Second, the sum of the M1 and M3 variables is equivalent to a vector that identifies members of the AM groups. Therefore, (M1 + M3) can be replaced by AM, which identifies participants who are members of the AM groups. Third, if one were to add the variables in Model 7.2b, that is, (M1 + M2) + (M1 + M3) + (M4 - M1), the resulting sum would be the four original dichotomous predictor variables (i.e., M1, M2, M3, and M4). Since these four original groups sum to a vector that contains only values of 1, which is equivalent to the unit vector U, one can include the unit vector U and eliminate one of the other vectors in Model 7.2b, say, (M4 - M1). Implementing these three transformations results in the following model:

$$Y = b_0U + b_1New + b_3AM + E_2 \qquad \text{(Model 7.2c)}$$

A review of the Full Model (Model 7.1) and the Restricted Model (7.2c) reveals the logic of the variables included in each model. A model that allows for an interaction effect should contain all of the groups, which is the case for Model 7.1—the Full Model. And the model that does not allow for an interaction effect should contain only the main effects (New and AM variables), which is the case for Model 7.2c—the Restricted Model. In general, the Full Model used to statistically test an interaction effect for a set of dichotomous variables will have a mean for each "cell" and the Restricted Model will only have main-effects information, that is, the A main effect and the B main effect.

To determine if the interaction effect represented by the Full Model (Model 7.1) was statistically significant, researchers would conduct an F test of the differences between the R^2 values of the Full Model and the Restricted Model (Model 7.2c). Since one restriction was placed on the Full Model to form the Restricted Model, and the Full Model contained four pieces of information, the numerator and denominator degrees of freedom values of the F test are 1 and N - 4, respectively. In general the numerator and denominator degrees of freedom values of the F test will be a function of the number of levels on each variable, where A equals the number of levels on variable A, and B equals the number of levels on variable B. Specifically, the numerator degrees of freedom (df_n) value will equal $((A - 1) * (B - 1))$, and the denominator degrees of freedom (df_d) value will equal $(N - (A * B))$.

GENERAL HYPOTHESIS 7.1

The discussion of General Research Hypothesis 7.1 (GRH 7.1) may serve as a guide for researchers who pose an interaction Research Hypothesis that involves dichotomous predictor variables. In this section, it is assumed that both the A main effect and the B main effect consist of two levels. The two variables used to represent the two levels of the A main effect (Treatment 1 and Treatment 2) are T1 and T2, while the two variables used to represent the two levels of the B main effect (pretest levels) are Level 1 and Level 2. The criterion variable is represented by Y. An interaction Research Hypothesis, such as GRH 7.1, could be stated as either a directional or a nondirectional Research Hypothesis as follows:

Directional GRH 7.1: For a given population, the positive difference between the means of Treatment 2 and Treatment 1 on criterion Y will be greater for Level 1 than for Level 2.

Nondirectional GRH 7.1: For a given population, the difference between the means of Treatment 1 and Treatment 2 on the criterion Y will differ for Level 1 and Level 2.

Note that directional GRH 7.1 specifically specifies an *ordinal interaction*. That is, Treatment 2 will exceed Treatment 1 for Level 1 and Level 2 but the difference will be greater for Level 1.

The corresponding General Statistical Hypotheses (GSH 7.1) are as follows:

Directional GSH 7.1: For a given population, the positive difference between the means of Treatment 2 and Treatment 1 on criterion Y will *not* be greater for Level 1 than for Level 2.

Nondirectional GSH 7.1: For a given population, the difference between the means of Treatment 1 and Treatment 2 on the criterion Y will *not* differ for Level 1 and Level 2.

This analysis involves two main effects. One main effect, labeled treatment, consisted of two treatment variables labeled T1 and T2. The other main effect, referred to as pretest, is represented by two levels with the variables labeled L1 and L2. The interaction being tested in this section involves the following four groups: (a) Treatment 1 Level 1—T1L1, (b) Treatment 2 Level 1—T2L1, (c) Treatment 1 Level 2—T1L2, and (d) Treatment 2 Level 2—T2L2.

In the previous section, the Full Model (Model 7.1) did not contain a unit vector, but it did include the four predictor variables (groups). In this section, we will include a unit vector (U) in the Full Model, which will be the case if one uses SPSS for Windows or Microsoft Excel to analyze the data. Since a unit vector is included in the Full Model and the four predictor variables (i.e., T1L1, T2L1, T1L2, and T2L2) are linearly dependent, one of the four variables will not be entered into the model. Since we chose not to include variable T1L1 (we could have chosen any one of the four predictor variables) the Full Model for GRH 7.1 is constructed as follows:

$$Y = a_0U + a_1T2L1 + a_2T1L2 + a_3T2L2 + E_1 \qquad \text{(Model 7.3)}$$

It is important to note that since the variable T1L1 was not included in this Full Model (Model 7.3) and the unit vector (U) was included, the coefficients a_1, a_2, and a_3, will not equal the means of their respective groups. Rather, they will equal the difference between the mean of their respective group and the T1L1 group (i.e., the group not explicitly included in the model). Due to this fact, the restriction that must be placed on the Full Model (Model 7.3) to satisfy the statement in GSH 7.1—regardless of whether it is directional or nondirectional—will appear different from the restriction presented in the previous section.

The restriction placed on the Full Model (Model 7.3) is as follows:

$a_1 = a_3 - a_2$ (1 restriction)

To understand this restriction, it is helpful to refer to the diagram in Figure 7.3. If no interaction effect exits, the distance between the means of T2L1 and T1L1 should equal the distance between the means of T2L2 and T1L2. Remember that the coefficients in the Full Model (Model 7.3) are equal to the differences between their respective means and the mean of the group not explicitly included in the Full Model (Model 7.3), which is the T1L1 group. Thus, a_1 will indicate how much higher or lower the mean of the T2L1 group is versus the mean of the T1L1 group, while $a_3 - a_2$ will indicate how much higher or lower the mean of the T2L2 group is versus the mean of the T1L2 group. Note that a_2 was subtracted from a_3. This order of subtraction was used to produce a difference value conceptually similar to the one produced by a_1 (T2L1 minus T1L1), that is, T2L2 minus T1L2.

As discussed in the previous section, once this restriction is placed in the Full Model (Model 7.3), the resulting Restricted Model can be represented by a model that contains the main effects. The treatment main effect can be represented by a variable labeled T, which is equal to the sum of variables T1L1 and T1L2; while the pretest main effect can be represented by a variable labeled L (for levels), which is equal to the sum of variables T1L1 and T2L1. (Note the variable T could also have been formed by summing the variables T2L1 and T2L2, and the variable L could have been formed by summing variables T1L2 and T2L2.) Thus the Restricted Model is as follows:

$$Y = a_0 + a_1 T + a_2 L + E_2 \qquad\qquad\qquad\text{(Model 7.4)}$$

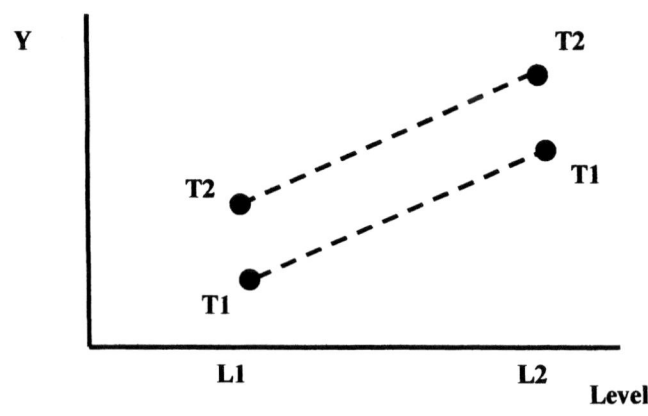

Figure 7.3.
Diagram for a case where an interaction effect does not exist.

The general F test, which was presented in Equation 4.1 of chapter 4, will be used to statistically test the difference between the R^2 values of the Full Model (Model 7.3) and the Restricted Model (Model 7.4). If the nondirectional version of the GRH 7.1 is used, the corresponding Statistical Hypothesis (GSH 7.1) would be rejected if the two-tailed probability of this F test is less than the established alpha level.

If the directional version of GRH 7.1 is used, the researchers would compare the one-tailed probability value of the F test to the established alpha level. If the F test indicated that the interaction effect was significant, it is not a sufficient finding, however, to suggest that directional GRH 7.1 is supported by the data. Before the researchers would be willing to make such a statement, they need to compare the signs of the coefficients generated for the Full Model to the signs required by the directional Research Hypothesis. Thus the researchers must identify the procedure we call the "coefficient criteria" (i.e., the criteria that the coefficients must meet to satisfy the Research Hypothesis).

Recall that the directional version of GRH 7.1 called for the *positive* difference between the mean of Treatment 2 and the mean of Treatment 1 on criterion Y to be *greater* for Level 2 than for Level 1. For this statement to be supported by the data, the positive difference between T2L1 and T1L1 must exceed the positive difference between T2L2 and T1L2. That is, a specific *ordinal* interaction effect is required to support GRH 7.1. If the type of ordinal interaction described in GRH 7.1 exists, the values for a_1 and $a_3 - a_2$ must be positive, and the value for a_1 must exceed the value for $a_3 - a_2$.

Two points should be noted regarding the use of directional interaction research questions. First, when a directional interaction research question is posed and statistically tested, the coefficient criteria may not support the specific interaction contained in the research question, but that does not necessarily indicate that no interaction effect exists. An interaction effect may exist, but it simply does not match the one specified in the research question.

Second, the coefficient criteria established to judge whether the directional interaction Research Hypothesis is supported depends on the specific type of interaction effect being statistically tested. For example, assume the researchers expected the mean of Treatment 2 to exceed the mean of Treatment 1 for Level 1 but the opposite condition to exist for Level 2, that is, the interaction effect is *disordinal.* For this type of interaction effect (the dashed lines in Figure 7.3 would cross) the researchers would expect, in addition to a significant F test of the difference between the R^2 values of the Full and Restricted Models, a *positive* value for a_1 and a *negative* value for $a_3 - a_2$.

APPLIED RESEARCH HYPOTHESIS 7.1

Applied Research Hypothesis 7.1 (ARH 7.1) illustrates the type of analysis presented in the General Hypothesis 7.1 section, which tested for an interaction effect between two sets of dichotomous variables. In this section, it is assumed the researchers were interested in the performance of students on a math examination (variable X9) when exposed to either of two methods of instruction, which were labeled Method A and Method B. The dichotomous variables used to identify which method of instruction a given participant received were variables X10 (Method A) and X11 (Method B). The students were also classified according to the type of setting in which their schools were located—City (Location 1) and Non-City (Location 2). The dichotomous variables used to identify these two locations were X12 (City) and X13 (Non-City). The methods of instruction and the school locations were the main effects in this study. To identify the four specific groups needed to test for an interaction effect between Method and Location, the following variables need to be generated:

1. X22 = X10*X12, which identifies the participants in Method A and Location 1.
2. X23 = X10*X13, which identifies the participants in Method A and Location 2.

3. X24 = X11*X12, which identifies the participants in Method B and Location 1.
4. X25 = X11*X13, which identifies the participants in Method B and Location 2.

A directional Applied Research Hypothesis (ARH 7.1) will be tested, which is as follows:

> Directional ARH 7.1: For a given population, the positive difference between means of Method B and Method A on the criterion of interest (X9) will be greater for Location 1 than for Location 2.

Note that this directional Research Hypothesis does specify which mean in each location will be higher. The corresponding Applied Statistical Hypothesis (ASH 7.1) is as follows:

> Directional ASH 7.1: For a given population, the positive difference between the means of Method B and Method A on the criterion of interest (X9) will be the same for Location 1 and Location 2.

Full and Restricted Models

Once the four predictor variables (i.e., X22, X23, X24, and X25) are generated, the following Full Model is constructed:

$$X9 = a_0U + a_{23}X23 + a_{24}X24 + a_{25}X25 + E_1 \qquad \text{(Model 7.5)}$$

Note that due to the fact that the unit vector was included in this Full Model, one of the four linearly dependent predictor variables, namely X22, was not placed in the model.

The restriction required by ASH 7.1 is as follows:

$$a_{24} = a_{25} - a_{23}$$

Note that this restriction reflects the specific restriction required by ASH 7.1. That is, the difference between the means of Method B in Location 1 and Method A in Location 1 (a_{24}) does not exceed the difference between the means of Method B in Location 2 and Method A in Location 2 ($a_{25} - a_{23}$).

As discussed previously in this chapter, this restriction is placed into the Full Model (Model 7.5), which results in a Restricted Model that contains only the main effects (i.e., Method and Location). In this Restricted Model, the Method main effect can be represented by either variable X10 or variable X11, and Location main effect can be represented by either variable X12 or variable X13. We will assume the researchers chose to enter into the Restricted Model variables X10 and X12 to represent the Method and Location main effects, respectively. Thus the Restricted Model is as follows:

$$X9 = a_0 U + a_{10}X10 + a_{12}X12 + E_2 \qquad \text{(Model 7.6)}$$

To reject the statistical hypothesis ASH 7.1, which is directional, the researchers would first determine whether the F test one-tailed probability value of the difference between the R^2 values of the Full and Restricted Models was less than the established alpha level of .05. If the difference between the two R^2 values was statistically significant, the researchers would next verify whether the coefficient criteria were met. The coefficient criteria are based on the type of interaction specified by the directional Research Hypothesis. ARH 7.1 specifies that the *positive* difference between the means of Method B in Location 1 and Method A in Location 1 *exceeds* the *positive* difference between the means of Method B in Location 2 and Method A in Location 2. Thus, the coefficient criteria require the values of a_{24} and a_{25} - a_{23} to be positive and the value of a_{24} to exceed the value of a_{25} - a_{23}.

Analysis ARH 7.1 with SPSS for Windows

Once the SPSS data file entitled "GLM DATA SPSS FORMAT" has been accessed from the internet site (see Appendix A) and the commands needed to access the Linear Regression menu (see chapter 4) have been completed, the Linear Regression menu presented in Exhibit 7.2 will be displayed.

The criterion variable (X9) and the predictor variables (X23, X24, and X25) contained in Full Model (Model 7.5) must be entered into the Linear Regression menu. As illustrated in Exhibit 7.2, these variables are entered as follows:

1. Click on the X9 variable (see Oval 1) and click on the arrow key next to the "Dependent" box. This will identify X9 as the criterion variable.
2. While holding down the control key on the keyboard, click on the X23, X24, and X25 variables (see Oval 2). Release the control key and click on the arrow key next to the "Independent(s)" box. This will identify X23, X24, and X25 as the predictor variables.
3. Click on the "OK" button (see Oval 3).

After these three steps are completed, the regression analysis output contained in Exhibit 7.3 will be displayed. The key piece of information on the output window for the Full Model (Model 7.5) is the R^2 value of .050 (see Oval 1). Note that the F test listed in the "ANOVA" table on the output (see Oval 2) is not the appropriate test for ARH 7.1. This F test would be applicable only when the Restricted Model is the Null Model, which is not the case here.

Exhibit 7.2.
SPSS Linear Regression menu—Full Model (Model 7.5) for ARH 7.1.

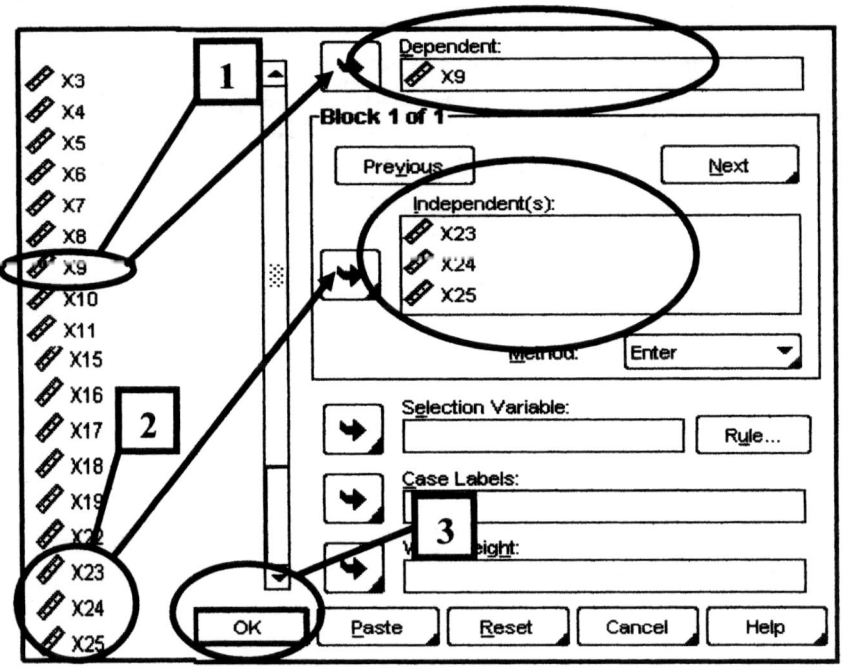

To obtain the Restricted Model (Model 7.6), the SPSS for Windows Linear Regression menu should once again be accessed. The Linear Regression menu will appear as displayed in Exhibit 7.4. Once this menu is displayed, complete the following:

1. Click on the "Reset" box (see Oval 1).
2. Enter the criterion variable X9 (see Oval 2) and the predictor variables X10 and X12 (see Oval 3) in the manner previously described for the Full Model.
3. Click on the "OK" box (see Oval 4).

After these three steps are completed, the regression analysis output contained in Exhibit 7.5 will be displayed.

The key piece of information on the output window for the Restricted Model (Model 7.6) is the R^2 value of .015 (see Oval 1). Note that once again the F test listed in the "ANOVA" table on the output (see Oval 2) is not the appropriate test for ASH 7.1.

Exhibit 7.3.
SPSS regression output—Full Model (Model 7.5) for ARH 7.1

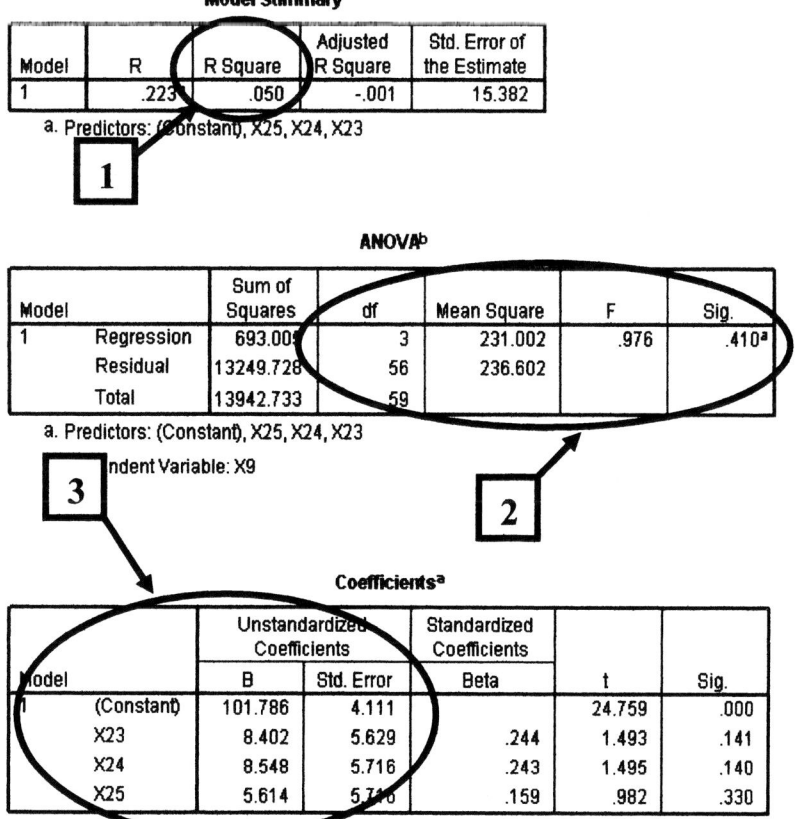

Model Summary

Model	R	R Square	Adjusted R Square	Std. Error of the Estimate
1	.223	.050	-.001	15.382

a. Predictors: (Constant), X25, X24, X23

1

ANOVA[b]

Model		Sum of Squares	df	Mean Square	F	Sig.
1	Regression	693.005	3	231.002	.976	.410[a]
	Residual	13249.728	56	236.602		
	Total	13942.733	59			

a. Predictors: (Constant), X25, X24, X23
b. Dependent Variable: X9

3

2

Coefficients[a]

Model		Unstandardized Coefficients		Standardized Coefficients	t	Sig.
		B	Std. Error	Beta		
1	(Constant)	101.786	4.111		24.759	.000
	X23	8.402	5.629	.244	1.493	.141
	X24	8.548	5.716	.243	1.495	.140
	X25	5.614	5.716	.159	.982	.330

a. Dependent Variable: X9

The F test formula needed to test ASH 7.1 is the one contained in Equation 4.1, which is listed in chapter 4. The appropriate F test and its corresponding probability value can be obtained by using the SPSS for Windows computer software. To use SPSS for Windows to conduct this test, however, the variables in the Full Model (Model 7.5) need to be changed to form an equivalent Full Model. An equivalent model to Model 7.5 can be formed as follows:

$$X9 = a_0U + a_{10}X10 + a_{12}X12 + a_{22}X22 + E_1 \qquad \text{(Model 7.5a)}$$

Exhibit 7.4.
SPSS Linear Regression menu—Restricted Model (Model 7.6) for ASH 7.1

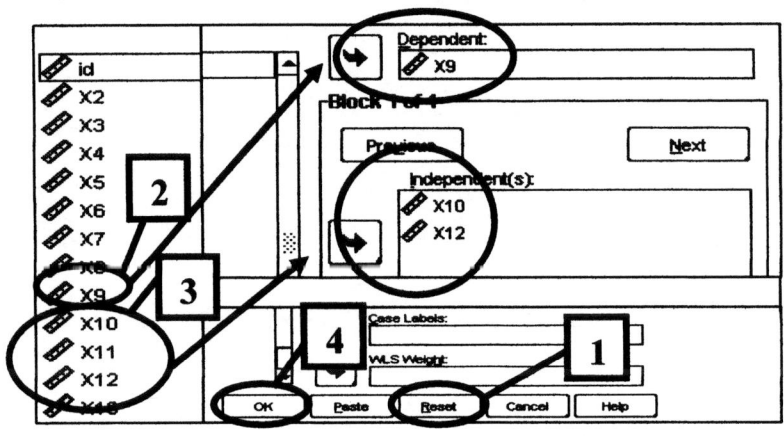

Note that this Full Model contains the two main effect variables (i.e., X10 and X12) plus the variable that represents the Method A Location 1 group (i.e., X22). This Full Model (Model 7.5a) will produce the same R^2 value as the original Full Model (Model 7.5). It is interesting to note that any one of the other group variables (i.e., X23, X24, or X25) could be placed in this Full Model (Model 7.5a) rather than X22, and the R^2 value would remain the same.

To allow the SPSS for Windows software to conduct the appropriate F test, the Full Model must contain the variables in the Restricted Model. Note that both of the variables in the Restricted Model (Model 7.6), namely, X10 and X12, are also part of the Full Model (Model 7.5a).

Once this new Full Model (Model 7.5a) is formed, the F test and its corresponding probability can be obtained by returning to the Linear Regression menu (see Exhibit 7.6). First, enter the Restricted Model (Model 7.6) as follows:

1. Click on the "Reset" box (see Oval 1).
2. Click on the X9 variable (see Oval 2) and click on the arrow key next to the "Dependent" box.
3. While holding down the control key on the keyboard, click on the X10 and X12 variables (see Oval 3). Release the control key and click on the arrow key next to the "Independent(s)" box.
4. Click on the "Next" box (see Oval 4).

After completing these steps the Linear Regression menu will reappear (see Exhibit 7.7).

Exhibit 7.5.
SPSS regression output—Restricted Model (Model 7.6) for ASH 7.1

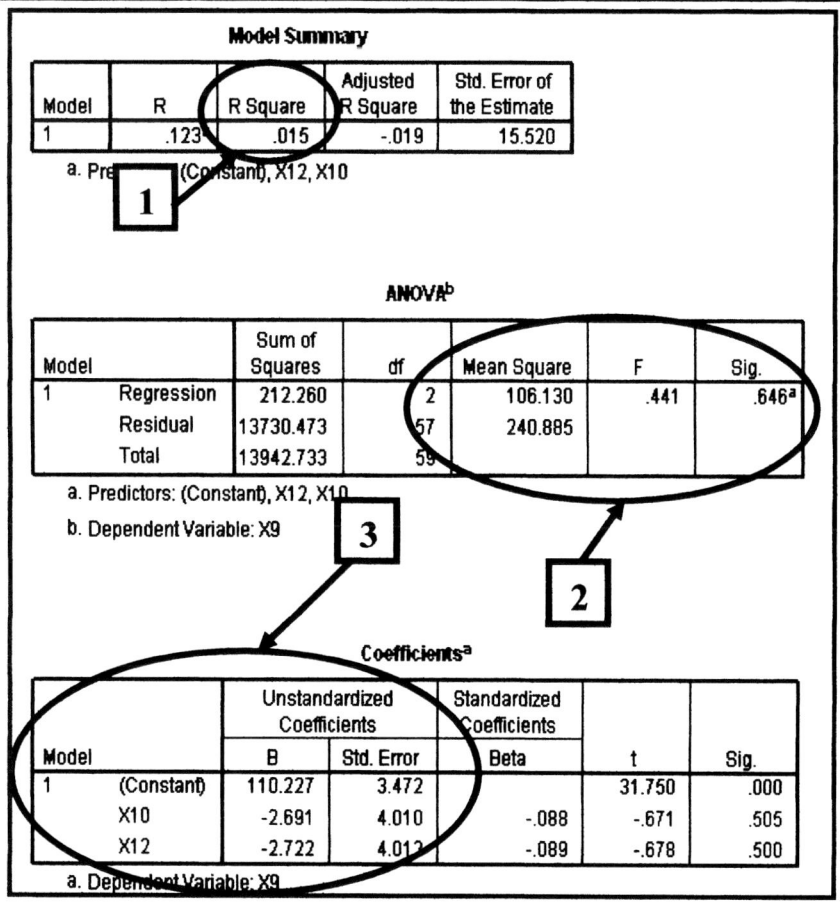

Once the Linear Regression menu is displayed, complete the following two steps:

1. Click on variable X22 (see Oval 1) and click on the arrow key next to the "Independent(s)" box.
2. Click on the "Statistics" box (see Oval 2).

Upon completion of these two steps the statistics menu will be displayed as presented in Exhibit 7.8. Click on the boxes in front of "R squared change," "Model Fit," and "Estimates" (see Oval 1) and click on the "Continue" (see Oval 2) key. The Linear Regression menu will reappear as displayed in Exhibit 7.7. Click on the "OK" key (see Oval 3). The output listing the F test of the difference between the R^2 values of the Full Model (Model 7.5a), which is equivalent to Model 7.5, and the Restricted Model (Model 7.6), along with its probability value, is displayed in Exhibit 7.9.

Exhibit 7.6.
SPSS Linear Regression menu—Restricted Model (Model 7.6) and Full Model (Model 7.5a) for the *F* Test of ASH 7.1.

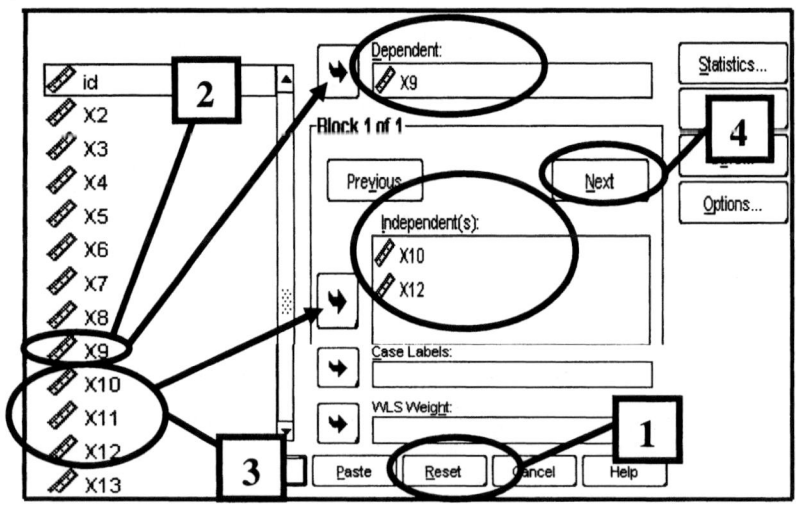

The key information contained in the output window (see Exhibit 7.9) is as follows:

1. The Restricted Model (Model 7.6) is labeled Model 1 (see Oval 1), and its R^2 value is .015, which, of course, matches the R^2 value listed for the Restricted Model in Exhibit 7.5.
2. The Full Model (Model 7.5a) is labeled Model 2 (see Oval 1), and its R^2 value is .050. Note that this value also matches the R^2 value for the Full Model (Model 7.5) in Exhibit 7.3.
3. The difference between the R^2 values of Model 1 (i.e., the Restricted Model—Model 7.6) and Model 2 (i.e., the Full Model—Model 7.5a) is .034 (see Oval 2). Note that the value of .034 is not the same as .050 - .015. This is due to the fact that the R^2 values of the Full and Restricted Models are not exactly .050 and .015, respectively. The researchers could appropriately report, however, that the difference between the R^2 values is equal to .035.
4. The *F*-test value of the difference between the R^2 values of the Full and Restricted Models is 2.032 (see Oval 3).

Exhibit 7.7.
SPSS Linear Regression menu—Full Model (Model 7.5a) for the F Test of ASH
7.1.

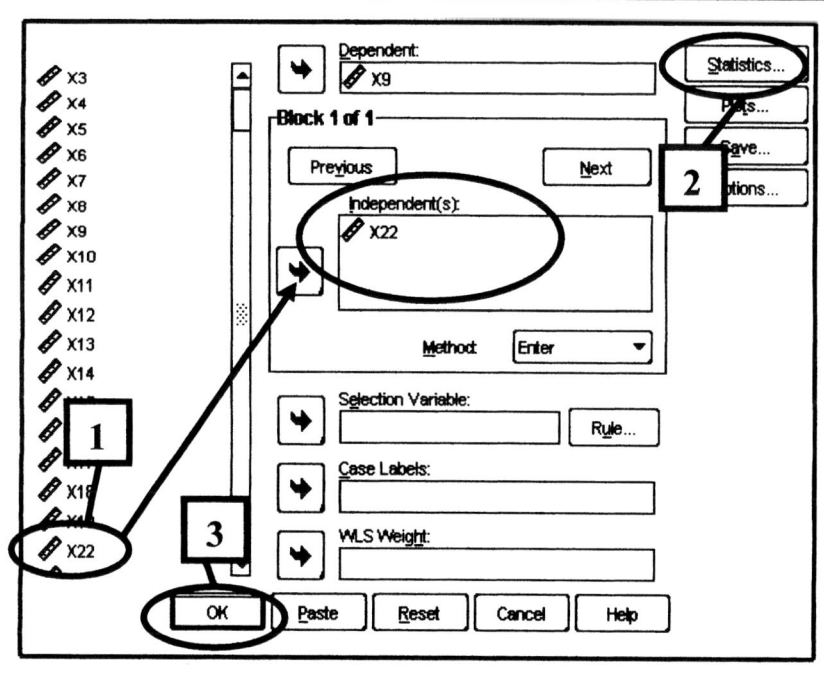

Exhibit 7.8.
SPSS Statistics menu—R Square Change for the F Test of ASH 7.1.

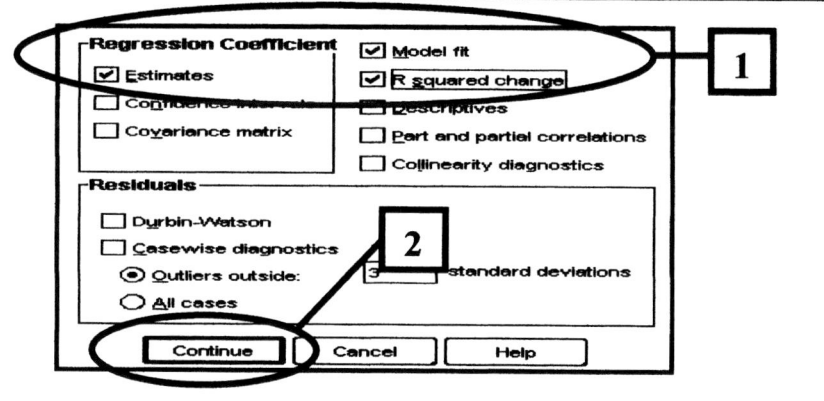

5. The numerator degrees of freedom value (dfn) value (i.e., 1) for the F test is found under the column heading "Change Statistics df1" in the row containing the information for Model 2 (see Oval 3). The denominator degrees of freedom value (df_d) value (i.e., 56) for the F test is found under the column heading "Change Statistics df2" in the row containing the information for Model 2 (see Oval 3).
6. The one-tailed probability value of .08 is one half of .160, the probability value listed under the heading "Sig. F Change" (see Oval 3).

Exhibit 7.9.
SPSS regression output—Restricted Model (Model 7.6), Full Model (Model 7.5a), R^2 Change, and F Test of ASH 7.1.

Model Summary

Model	R	R Square	Adjusted R Square	Std. Error of the Estimate	R Square Change	F Change	df1	df2	Sig. F Change
							Change Statistics		
1	.123ᵃ	.015	-.019	15.52	.015	.441	2	57	.646
2	.223ᵇ	.050	-.001	15.382	.034	2.032	1	56	.160

a. Predictors: (Constant), X12, X10
b. Predictors: (Constant), X12, X10, X22

Recall that the Statistical Hypothesis ASR 7.1 will be rejected only if two conditions are satisfied. First, the one-tailed probability of the F test of the difference between the R^2 values of the Full and Restricted Models must be less than the established alpha level (i.e., .05). A review of the information provided in the output window (see Oval 3 in Exhibit 7.9) indicates that the one-tailed probability value (.08, i.e., .16 / 2) of the F test is not less than the alpha level of .05. Thus, the interaction effect between the Methods and Location is not statistically significant. Based on this result, the researchers would not reject the ASH 7.1. Thus, the researchers would conclude that for a given population, the positive difference between means of Method B and Method A on the criterion of interest (X9) was not greater for Location 1 than for Location 2.

In spite of this non-significant finding, the researcher may want to examine whether the coefficient criteria were met as a means of providing insight for future research. Although this analysis utilized Model 7.5a as the Full Model, the values of the coefficients generated for Model 7.5 in the first analysis are far easier to interpret (see Oval 3 of Exhibit 7.3).

The values for the coefficients a_{23}, a_{24}, and a_{25}, which were generated for the predictor variables in Model 7.5, were 8.40, 8.55, and 5.61, respectively. Thus, the values for a_{24} and a_{25} - a_{23} were 8.55 and -2.79 (i.e., 5.61 – 8.40), respectively. Although the positive sign for a_{24} matched the sign required by ARH 7.1, the negative sign for a_{25} - a_{23} was not the expected sign. Thus, the researchers discovered that not only was the interaction effect not significant, the interaction contained in the sample was not the type of interaction specified in ARH 7.1. This information may be useful for the formulation of hypotheses for future studies.

Writing the Results for ARH 7.1 in APA Style

The results for ARH 7.1 could be written in APA style as follows:

The mean math examination scores of the four groups of students were as follows: (a) Method A in city locations, 101.79; (b) Method A in non-city locations, 110.19; (c) Method B in city locations, 110.34; and (d) Method B in non-city locations, 107.40. The difference between the means of those students taught by Method B in city locations and those students taught by Method A in city locations was 8.55; while the difference between the means of those students taught by Method B in non-city locations and those students taught by Method A in non-city locations was -2.79. The interaction effect contained in the sample was not statistically significant (F (1, 56) = 2.03, $p / 2$ = .08) at the one-tailed alpha level of .05. In addition, the expected ordinal interaction was not reflected in the model parameter estimates.

Note that in the APA style report of the results, the mean mathematics test scores were reported for each of the four groups. The values of the four group means were obtained from the output of the Model 7.5 located in Exhibit 7.3. Since the variable representing the students taught by Method A in city locations (X22) was not included, the mean for this group was equal to the coefficient for the unit vector (101.79), which is the coefficient listed for the constant term of Model 7.3 (see Oval 3 in Exhibit 7.3). The mean of the students taught by Method A in non-city locations is equal to the value of the coefficient for variable X23 plus the value for the constant (i.e., 8.40 + 101.79 = 110.19). The mean of the students taught by Method B in city locations is equal to the value of the coefficient for variable X24 plus the value for the constant (i.e., 8.55 + 101.79 = 110.34). And the mean of the students taught by Method B in non-city locations is equal to the value of the coefficient for variable X25 plus the value for the constant (i.e., 5.61 + 101.79 = 107.40).

We encourage researchers to include a diagram of any statistically significant interaction. If the interaction effect is not statistically significant, which is the case for this study, a hypothetical diagram of the type of interaction specified in the Research Hypothesis would possibly aid the readers. The diagram could take the form of one of the diagrams contained in Figure 7.1. Such a diagram will facilitate the reader's understanding of the nature of the specified interaction effect. Remember, a picture is worth a thousand words.

INTERACTION EFFECTS WITH DICHOTOMOUS PREDICTOR VARIBLES WITH MORE THAN TWO LEVELS

Consider a two-way design with four levels of IQ and three drugs. The interaction question is only one of the infinite number of possible Research Hypotheses relevant to this design. There are many sets of multiple comparisons possible with this design, and traditional interaction is a part of some of those comparisons. The Research Hypothesis for interaction could be stated as:

Research Hypothesis: The differences between the four IQ levels are not constant across the three drug treatments.

This Research could be stated more explicitly as follows:

Research Hypothesis: The differences between the four IQ levels given Drug 1, and the differences between the four IQ levels given Drug 2, and the differences between the four IQ levels given Drug 3 are not all equal.

The corresponding Statistical Hypothesis is as follows:

Statistical Hypothesis: The differences between the four IQ levels are constant across the three drug treatments.

The regression formulation for testing the significance of the interaction has been well documented (Bottenberg & Ward, 1963; Jennings, 1967; Kelly, Beggs, McNeil, Eichelberger, & Lyon, 1969). A Full Model that allows for an interaction effect, must permit the 12 cell means to manifest themselves. Then a Restricted Model, allowing only the row (IQ) and column (Drug) means to manifest themselves, can be compared to the Full Model via the generalized F test. The following Full Model, which does not include a unit vector, allows the 12 means to manifest themselves as 12 linearly independent vectors, or 12 pieces of information:

$$Y1 = a_1C1 + a_2C2 + a_3C3 + a_4C4 + a_5C5 + a_6C6 + a_7C7 + a_8C8 + a_9C9 + a_{10}C10 + a_{11}C11 + a_{12}C12 + E_1 \qquad \text{(Model 7.7)}$$

where:
1. $Y1$ = the criterion score;
2. $C1$ = 1 if IQ level 1 and Drug 1, 0 otherwise;
3. $C2$ = 1 if IQ level 1 and Drug 2, 0 otherwise;
4. $C3$ = 1 if IQ level 1 and Drug 3, 0 otherwise;

5. C4 = 1 if IQ level 2 and Drug 1, 0 otherwise;
6. C5 = 1 if IQ level 2 and Drug 2, 0 otherwise;
7. C6 = 1 if IQ level 2 and Drug 3, 0 otherwise;
8. C7 = 1 if IQ level 3 and Drug 1, 0 otherwise;
9. C8 = 1 if IQ level 3 and Drug 2, 0 otherwise;
10. C9 = 1 if IQ level 3 and Drug 3, 0 otherwise;
11. C10 = 1 if IQ level 4 and Drug 1, 0 otherwise;
12. C11 = 1 if IQ level 4 and Drug 2, 0 otherwise;
13. C12 = 1 if IQ level 4 and Drug 3, 0 otherwise; and
14. a_1, a_2, a_3, . . ., a_{12} are least squares weighting coefficients calculated to minimize the sum of the squared values in the error vector, E_1.

This Full Model (Model 7.7) allows each of the 12 means to manifest itself and is called by some the "cell means model." Model 7.7 could be written in an equivalent form, which we believe more clearly reveals the allowance for interaction between IQ and Drugs. This equivalent Full Model is as follows:

$$Y1 = a_1(Q1 * D1) + a_2(Q1 * D2) + a_3(Q1 * D3) + a_4(Q2 * D1) + a_5(Q2 * D2) + a_6(Q2 * D3) + a_7(Q3 * D1) + a_8(Q3 * D2) + a_9(Q3 * D3) + a_{10}(Q4 * D1) + a_{11}(Q4 * D2) + a_{12}(Q4 * D3) + E_1 \quad \text{(Model 7.7a)}$$

The multiplication signs contained in this Full Model (Model 7.7a) indicate that the model allows for interaction effects, as these signs will whenever they appear in subsequent models. Since Model 7.7 and Model 7.7a are equivalent models, forcing the restrictions of no interaction on either model will result in the same Restricted Model. The restrictions specifying no interaction are as follows:

$(a_1 - a_2) = (a_4 - a_5) = (a_7 - a_8) = (a_{10} - a_{11})$ (Restriction Set 1—3 restrictions)
$(a_2 - a_3) = (a_5 - a_6) = (a_8 - a_9) = (a_{11} - a_{12})$ (Restriction Set 2—3 restrictions)

See Figure 7.4 for a depiction of what these restrictions require regarding the differences among the 12 group means. As shown in Figure 7.4, the differences between the 12 means are such that the dotted line connecting the 4 means for Drug 1 across the 3 IQ levels must be parallel to the dotted line connecting the 4 means for Drug 2 across the 3 IQ levels, which in turn must be parallel to the dotted line connecting the 4 means for Drug 3 across the 3 IQ levels.

Forcing these restrictions onto either Full Model would result in a set of very strange looking variables in the Restricted Model. But, using the information provided in the previous sections, such a Restricted Model is equivalent to a Restricted Model that has only information about the "main effects." Recall that since the sum of Q1, Q2, Q3, and Q4 is equal to the unit vector; and the sum of D1, D2, and D3 is also equal to the unit vector, the equivalent Restricted Model that contains only the main effects is constructed as follows:

$$Y1 = a_0U + q_2Q2 + q_3Q3 + q_4Q4 + d_2D2 + d_3D3 + E_2 \qquad \text{(Model 7.8)}$$

where:
1. Y1 = the criterion score;
2. U = the unit vector
3. Q2 = 1 if IQ level 2, 0 otherwise;
4. Q3 = 1 if IQ level 3, 0 otherwise;
5. Q4 = 1 if IQ level 4, 0 otherwise;
6. D2 = 1 if Drug 2, 0 otherwise; and
7. D3 = 1 if Drug 3, 0 otherwise.

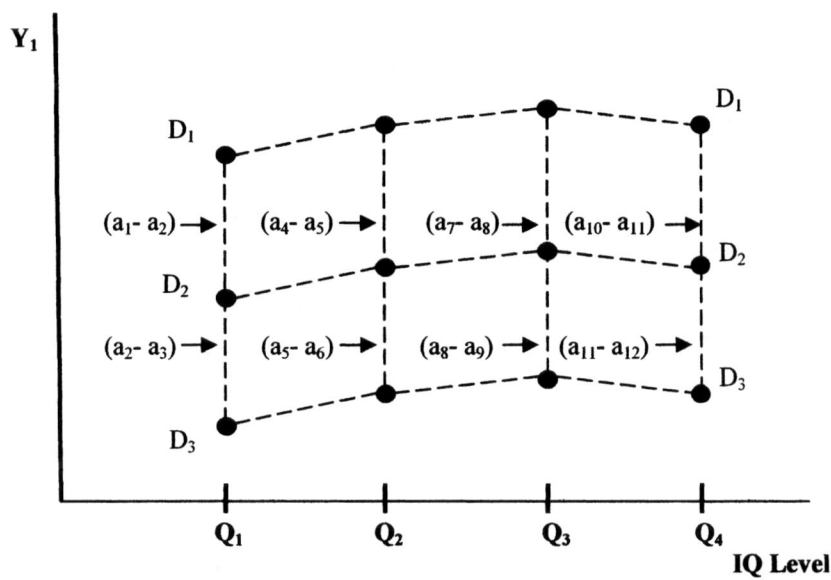

Figure 7.4.
No interaction effect between IQ and Drug.

Most computer programs, including SPSS for Windows and Microsoft Excel, will automatically provide the unit vector. Since computer software will be used to estimate the models and because researchers are less likely to make an error in determining the number of linearly independent vectors when the unit vector is included, we believe it should be included. With the inclusion of the unit vector in the Full Model (i.e., Model 7.7 or Model 7.7a), one of the interaction vectors is not a new piece of information; and therefore, must be eliminated, such as variable (Q1 * D1). The following Full Model (Model 7.7b) contains the unit vector and 11 of the 12 interaction variables contained in Models 7.7 and 7.7a:

$$Y1 = a_0U + a_2(Q1 * D2) + a_3(Q1 * D3) + a_4(Q2 * D1) + a_5(Q2 * D2) +$$
$$a_6(Q2 * D3) + a_7(Q3 * D1) + a_8(Q3 * D2) + a_9(Q3 * D3) + a_{10}(Q4 *$$
$$D1) + a_{11}(Q4 * D2) + a_{12}(Q4 * D3) + E_1 \qquad \text{(Model 7.7b)}$$

Two important characteristics of this model should be understood. First, in Models 7.7 and 7.7a, the coefficients (i.e., a_2, a_3, etc.) will equal the mean of their respective groups. In Model 7.7b, however, with the inclusion of the unit vector and the elimination of $Q1 * D1$, the coefficient for the unit vector (a_0) will equal the mean of the eliminated group (i.e., the $Q1 * D1$ group); and each of the other coefficients (i.e., a_2, a_3, etc.) will equal the difference between the mean of its respective group and the mean of the $Q1 * D1$ group. Second, although the $Q1 * D1$ interaction variable was not included in Model 7.7b; any one of the 12 interaction terms could have been eliminated. It is important to note that regardless of which interaction term is eliminated, the Full Model will have the same R^2 value.

The Restricted Model represented by 7.8, which includes the unit vector can be used in conjunction with a Full Model containing the interaction terms without the unit vector (i.e., Models 7.7 or 7.7a) or without the unit vector (i.e., Model 7.7b). The important point to note regarding this Restricted Model (Model 7.8) is that it contains six pieces of information, that is, the unit vector, three IQ vectors, and two drug vectors. Thus, Model 7.8 contains six linearly independent pieces of information.

Simple Effects Tests with a Restricted Sample Size

If the interaction is significant, most statistical authors indicate that the main-effects question is not appropriate, and that simple effects should be investigated. Each of the simple effects is simply a one-way ANOVA at any one level. The regression formulation for the simple effect of IQ level 1 would require the following Full Model that includes the unit vector:

$$Y1 = a_0U + c_2(Q1 * D2) + c_3(Q1 * D3) + E_1 \qquad \text{(Model 7.9)}$$

It is important to understand that only the data for the subjects in IQ level 1 are included in the analysis of this Full Model (Model 7.9).

The restriction implied by the simple effects is as follows:

$$0 = c_2 = c_3 \quad \text{(2 restrictions)}$$

Forcing these restrictions into the Full Model (Model 7.9) yields the following Restricted Model:

$$Y1 = a_0U + E_2 \qquad \text{(Model 7.10)}$$

Once again, only the data for the subjects in IQ level 1 are included in the analysis of this Restricted Model (Model 7.10).

The Full Model (Model 7.9) contained three pieces of information (i.e., the unit vector and the Q1* D2 and Q1* D3 vectors), while the Restricted Model (Model 7.10) contained only one (i.e., the unit vector). Thus, the numerator and denominator degrees of freedom values for the F test of the difference between the R^2 values of the two models would be (3 - 1) and (N - 3), respectively. Recall that the Full and Restricted Models deal only with subjects at IQ level 1. Due to this fact, N for this analysis is equal to the number of subjects in IQ level 1 only—not the total number of subjects in the study.

Simple Effects Tests with the Total Sample Size

A more stable estimate of the within-group variance is available by considering all the cells, and therefore the total sample size, not just those three cells directly considered in the Full Model represented by Model 7.9 and the Restricted Model represented by Model 7.10. When utilizing the total sample, the Full Model would be the same as Model 7.7b—assuming the unit vector was included in the model and the group represented by D1* Q1 was eliminated.

The restrictions placed on this Full Model (Model 7.7b) would be as follows:

$$0 = a_2 = a_3 \quad \text{(2 restrictions)}$$

Placing these restrictions on the Full Model (Model 7.7b) produces a Restricted Model with two fewer predictor variables (i.e., Q1*D2 and Q1*D3). The resulting Restricted Model is as follows:

$$Y1 = a_0U + a_4(Q2 * D1) + a_5(Q2 * D2) + a_6(Q2 * D3) + a_7(Q3 * D1) +$$
$$a_8(Q3 * D2) + a_9(Q3 * D3) + a_{10}(Q4 * D1) + a_{11}(Q4 * D2) + a_{12}(Q4 *$$
$$D3) + E_1 \qquad \qquad \text{(Model 7.11)}$$

Note that this Restricted Model (Model 7.11) does not contain information regarding IQ level 1, which was also the case for the Restricted Model represented by Model 7.10. However, unlike the analysis of Models 7.9 and 7.10, which involved only the subjects in the IQ level 1, the analyses of Models 7.7b and 7.11 utilized all of the subjects in the study. Thus, the value of N used in the F test of the difference between the R^2 values of these models is equal to the total sample size. This increase in the number of subjects will often provide additional power to the F test not only due to the increased sample size but also an increased R^2 value for the Full Model.

Testing for a Main Effect when the Interaction Effect is not Significant

Even when there is significant interaction among the variables, we would argue that one might still want to examine the main effects when the interaction effect is ordinal. Ordinal interaction occurs when the rank order of the categories of one variable based on their criterion scores is the same within each category of the second predictor variable (i.e., the lines do not cross in the interaction diagram). When significant ordinal interaction exists in the data, the magnitude of the main effects is not indicative of the effects at each level, but it is the case that one level is uniformly superior, as depicted in Figure 7.1a. On the other hand, Figure 7.1c indicates a disordinal interaction, that is, the lines crossed.

Testing for main effects can be done in two ways: (a) one in which the within term is used as the best estimate of expected variance and (b) the other in which the within source is pooled with the non-significant interaction and used as a new estimate of expected variance. These notions are more fully delineated by Jennings (1967).

Not pooling the interaction term. Jennings (1967) presented the regression models appropriate for the case in which the sources are not pooled. If the main effect for IQ were to be tested, the following restriction would be made on Model 7.7 and Model 7.7a if the cell numbers ($N1$, $N2$, etc.) were not proportional:

$$\frac{N_1 a_1 + N_2 a_2 + N_3 a_3}{N_1 + N_2 + N_3} = \frac{N_4 a_4 + N_5 a_5 + N_6 a_6}{N_4 + N_5 + N_6} =$$

$$\frac{N_7 a_7 + N_8 a_8 + N_9 a_9}{N_7 + N_8 + N_9} = \frac{N_{10} a_{10} + N_{11} a_{11} + N_{12} a_{12}}{N_{10} + N_{11} + N_{12}}$$

(Equation 7.1)

Placing these restrictions on either of the Full Models (i.e., Model 7.7 or Model 7.7b) results in an extremely complicated Restricted Model—a model with little conceptual value. The pooling analog, which is discussed in the next section, is much more conceptually pleasing.

Pooling of the interaction term. When the interaction source of variance is not significant, one can consider this variance to be essentially random and thus pool that variance with the variance from the within source. The argument for this procedure (when there is little sample interaction) is that the resulting variance estimate is a more stable estimate of the expected chance variance. Winer (1971) developed a guideline stating that, if the significance of the interaction source had a p value greater than .30, then one could pool the interaction and within sources of variance.

The Restricted Model represented by Model 7.8 depicted a state of affairs in which interaction was not allowed. If this Restricted Model is not significantly less predictable than the Full Model represented by Model 7.7b, then Model 7.8

can be thought of as providing a more stable variance estimate of the expected chance variance. Thus, Model 7.8 becomes the Full Model, which is, once again, as follows:

$$Y1 = a_0U + q_2Q2 + q_3Q3 + q_4Q4 + d_2D2 + d_3D3 + E_2 \qquad \text{(Model 7.8)}$$

It is important to note that the coefficients q2, q3, and q4 are equal to the differences between their respective IQ level group means and the mean of IQ level 1, which was the IQ group not included in the model. Since these restrictions require the three coefficients for the IQ levels to be equal, the coefficient q1 can be substituted for coefficients q2, q3, and q4.

If the researchers use Model 7.8 as the Full Model, testing for the IQ main effect can be accomplished by placing the following restrictions on this model:

$$q_2 = q_3 = q_4 = 0 \quad \text{(3 restrictions)}$$

Placing this restriction on the Full Model (Model 7.8) produces the following Restricted Model:

$$Y1 = q_1Q1 + q_1Q2 + q_1Q3 + q_1Q4 + d_1D1 + d_2D2 + d_3D3 + E_2$$
$$\text{(Model 7.12)}$$

This model can be expressed as follows:

$$Y1 = q_1(Q1 + Q2 + Q3 + Q4) + d_1D1 + d_2D2 + d_3D3 + E_2 \quad \text{(Model 7.12a)}$$

The sum of vectors Q1, Q2, Q3, and Q4 produces the unit vector. The sum of vectors D1, D2, and D3 also produces the unit vector, so one of those D vectors is redundant with the unit vector, resulting in Model 7.11:

$$Y1 = a_0U + d_2D2 + d_3D3 + E_3 \qquad \text{(Model 7.12b)}$$

The Full Model (Model 7.8) contains six pieces of information and the Restricted Model (Model 7.12b) includes three pieces of information. Thus, the F value used to test the difference between the R^2 values of these two models would have a numerator degrees of freedom value and a denominator degrees of freedom value of (6 - 3) and (N - 6), respectively.

One can think of the IQ main-effect question as having information about IQ and Drugs and simply giving up information about IQ. Alternatively, one can ask: Does knowledge of IQ account for criterion variance, over and above knowledge of drugs? The drug main-effects question would be tested by forcing the following restrictions on the Full Model (Model 7.8):

$$d_1 = d_2 = d_3 \quad \text{(2 restrictions)}$$

The following Restricted Model is produced:

$$Y1 = a_0U + q_2Q2 + q_3Q3 + q_4Q4 + E_4 \qquad \text{(Model 7.13)}$$

The numerator and denominator degrees of freedom values for the F test used to statistically test the difference between the R^2 values of the Full Model (Model 7.8) and the Restricted Model (Model 7.13) are equal to 2 and $N - 6$, with N being equal to the total sample size.

INTERACTION BETWEEN A DICHOTOMOUS PREDICTOR VARIABLE AND A CONTINUOUS PREDICTOR VARIABLE

Ambiguous results regarding the effects of specific treatments are often found in various research domains. For example, in the field of physical conditioning and strength improvement, some studies report that isometric procedures are more effective in increasing strength than isotonic procedures, while others report the opposite outcome. Isometrics are exercises in which the individual tenses the muscle to be strengthened against a static object; isotonic exercises use the muscle in such activities as weight lifting and push-ups. [The following material dealing with isotonic and isometric conditioning is adapted from an article by Bender, Kelly, Pierson, and Kaplan (1968).] When studies consistently provide contradictory results, one may suspect that another, undetected variable is operating to cause one method to work best for one type of individual and another method may work best for another type of individual. (It is assumed that the researchers are competent and conscientious; thus, one hopes experimenter bias is eliminated and the data are good.) The subjects that have been previously researched may represent different populations with respect to this undetected variable.

As an example, assume Figure 7.5 represents the data regarding two such contradictory studies on improving arm strength. A review of Study 1 in Figure 7.5a reveals that isotonic exercises resulted in a higher mean posttest arm strength, while an examination of Study 2 in Figure 7.5b indicates that isometric exercises resulted in a higher mean posttest arm strength.

The search for the underlying variable that can clarify the issue would most likely start with an examination of the details of the contradictory studies to determine if they represent different populations. Suppose that in the literature review it was found that most of the studies that indicated superiority for isometric procedures used skilled athletes, who were relatively strong initially, and that most of the studies supporting the superiority of isotonic procedures used subjects with average pretest strength levels. Some studies may have reported the initial arm strength of the subjects.

A subsequent researcher could incorporate that information and modify the diagrams contained in Figure 7.5 to produce the results contained in the two diagrams contained in Figures 7.6. A review of Figure 7.6a indicates that isotonic procedures result in higher posttest arm strength for all initial arm strength values studied except a few at the higher end; whereas an examination of Figure 7.6b reveals that isometric procedures result in higher posttest arm strength for all initial arm strength values studied except a few at the lower end.

An examination of the hypothetical data represented in Figures 7.6a and 7.6b reveals several important considerations. The first thing one should note is that Study 1 (Figure 7.6a) included subjects with a range of initial arm strength values from 60 pounds to 85 pounds, whereas the pretest strength range for subjects in Study 2 (Figure 7.6b) was from 80 pounds to 120 pounds. It appears, then, that these two studies included subjects from different populations with respect to pretest arm strength.

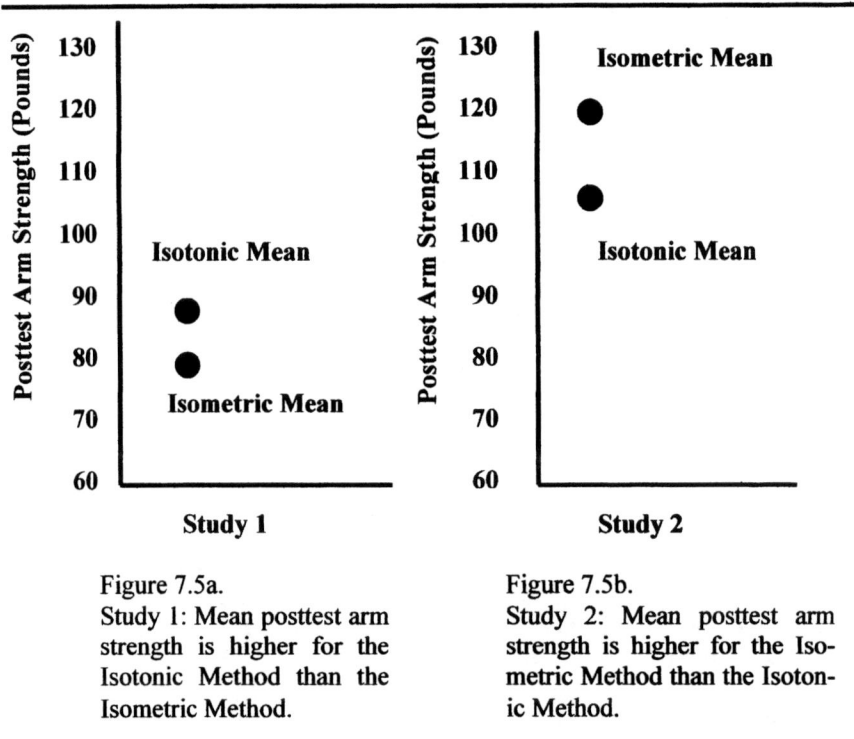

Figure 7.5a.
Study 1: Mean posttest arm strength is higher for the Isotonic Method than the Isometric Method.

Figure 7.5b.
Study 2: Mean posttest arm strength is higher for the Isometric Method than the Isotonic Method.

Figure 7.5.
Two studies of posttest arm strength of participants in the Isotonic and Isometric Methods.

In addition, one might note another consideration by examining the lines for the isotonic method in both Figures 7.6a and 7.6b. In Figure 7.6a, the posttest score is higher for the isotonic method at all levels of observed initial arm strength values, indicating that the subjects' strength levels improved. A review

of Figure 7.6b, the line for the isotonic method reveals that the difference between initial arm strength values and post-treatment arm strength values decreases until at an initial arm strength value of 120 pounds the post-treatment
arm strength value is also 120 pounds, indicating no improvement. It appears
that if a person is originally relatively weak, isotonic procedures are better than
isometric for improvement; and if a person is already relatively strong, isometric
procedures are more effective than isotonic. Suppose the researchers review the
literature and the observed trend is found repeatedly. Such a finding may lead
the researchers to conclude they have isolated the variable that explains the contradictory results reported in the literature.

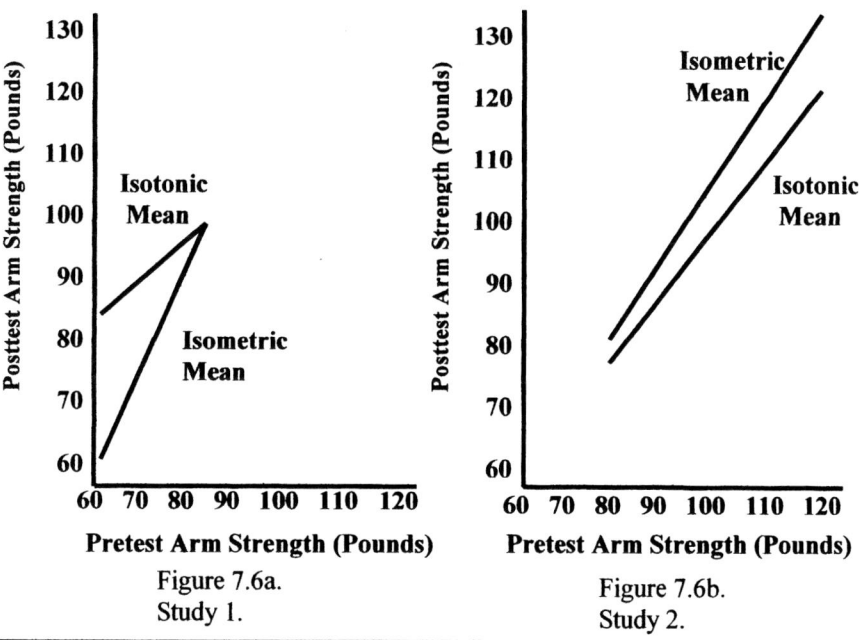

Figure 7.6.
Lines of best fit for data from Isotonic and Isometric Methods, when viewed
across levels of pretest arm strength—Study 1 and Study 2.

These researchers may pose the following directional Research Hypothesis:

Research Hypothesis: Isometric exercises are superior to isotonic exercises
in terms of post-treatment arm strength for men who initially score high on
arm strength, while isotonic exercises are superior to isometric exercises for
men who initially score low on arm strength.

The corresponding Statistical Hypothesis is as follows:

Statistical Hypothesis: The differential effects of isotonic and isometric methods on post-treatment arm strength, if any, are the same across the range
of initially observed arm strength values.

Since very weak men (below a pretest score of 80) are not included in Study
2, and very strong men (above 85) are not included in Study 1, a new study is
required to test the Research Hypothesis against the Statistical Hypothesis to
determine if the two studies differed only in pretest arm strength. Suppose the
researchers have the resources to select and measure 60 men who range in
strength from 60 pounds through 110 pounds. They want to select two groups,
one to be given isometric treatment and the other isotonic treatment. Random
selection is needed, but they also want to be sure that all levels of strength are
included in both treatments. The researchers may take strength intervals of 10
pounds (e.g., from 60 to 70 pounds) and randomly assign one half of the individuals within that range to each treatment group. They can do this for all 10-
pound intervals to be sure the two treatment groups represent the population.

Testing for Interaction

The Research Hypothesis requires that the data for both treatments be compared over the same range of interest on initial arm strength. To illustrate this
point, refer to Figure 7.7.

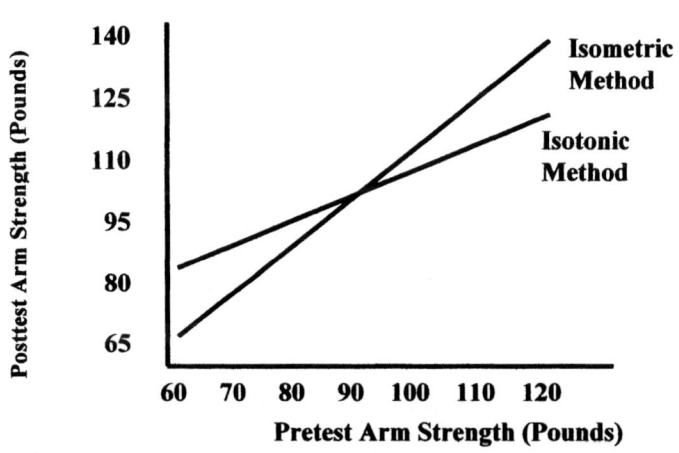

Figure 7.7.
The two lines of best fit— Isometric and Isotonic Methods—for the relationship
between pretest and posttest arm strength measures.

It would not be appropriate to estimate and examine a line of best fit for only the isotonic procedure or to estimate and examine a line of best fit for only the

isometric procedure. Two lines of best fit (one for each method) must be found simultaneously, and then their slopes must be compared. Figure 7.7 represents the two lines of best fit found simultaneously for the data obtained from the 60 subjects.

An inspection of Figure 7.7 suggests that the Research Hypothesis is supported because weak men improve more with isotonic exercise, and strong men improve more with isometric exercise. But how likely is it that the observed interaction is due to sampling error? To statistically test the Research Hypothesis stated by the researchers, they must first design a Full Model that estimates the two lines simultaneously and a Restricted Model that requires the slopes of the two lines to be equal. Next the researchers will need to test the differences between the R^2 values of these two models.

Full Model. One linear model must be constructed to reflect the two lines with differing slopes, as indicated in the Research Hypothesis. A Full Model that allows for different slopes of the two regression lines—one for the isometric group and another for the isotonic group—is constructed as follows (note that this model does not contain the unit vector):

$$Y1 = a_1T1 + a_2T2 + b_1L1 + b_2L2 + E_1 \qquad \text{(Model 7.14)}$$

where:

1. $Y1$ = post-treatment arm strength values;
2. $T1 = 1$ if the post-treatment score is from a man given isotonic exercises, 0 otherwise;
3. $L1$ = initial arm strength if the post-treatment score is from a man given isotonic exercise, 0 otherwise;
4. $T2 = 1$ if the post-treatment score is from a man given isometric exercise, 0 otherwise;
5. $L2$ = initial arm strength if the score on the criterion is from a man given isometric exercise, 0 otherwise; and
6. a_1, a_2, b_1, and b_2 are least squares weights calculated to minimize the sum of the squared elements in E_1.

The data file for Model 7.14 contains the vectors listed on the following page.

Subject identification numbers are at the left-hand side of the vector display for each of the 60 male subjects. To illustrate how to read this vector display, consider the two participants assigned identification (ID) numbers 1 and 31. The person with ID number 1 has a criterion score of 78 (the first element in vector Y1). This male's corresponding value in T1 is 1 because he participated in the isotonic exercise group. His initial arm strength score (L1) was 60 and his values for T2 and L2 are 0 because he was not in the isometric group. The person with ID number 31 has a Y1 score of 84; and he has a 0 in vectors T1 and L1, which means he was not given isotonic exercise. The value for person 31 in T2 is 1 and in L2 is 80, thus he was given isometric exercise and his pretest score was 80 pounds.

Individual $\quad$ Y1 $\quad = a_1 T1 \quad + \quad a_2 T2 \quad + \quad b_1 L1 \quad + \quad b_2 L2 \quad + E_1$

Individual	Y1	$= a_1$ T1	$+ a_2$ T2	$+ b_1$ L1	$+ b_2$ L2	$+$ E1
1	78	1	0	60	0	?
2	99	1	0	100	0	?
3	91	1	0	80	0	?
.	.	1	0	.	0	?
.	.	1	0	.	0	?
.	.	1	0	.	0	?
30	120	1	0	120	0	?
31	84	0	1	0	80	?
.	.	0	1	0	.	?
.	.	0	1	0	.	?
.	.	0	1	0	.	?
60	120	0	1	0	122	?

It is important to note that this Full Model (Model 7.14) has four linearly independent predictor vectors. These vectors allow for two separate lines because they allow for two slopes (b_1 and b_2) and for two Y intercepts (a_1 and a_2). To support the Research Hypothesis under study, the plot of the two lines of best fit must show that the isometric treatment yields higher post-treatment arm strength than the isotonic treatment for men with high initial arm strength; while the isotonic treatment yields higher post-treatment arm strength than the isometric treatment for men with low initial arm strength. The algebraically equivalent way to express this hypothesis is to expect that (a) the slope b_2 is greater than the slope b_1 and (b) these lines will cross within the relevant range of the initial arm strength values, that is, a disordinal interaction effect will exist.

Restricted Model. Solving for weights a_1, a_2, b_1, and b_2, one obtains an R_F^2 value that specifies the proportion of variance in the criterion variable accounted for by the four predictor pieces of information. The Research Hypothesis stipulated that b_2 is greater than b_1. This condition can be statistically tested by constructing a Restricted Model that forces b_1 to equal b_2. Thus the restriction placed on the Full Model (Model 7.14) is as follows:

$b_1 = b_2 \qquad$ (1 restriction)

This restriction forces the two regression lines—one estimated for each of the treatments—to have the same slope. It is important to note, however, that the Research Hypothesis does not require the Y intercept values (i.e., a_1 and a_2) be equal.

Since b_1 and b_2 are required to be equal, b_1 can be substituted for b_2 in the Full Model (Model 7.14) producing the following Restricted Model (Model 7.15):

$Y1 = a_1 T1 + a_2 T2 + b_1 L1 + b_1 L2 + E_2 \qquad\qquad$ (Model 7.15)

Once again, note that this Restricted Model does not contain the unit vector.

Collecting like terms in this Restricted Model results in the following Restricted Model:

$$Y1 = a_1 T1 + a_2 T2 + b_1(L1 + L2) + E_2 \qquad \text{(Model 7.15a)}$$

When variables L1 and L2 are summed the result is a vector, which is labeled LALL (which refers to LEVELS ALL) that contains the initial arm strength score for all of the 60 males. Substituting LALL for (L1 + L2) results in the following Restricted Model:

$$Y1 = a_1 T1 + a_2 T2 + b_3 LALL + E_2 \qquad \text{(Model 7.15b)}$$

As previously stated, this Restricted Model (model 7.15b) forces the slopes of the two regression lines—one for each of the treatments—to be equal. It is important to note, however, that Model 7.15b allows the Y intercepts of the two regression lines to differ (i.e., a_1 and a_2 are not forced to be equal). Thus this Restricted Model (Model 7.15b) "gives up" only one piece of information from the Full Model (Model 7.14), which is, the knowledge of which treatment group is associated with given initial arm strength values.

Statistical test of different slope values. If there is a difference in slopes, then the sum of the squared values in the error vector for the Restricted Model (E_2) will be larger than those values in the Full Model (E_1). Thus the R^2 value of the Restricted Model (R_R^2), which used only three pieces of information, will be smaller than the R^2 value of the Full Model (R_F^2), which contained four pieces of information. The F test will indicate how likely it is that the difference between R_F^2 (using Model 7.14) and R_R^2 (using Model 7.15b) is due to sampling error. Again, the same general F test, which was presented in Equation 4.1 of chapter 4, can be used to answer this Research Hypothesis. Once again, the F test formula is as follows:

$$F\,(df_n,\,df_d) = \frac{(R_F^2 - R_R^2)/(df_n)}{(1 - R_F^2)/(df_d)}$$

$$\text{(Equation 4.1)}$$

where:

1. R_F^2 = the R^2 value of the Full Model (Model 7.14).
2. R_R^2 = the R^2 value of the Restricted Model (Model 7.15b).
3. df_n = the number of linearly independent vectors in the Full Model minus the number of linearly independent vectors in the Restricted Model.
4. df_d = the sample size minus the number of linearly independent vectors in the Full Model.

It should be recalled that $((R_F^2 - R_R^2)/df_n)$ is a proportional estimate of the population variance and in this application of the general F test, rather than being an among-group estimate, it is an estimate of the population variance using interaction. Likewise, $((1 - R_F^2)/df_d)$ is the best proportional estimate of the population variance using all the information in the Full Model. Since the number of linearly independent vectors in the Full Model (Model 7.14) and Restricted Model (Model 7.15b) are 4 and 3, respectively, the df_n value is equal to 1 (i.e., 4 - 3). In addition, since the analysis involves 60 subjects, the df_n value is equal to 56 (i.e., 60 - 4). Substituting these degrees of freedom values along with the R^2 values of the Full Model (Model 7.14) and the Restricted Model (Model 7.15b) into Equation 4.1 will produce the appropriate F test value. Computer software, such as SPSS for Windows and Microsoft Excel can be used to calculate this F value and its corresponding probability value.

Remember that this analysis involved a directional Research Hypothesis. Thus, if this F test indicates that the interaction effect is statistically significant, the researchers must review the nature of this interaction effect. The Research Hypothesis would be supported *only* if (a) b_2 is indeed greater than b_1 for the coefficients generated for the Full Model (Model 7.14) and (b) the interaction effect is disordinal within the relevant range of initial arm strength values. When b_2 exceeds b_1, the hypothesized type of the interaction exists (i.e., disordinal); however, before the Research Hypothesis can be accepted, the lines of best fit must cross within the range of observed initial arm strength values. Thus, the plot of the lines of best fit as illustrated in Figure 7.7 must be inspected closely before the Research Hypothesis can be supported.

These lines can be plotted by hand by relying on the weighting coefficients. Two points on each regression line need to be calculated in order for the line to be drawn. To calculate these two points for each regression line we need to select two values for the continuous predictor variables (i.e., L1 and L2—the variables containing the initial arm strength values) and the dichotomous treatment variables (i.e., T1 and T2). We will select the two arm strength values at the minimum and maximum values found in variables L1 and L2, which for this example we will assume are 60 and 122, respectively. The values for T1 and T2 must also be set to represent the appropriate group for a given regression line. Thus the values corresponding to the variables contained in the Full Model (Model 7.14) are as follows for the two regression lines:

1. Isometric Group Regression Line:
    a. For Point 1 T1 = 1, T2 = 0, L1 = 60, and L2 = 0.
    b. For Point 2 T1 = 1, T2 = 0, L1 = 122, and L2 = 0.
2. Isotonic Group Regression Line:
    a. For Point 1 T1 = 0, T2 = 1, L1 = 0, and L2 = 60.
    b. For Point 2 T1 = 0, T2 = 1, L1 = 0, and L2 = 122.

These values are substituted into the Full Model (Model 7.14) as follows:

1. Isometric Group Regression Line:
    a. $Y1_{T1Min} = a_1(1) + a_2(0) + b_1(60) + b_2(0)$
    b. $Y1_{T1Max} = a_1(1) + a_2(0) + b_1(122) + b_2(0)$
2. Isotonic Group Regression Line:

a. $Y1_{T2Min} = a_1(0) + a_2(1) + b_1(0) + b_2(60)$
b. $Y1_{T2Max} = a_1(0) + a_2(1) + b_1(0) + b_2(122)$

Once the coefficients (i.e., a_1, b_1, a_2, and b_2) are estimated for the Full Model (Model 7.14), the two values for each of the regression lines can be calculated.

The Y axis for the interaction diagram is scaled for the post-treatment arm strength values, and the X axis is scaled for the initial arm strength values. The regression line for the isometric group will be formed by plotting the $Y1_{T1Min}$ and $Y1_{T1Max}$ values, and connecting those two points with a solid line. In like manner, the regression line for the isotonic group will be formed by plotting the $Y1_{T2Min}$ and $Y1_{T2Max}$ values, and connecting those two points with a solid line. (Note that a solid line is used to connect the two points when the interaction effect involves a continuous variable, but when the interaction effect involves two dichotomous variables, such as the interactions presented in the previous sections, a dotted line is used to connect the two points.)

Once the two regression lines are plotted on the interaction diagram, the researchers need to determine if the regression lines cross in the relevant range of initial arm strength values. The regression lines do cross in the relevant range in Figure 7.7, but they do not cross in the relevant range in Figure 7.8.

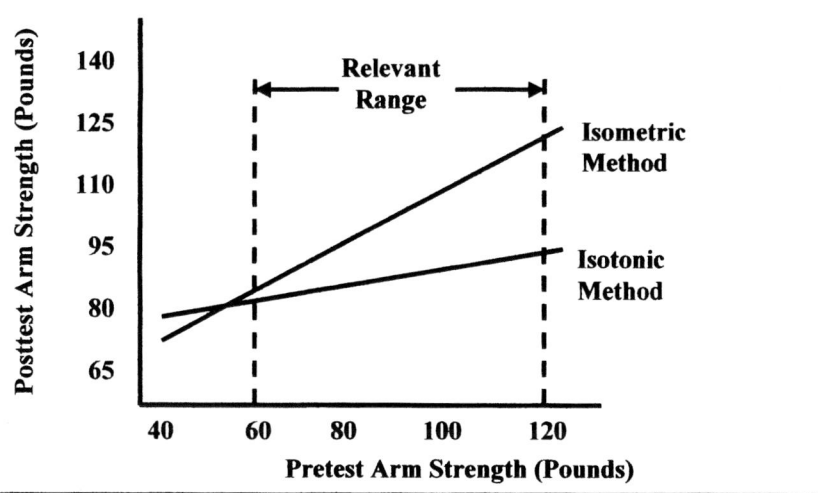

Figure 7.8.
A treatment-by-level interaction—The lines cross beyond the range of interest.

The importance of inspecting the interaction diagram. Inspection of the lines of best fit is crucial because several data configurations can generate a large F value. Figure 7.8 is one such situation. The Research Hypothesis indicated that, among other things, initially weak men will benefit more from isotonic exercise. Yet an inspection of Figure 7.8 reveals that the regression lines cross at 80 pounds on the initial arm strength scale. The weakest men in the study had initial arm strength values of 90 pounds. If results such as in Figure 7.8 occur, then the Research Hypothesis is not supported. An examination of the results in Figure 7.8 indicate that isometric is best for all initial arm strength levels sampled; however the stronger the man, the relatively more effective is the isometric treatment. Such a finding provides data that leaves the original problem unresolved; the relevant variable that accounts for the previously reported contradictory findings would not yet have been found.

If the data are accurately reflected in Figure 7.7 (the regression lines crossing in the range of interest), the Research Hypothesis can be supported, and the data are worth reporting because it looks like at least one source of "contamination" contributing to contradictory results has been found. If the researcher had done a good job in analyzing past research and the Research Hypothesis was based upon past findings, it would be surprising if the results did not conform to the situation shown in Figure 7.7.

There is a procedure available (the Johnson-Neyman technique) that allows one to determine the regions where the differences between groups are significant. Fraas and Newman (1997) provide an SPSS syntax program designed to calculate Johnson-Neyman confidence limits. The approach taken in this text is to pay attention to the whole range. We acknowledge that, with respect to Figure 7.7, the isometric procedure may not be significantly better than the isotonic procedure for some subjects with pretest arm strength greater than 80. But if the two procedures are equally costly and one has the option of giving either treatment, then one would assign the isometric treatment to subjects who have a pretest arm strength greater than 80, whereas the isotonic treatment would be assigned to those whose pretest arm strength was less than 80.

Summary of the Procedure Used to Test for an Interaction Effect

Contradictory results were reported in the literature regarding the relative effectiveness of two treatment methods. The examination of the literature led to the expectation that isometric treatment would yield a steeper slope across initial arm strength levels when compared with the slope associated with men given isotonic exercise. Furthermore, the researchers were led to expect that the starting point (the Y intercept) would be higher for the isotonic treatment because previous data indicated that weaker men performed better under the isotonic exercise treatment. The researchers could model these expectations through the use of variables that allowed for an interaction effect between the dichotomous treatment variable and the continuous variable that included initial arm strength. The steps used to investigate this interaction effect were as follows:

1. A Research Hypothesis was stated that expressed the expectation that two straight lines with differing slopes would be needed to reflect the data. From this Research Hypothesis, a Full Model was constructed that allowed for two Y intercepts and two slopes.

2. A Statistical Hypothesis was stated that expressed the expectation that two straight lines with a single common slope would reflect the data. From the Statistical Hypothesis a Restricted Model was constructed that allowed for two Y intercepts, but only one common slope.
3. The researchers used an F test to determine whether the interaction effect was statistically significant. If the F test was significant, the researchers would use the coefficients generated by the Full Model to plot the interaction effect. If they judged that the statistically significant interaction exhibited the characteristics stated in the Research Hypothesis, they would reject the Statistical Hypothesis and suggest that the data support the Research Hypothesis.

GENERAL RESEARCH HYPOTHESIS 7.2

This section presents a procedure that researchers can follow when addressing a research question that involves an interaction effect between dichotomous variable and continuous variable. Specifically, this presentation focuses on the interaction between two groups, which are identified as Method A (G_A) and Method B (G_B), and a continuous variable that represents pretest scores (Pretest). The criterion variable consists of the participants' posttest scores (Posttest).

The research question posed by the researchers is: Does the relative superiority of Method A over Method B increase as the pretest scores increase? Since this is a directional research question, the General Research Hypothesis 7.2 (GRH 7.2) is stated as a directional hypothesis as follows:

Directional GRH 7.2: For a given population, as the pretest scores increase, the relative superiority of Method A over Method B on the posttest scores increases.

The corresponding General Statistical Hypothesis (GSH 7.2) is stated as follows:

Directional GSH 7.2: For a given population, as the pretest scores increase, the relative superiority of Method A over Method B on the posttest scores does *not* increase.

Using Microsoft Excel to test GSH 7.2. In this section we will assume the researchers will use Microsoft Excel to statistically test GSH 7.2. In the following section we will discuss how the analysis changes when SPSS for Windows is used.

Before the Full Model, which reflects the condition stated in GRH 7.2, can be constructed, two variables must be generated to allow the model to include the stated interaction effect. One such variable, which is labeled G_APretest, is the product of the G_A and Pretest variables. The other variable, which is labeled

G_BPretest, is the product of the G_B and Pretest variables. Once these variables are generated, the Full Model is constructed as follows:

$$\text{Posttest} = a_0U + a_1G_A + a_2G_A\text{Pretest} + a_3G_B\text{Pretest} + E_1 \qquad \text{(Model 7.16)}$$

Two characteristics of this Full Model should be noted. First, it contains the unit vector, which the models presented in the previous section did not include. Second, since this model contains the unit vector, only one of the group dichotomous variables is included (G_A) in the Full Model. Remember, with the inclusion of the unit vector and variable G_A, the information provided by variable G_B would be redundant.

The restriction required by GSH 7.2 is that the slopes of the two regression lines must be equal. Thus for Model 7.16 this requirement is stated as follows:

$$a_2 = a_3 \quad \text{(1 restriction)}$$

Replacing coefficient a_3 with coefficient a_2 in the Full Model (Model 7.16) results in the following Restricted Model:

$$\text{Posttest} = a_0U + a_1G_A + a_2G_A\text{Pretest} + a_2G_B\text{Pretest} + E_2 \qquad \text{(Model 7.17)}$$

Collecting like terms produces the following Restricted Model:

$$\text{Posttest} = a_0U + a_1G_A + a_2(G_A\text{Pretest} + G_B\text{Pretest}) + E_2 \qquad \text{(Model 7.17a)}$$

Since adding the variables G_APretest and G_BPretest produces the Pretest variable the Restricted Model used in the analysis is as follows:

$$\text{Posttest} = a_0U + a_1G_A + a_4\text{Pretest} + E_2 \qquad \text{(Model 7.17b)}$$

An examination of the Full Model (Model 7.16) and the Restricted Model (Model 7.17b) reveals that the Full Model contains four pieces of information (m1 = 4) and the Restricted Model contains one less piece of information (m2 = 3). Remember, only when the difference between the values for m1 and m2 is equal to 1, can a directional research question be stated.

To statistically test GSH 7.2 an F test of the differences between the R^2 values of the Full and Restricted Models is conducted using Equation 4.1, which was reprinted in a previous section of this chapter. In this F test the R_F^2 and the R_R^2 values are the R^2 value of the Full and Restricted Models, respectively. The df_n value is equal to m1 minus m2, and df_d is equal to N minus m1.

Recall that GRH 7.2 is a directional Research Hypothesis. Thus, in addition to conducting the F test, the researchers must determine whether certain stipulations in GRH 7.2 regarding the interaction effect have been met in the Full Model (Model 7.16). Those stipulations regarding the interaction effect require that Method A be superior to Method B, and the interaction effect must be ordinal (i.e., the regression lines do not cross) within the relevant range of the pretest scores.

If Method A is superior to Method B in the interaction effect, the slope of Method A must exceed the slope of Method B. In Model 7.16 the coefficients a_2 and a_3 are the slopes of Method A and Method B, respectively. Thus, when Model 7.16 is used as the Full Model a_2 must exceed a_3. To determine if the interaction effect is ordinal within the relevant range of the pretest scores, the researchers need to plot the interaction effect by completing the following steps:

1. Select two pretest values. The selected values should be the minimum and maximum pretest values in the sample data.

2. The first point used to form the regression line for Method A is calculated by substituting into the model (a) the value of 1 for U, (b) the values of 1 and 0 for the G_A and G_B variables, respectively, (c) the minimum pretest value for the G_APretest variable, and (d) the value of 0 for the G_BPretest variable. Next multiply these values by their respective coefficient values and add these products. To calculate the second point, follow the same procedure except substitute the maximum pretest value for G_APretest.

3. Calculate the first point used to form the regression line for Method B by substituting into the model (a) the value of 1 for U, (b) the values of 0 and 1 for the G_A and G_B variables, respectively, (c) the value of 0 for the G_APretest variable, and (d) the minimum pretest value for the G_BPretest variable. Next multiply these values by their respective coefficient values and add these products. To calculate the second point, follow the same procedure except substitute the maximum pretest value for G_BPretest.

4. Construct an interaction diagram with the Y axis scaled for the posttest scores and the X axis scaled for the pretest scores. Plot the two points for Method A and connect the points with a dotted line. Do the same for the two points for Method B.

The researchers would examine this diagram of the interaction effect to determine whether the interaction effect was indeed ordinal. Assuming the F test of the interaction effect was statistically significant, the slope of Method A exceeded the slope of Method B, and the interaction effect was ordinal, the researchers would reject GSH 7.2 and state that the data support GRH 7.2. That is, their analysis indicates that for a given population, as the pretest scores increase, the relative superiority of Method A over Method B on the posttest scores increases.

Using SPSS for Windows to test GSH 7.2. If researchers choose to use SPSS for Windows to statistically test GSH 7.2, a Full Model equivalent to Model 7.16 needs to be constructed. This equivalent Full Model is as follows:

$$\text{Posttest} = a_0 U + a_1 G_A + a_5 G_A \text{Pretest} + a_6 \text{Pretest} + E_1 \qquad \text{(Model 7.16a)}$$

This Full Model (Model 7.16a), which does not require the variable G_BPretest be generated, would produce the same R^2 value as Model 7.16. The question one might raise is: Why construct this equivalent Model? When SPSS for Windows is used to analyze the difference between the Full Model and the Restricted Model the equivalent Full Model (Model 7.16a) allows the "R square change" function to be used; where the original Full Model (Model 7.16) does not.

Although Models 7.16 and 7.16a are equivalent, note that the subscript for the G_APretest variable was changed from a_2 in Model 7.16 to a_5 in Model 7.16a. The reason for this change is that in Model 7.16 the a_2 value is equal to the slope of the regression line for Method A; while the value for a_5 in Model 7.16a is equal to the difference between the slopes of the regression lines for Method A and Method B. In addition, in Model 7.16a the coefficient a_6 for the Pretest variable is equal to the slope of the regression line for Method B.

To require the slopes of the regression lines for Methods A and B to be equal, the following restriction is placed on the Full Model (Model 7.16a):

$a_5 = 0$ (1 restriction)

Placing this restriction on the Full Model (Model 7.16a) produces the following Restricted Model, which is the same one constructed for the analysis that used Microsoft Excel:

$$\text{Posttest} = a_0 U + a_1 G_A + a_4 \text{Pretest} + E_2 \qquad \text{(Model 7.17b)}$$

Note that the coefficient for the Pretest variable is a_4 because with the deletion of G_APretest variable, the coefficient will now equal the estimated common slope of Methods A and B.

As before, the researchers would statistically test the differences between the R^2 values of the Full Model (Model 7.16a) and the Restricted Model (Model 7.17b) with an F test. As previously discussed in this chapter, when researchers use SPSS for Windows to conduct an F test between a Full Model and a Restricted Model and the Restricted Model is not the Null Model, the F test and its corresponding probability value will be produced by utilizing the Statistics menu in the Linear Regression menu (see the section entitled "Applied Research Hypothesis 7.1").

If the interaction effect tested by this F test is statistically significant the researchers would plot the interaction effect using the same steps discussed in the previous section. However, when Model 7.16a is used as the Full Model, the steps are completed as follows:

1. Select two pretest values. The selected values should be the minimum and maximum pretest values in the sample data.
2. The first point used to form the regression line for Method A is calculated by substituting into the model (a) a value of 1 for both U and the G_A variable and (b) the minimum pretest value for the G_APretest and

Pretest variables. Next multiply these values by their respective coefficient values and add these products. To calculate the second point, follow the same procedure except substitute the maximum pretest value for G_APretest and Pretest.

3. The first point used to form the regression line for Method B is calculated by substituting into the model (a) a value of 1 for U, (b) a value of 0 for the G_A variable, (b) a value of 0 for the G_APretest variable, and (c) the minimum pretest value for the Pretest variable. Next multiply these values by their respective coefficient values and add these products. To calculate the second point for Method B, follow the same procedure except substitute the maximum pretest value for the Pretest variable.

4. Construct an interaction diagram with the Y axis scaled for the posttest scores and the X axis scaled for the pretest scores. Plot the two points for Method A and connect the points with a solid line. Do the same for the two points for Method B.

As previously discussed, the researchers would examine this diagram of the interaction effect to determine whether the interaction effect was indeed ordinal. Assuming the F test of the interaction effect was statistically significant, the slope of Method A exceeded the slope of Method B, and the interaction effect was ordinal, the researchers would reject GSH 7.2

APPLIED RESEARCH HYPOTHESIS 7.2

The analysis of Applied Research Hypothesis 7.2 (ARH 7.2) provides an empirical illustration of the steps researchers can follow to statistically test for an interaction effect between dichotomous and continuous predictor variables. In this section, we are assuming that the researchers are interested in the same type of interaction effect research question as stated in the previous section. The specific research question is stated as follows:

Does the relative superiority of Treatment A to Treatment B, with respect to post-treatment self-confidence, increase as pre-treatment confidence levels increase?

To address this research question, the researchers collected pre-treatment (X8) and post-treatment (X9) self-confidence scores. Variable X8 was a continuous predictor variable, while variable X9 served as the continuous criterion variable. Each of the study's participants was randomly placed into one of two methods—Method A or Method B. Variable X10 was a dichotomous variable in which a value of 1 identified a participant as a member of Method A and a value of 0 indicated the participant was not a member of Method A but rather a member of Method B. In a similar fashion, variable X11 was a dichotomous variable in which a value of 1 identified a participant as a member of Method B and a value of 0 indicated the participant was not a member of Method B but rather a member of Method A.

Two variables were generated to allow the interaction effect between treatment and pre-treatment self-confidence scores to be estimated and statistically tested. Variable X26 is the product of X8 (pre-treatment self-confidence scores) and X10 (Method A); while X27 is the product of X8 (pre-treatment self-confidence scores) and X11 (Method B).

Since a directional research question was posed, ARH 7.2 was stated as a directional Research Hypothesis as follows:

Directional ARH 7.2: For a given population, as the pre-treatment self-confidence scores increase, the relative superiority of Method A over Method B increases with respect to the post-treatment self-confidence scores.

The corresponding Applied Statistical Hypothesis (ASH 7.2) is stated as follows:

Directional ASH 7.2: For a given population, as the pre-treatment self-confidence scores increase, the relative superiority of Method A over Method B does *not* increase with respect to the post-treatment self-confidence scores.

Full and Restricted Models

As discussed in the General Research Hypothesis 7.2 section, the Full Model that reflects the condition stipulated in ARH 7.2 could be constructed in two equivalent forms as follows:

$$X9 = a_0U + a_{10}X10 + a_{26}X26 + a_{27}X27 + E_1 \qquad \text{(Model 7.18)}$$

$$X9 = a_0U + a_{10}X10 + a_8X8 + a_{26}X26 + E_1 \qquad \text{(Model 7.18a)}$$

If the researchers use SPSS for Windows to analyze the Full Model and its corresponding restricted model, Model 7.18a allows for the researchers to use the "R squared change" command. When using Microsoft Excel, the researchers will not find any difference between the utilization of either model with respect to ease of analysis except they will not have to calculate variable X27 if Model 7.18a is used as the Full Model. In this illustration, we will assume the researchers chose to use Model 17.8a as the Full Model.

Since ASH 7.2 (the Statistical Hypothesis) requires the slopes of the two regression lines—one for Method A and one for Method B—to be equal and Model 7.18a is used as the Full Model, the restriction placed on this Full Model is as follows:

$a_{26} = 0$ (1 restriction)

Placing this restriction into the Full Model (Model 7.18a) will produce the following Restricted Model:

$$X9 = a_0U + a_{10}X10 + a_8X8 + E_2 \qquad \text{(Model 7.19)}$$

Analysis of ARH 7.2 with Microsoft Excel

The data contained in the GLM DATA EXCEL FORMAT file can be accessed from the internet site (see Appendix A). After this file has been accessed with the Excel computer software, the X26 variable must be generated. (Note if Model 7.18 is used, variable X27 would also need to be generated.) As noted in the previous section, multiplying X10 and X8 forms the variable X26.

Before the Regression menu is accessed, the predictor variables contained in the Full Model (Model 7.18a) must be placed next to each other in the data file. Thus the researchers would cut and paste the X8, X10, and X26 variables in the data set so that they are adjacent to each other in the data file. Once these predictor variables are placed next to each other in the data file and the commands needed to access the Regression menu (see chapter 4) have been completed, the Regression menu presented in Exhibit 7.10 will be displayed.

Exhibit 7.10.
Microsoft Excel Regression menu—Full Model (Model 7.18a) for ARH 7.2.

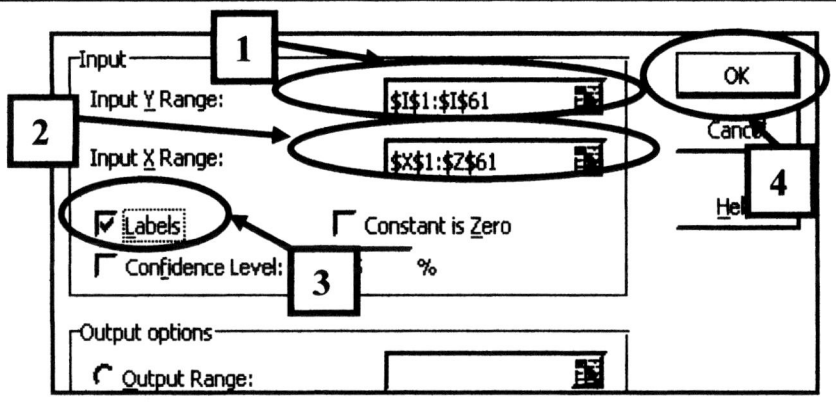

To obtain the output for the Full Model (Model 7.18a) used to statistically test ASH 7.2, which is the statistical hypothesis corresponding to ARH 7.2, the following four steps are completed in the Regression menu (see Exhibit 7.10):

1. Click on the box next to "Input $\underline{Y}$ Range" (see Oval 1). Next, click and drag on the cells located in the column in which X9 is located—including the name (label) of the variable.
2. Click on the box next to "Input $\underline{X}$ Range" (see Oval 2). Next, click and drag on the cells containing the predictor variables X8, X10, and X26—including their names (labels). Remember these variables must be located next to each other in the data file. In this example variables X8, X10, and X26 were placed in columns X, Y, and Z.
3. Click on the box in front of "Labels" (see Oval 3).
4. Click on the "OK" button (see Oval 4).

After these four steps are completed, the output window for the Full Model (Model 7.18a) will appear as displayed in Exhibit 7.11.

Exhibit 7.11.
Microsoft Excel regression output—Full Model (Model 7.18a) for ARH 7.2.

Regression Statistics	
Multiple R	0.9909
R Square	0.9820
Adjusted R Square	0.9810
Standard Error	2.1188
Observations	60

ANOVA

	df	SS	MS	F	Significance F
Regression	3	13691.3315	4563.7772	1016.5856	0.0000
Residual	56	251.4019	4.4893		
Total	59	13942.7333			

	Coefficients	Standard Error	t Stat	P-value	Lower 95%
Intercept	38.3032	2.8460	13	0.0000	32.6021
X8	0.6655	0.0266	25.0264	0.0000	0.6122
X10	-72.6810	4.0606	-17.8992	0.0000	-80.8153
X26	0.6943	0.0384	18.0638	0.0000	0.6173

When reviewing the output for the Full Model (Model 7.18a) it is important to remember that the Restricted Model (Model 7.19) contains two predictor variables. Thus, the Restricted Model is not the Null Model. Since the Restricted Model is not the Null Model, the F test listed in the portion of the output window entitled "ANOVA Table" is not the appropriate F test. The appropriate F test must be conducted with the use of Equation 4.1, which was previously presented in chapter 4 and repeated in this chapter.

Three values contained in the Full Model output window displayed in Exhibit 7.11 are used in the calculation of the F test. The first piece of information is the R^2 value of .982 (see Oval 1). The second and third key values are the regression degrees of freedom value of 3 and the residual degrees of freedom value of 56 (see Oval 2). The two other pieces of information must be obtained from the Restricted Model output.

The output for the Restricted Model (Model 7.19) is obtained by, first, accessing the Regression menu once again (see Exhibit 7.12). (The researchers need to be sure to have the data sheet displayed before they access the Regression menu.) The output for the Restricted Model is obtained by completing the following steps:

1. Click on the box next to "Input $\underline{Y}$ Range" (see Oval 1). Next, click and drag on the cells in the column in which variable X9 is located.
2. Click on the box next to "Input $\underline{X}$ Range" (see Oval 2). Next, click and drag on the cells in the columns containing the variables X8 and X10. Remember, these variables must be next to each other in the data file.
3. Click on the box in front of "Labels" (see Oval 3).
4. Click on the "OK" button (see Oval 4).

After these four steps are completed, the output window for the Restricted Model (Model 7.19) will be produced as displayed in Exhibit 7.13.

Exhibit 10.12
Microsoft Excel Regression menu—Restricted Model (Model 4.24c) for ASH 7.2.

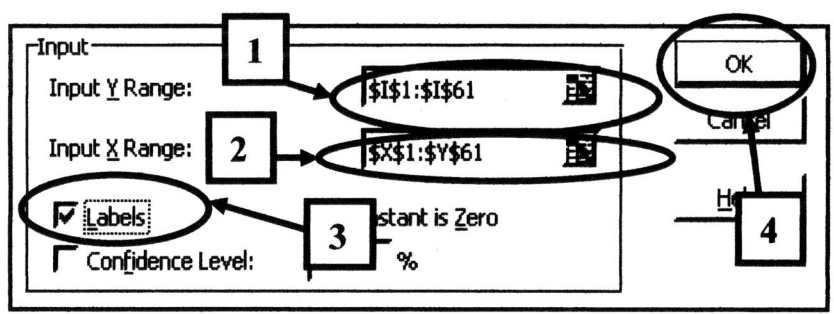

The two values obtained from the Restricted Model (Model 7.19), which is displayed in Exhibit 7.13, are (a) the R^2 value of .877 (see Oval 1) and (b) the regression degrees of freedom value of 2 (see Oval 2). Remember, the Microsoft Excel output windows for the Full and Restricted Models do not provide the appropriate F test for the difference between the R^2 values of these two models designed for the analysis of ARH 7.2.

Exhibit 7.13.
Microsoft Excel regression output—Restricted Model (Model 7.19) for ASH 7.2

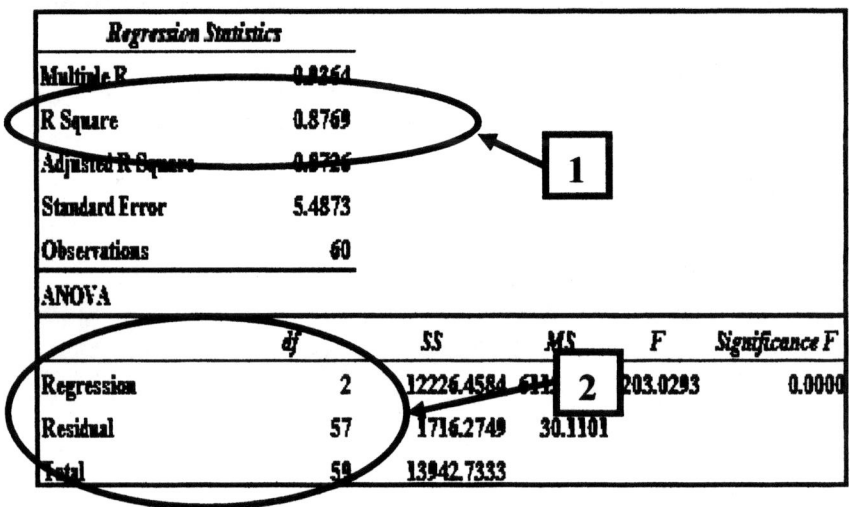

The appropriate F test value can be calculated with the use of the Microsoft Excel file entitled "F Test Calculation," which is contained in the internet site for the text (see Appendix A). The commands required to generate this program are listed in Appendix D. After the "F Test Calculation" program is accessed, the following values are entered into the file (see Oval 1 in Exhibit 7.14):

1. The R^2 value of the Full Model (.982) is entered into row 1 of column B.
2. The R^2 value of the Restricted Model (.877) is entered into row 2 of column B.
3. The regression degrees of freedom value for the Full Model (3) is entered into row 3 of column B.
4. The regression degrees of freedom value for the Restricted Model (2) is entered into row 4 of column B.
5. The residual degrees of freedom value for the Full Model (56) is entered into row 5 of column B.

Once these values are entered into the F test program, the F test value (326.67) and its corresponding probability value (.000) will be displayed (see Oval 2 of Exhibit 7.14).

Exhibit 7.14.
Microsoft Excel *F*-test calculation of ASH 4.3.

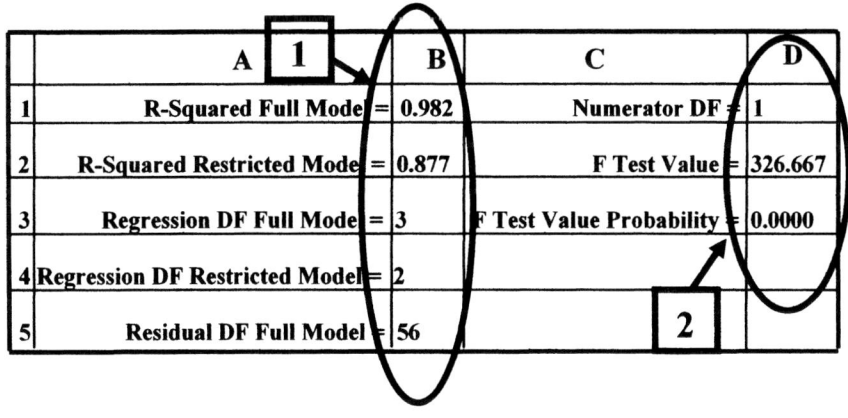

A 1	B	C	D
1	R-Squared Full Model =	0.982	Numerator DF = 1	
2	R-Squared Restricted Model =	0.877	F Test Value = 326.667	
3	Regression DF Full Model =	3	F Test Value Probability = 0.0000	
4	Regression DF Restricted Model =	2		
5	Residual DF Full Model =	56	2	

The researchers would express the .000 probability value as *p* < .001. Since ARH 7.2 is a directional hypothesis, this probability value would be divided by 2, which would expressed as *p* / 2 < .001. Since the one-tailed probability value (*p* / 2 < .001) is less than the established alpha level of .05, the Statistical Hypothesis (ASH 7.2) is rejected.

Now the researchers need to verify if this statistically significant interaction reflects the type of interaction effect ARH 7.2 requires. The type of interaction effect contained in ARH 7.2 requires two criteria be met. The first criterion is the value of the coefficient for variable X26 in the Full Model (Model 7.18), which is equal to the slope of Treatment A minus the slope of Treatment B, must be positive. Since this coefficient value for variable X26 is 0.694 (see Oval 3 in Exhibit 7.11), this criterion is met.

The second criterion requires that the interaction effect be ordinal within the relevant range of the pre-treatment scores. To determine if this criterion is met, the researchers will plot the two regression lines—one line for Treatment A and another line for Treatment B. To complete this task the researchers will complete the following steps:

1. Select two pre-treatment self-confidence scores. The selected values, which are 82 and 130, represent the minimum and maximum values, respectively.

2. The one point used to plot the regression line for Treatment A requires that values for the unit vector and the variables be set as follows: (a) U = 1, (b) X8 = 82, (c) X10 = 1, and (d) X26 = 82 (i.e., 1 multiplied by 82). For the other point the values are set as follows: (a) U = 1, (b) X8 = 130, (c) X10 = 1, and (d) X26 = 130 (i.e., 1 multiplied by 130).

3. The one point used to plot the regression line for Treatment B requires that values for the unit vector and the variables be set as follows: (a) U = 1, (b) X8 = 82, (c) X10 = 0, and (d) X26 = 0 (i.e., 0 multiplied by 82). For the other point the values are set as follows: (a) U = 1, (b) X8 = 130, (c) X10 = 0, and (d) X26 = 0 (i.e., 0 multiplied by 130).
4. The values identified in Steps 2 and 3 are multiplied by their respective regression coefficients generated for the Full Model (Model 7.18). A review of the coefficients listed in the output for the Full Model (see Oval 3 in Exhibit 7.12) reveals the values are: (a) a_0 = 38.303, (b) a_{10} = -72.681, (c) a_8 =.665, and (d) a_{26} = .694. Thus the post-treatment self-confidence values are calculated as follows:

Treatment A

$$a_0 \ U \ + \ a_{10} * X10 \ + \ a_8 * X8 \ + \ a_{26} * X26 \ = \ X9$$
$$38.303(1) + (-72.681)(1) + 0.665(82) \ + 0.694(82) \ = \ 77.1$$
$$38.303(1) + (-72.681)(1) + 0.665(130) + 0.694(130) = \ 142.3$$

Treatment B

$$a_0 \ U \ + \ a_{10} * X10 \ + \ a_8 * X8 \ + \ a_{26} * X26 \ = \ X9$$
$$38.303(1) + (-72.681)(0) + \ 0.665(82) \ + \ 0.694(0) \ = \ 92.8$$
$$38.303(1) + (-72.681)(0) + \ 0.665(130) + \ 0.694(0) \ = 124.8$$

5. The diagram for this interaction effect is constructed as displayed in Figure 7.9. The Y axis and X axis are scaled for the post-treatment self-confidence scores and the pre-treatment self-confidence scores, respectively. The two points for Treatment A (77.1 and 142.3) are plotted and connected with a solid line. In a similar manner, the two points for Treatment B (92.8 and 124.8) are plotted and connected with a solid line.

A review of this interaction effect reveals that this interaction effect is disordinal, that is, the lines cross in the relevant range of pre-treatment self-confidence scores. Thus, ARH 7.2 is not supported. It may be important, however, for the researchers to determine the point of intersection of the two lines in the plot of the interaction effect. This type of information may prove useful in the designing of future studies.

The value for the pre-treatment self-confidence score (i.e., the value for X8) located at the point of intersection of the two regression lines can be calculated as follows:

1. The researchers must establish for each regression line the values for U and the variable X10. The value for U is 1 in both regression lines. The value of X10 is 1 (X10 is equal to 1 for Treatment A) for the Treatment A regression line; while for the Treatment B regression line the value of X10 is 0 (X10 is equal to 0 for Treatment B).

2. Since the value of X10 is 1 for the Method A regression line, X26, which is the product of X8 and X10, becomes just X8. In addition, since the value of X10 is 0 for the Method B regression line and X26 is the product of X8 and X10, X26 becomes 0.
3. The two lines are set equal to each other and solved for the value of X8 as follows:

Regression Line for Treatment A
$38.303(1) + (-72.681)(1) + 0.665X8 + 0.694(X8)$ =

Regression Line for Treatment B
$38.303(1) + (-72.681)(0) + 0.665X8 + 0.694(0)$

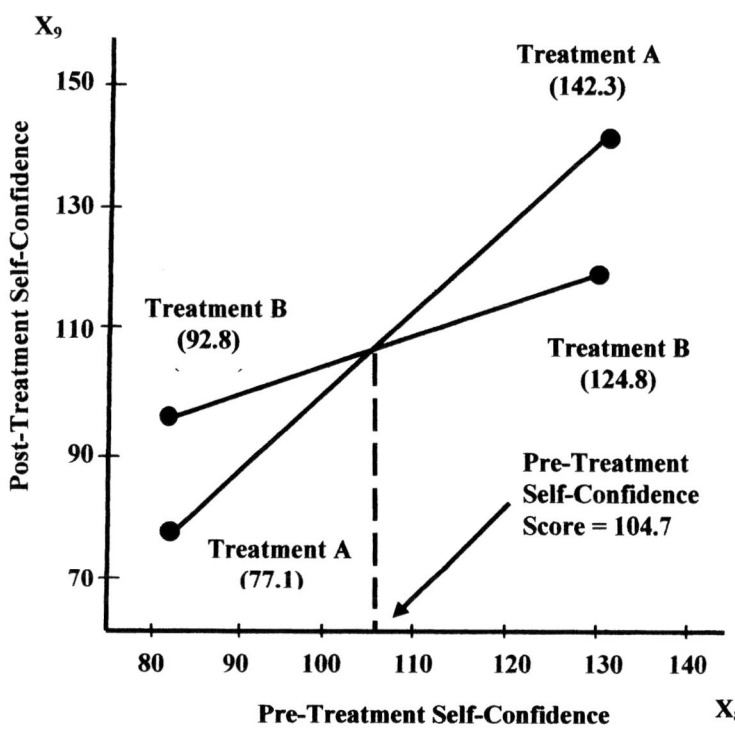

Figure 7.9.
Diagram of the interaction effect estimated by the Full Model (Model 7.18) for ARH 7.2.

Solving this equality for X8, results in a pre-treatment self-confidence score of 104.7 or approximately 105 (see Figure 7.9). Thus Treatment A is superior to

Treatment B for participants who have pre-treatment self-confidence scores of 105 or higher. In addition, it is important to note that since the slope of the regression line for Treatment A is greater than the slope for Treatment B, the post-treatment scores of those participants in Treatment A exceed those in Treatment B at an increasing rate for those participants who have pre-treatment self-confidence scores of at least 105. Due to the disordinal interaction, the researchers would conclude that ARH 7.2 is not supported for a given population. However, the fact that the pre-treatment self-confidence scores increase *above* the 105 level, the relative superiority of Method A over Method B increases with respect to the post-treatment self-confidence scores may prove useful for future studies.

We believe that the results produced for the statistical testing of ARH 7.2 provide an excellent example of what researchers should and should not do. Researchers should not *blindly* apply statistical tests to their hypotheses, but rather they should carefully review all the information the results reveal about the relationships among the variables they are investigating before they state their conclusions.

Writing the Results for ARH 7.2 in APA Style

When reporting the findings regarding the statistical test results for ARH 7.2, the researchers should provide: (a) a table containing the regression information for the Full and Restricted Models, which is identified for this APA presentation as Table 1 and (b) a figure that displays the interaction effect, which is identified as Figure 1 for this APA presentation, but labeled Figure 7.9 in this chapter. Assuming the researchers included such a table and figure, they could report their findings in APA style as follows:

> As indicated by the results reported in Table 1, the statistical test of the difference between the R^2 values of the Full Model ($R_F^2 = .98$) and the Restricted Model ($R_R^2 = .88$), which tested the interaction effect between the treatments and pre-treatment self-confidence levels, was statistically significant at the one-tailed .05 alpha level ($F(1, 56) = 326.67, p / 2 < .001$). A review of Figure 1 reveals that this interaction effect, which was disordinal, did not reflect the interaction effect specified in the research question. Only when the pre-treatment self-confident scores exceeded 105 were the post-treatment scores of those participants who received Treatment A higher than the post-treatment scores of those who received Treatment B. In addition those differences in the post-treatment self-confidence scores increased as the pre-treatment self-confidence scores increased. The amount of unique variation accounted for by the interaction effect between the treatments and the pre-treatment self-confidence scores was .10, which is classified as a large effect size according to the criteria established by Cohen (1988).

PARTIAL INTERACTION

The notion of interaction can easily be extended to more than two groups and more than one continuous predictor variable. When more than two groups are considered, interaction can occur in only certain segments of the design. We suggest that one should consider models that allow interaction to occur only in those parts of the design where interaction is suspected of occurring. Often a researcher will suspect interaction is occurring in only one segment of the design and, rather than test this specific question, will unfortunately test the overall (omnibus) interaction question. Andrews, Morgan, and Sonquist (1967) presented some notions that are supportive of investigating interaction in specific aspects of the design. Figure 7.10 depicts an extreme case.

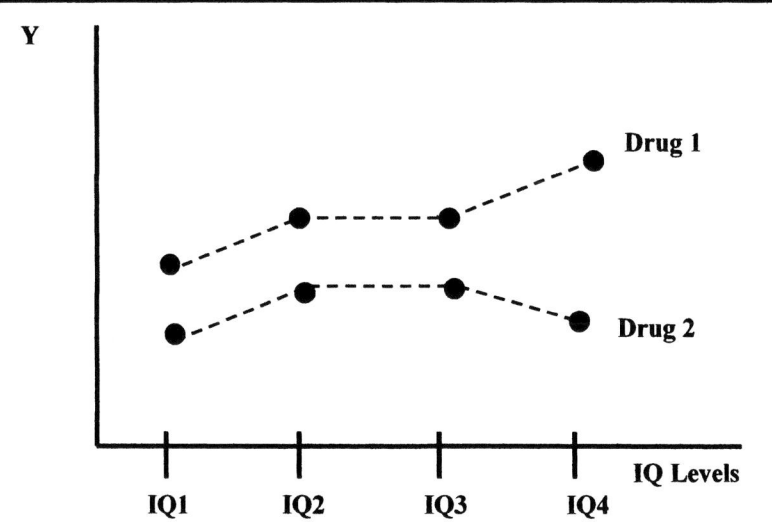

Figure 7.10.
Interaction occurring at only one level of IQ.

As indicated by the Drug 1 and Drug 2 lines in Figure 7.10, the distance between the two levels of this one independent variable at one point (IQ4) is different from that of other points (IQ1, IQ2, and IQ3). What is happening at IQ4 is clearly different from what is happening at IQ1, IQ2, and IQ3. If the omnibus interaction question had been calculated, the exact source of significance would not have been found, nor could a directional interpretation have been made. But if interaction is tested by equating the difference at IQ4 with the average of the differences at IQ1, IQ2, and IQ3, then, because only one restriction is being made, an unambiguous interpretation can be made on a significant F test. (Again, unambiguous only if the direction of the interaction had been specified before analysis.) The careful reader will note that this is very similar to the notion presented in the previous section—a specific hypothesis and the testing of one interaction variable. Fraas and Drushal (1987) provided an interesting application of the partial interaction question.

Directional Partial Interaction

The notions of directional hypothesis testing are as applicable to interaction hypotheses as to any other kind of hypothesis. In the previous example illustrated by Figure 7.10, the researcher expected the difference at IQ4 to be different from the average of IQ1, IQ2, and IQ3. In addition, the researcher may have expected the superiority of Drug 1 over Drug 2 at IQ4 to be *greater* than at the average of IQ1, IQ2, and IQ3. This specific expectation calls for a directional test of significance. For both the directional and nondirectional hypotheses, the Full Model, restrictions, and Restricted Model are the same. The only difference is the emphasis that such directional Research Hypotheses place on the signs of the relationship among the coefficients in the Full Model.

To illustrate this point consider the Full Model required to test the interaction just discussed, which is as follows:

$$Y1 = a_1C1 + a_2C2 + a_3C3 + a_4C4 + a_5C5 + a_6C6 + a_7C7 + a_8C8 + E_1$$
(Model 7.20)

where:
1. $Y1$ = the criterion variable;
2. $C1 = 1$ if Drug 1 and IQ Level 1, 0 otherwise;
3. $C2 = 1$ if Drug 2 and IQ Level 1, 0 otherwise;
4. $C3 = 1$ if Drug 1 and IQ Level 2, 0 otherwise;
5. $C4 = 1$ if Drug 2 and IQ Level 2, 0 otherwise;
6. $C5 = 1$ if Drug 1 and IQ Level 3, 0 otherwise;
7. $C6 = 1$ if Drug 2 and IQ Level 3, 0 otherwise;
8. $C7 = 1$ if Drug 1 and IQ Level 4, 0 otherwise; and
9. $C8 = 1$ if Drug 2 and IQ Level 4, 0 otherwise.

The researchers expect the following relationship to exist among the coefficients in this Full Model:

$$(a_7 - a_8) > ((a_1 - a_2) + (a_3 - a_4) + (a_5 - a_6)) / 3$$

The antithesis of this restriction is as follows:

$$(a_7 - a_8) = ((a_1 - a_2) + (a_3 - a_4) + (a_5 - a_6)) / 3 \quad \text{(1 restriction)}$$

Imposing this restriction on the Full Model results in an extremely complicated Restricted Model and will not be presented in this text. The directional Research

Hypothesis, of course, demands that (a_7 - a_8) be greater than the quantity on the right-hand side of the restriction.

Use of directional Research Hypotheses that deal with interaction. If a researcher is going to treat interaction as an interesting phenomenon in its own right, then the interaction should be expected to be occurring in a certain specified way. Many researchers have expectations about how the interaction will occur. In such cases, directional Research Hypotheses should be stated and statistically tested.

Researchers should be sure to understand two points related to the testing of directional Research Hypotheses. First, remember that the F test probabilities generated by most statistical software, including SPSS for Windows and Microsoft Excel, are two-tailed probabilities. Such probability values should be divided by 2 if directional Research Hypotheses are tested. Second, directional Research Hypotheses are limited to a single restriction on a Full Model, which means that there will be only one degree of freedom for the numerator of the F ratio. This is why directional Research Hypotheses are sometimes referred to as "one degree of freedom questions." If the degrees of freedom value in the numerator of the F ratio is greater than one, then a directional research hypothesis has not been tested.

Interaction terms used as covariates. Any variable may be used as a covariate, as long as that variable is not influenced by the treatments. If one considers interaction as a phenomenon in its own right, then interaction terms could be used as covariates. Some rationale should be used for including this covariate, otherwise many interactions (second-degree, third degree, etc.) might be included that would drain the degrees of freedom and spuriously generate nonsignificant findings. The regression models for the covariance analogue are presented in chapter 8. Basically, the covariates appear in both the Full Model and the Restricted Model, with the treatment vectors appearing in only the Full Model.

INTERACTION BETWEEN TWO CONTINUOUS PREDICTORS

Normally when one uses a continuous variable as a predictor, one uses the originally scaled values of that predictor in a model, such as the following model, which actually expresses the linear correlation between Y and P:

$$Y = a_0U + a_1P + E_1 \qquad\qquad\qquad \text{(Model 7.21)}$$

This relationship is shown in Figure 7.11 as a straight line. When multiple predictors are used to predict the criterion, most researchers still choose to use the originally scaled variables. With more than one predictor, the graphic representation of the relationship between the criterion and predictors may require more than two dimensions. Several such situations are discussed in this section.

It must be remembered that the research question dictates the model and graphic representation of the model. There is no need to center continuous variables by subtracting the mean from each score, as with dichotomous interaction discussed earlier. Kromrey and Foster-Johnson (1996) conducted a Monte Carlo study and concluded that centering and non-centering produced the same R^2 values. Some of the coefficients will be different, however, which might facilitate interpretations of the centering data.

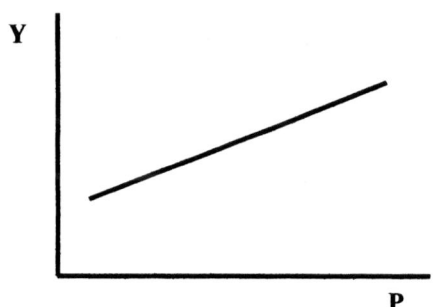

Figure 7.11.
A single straight line fit by Model 7.21.

Model 7.21 was represented in Figure 7.11 as a two-dimensional straight line. It also could be represented in three dimensions. The following model (Model 7.22) shows this possibility:

$$Y = a_0U + a_1P + a_2Q + E_2 \qquad\qquad \text{(Model 7.22)}$$

If the relationship between Y and P is expected to be the same for all values of Q, then the model needed to express this relationship is still Model 7.21 and to show the constancy of the relationship over values of Q, Figure 7.12 would be used. The criterion, Y, is on the vertical axis, and the values for P and Q are along the sides of the cube. The predicted scores on Y form a flat plane defined by the values of Y, P, and Q. Figure 7.12 illustrates that Q is not needed as the plane is at the same height on Y for all values of Q (for any value of P). That is, at the front right side of the cube, the plane is parallel to the Q axis. Also, at the back left side of the cube, the plane is parallel to the Q axis.

If one were to model the relationships depicted in Figure 7.12 with only two dimensions, the coefficient for Q (a_2) would be zero. This value would indicate that the slope of the plane along Q is zero; thus Q would not add to the predictability of Y. That is, the Statistical Hypothesis, which requires the a_2 coefficient to be equal to 0, could not be rejected.

Suppose, however, that the relationship between Y and P is not expected to be constant across values of Q as depicted in Figure 7.13. Then Q also would have to be included in the model as a predictor and the a_2 coefficient would not be 0.

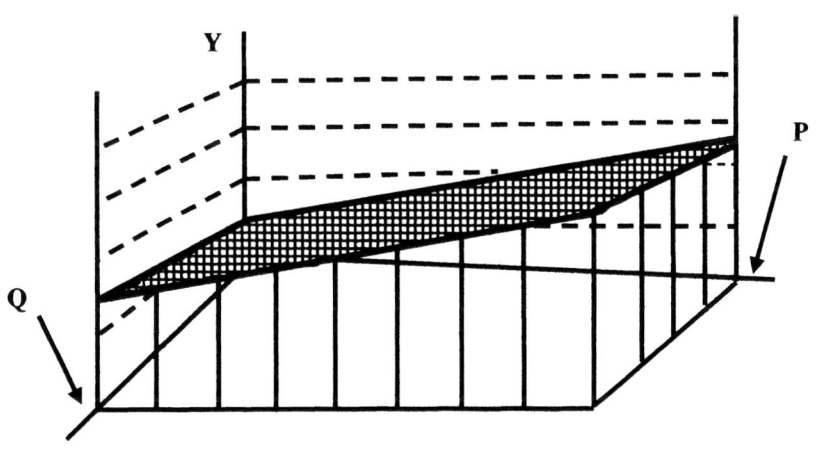

Figure 7.12.
A plane fit by Model 7.22 that does not vary over the range of variable Q $(a_2 = 0)$.

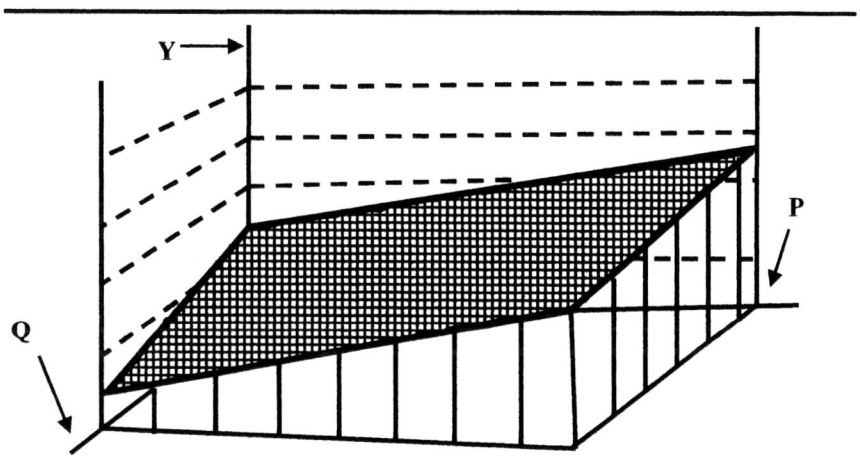

Figure 7.13.
A plane fit by Model 7.21 that does vary over the range of variable Q $(a_2 \neq 0)$.

A review of Figure 7.13 reveals that Model 7.22 allows the predicted Y scores to form a flat place that is tilted such that the criterion Y value for a given

P value is different for each Q value. This point can also be seen by noting that at each corner of the cube the plane has a different Y value.

As was discussed in a previous section, one may have reason not to want to use the original values of the predictor variables. As depicted in Figure 7.14, one may, for example, expect a particular relationship (say, a flat plane) to exist for values below a particular point (R) on the Q variable and expect another relationship (say, a flat plane) to exist for values at and above R on variable Q. The following model (Model 7.23) would allow this expectation:

$$Y = a_0U + a_1W1 + a_2(P * W1) + a_3(Q * W1) + a_4W2 + a_5(P * W2) +$$
$$a_6(Q * W2) + E_3 \qquad \qquad \text{(Model 7.23)}$$

where:
1. $W1 = 1$ if the score on Q is below the value of R, 0 otherwise; and
2. $W2 = 1$ if the score on Q is at or above R, 0 otherwise.

Model 7.23 allows for two planes, one for Q values below R and another flat plane for Q values at and above R. These two planes would not have to intersect at R as depicted in Figure 7.14, although if the data were systematic, the intersection would likely be at R. Notice that each of the interactions in the Model 7.23 is again represented by multiplication.

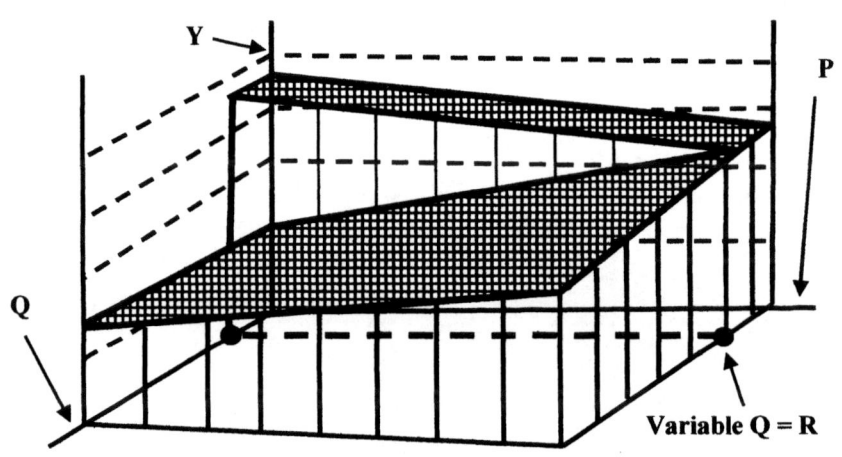

Figure 7.14.
A plane fit below the R value on variable Q and another plane above the R value both fit by Model by 7.23.

If the researcher expects that a interaction exists between P and Q as they relate to Y, a model such as Model 7.24 could be used, which is as follows:

$$Y = a_0 U + a_1 P + a_2 Q + a_3 (P * Q) + E_4 \qquad \text{(Model 7.24)}$$

The relationship in Model 7.24 does not represent a flat plane or a combination of flat planes, but instead a twisted plane of the type pictured in Figure 7.15. Each edge of the plane is straight, and each line drawn across the values of Q is straight, yet the plane is twisted such that the twisted plane intersects the four corners of the cube at values different from those that would be possible if the plane were flat.

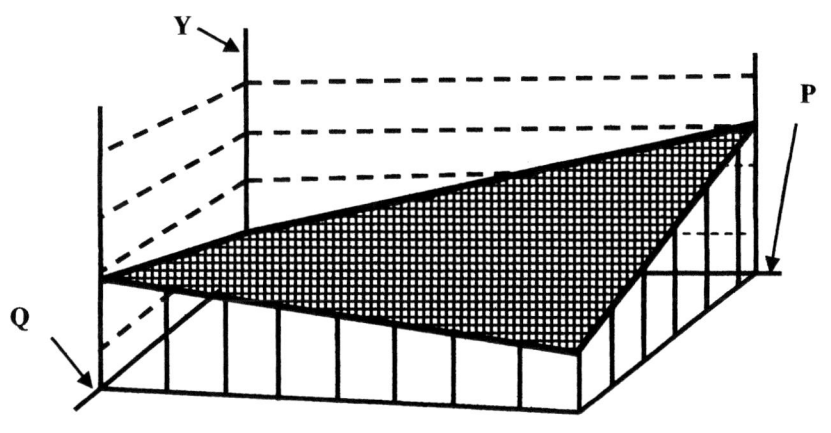

Figure 7.15.
A twisted plane depicting an interaction between the P and Q continuous predictor variables fit by Model 7.24.

COMPARISON OF CATEGORICAL AND CONTINUOUS INTERACTION

One major weakness of ANOVA analyses is the artificial categorization of continuous variables. Researchers can usually obtain continuous data and usually would prefer to infer along some continuum, but they often categorize their data before analyzing it. Indeed, phenomena in the real world usually follow systematic functions, rather than discrete leaps and bounds. The GLM approach readily allows one to investigate continuous variables and specifically to investigate the interaction between categorical variables and continuous variables.

Suppose that one wanted to treat IQ as a continuous variable rather than artificially categorizing it into four levels. Also assume that Figure 7.16 presents the possible state of affairs regarding a possible interaction between IQ and Drug.

Figure 7.16a is the pictorial representation of the suspected interaction, and Figure 7.16b is the state of affairs allowing no interaction between drugs and IQ as they affect the criterion. The regression formulation allowing for interaction would be as follows:

$$Y1 = a_0U + a_2D2 + b_1S1 + b_2S2 + E_1 \qquad \text{(Model 7.25)}$$

where:
1. Y1 = criterion score;
2. U = the unit vector;
3. D2 = 1 if have Drug 2, 0 otherwise;
4. Q5 = IQ for each subject;
5. S1 = D1 * Q5 (IQ if Drug 1 is used, 0 otherwise); and
6. S2 = D2 * Q5 (IQ if Drug 2 is used, 0 otherwise).

It should be noted that since the unit vector (U) is included in this model that variable D1 is excluded. The slope of the Drug 1 line would be b_1, and the Y intercept would be a_0. The slope of the Drug 2 line would be b_2 and the Y intercept would be equal to the sum of a_0 and a_1.

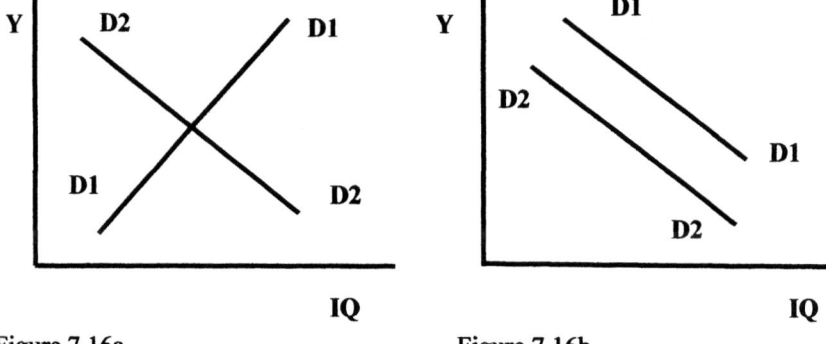

Figure 7.16a.
A possible interaction between IQ and Drug treatment.

Figure 7.16b
No interaction between IQ and Drug treatment.

Figure 7.16.
Regression lines for the two Drug groups.

This Full Model (Model 7.25) allows for an interaction effect (see Figure 7.16a), which indicates that the regression lines for the two different drug treatments will not be parallel. If these two lines do not reflect an interaction effect (see Figure 7.16b), they will be parallel, which means that the two slopes must be equal, resulting in the following restriction:

$$b_1 = b_2 \qquad \text{(1 restriction)}$$

Setting the two slopes equal to a common slope (b_3), results in the following Restricted Model:

$$Y1 = a_0U + a_2D2 + b_3IQ + F_2 \qquad \text{(Model 7.26)}$$

The F test of significance between Models 7.25 and 7.26 would have a df_n value equal to 1 (i.e., 4 - 3) and df_d equal to N - 4. Note that the vectors that allow for interaction can be formed by multiplying the drug vector by the continuous IQ vector.

Suppose now that the study involved a third drug, as illustrated in Figure 7.17. The regression formulation of the Full Model that allows for an interaction effect among the three drugs would be as follows:

$$Y1 = a_0U + a_2D2 + a_3D3 + b_1S1 + b_2S2 + b_3S3 + E_3 \qquad \text{(Model 7.27)}$$

where:
  1. $Y1$ = criterion score;
  2. U = the unit vector;
  3. $D2 = 1$ if have Drug 2, 0 otherwise;
  4. $D3 = 1$ if have Drug 3, 0 otherwise;
  5. $Q5 = IQ$ for each subject;
  6. $S1 = D1 * Q5$ (IQ if Drug 1 was used, 0 otherwise);
  7. $S2 = D2 * Q5$ (IQ if Drug 2 was used, 0 otherwise); and
  8. $S3 = D3 * Q5$ (IQ if Drug 3 was used, 0 otherwise).

The slope of the Drug 3 line would be b_3 and the Y intercept would be equal to the sum of a_3 and a_0.

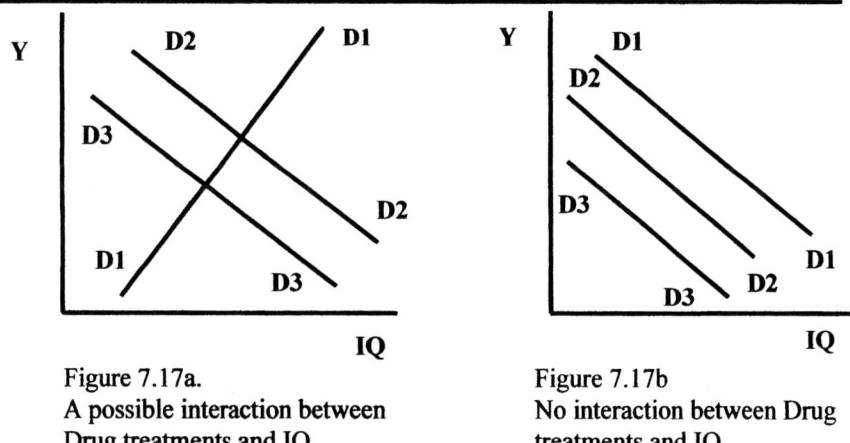

Figure 7.17a.
A possible interaction between
Drug treatments and IQ.

Figure 7.17b
No interaction between Drug
treatments and IQ.

Figure 7.17.
Regression lines for the three Drug treatments.

Once again, this Full Model (Model 7.27) allows for an interaction effect (see Figure 7.17a), which indicates that the regression lines for the three Drugs will not be parallel. If these three lines do not reflect an interaction effect (see Figure 7.17b), they will be parallel, which means that the three slopes must be equal, resulting in the following restriction:

$b_1 = b_2 = b_3$ (2 restrictions)

Setting the three slopes equal to a common slope, say b_5, results in the following Restricted Model:

$$Y1 = a_0U + a_2D2 + a_3D3 + b_5Q5 + E_4 \qquad \text{(Model 7.28)}$$

For the F test of significance between the Full Model (Model 7.27) and the Restricted Model (Model 7.28) the df_n and df_d values would equal (6 - 4) and $(N - 6)$, respectively. Once again, note that the vectors that allow for interaction are formed by multiplying the drug variables by the continuous IQ variable.

Now suppose that the three drugs were actually three different dosage levels of the same drug, and we now wish to treat the drug variable as a continuous variable (labeled D4), rather than artificially categorizing it into levels. The interaction Research Hypothesis now would be as follows:

Research Hypothesis: There is an interaction between IQ and drug dosage level, over and above the separate linear effects of IQ and drug dosage level.

A more precise and insightful way of stating the Research Hypothesis would be as follows:

Research Hypothesis: The combined effect of IQ and drug dosage level is needed to predict the criterion, over and above the separate linear effects of IQ and drug dosage level.

The interaction between the continuous variable of IQ and the drug dosage level could be tested through the use of the Full Model represented by Model 7.29 and the Restricted Model represented by Model 7.30. The Full Model is constructed as follows:

$$Y1 = a_0U + a_4D4 + b_5Q5 + c_4(D4 * Q5) + E_5 \qquad \text{(Model 7.29)}$$

where:
1. $Y1$ = criterion score;
2. U = unit vector;
3. $D4$ = drug dosage level for each subject;

4. $Q5 = IQ$ for each subject; and

5. $D4 * Q5 =$ product of drug dosage level and IQ for each subject.

The product term $(D4 * Q5)$ allows for the linear interaction between drug dosage level and IQ. The restriction on Model 7.29 that would not allow interaction to occur would be as follows:

$c_4 = 0$ (1 restriction)

Placing this restriction into the Full Model (Model 7.29) produces the following Restricted Model:

$$Y1 = a_0U + a_4D4 + b_5Q5 + E_6 \qquad \text{(Model 7.30)}$$

Note that the unit vector appears in both Full Model (Model 7.29) and the Restricted Model (Model 7.30). When all variables in the regression model are continuous, the unit vector is not generally a redundant piece of information and therefore must be counted as one of the linearly independent vectors. Furthermore, the interaction term $(D4 * Q5)$ in the Full Model 7.29 is not *linearly* dependent upon the other vectors in the model. The interaction term is a function of these variables, but it is not a *linear* function; therefore it is *linearly* independent. There are four linearly independent vectors in the Full Model (Model 7.29) and three in the Restricted Model (Model 7.30), hence for the F test used to statistically test the difference between the R^2 values of the Full and Restricted Models, the df_n value equal to 1 (i.e., 4 - 3) and df_d equal to N - 4.

The product of two continuous variables has been labeled the *moderator variable* in the literature (Saunders, 1956). The multiplication of two continuous variables is simply an extension of the multiplication of two categorical variables (already referred to in the literature as *interaction*), so it is unfortunate that new terminology, (moderator variable) was introduced. For heuristic purposes, the single concept of interaction should suffice, although any variable that increases predictability has a legitimate place in a model, whether it has a special name or not.

Some authors (Goldberg, 1972; McClelland & Judd, 1993) reported that few interactions between continuous variables have been found. Goldberg's work (he called them *configural variables*) was with highly correlated test items, which one would not necessarily expect to interact with one another. Others (DuCette & Wolk, 1972; Wood & Langevin, 1972) have found limited, selected instances in which a moderator variable increased prediction. Furthermore, McNeil and Newman (1996) pointed out that whether one searches for categorical or continuous interactions should be a function of the (a) design of the study, (b) measurement of the predictor variable(s), and (c) nature of the question asked. Once again, the research hypothesis should dictate the kind of analysis and the nature of the variables in that analysis.

Table 7.1 indicates the advantages of treating variables as continuous variables rather than artificially categorized variables. When variables are treated as continuous variables, more degrees of freedom are retained in the denominator, and at the same time fewer degrees of freedom exist in the numerator of the F test used to statistically test the difference between the R^2 values of the Full and Restricted Models.

Table 7.1.
Comparison of Interaction Possibilities.

	Models from Text	Specific		General	
		dfn	dfd	dfn	dfd
Both predictor variable Categorical	7.7 vs. 7.8	6	$N - 12$	$(r - 1)(c - 1)$	$N - (c * r)$
One predictor variable continuous and one categorical	7.27 vs. 7.28	2	$N - 6$	$(k - 1)$	$N - (k * 2)$
Both predictor variables Continuous	7.29 vs. 7.30	1	$N - 4$	1	$N - 4$

Note: N = number of subjects; r = number of categories of one predictor variable; c = number of categories of the other predictor variable; and k = number of categories of the categorical variable.

The extreme situation appears when both variables are treated as continuous, resulting in one degree of freedom in the numerator, corresponding to the one restriction made on the Full Model (Model 7.29). So, if the F test of the difference between the R^2 values of the Full Model (Model 7.29) and the Restricted Model (Model 7.30) is statistically significant, one knows which restriction is generating the significance. Inspection of the regression coefficient for that source will then indicate if the results are in the hypothesized direction. But when more than one restriction is made, as is the case when the difference between the R^2 values of Model 7.27 and Model 7.28 is statistically tested, and significance is found, the source of the significance cannot be pinpointed. These comments are applicable to any test in which more than one restriction is made—not solely to a problem with interaction hypotheses.

One limitation regarding the interaction of continuous variables is that many more subjects are required to get stable estimates. Essentially, the number of "inside" cells is equal to the product of the range of scores on the one variable and the range of scores on the other variable. Thus if Variable A has 20 different score values and Variable B has 14 different score values, the product would be 280. The traditional rule of thumb of 30 subjects per cell would require 8,400 subjects. We argue that because the variables are probably somewhat continuously related, fewer than 8,400 subjects are necessary. But the point is more subjects are necessary when considering continuous data than when considering categorical data. We discuss the number of subjects needed in a particular analysis more fully at the end of chapter 8.

CHAPTER 8

STATISTICAL CONTROL OF POSSIBLE CONFOUNDING VARIABLES

One goal of research is to isolate the unique effects of a variable. To reveal that a particular variable is influencing the criterion in a particular way, the influence of other competing variables or confounding variables must be eliminated. Competing variables can be eliminated, to various degrees, by theoretical (logical) arguments, by research design, or by statistical control. Theoretical argument is usually the weakest defense, whereas research design is usually the strongest. Unfortunately, it is not always possible in the behavioral sciences to either "argue away" or "design away" the competing variables. Statistical control, through the analysis of covariance, can be a very useful tool for assisting researchers in their attempts to isolate the effects of explanatory variables.

CONTROL OF CONFOUNDING VARIABLES THROUGH THEORY

One might argue based on theory, or common sense, or logic that a certain variable cannot be a confounding variable. For example, researchers might argue that the color of a person's hair, the number of letters in the person's name, and the person's Social Security number are unrelated to that individual's math achievement; and thus they could then ignore these variables when analyzing math achievement. Usually, though, researchers do not go through such a thought process—they just choose to investigate certain variables and by default ignore others. Whether those ignored variables are relevant or not is ultimately an empirical question, but for the time being the variables are ignored.

CONTROL OF CONFOUNDING VARIABLES THROUGH DESIGN

Research conducted in the laboratory is directed toward determining the effects of a particular independent variable upon some outcome state. Some examples of laboratory research questions are:

1. Is titanium a better luminescent filament than carbonized cotton?
2. Do pigs fed daily with 50% of the feed consisting of protein weigh more after 60 days than pigs fed with 25% of the feed consisting of protein?
3. Do dogs raised in isolation for 80 days after birth respond to noxious stimuli less effectively than dogs raised in a natural kennel environment?

To answer these questions, the researchers attempt to control for all known contaminants in the research design. In the dog question, breed differences surely may contaminate the results. If the isolated dogs were Doberman Pinschers and the control dogs were mongrels picked up at the pound, the observed differences, if any, in sensitivity to noxious stimuli might be contaminated by breed differences in sensory responsiveness.

Randomization

The researchers may control for the effects of a contaminating variable, such as breed, by choosing only one "level" of that variable (one breed) or by choosing several levels (breeds) and then, in either case, randomly assigning one half the members of each breed to each treatment. If only one level (breed) is selected, that is the only level to which the results can be safely generalized. In the dog question, the researchers may select a specific breed and randomly split litters of puppies into experimental and control groups. The ANOVA on the sensory response criterion would then yield an F value that gives the researchers an estimate of how likely it is that the observed differences are due to sample variation within that breed. By using the split litter of a specific breed, the population is that specific breed, and thus large response deviations between the two treatment samples would not be contaminated by the possibility of the dogs having originally come from two different breed populations.

In laboratory research, the researchers may have most of the relevant (potentially contaminating) variables under physical control. Such control also may be achieved in the laboratory for complex manipulations of more than one variable. Applied research in the behavioral sciences and education, though, is frequently conducted in natural settings where physical control of contaminants is either costly or impossible.

Matching

Consider the case of researchers who want to test the effectiveness of a new curriculum designed to improve reading ability. If students are taught using the proposed (new) curriculum, the researchers expect the students to perform better than they would when taught using the current curriculum. Suppose the researchers have a sample of 200 students available for random assignment. Past research may suggest that, in relation to the students' criterion score of reading, several variables are known to be related to performance, such as (a) female students score better than male students, (b) students from middle-class homes tend to perform better, (c) high-IQ students tend to have higher reading scores, and (d) the students' past reading abilities are positively related to their reading scores. These four variables are possible sources of contamination if subjects are not assigned randomly to treatments or if the random assignments, by chance, place more of one kind of student in one of the two treatment groups.

To avoid such contaminating effects (i.e., non-equivalent random samples), one may attempt a matched-pairs assignment procedure by first selecting two females from middle-class homes with high IQ scores and high initial reading ability. Next one of the females would be randomly assigned to use the new curriculum and the other would be assigned to the currently used curriculum. If such a procedure is followed rigorously for all ranges of the possible contaminants for all of the study's participants, the researcher will have two groups of matched pairs. With the limited original sample of 200 children, however, one would typically find that, after obtaining, say, 30 pairs, no "real" pairs are left. Some of the remaining students could possibly be paired with respect to one or two of the variables, but they would not match on the other variables. The mind boggles at the effort to form matched pairs for more than one or two variables, and successful matching on all 200 subjects would surely be unlikely.

Given that matching is accomplished for 30 pairs, one may conduct the study, but then the results could be generalized only to the peculiar population that the 30 sample pairs represent. Experimental control such as this parallels the rigor of the laboratory, but does it really solve these researchers' question because of the initial 200 students only 60 students were used, and these 60 students are not necessarily the same on the other relevant variables as those students who could not be matched. The population to which the results can be generalized is not readily apparent. If the 200 students represent the population to which the researchers intended to generalize, the selected matched sample does not really represent that population. Therefore, the results of the study will not allow the researchers' question to be satisfactorily addressed.

Once researchers obtain their matched samples, they usually proceed as if those samples were randomly assigned; and they would use a t test to determine if the post-treatment reading mean of the students taught using the new curriculum exceeds the post-treatment reading mean of the students taught using the current curriculum. More is known about the subjects, however, than that students were taught using different curricula. The information that matched pairs were used in the analysis can be used by the researchers. The resulting design is analogous to the dependent-groups design in which the "person vectors" are the "pair vectors."

Controlling Competing Explanations through the Research Design

Another way to control for possible competing explanations for the observed changes in the criterion variable is to design the study in such a manner that the variables represent these competing explanations. For example, if IQ is known to influence the criterion variable, the effects of IQ on this criterion variable can be controlled. For the curriculum study discussed in the preceding section, the inclusion of IQ as a blocking variable would change the design from a one-way ANOVA design (two-curriculum methods) to a two-way ANOVA design (with the main effects being curriculum types and IQ). The analysis procedure discussed for a two-way ANOVA design in chapter 4 would be applicable to such a study. Note, however, as discussed in chapter 4, that IQ "groups," "levels," or "blocks" would be created, thus "lumping" together many different IQ scores and considering their effects to be similar. The covariance design, to be discussed in this chapter, does not lump all IQ scores together but treats the possible contaminating variable as one continuous variable.

CONFOUNDING VARIABLES
UNDER STATISTICAL CONTROL

Rationale

An alternate approach to controlling alternative explanations for the changes in the criterion variable values in the proposed curriculum study would be to form two groups that are as nearly equivalent as the researchers can manage to arrange and then statistically control for the contamination that was not under their control. Given a rough equivalency between the experimental and control groups, one may find that one group has a few more high IQ students, fewer girls, fewer middle-class children, and a slight difference in initial reading ability when compared with the other group. These differences are contaminants whose magnitude of effect on the criterion would be unknown if not accounted for in the research question and subsequent analysis. In such a situation the researchers are asking: Over and above the influences of IQ, gender, entering reading ability, and social class, is the new curriculum superior to the current curriculum as measured by posttest reading-ability scores?

The question can be cast into the following Research Hypothesis:

Research Hypothesis: Over and above the influences of IQ, gender, entering reading ability, and social class, the new curriculum is superior to the current curriculum as measured by the posttest reading-ability scores for the population of interest.

The competing Statistical Hypothesis would be as follows:

Statistical Hypothesis: Over and above the influences of IQ, gender, entering reading ability, and social class, the new curriculum is equal to the current curriculum as measured by the posttest reading-ability scores for the population of interest.

Full Model, Restricted Model, and F test

The Full Model, which reflects the Research Hypothesis, utilizes the post-treatment reading scores as the criterion variable, and it includes a predictor variable that represents the "treatments." In addition, this Full Model contains variables that represent each student's IQ level, gender, pre-treatment reading score, and social class. These variables serve as the covariates, which allows the difference between the post-treatment reading scores of the two groups of students (i.e., new curriculum and current curriculum groups) to be estimated *over and above* the influences of the student characteristics the variables represent. This Full Model is as follows:

$$Y = a_0U + a_1NC + c_1IQ + c_2G + c_3Pre + c_4SC1 + c_5SC2 + E_1 \quad \text{(Model 8.1)}$$

where:
1. $Y1$ = post-treatment reading score;
2. U = the unit vector;
3. $NC = 1$ if $Y1$ is from a member of the new curriculum group, 0 if $Y1$ is from a member of the current curriculum group;
4. IQ = IQ score of each individual represented in Y;
5. $G = 1$ if the subject is female, 0 otherwise;
6. Pre = Pre-treatment reading score;
7. $SC1 = 1$ if the numerical value on Social Class Index was 1, 0 otherwise;
8. $SC2 = 1$ if the numerical value on Social Class Index was 2, 0 otherwise;
9. $SC3 = 1$ if the numerical value on Social Class Index was 3, 0 otherwise;
10. E_1 = the error vector, the difference between the observed criterion and the predicted criterion $(Y - \hat{Y})$; and
11. $a_0, a_1, c_1, c_2, c_3, c_4$ and c_5 are regression coefficients calculated to minimize the sum of the squared elements in vector E_1.

Four characteristics of this Full Model should be noted. First, it contains the unit vector. Almost all scales in the social sciences are arbitrarily scaled, so the regression constant is almost always employed to adjust all the predicted scores up or down to have the same mean as the criterion.

Second, it is assumed that the Social Class Index contains three values (i.e., 1, 2, and 3). A series of three dummy variables were constructed to represent the students' social classes as described in the variable descriptions for the Full Model. Since this Full Model (Model 8.1) contained the unit vector, only two of the three dummy variables are included in the model (refer to chapter 4). For this Full Model variable SC3 was not included. (Remember, it does not matter which of the three dummy variables is the one excluded.)

Third, the coefficients for the covariates are labeled c_1, c_2, c_3, c_4 and c_5. One could use any letter to represent these weights, yet some researchers may find that for mnemonic purposes the label c helps to suggest that these weights are associated with the covariates.

The fourth characteristic of this model is very important to understand. That is, the coefficient for the treatment variable (NC) is the *adjusted* post-treatment reading mean of the new curriculum minus the *adjusted* post-treatment reading mean of the current curriculum group. It is important to note that the difference measured by this coefficient is an adjusted mean. That is, the means have been adjusted for differences between the two treatment groups with respect to the covariates. Unless each covariate is uncorrelated with the group variable (NC), which is very unlikely in the settings encountered by social scientists and educators, the adjusted mean of each group will not match its corresponding unadjusted mean.

The following illustrates how the Full Model might look in vector form:

$$\text{Student } Y = a_0 U1 + a_1 NC + c_1 IQ + c_2 G + c_3 Pre + c_4 SC1 + c_5 SC2 + E_1$$

Student	Y	U1	NC	IQ	G	Pre	SC1	SC2	E
1	95	1	1	120	1	70	1	0	?
2	75	1	0	115	0	66	0	0	?
3	54	1	0	110	0	39	0	0	?
4	80	1	1	120	0	70	0	1	?
5	78	1	1	120	0	64	0	1	?
.	.	.	.	.	.	.	.	.	?
.	.	.	.	.	.	.	.	.	?
.	.	.	.	.	.	.	.	.	?
200	84	1	1	105	1	69	1	0	?

A review of row one, the data for Student 1, reveals that the student (a) has a post-treatment reading score of 95, (b) is a member of the new curriculum group—the value for the NC variable is 1, (c) has an IQ score of 120, (d) is a female—the value for the G variable is 1, (e) has a pre-treatment reading score of 70, and (f) is a member of the social class with a Social Class Index number of 1—the value for the SCI variable is 1.

For Student 2, a review of the vectors reveals that the student (a) has a post-treatment reading score of 75, (b) is a member of the current curriculum group—the value for the NC variable is 0, (c) has an IQ score of 115, (d) is a male—the value for the G variables is 0, (e) has a pre-treatment reading ability pretest score of 66, and (f) is a member of the social class with a Social Class Index number of 3—both of the values for the SC1 and SC2 variables are 0.

Since the coefficient for the treatment variable (NC) estimates the adjusted post-treatment reading mean of the new curriculum minus the adjusted post-treatment reading mean of the current curriculum group, and the Statistical Hypothesis requires the adjusted means of the two treatment groups to be equal, the restriction placed on the Full Model (Model 8.1) is as follows:

$a_1 = 0$ (1 restriction)

Placing this restriction in the Full Model results in the following Restricted Model:

$$Y1 = a_0U + c_1IQ + c_2G + c_3Pre + c_4SC1 + c_5SC2 + E_2 \qquad \text{(Model 8.2)}$$

This Restricted Model forces both treatment groups to have the same adjusted mean post-treatment reading score and thus any predictability is due solely to the covariates and the regression constant (a_0). Strictly speaking, the unit vector in the Restricted Model is a covariate and should be specified in the Research Hypothesis. Customary usage has led researches to place the unit vector in the Full and Restricted Models.

The R_F^2 (i.e., the R^2 value of the Full Model—Model 8.1) is the proportion of the observed criterion variance accounted for by treatment-group membership and the covariates. The R_R^2 (i.e., the R^2 value of the Restricted Model—Model 8.2) is the proportion of the observed criterion variance accounted for by the covariates alone. Any loss in the R^2 value between the Full and Restricted Models is identified as the proportion of *unique* criterion variance accounted for by the knowledge of which treatment each student received (i.e., over and above the variance in the criterion accounted for by the covariates).

The difference between the R_F^2 and R_R^2 values is statistically tested with Equation 4.1, which was presented in chapter 4. To assist our discussion, Equation 4.1 is once again presented.

$$F(df_n, df_d) = \frac{(R_F^2 - R_R^2)/(df_n)}{(1 - R_F^2)/(df_d)} \qquad \text{(Equation 4.1)}$$

The number of linearly independent vectors in the Full Model is seven (i.e., one for the unit vector and one for each of the six predictor variables) and six in the Restricted Model (i.e., one for the unit vector and one each for the five covariate variables. Thus, the numerator degrees of freedom (df_n) value is 1 (i.e., 7 - 6); and with 200 students being included in the analysis, the denominator degrees of freedom (df_d) value is 193 (i.e., 200 - 7). If R_F^2 is equal to .53 and R_R^2 is equal to .45, the F test value would be calculated as follows:

$$F(1, \ 194) = \frac{(.53 - .45)/(1)}{(1 - .53)/(193)} = \frac{.08/1}{.47/193} = 32.85$$

The value of .08 (i.e., .53 - .45) is the proportion of the *sample* criterion variance uniquely due to knowledge of group membership, over and above the covariate knowledge. When this is divided by the numerator degrees of freedom, the result is a proportional *estimate* of the *population* criterion variance using the unique knowledge of group membership. One may want to label this $\hat{v}_u$, where u indicates the proportionally estimated population variance is due to *unique* knowledge of group membership.

The value of .47, which is (1 - .53) and located in the denominator of the F test, is the proportion of the *sample* criterion variance unaccounted for by the variables in the Full Model, and is called *error variance*. When this value is divided by the denominator degrees of freedom, one obtains a proportional *estimate* of the *population* criterion variance that is the most stable estimate. (See chapter 2 for a review.)

One can determine how often an F value of 32.85 or larger is observed due to sampling variation by examining the F test's corresponding probability value. The corresponding probability value for the F value of 32.85 is less than .001. Remember, this analysis involved a directional Research Hypothesis. Thus, the probability value would be divided by two to obtain the one-tailed probability value, which is, of course, still less than .001. This one-tailed probability value would be compared to an alpha value set by the researchers, say .05. Since this one-tailed probability value was less than .05, the researchers would judge the difference between the adjusted mean post-treatment reading scores of the students taught by the new curriculum and the adjusted mean post-treatment reading scores of the students taught by the current curriculum to be statistically significant.

Before the Statistical Hypothesis can be rejected, however, the researchers must verify that the adjusted mean post-treatment reading score of the new curriculum group exceeded the adjusted mean post-treatment reading score of the current curriculum group, as required by the directional Research Hypothesis. The researchers could verify if this was the case by examining the sign of the coefficient for the NC variable (i.e., a_1). Since, as previously discussed, the a_1 was equal to the adjusted mean post-treatment reading score of the new curriculum group minus the adjusted mean post-treatment reading score of the current curriculum group, the sign of a_1 must be positive to meet the requirement stated in the directional Research Hypothesis. If the one-tailed probability of the F test is less than the established alpha level *and* the sign of the a_1 coefficient is positive, the Statistical Hypothesis would be rejected, and the researchers would have evidence that supports the Research Hypothesis.

A t Test of the Group Coefficient in the Full Model

The previous section discussed a procedure that involved an F test of the difference between the R^2 values of the two models. Since the numerator degrees of freedom value used in the F test was equal to one, that is, one restriction was placed on the Full Model, the researchers could t test the group variable coefficient contained in the Full Model to obtain the same results. In this section we present how researchers would implement this statistical testing procedure.

When using the t test of the treatment group variable to determine whether the difference between the adjusted means of the groups was statistically significant, researchers would, first, construct the Full Model (Model 8.1), which, again, is as follows:

$$Y1 = a_0U + a_1NC + c_1IQ + c_2G + c_3Pre + c_4SC1 + c_5SC2 + E_1 \quad \text{(Model 8.1)}$$

Next, the researchers would examine the t test of the coefficient for the NC (group) variable and its corresponding (nondirectional) probability value, which is provided routinely by the SPSS for Windows and Microsoft Excel programs. Since a directional Research Hypothesis was posed, the researchers would divide this probability value by two to obtain the one-tailed probability value. Two conditions must be verified by the researchers to enable them to reject the Statistical Hypothesis. First, the one-tailed t test probability value must be less than the established alpha level. Second, the coefficient for the NC variable (a_1), which is equal to the adjusted mean post-treatment reading score of the new curriculum group minus the adjusted mean post-treatment reading score of the current curriculum group, must be positive.

One important value that is not directly provided by this procedure is the amount of unique variance in the criterion variable accounted for by the treatment-group variable, which can be reported as the effect size. This value can be obtained, however, by utilizing Equation 6.4, which was discussed in chapter 6. Equation 6.4 is as follows:

$$\Delta R^2 = \frac{(t)^2 \, (1 - R^2)}{df_d}$$

(Equation 6.4)

where:
1. The ΔR^2 symbol represents the change in the R^2 value associated with the inclusion of the treatment-group variable in the model.
2. The t^2 symbol represents the square of the t-test value for the treatment-group variable.
3. The quantity $(1 - R^2)$ is the proportion of the variation in the criterion variable not accounted for by the covariate variables in the model.
4. The df_d symbol is the degrees of freedom of the denominator for the model, which is equal to the sample size minus a quantity consisting of the number of predictor variables (i.e., the treatment-group variable and covariate variables) in the model plus 1.

To calculate the amount of unique variance in the post-treatment scores (i.e., the criterion variable) that is accounted for by the treatment-group variable (ΔR^2), three values must be known: (a) the t test value for the coefficient of the treatment-group variable, (b) the R^2 value for the Full Model (Model 8.1), and (c) the residual or the denominator degrees of freedom (df_d) value for the Full Model. Nearly all computer statistical software, including SPSS for Windows and Microsoft Excel, will provide these three values.

In the previous section the R^2 and the df_d values were listed as .53 and 193, respectively. Since the F test value was 32.85 and the value for a_1 (the coefficient for the group variable) was positive, the t test value for the coefficient of the treatment-group variable would be 5.75. (How do we know the t test value is 5.75? If the numerator degrees of freedom value in the F test is 1, the positive square root value of the F value is equal to the absolute value of the t test value; and the positive value of a_1 indicates that the t value is positive.)

Substituting the t test value (5.75), the R^2 value (.53), and the df_d value (194) into Equation 6.4 results in the following:

$$\Delta R^2 = \frac{(5.75)^2 \ (1 - .53)}{194} = .08$$

This .08 value, which matches the corresponding value listed in the previous section, indicates the proportion of unique variation in the criterion variable that was accounted for by the treatment-group variable.

Utility of Analysis of Covariance

It is important to note that in this text we take a position somewhat different from others regarding the legitimacy of the covariance analysis. Some statisticians take the position that the lack of random assignment disallows a meaningful conclusion. Our position is that research and decisions must be made in the real world. Random assignment of groups is ideal, but insight can be gained when this is not possible. Our emphasis on replication (discussed in full in chapter 12) is a check on any bias that might occur from not having equivalent groups. *Analysis of covariance cannot completely fix a badly designed study but controlling for confounding variables is better than ignoring them.*

In the previous curriculum example the reader should realize that if the coefficient for the NC variable (a_1) in Model 8.1 is negative, which indicates that the new curriculum group had a lower adjusted mean post-treatment reading score than did the current curriculum group—opposite from what was hypothesized—the researchers should not report the results as "significant in the opposite direction" because that was not the question under investigation. (Review our discussion of this point in chapter 4 in the "Analysis of ARH 4.1 with Microsoft Excel.") Indeed, given the apparent debilitating influence of the new curriculum, the researchers should be surprised because their careful planning, based upon past knowledge, was *not* supported. Several questions may be worth pursuing, such as the following:

1. Is the criterion measure appropriate?
2. Did the teachers sabotage the program?
3. Is the method interacting with one or more of the covariates?
4. Is the innovative treatment really not that good after all?
5. Were the data collected and analyzed properly?

If, upon tracking down possible contaminants, the researchers find no explanation, the researchers may then replicate the study (with the new Research Hypothesis opposite to the original). Given that the new Research Hypothesis can be accepted, they should publish their results to alert the reading research community to a possible flaw in the theoretical knowledge in their field, assuming that anyone cares that the new curriculum is significantly worse than the current curriculum.

What Does the Over and Above Analysis Do?

The hypothetical study just presented is rather complex, but it is just such complexity of applied research that makes the analysis of covariance useful. For a conceptual understanding of covariance, consider the following less complex hypothetical study.

Suppose in a given study the researchers wanted to test the influence of an innovative reading program relative to the current reading program. In this study 20 male students and 20 female students were randomly divided into the two methods. The researchers believed that initial reading ability would be a factor in the students' posttest (i.e., post-treatment) reading scores (i.e., the criterion variable). If the two treatment groups used by the researchers have moderately different means on the pretest reading scores, any observed posttest difference between groups is likely to be influenced by those pretest mean differences.

Note two items regarding this study. First, each group contains an equal number of male and female students, thus gender cannot be a competing explanation of a possible difference between posttest means of the groups. Second, the study utilized a relatively small sample size. This may be a possible concern to the researchers, which we will discuss in detail later in this chapter.

Graphical depiction of the analysis. Figures 8.1 and 8.2 provided graphical representations of the analysis of the hypothetical study. An inspection of Figure 8.1 reveals that on the pretest reading scores Method 1 (i.e., the new reading instructional method) had a mean of 4.9 and Method 2 (i.e., the currently used reading instructional method) had a mean of 4.2. Method 1 also had a higher posttest mean ($\overline{Y}$ = 5.7) than did Method 2 ($\overline{Y}$ = 4.9), but notice that both groups improved.

If pretest differences were not statistically taken into account, the researchers may conclude that Method 1 is .8 of a point better than Method 2; however, Method 1 was initially .7 of a point better on the pretest. On inspection of the two lines of best fit one should note that across the range of scores where the two groups overlap only about a tenth of a point (i.e., 0.1) difference exists between the methods, which is in favor of Method 1. These pretest group differences should be statistically controlled to determine if Method 1 is superior to Method 2 over and above the criterion variance accounted for by the pretest scores.

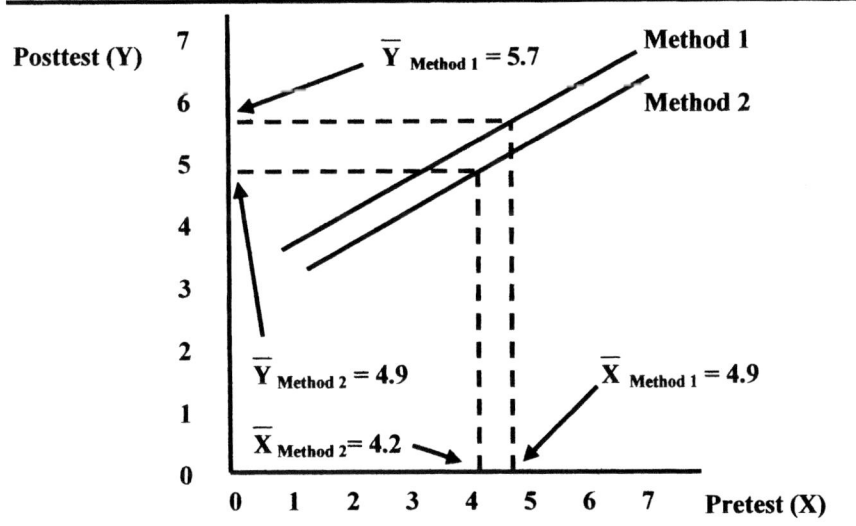

Figure 8.1.
The lines of best fit representing Method 1 and Method 2 across the range of pretest scores on reading as related to the posttest scores.

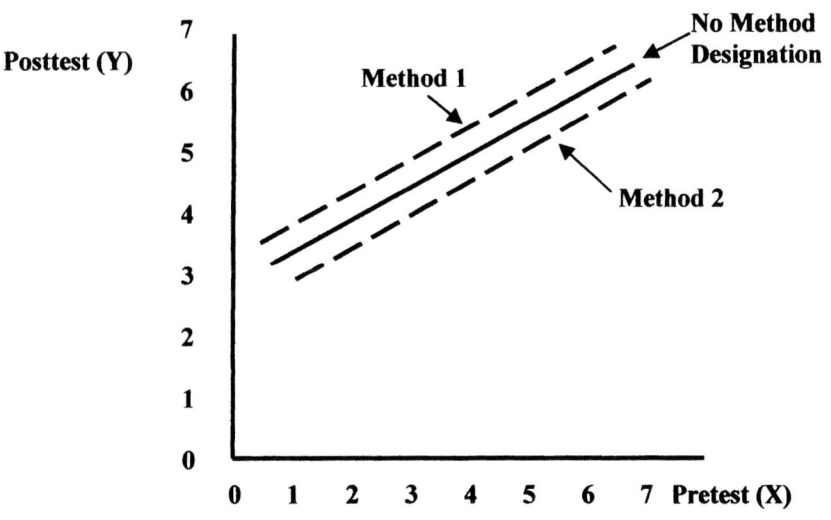

Figure 8.2.
The line of best fit without knowledge of group membership—The dashed lines are the superimposed lines of best fit shown in Figure 8.1.

The solid line in Figure 8.2 depicts the relationship of the pretest reading scores with the posttest reading scores without knowledge of the treatment-group membership. The dashed lines are taken from Figure 8.1. It should be obvious that the moderate pretest differences are accounting for the observed posttest group differences, therefore knowledge of treatment-group membership is almost totally redundant with knowledge of pretest scores. When this is observed, one can conclude that, over and above pretest scores, observed group differences are chance differences due to sampling variation.

Full and Restricted Models used to analyze the data. Now let's turn our attention to how the researchers would statistically analyze the data for this study. To statistically test whether Method 1 is superior to Method 2 with respect to the posttest reading scores adjusting for pretest reading score differences between the methods, the following Research Hypothesis could be stated:

Research Hypothesis (Version 1): For the population, Method 1 is superior to Method 2 on posttest reading, over and above pretest scores.

This Research Hypothesis can be stated in several ways. The following two alternate wordings are equivalent to the first Research Hypothesis:

Research Hypothesis (Version 2): For the population, Method 1 is superior to Method 2 on posttest reading when the pretest differences are statistically controlled.

Research Hypothesis (Version 3): For the population, Method 1 is superior to Method 2 on posttest reading when the pretest differences are covaried.

The corresponding Statistical Hypotheses are as follows:

Statistical Hypothesis (Version 1): For the population, Method 1 is *not* superior to Method 2 on posttest reading, over and above the pretest scores.

Statistical Hypothesis (Version 2): For the population, Method 1 is *not* superior to Method 2 on posttest reading when the pretest differences are statistically controlled.

Statistical Hypothesis (Version 3): For the population, Method 1 is *not* superior to Method 2 on posttest reading when the pretest differences are covaried.

To statistically test any one of the versions of the Statistical Hypotheses (we will assume the researchers elected to use Version 1), the following Full Model is constructed and estimated:

$$Y1 = a_0U + a_1M + c_1Pre + E_1 \qquad\qquad\qquad (\text{Model 8.3})$$

where:
1. $Y1$ = posttest reading score;
2. U = the unit vector;
3. $M = 1$ if the posttest reading score is from an individual in Method 1, 0 otherwise;
4. Pre = pretest score; and
5. a_0, a_1, and c_1 are regression coefficients calculated so as to minimize the sum of the squared elements in vector E_1.

Since the Research Hypothesis requires the adjusted mean posttest reading score of Method 1 to exceed the adjusted mean posttest reading score of Method 2, the value for the method variable (M) coefficient (a_1) must be positive to satisfy this condition.

The Statistical Hypothesis states that the adjusted mean posttest reading score of Method 1 does not exceed the adjusted mean posttest reading score of Method 2, thus the restriction placed on the Full Model (Model 8.3) is as follows:

$a_1 = 0$ (1 restriction)

Placing this restriction on the Full Model results in the following Restricted Model:

$$Y1 = a_0U + c_1Pre + E_2 \qquad\qquad\qquad (\text{Model 8.4})$$

There are three linearly independent vectors in the Full Model (U, M, and Pre) and two linearly independent vectors in the Restricted Model (U and Pre).

Statistical test. The Statistical Hypothesis can be tested with an *F* test of the difference between the R^2 value of the Full Model (R_F^2) and the R^2 value of the Restricted Model (R_R^2). Assuming the R_F^2 and R_R^2 values are .60 and .575, respectively; and the sample size is 40, the *F* test of the differences between these R^2 values would be calculated as follows:

$$F = \frac{(.600 - .575)/(3 - 2)}{(1 - .600)/(40 - 3)} = 2.31$$

The probability corresponding to this *F* value, which the computer software would provide to the researchers, would be .14. Since the Research Hypothesis was directional, the researchers would divide the probability value of .14 by 2 to obtain the one-tailed probability of .07. With the one-tailed probability being greater than the established alpha level, say, of .05, the researchers would fail to reject the Statistical Hypothesis (Method 1 is *not* superior to Method 2 on posttest reading, over and above the pretest scores). The researchers might still maintain a belief, appropriately, that the Research Hypothesis is true—but the relatively small treatment effect examined with a sample of only modest size led to a relatively small F value and, therefore, a high one-tailed probability value. Indeed, the Research Hypothesis may yet be true; but based upon the study's findings the Statistical Hypothesis cannot be refuted. As previously stated, we will address this concern later in this chapter when the concept of statistical power is examined.

The analysis of the difference between the two treatments while statistically controlling for the effects of one confounding variable is presented in the following sections. The section entitled "General Research Hypothesis 8.1" provides a general guide to this type of analysis, while the section entitled "Applied Research Hypothesis 8.1" presents a numerical example.

GENERAL RESEARCH HYPOTHESIS 8.1

General Research Hypothesis 8.1 requires an analysis that involves the use of covariates. The criterion variable is a continuous variable, which contained posttest scores and was labeled Y1. One of the predictor variables is a dichotomous variable that identifies which of the two treatment groups each study participant was a member. In this grouping variable, which was labeled T, a value of 1 represented the members of the Treatment 1 and a value of 0 represented members of the Treatment 2. The participants were not randomly assigned—a common characteristic of studies in education and the social sciences.

To statistically adjust for possible differences between the two treatment groups, three covariates were included in the analysis. The first covariate represented the gender of each participant. In this gender variable, which was labeled G, a value of 1 represented the females and a value of 0 represented males. The second covariate, which was labeled Pre, consisted of the participants' pretest scores. The third covariate, which consists of a series of three dichotomous variables (EM1, EM2, and EM3), represented the employment status of the participants. The dichotomous variables of EM1, EM2, and EM3 represented no employment, part-time employment, and full-time employment, respectively. Note that the GLM approach can handle the three different types of covariates used in this study—a dichotomous variable (gender), a continuous variable (pretest scores), and a series of dichotomous variables (employment status).

The research question being address is: Does the mean posttest score of Treatment 1 exceed the posttest score of Treatment 2 adjusting for differences in the participants' gender, pretest scores, and employment status? The directional General Research Hypothesis 8.1 (GRH 8.1) and its corresponding General Statistical Hypothesis 8.1 (GSH 8.1) are as follows:

Directional GRH 8.1: For a given population, the posttest mean of Treatment 1 will be higher than the posttest mean of Treatment 2 over and above the influences of gender, pretest scores, and employment status.

Directional GSH 8.1: For a given population, the posttest mean of Treatment 1 will *not* be higher than the posttest mean of Treatment 2 over and above the influences of gender, pretest scores, and employment status.

The Full Model required to reflect the relationships stated in GRH 8.1 is as follows:

$$Y1 = a_0U + a_1T + c_1G + c_2Pre + c_3EM2 + c_4EM3 + E_1 \qquad \text{(Model 8.5)}$$

Note that this Full Model contains, in addition to the unit vector, the treatment variable (T) and the information required to represent the four covariates. That is, variables G (gender), Pre (pretest scores), EM2 and EM3 (employment status). Recall that when a model contains the unit vector, which is the case for this Full Model (Model 8.5), only two of the three employment variables can be entered into the model—the remaining employment variable (EM1) is redundant information.

Since the Statistical Hypothesis (GSH 8.1) requires the posttest means of the two treatment groups not to differ, the restriction placed on the Full Model (Model 8.5) is as follows:

$$a_1 = 0 \quad \text{(1 restriction)}$$

Placing this restriction into the Full Model (Model 8.5) produces the following restricted model.

$$Y1 = a_0U + c_1G + c_2Pre + c_3EM2 + c_4EM3 + E_1 \qquad \text{(Model 8.6)}$$

Since a_1 was set equal to 0, the treatment variable has been eliminated from this Restricted Model, but the Restricted Model retains the covariates.

The difference between the R^2 values of the Full Model (Model 8.5) and the Restricted Model (Model 8.6) would be tested with an F test. This F test would have numerator and denominator degrees of freedom values of 1 (i.e., 6 - 5) and $N - 6$, respectively. Recalling that GRH 8.1 was a directional Research Hypothesis, two conditions must be met before the researchers would reject GSH 8.1. First, the one-tailed probability of the F test must be less than the established alpha level. Second, since the GRH 8.1 called for the adjusted posttest mean of Treatment 1 to be higher than the adjusted posttest mean of Treatment 2; and since the coefficient for the method variable (i.e., a_1) estimates how much higher or lower the adjusted posttest mean of Treatment 1 is as compared to the adjusted posttest mean of Treatment 2, the coefficient a_1 must be positive.

Based on the fact that only one restriction was placed on the Full Model (i.e., $a_1 = 0$), the researchers could also determine whether the GRH 8.1 should be rejected by using the t test of the treatment coefficient (a_1). Once again, two conditions must be satisfied before the researchers could reject GSH 8.1 and suggest that the data support GRH 8.1. First, the one-tailed probability of the t test of the treatment coefficient must be less than the established alpha level. Second, the value for the a_1 coefficient must be positive.

If the researchers reject GSH 8.1, they should use Equation 6.4, which was presented in chapter 6 and in a previous section of this chapter, to calculate the amount of unique variation in the criterion variable that was accounted for the methods. This value could serve as a measure of effect size for the treatment variable.

APPLIED RESEARCH HYPOTHESIS 8.1

Applied Research Hypothesis 8.1 (ARH 8.1) illustrates the type of analysis presented in the previous section, that is, an analysis of covariance with the GLM. In this section the criterion variable (X8) consists of post-treatment anxiety scores. The key predictor variable is a dichotomous variable that identifies whether a study participant received individual counseling sessions (Treatment 1) or group counseling sessions (Treatment 2). This treatment variable (X6) consisted of the values of 0 (Treatment 1) and 1 (Treatment 2). The researchers decided to statistically control for two characteristics of the participants. One characteristic was age, which was represented by a continuous variable (X4); and the other characteristic was the pre-treatment anxiety score (X9).

The directional research question posed by the researchers is as follows:

Is the mean post-treatment anxiety score for Treatment 2 less than the mean post-treatment anxiety score for Treatment 1 adjusting for age and pre-treatment anxiety scores?

The directional Applied Research Hypothesis (ARH 8.1) that reflects this directional research question is as follows:

Directional ARH 8.1: The mean post-treatment anxiety score for Treatment 2 is less than the mean post-treatment anxiety score for Treatment 1 adjusting for age and pre-treatment anxiety scores.

Corresponding to ARH 8.1 is the Applied Statistical Hypothesis 8.1 (ASH 8.1), which is as follows:

Directional ASH 8.1: The mean post-treatment anxiety score for Treatment 2 is *not* less than the mean post-treatment anxiety score for Treatment 1 adjusting for age and pre-treatment anxiety scores.

Full and Restricted Models

The Full Model that contains the relationships stated in ARH 8.1 is as follows:

$$X8 = a_0U + a_1X6 + a_2X4 + a_9X9 + E_2 \hspace{2cm} \text{(Model 8.7)}$$

Since ASH 8.1 requires the post-treatment anxiety means of Treatments 1 and 2 to be equal, the restriction required by ASH 8.1 is as follows:

$$a_1 = 0 \quad \text{(1 restriction)}$$

Forcing this restriction into the Full Model (Model 8.7) results in the following Restricted Model:

$$X8 = a_0U + a_2X4 + a_9X9 + E_1 \hspace{2cm} \text{(Model 8.8)}$$

An F test of the difference between the R^2 values of the Full Model (Model 8.7) and the Restricted Model (Model 8.8) will indicate whether the adjusted post-treatment anxiety means of Treatment 1 and Treatment 2 differ. (Note that a one-tailed F test probability will be used due to the fact the ARH 8.1 was directional.) Before ASH 8.1 can be rejected and ARH 8.1 can be supported, however, the adjusted post-treatment anxiety means of Treatment 2 must be less than the adjusted post-treatment anxiety means of Treatment 1. To satisfy this condition, the coefficient for the treatment variable (a_1) must be negative. If ASH 8.1 is rejected, the researchers will identify the proportion of variance in the post-treatment anxiety scores accounted for by the treatments, which will serve as a measurement of effect size.

Analysis ARH 8.1 with SPSS for Windows

The data file used in conjunction with ARH 8.1 is entitled "GLM DATA SPSS FORMAT," and it is located on the internet site (see Appendix A). Once this file is accessed and the commands needed to access the Linear Regression menu (see the chapter 4) have been completed, the menu presented in Exhibit 8.1 will be displayed.

The criterion variable (X8) and the predictor variables (X4 and X9) contained in Restricted Model (Model 8.8) must be entered into the Linear Regression menu. As illustrated in Exhibit 8.1, these variables are entered as follows:

1. Click on the X8 variable (see Oval 1) and click on the arrow key next to the "Dependent" box. This will identify X8 as the criterion variable.
2. While holding down the control key on the keyboard, click on the X4 and X9 variables (see Oval 1). Release the control key and click on the

arrow key next to the "Independent(s)" box. This will identify these variables as the predictor variables in the Restricted Model.
3. Click on the "Next" box (see Oval 3).

Exhibit 8.1
SPSS Linear Regression menu—Restricted Model (Model 8.8) for ASH 8.1.

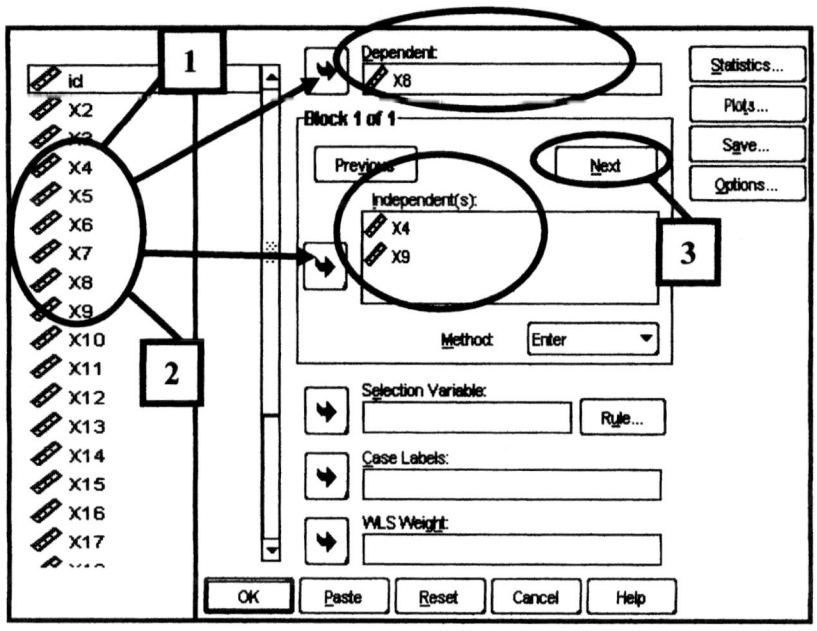

Once these three steps are completed, the Linear Regression menu will appear as displayed in Exhibit 8.2. Complete the following steps in this menu:
1. Click on variable X6 (see Oval 1) and click on the arrow key next to the
   "Independent(s)" box.
2. Click on the "Statistics" box (see Oval 2).
After these two steps are executed, the statistics menu will appear as displayed
in Exhibit 8.3. Complete the following steps in this menu:
1. Click on the boxes in front of "R squared change," "Model fit" and "Estimates" (see Oval 1).
2. Click on the "Continue" box (see Oval 2).
The Linear Regression menu will reappear as displayed in Exhibit 8.2. Click on
the "OK" key (see Oval 3) and the output for both the Full Model (Model 8.7)
and the Restricted Model (Model 8.8) will be generated as displayed in Exhibit
8.4.

Exhibit 8.2.
SPSS Linear Regression menu—Full Model (Model 8.7) for ARH 8.1 and the *F*
test of ASH 8.1

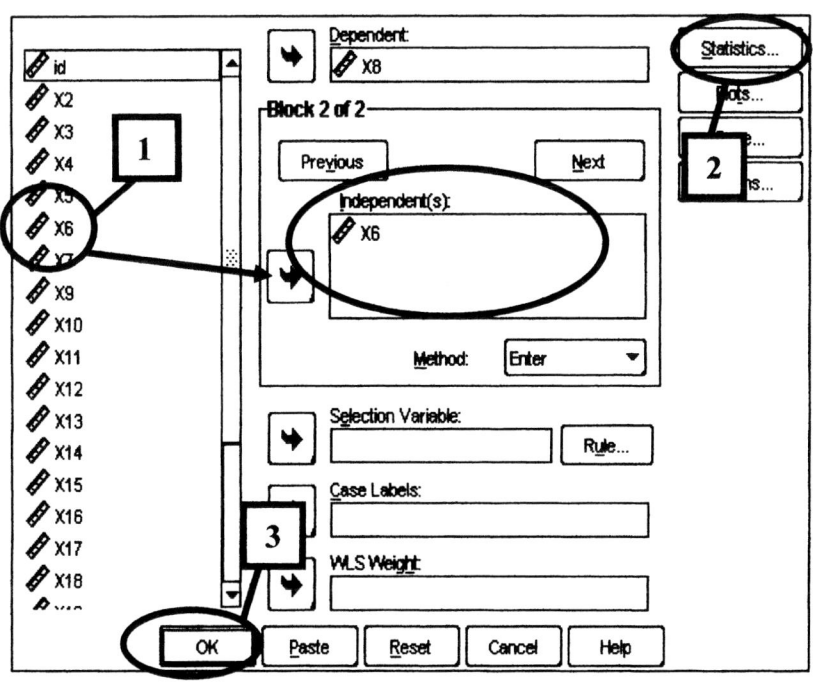

The key information contained in the output window (see Exhibit 8.4) is as
follows:

1. The Restricted Model (Model 8.8) is labeled Model 1 (see Oval 1), and
   its R^2 value is .890 (see Oval 2).
2. The Full Model (Model 8.7) is labeled Model 2 (see Oval 1), and its R^2
   value is .903 (see Oval 2).
3. The difference between the R^2 values of Model 2, that is the Full Model
   (Model 8.7), and Model 1, that is, the Restricted Model (Model 8.8), is
   .013 (see Oval 3).
4. The *F*-test value of the difference between the R^2 values of the Full and
   Restricted Models is 7.652 (see Oval 3).
5. The numerator degrees of freedom value (*dfn*) value of 1 for the *F* test is
   found under the column heading "Change Statistics df1" in the row containing the information for Model 2 (see Oval 3). The denominator degrees of freedom value (*dfd*) value of 56 for the *F* test is found under the

column heading "Change Statistics df2" in the row containing the information for Model 2 (see Oval 3).

6. The F-test probability value is .008, which is listed under the heading "Sig. F Change" for Model 2(see Oval 3). Since ARH 8.1 is directional, this probability value of .008 is divided by 2 to produce the one-tailed probability value of .004. This probability value will be expressed as $p / 2 < .01$ in the researchers' report.

7. The coefficient for the treatment variable (X6) is -3.356 (see Oval 4).

Exhibit 8.3.
SPSS Statistics menu—R squared change for the F test of ASH 8.1.

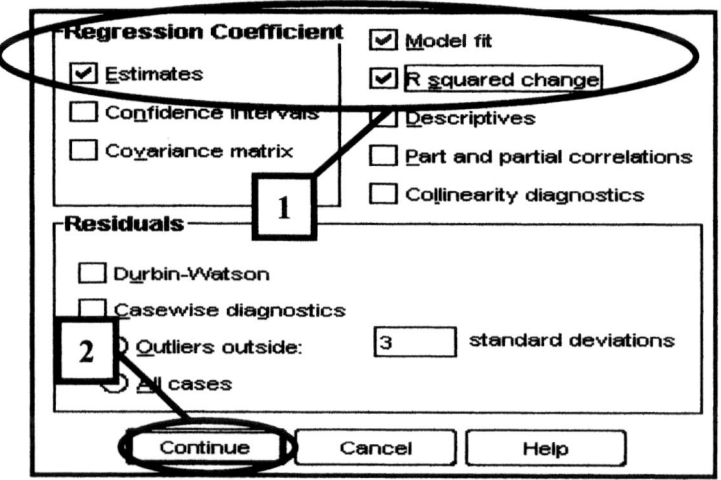

The Statistical Hypothesis ASH 8.1 will be rejected only if two conditions are satisfied. First, the one-tailed probability of the F test of the difference between the R^2 values of the Full and Restricted Models ($\Delta R^2 = .013$) must be less than the established alpha level (i.e., .05). A review of the information provided in the output window (see Oval 3 in Exhibit 8.4) indicates that the one-tailed probability value ($p / 2 < .01$) of the F test is less than the alpha level of .05. Thus, the difference between the R^2 values of the Full and Restricted Models ($\Delta R^2 = .013$) is statistically significant.

Second, the coefficient for the X6 variable (i.e., a_1) must be negative if the adjusted mean post-treatment anxiety score for Treatment 2 is less than the adjusted mean post-treatment anxiety score for Treatment 1. The value for a_1, which is -3.356, indicates that the adjusted mean post-treatment anxiety score for Treatment 2 is 3.356 points lower than the adjusted mean post-treatment anxiety score for Treatment 1.

Exhibit 8.4.
SPSS regression output—Restricted Model (Model 8.8) for ASH 8.1 and Full
Model (Model 8.7) for ARH 8.1

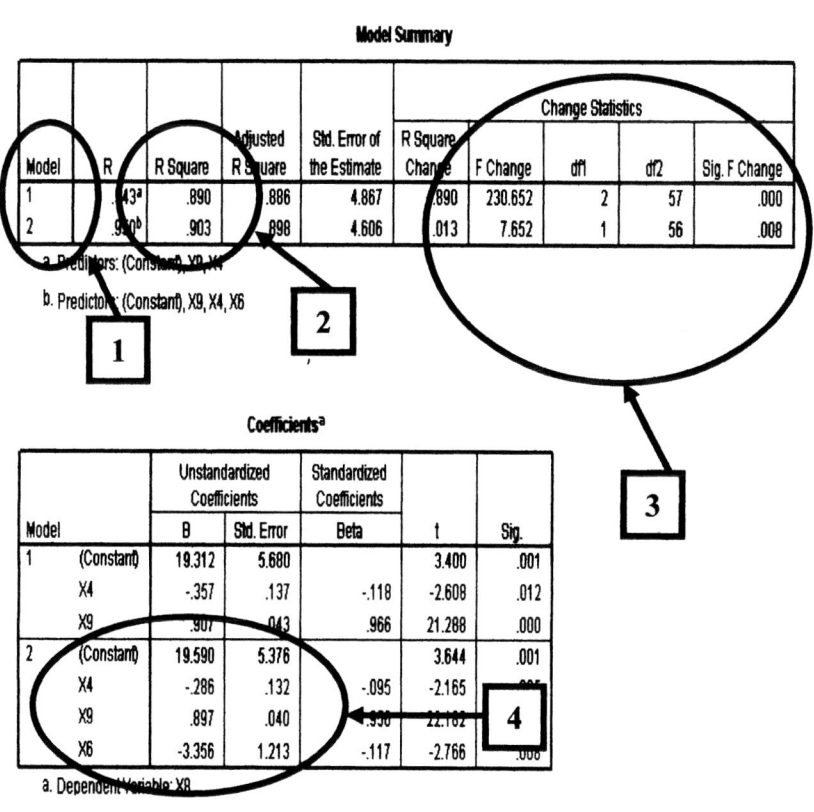

Model Summary

Model	R	R Square	Adjusted R Square	Std. Error of the Estimate	R Square Change	F Change	df1	df2	Sig. F Change
1	.943ᵃ	.890	.886	4.867	.890	230.652	2	57	.000
2	.950ᵇ	.903	.898	4.606	.013	7.652	1	56	.008

a. Predictors: (Constant), X9, X4
b. Predictors: (Constant), X9, X4, X6

Coefficientsᵃ

Model		Unstandardized Coefficients		Standardized Coefficients	t	Sig.
		B	Std. Error	Beta		
1	(Constant)	19.312	5.680		3.400	.001
	X4	-.357	.137	-.118	-2.608	.012
	X9	.907	.043	.966	21.288	.000
2	(Constant)	19.590	5.376		3.644	.001
	X4	-.286	.132	-.095	-2.165	.001
	X9	.897	.040	.950	22.162	.000
	X6	-3.356	1.213	-.117	-2.766	.008

a. Dependent Variable: X8

Since the one-tailed probability value of the F test was less than the alpha
level of .05 and the value for the treatment variable coefficient (a_1) was negative,
the researchers would reject the Statistical Hypothesis (i.e., ASH 8.1 is rejected);
and the researchers would be willing to support ARH 8.1, which states that the
adjusted mean post-treatment anxiety score for Treatment 2 is less than the adjusted mean post-treatment anxiety score for Treatment 1.

Writing the Results for ARH 8.1 in APA Style

The researchers could write in APA style the statistical test results for ARH
8.1 as follows:

The difference between the Full Model and Restricted Model R^2 values was .013. This value, which represented the amount of variation in the post-treatment anxiety scores uniquely accounted for by the treatments, was statistically significant at the .05 alpha level ($F(1, 56) = 7.65$, $p/2 < .01$). As indicated by the negative coefficient for the treatment variable in the Full Model (-3.36), the   adjusted mean post-treatment anxiety score for Treatment 2 was 3.36 points less than the adjusted mean post-treatment anxiety score for Treatment 1, which supports the Applied Research Hypothesis 8.1. The effect size, as measured by the amount of unique variation in the post-treatment anxiety means accounted for by the treatments (.013), is classified as a small effect size according to the criteria established by Cohen (1988).

Notice that, once again, the researchers' report contains the effect size along with the results of the statistical test.

ASSUMPTIONS OF THE ANALYSIS OF COVARIANCE

Analysis of covariance (the over and above question) makes one assumption in addition to those for the ANOVA. This added assumption is that the slope of the line for each of the k groups is the same across the range of the covariate. The assumption is imposed by the linear model and can easily be seen in the following linear equation, which does not contain a unit vector:

$$Y1 = a_1G1 + a_2G2 + a_3G3 + \ldots + a_kGk + c_1X1 + E_1 \qquad \text{(Model 8.9)}$$

There are k groups (i.e., G1, G2, . . ., Gk) and $k + 1$ coefficients (a_1, a_2, . . . , a_k, and c_1). Each group has its own Y-intercept, yet there is only one weight (c_1) associated with the covariate (X1). Since a one-unit increase on X1 will yield a c_1 increase in Y1, regardless of the group with which the score is associated, all lines by necessity are parallel (have the same slope), as illustrated in Figure 8.3 for four groups.

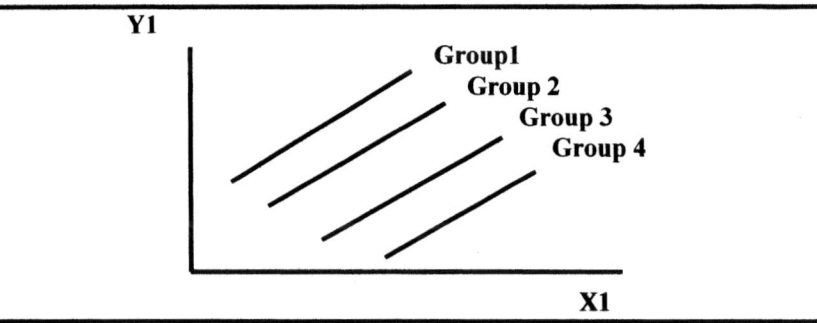

Figure 8.3.
Four parallel lines that represent homogeneous slopes across X1 (four groups).

The assumption just presented is referred to in statistical texts as *homogeneity of regression lines*. Although violations of ANOVA assumptions have been shown to be inconsequential when N is large, and thus the F test in the ANOVA is said to be robust, if the assumption of homogeneity of regression lines is violated when addressing an ANCOVA-type question, one should not pursue that question. While some statistical texts would have researchers "throw in the towel" at this point, we encourage researchers to carefully examine the resulting interaction and to treat the interaction as a valuable phenomenon in its own right. If the lines are not parallel, then interaction exists and can be tested with the procedures discussed in chapter 7 and further examined in the next section.

The Procedure Used to Test Groups that Assume Homogeneity of Regression Lines.

Consider the two diagrams contained in Figure 8.4. Suppose Method 1 received one treatment, and Method 2 received another treatment, and X1 was a potential covariate of interest. If one obtained the lines of best fit for each group independently, the models would appear as follows:

Method 1 Model
$Y1 = a_1 G1 + b_1 X1 + E_1$ (Model 8.10)

Method 2 Model
$Y1 = a_2 G2 + b_2 X1 + E_1$ (Model 8.11)

The plots of the regression lines generated by these two models may appear as depicted in Figure 8.4a. It would be apparent that the two slopes were not equal, that is, b_1 was not equal to b_2. The two groups do not have a common slope across the potential covariate.

If the assumption of the common slope is made, however, the Research Hypothesis would be as follows:

Research Hypothesis: For the population, the criterion mean for Group 1 is higher than the criterion mean for Group 2 across the range of values for variable X1.

The corresponding Statistical Hypothesis would be as follows:

Statistical Hypothesis: For the population, the criterion mean for Group 1 is *not* higher than the criterion mean for Group 2 across the range of values for variable X1.

It is important to note that the homogeneity of regression slopes is implicit in both the Research and Statistical Hypotheses, which is often overlooked by some researchers. The Full Model that reflects this Research Hypothesis is as follows:

$$Y1 = a_1G1 + a_2G2 + b_1X1 + E_1 \qquad\qquad \text{(Model 8.12)}$$

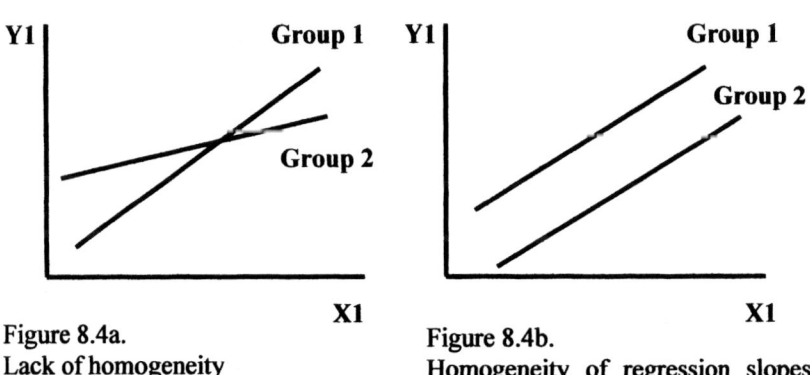

Figure 8.4a.
Lack of homogeneity
of regression slopes.

Figure 8.4b.
Homogeneity of regression slopes
with different Y intercepts (unequal
group means).

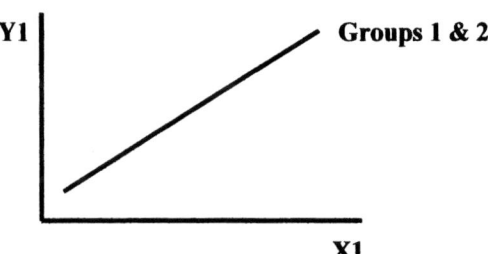

Figure 8.4b.
Homogeneity of regression slopes
with the same Y intercepts (equal
group means).

Figure 8.4.
Regression lines for Group 1 and Group 2.

This Full Model (Model 8.12) would yield coefficients (a_1, a_2, and b_1) for the condition contained in the Research Hypothesis and would yield lines of best fit as depicted in Figure 8.4b, which allows for two different Y-intercept points

(a_1 and a_2) but one common slope (b_1). Thus the Full Model (Model 8.12) allows for a difference between the group means that remains constant across the range of values for variable X1. It is important for researchers to understand that when they use such a model, the assumption of homogeneity of regression slopes is automatically a part of the analysis.

Since the Research Hypothesis requires that the criterion mean for Group 1 and Group 2 do not differ across the range of values of the X1 variable, the restriction placed on the Full Model is as follows:

$a_1 = a_2$ (1 restriction)

Placing this restriction into the Full Model (Model 8.12) produces the following Restricted Model:

$$Y1 = a_0 U + b_1 X1 + E_7 \qquad \text{(Model 8.13)}$$

In this Restricted Model, not only are the slopes of the two lines forced to be equal (i.e., the model contains the common slope of b_1), but the means of the two groups are forced to be equal also (i.e., the model contains the common Y intercept of a_0). The regression lines for the two groups, which reflect these restrictions, are depicted in Figure 8.4c. Since both regression lines have a common slope and a common Y intercept point (i.e., means), the one regression line is superimposed onto the other regression line.

Holding tenable the Statistical Hypothesis is reasonable if one is willing to consider the treatment effects averaged over the whole range of the covariate; but averaging the effects results in ignoring a very systematic state of affairs, as pictured in Figure 8.4a. If the condition depicted in Figure 8.4a exists, the assumption of homogeneity of regression slopes is not tenable, and therefore the data should be analyzed differently.

Some researchers shun the ANCOVA for the very reason just presented. The covariate, however, is not necessarily the cause of the "problem." Researchers also may fail to detect a difference between group means when an important covariate is overlooked or, in fact, any time the Full Model does not describe the data well (i.e., does not achieve a high R^2). Indeed, no analysis—including an ANCOVA—should be made without the researchers taking time to reflect prior to the actual data collection on the variables that should be included in the study and the nature of the relationships among those variables. The reader should note that Figure 8.4a represents a state of affairs that was discussed extensively in chapter 7 as interaction, and the test for interaction is just the test one needs to check the assumption of homogeneity of regression slopes.

A Procedure Used to Test Groups that Does Not Assume Homogeneity of Regression Lines

Suppose the researchers have a new treatment and they want to test the superiority of the new treatment over the currently used treatment. Furthermore, the researchers "know" initial ability influences posttest performance, and they want to ask the question: Over and above the influence of pretest ability, is Treatment 1 superior to Treatment 2 in relation to posttest performance? Once this question is posed, it should force the researchers to reflect on a second question: Is this really the best question? Reflection on this second question requires the researches to think of whether there is anything about Treatments 1 and 2 that could make one of the treatments superior for a certain range of the pretest scores but not for another range?

Upon reflection on the two treatments, the researchers may have a suspicion that the new treatment will "really" work better for the individuals who score high on the pretest but not so well for the individuals who score low on the pretest. If this condition is verified through their testing procedure, it indicates that the slopes of the regression lines are not homogenous, that is, an interaction effect exists; and it would be necessary for them to interpret this interaction effect. If the researchers' suspicion regarding the lack of consistency of the differences between the group means across the range of pretest values is not supported by their testing procedure, they may next want to know: Is the new treatment (Treatment 1) superior to the current treatment (Treatment 2), over and above pretest ability measures?

This type of thought process would lead researchers to conduct their analysis through a two-stage analytic procedure. These two stages are as follows:

1. Stage 1: Test for directional interaction. If interaction is found to exist, the researcher would then plot the lines of best fit to determine the tenability of the *directional* hypothesis.
2. Stage 2: If *no* directional interaction is found, then the researcher would test the directional hypothesis that Treatment 1 is superior to Treatment 2 when pretest ability is covaried.

The sequence of hypothesis testing for the two-group case follows.

Stage 1: Testing Stage 1 Research Hypothesis. The Stage 1 Research Hypothesis would be:

Directional Research Hypothesis—Stage 1: The differences between the posttest scores of Treatments 1 and 2 increase across the range of pretest scores with the posttest scores of Treatment 1 exceeding those of Treatment 2.

The corresponding Statistical Hypothesis would be as follows:

Directional Statistical Hypothesis—Stage 1: The differences between the posttest scores of Treatments 1 and 2 are constant across the range of pretest ability.

To test the Research Hypothesis for Stage 1, the Full Model must have two coefficients associated with pretest ability (one for each treatment). The Full Model, which in this presentation does not contain the unit vector, would be constructed as follows:

$$Y1 = a_1T1 + b_1(T1 * X1) + a_2T2 + b_2(T2 * X1) + E_1 \qquad \text{(Model 8.14)}$$

where:
1. $Y1$ = the criterion of posttest scores;
2. $T1 = 1$ if the score on the criterion is from a subject given Treatment 1, 0 otherwise;
3. $X1$ = the pretest score;
4. $T2 = 1$ if the score on the criterion is from a subject given Treatment 2, 0 otherwise;
5. $(T1 * X1)$ = the pretest score if the individual is from Treatment 1, 0 otherwise;
6. $(T2 * X1)$ = the pretest score if the individual is from Treatment 2, 0 otherwise;
7. E_1 = the error vector; and
8. a_1, a_2, b_1 and b_2 are regression coefficients calculated to minimize the sum of the squared values in E_1.

To reflect the Statistical Hypothesis, the two lines of best fit must be forced to be parallel. This can be done by setting the two slopes equal to each other, $b_1 = b_2$, or equivalently both equal to a common weight, b_3, resulting in the following Restricted Model for Stage 1:

$$Y1 = a_1T1 + a_2T2 + b_3X1 + E_2 \qquad \text{(Model 8.15)}$$

If difficulty is encountered in understanding these models, the reader should review chapter 7 regarding interaction.

An F test is conducted to statistically test the difference between the R^2 values of the Full Model (Model 8.14) and the Restricted Model (Model 8.15). The numerator degrees of freedom value is equal to the number of linearly independent vectors in the Full Model (m1) minus the number of linearly independent vectors in the Restricted Model (m2), which is equal to 1 (i.e., 4 - 3). The denominator degrees of freedom value is equal to the sample size (N) minus m1 (i.e., $N - 4$).

Graphically, if the Research Hypothesis associated with Stage 1 is supported by the data, the regression lines of the two treatments may appear as plotted in either of the graphs contained in Figure 8.5. In Figure 8.5a, the lines of best fit do *not* cross in the range of interest, but an interaction exists because the higher the score on X1, the greater the difference between Treatment 1 and Treatment 2 on the criterion, favoring Treatment 1. Figure 8.5b shows the case where the lines of best fit cross within the range of interest. If the individual had an X1 score of better than 4, Treatment 1 yields higher criterion scores; but if an individual's X1 score is below 4, Treatment 2 yields higher criterion scores.

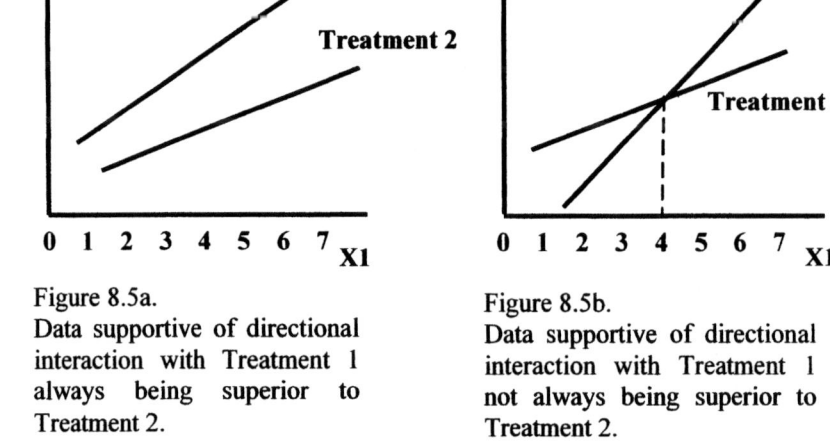

Figure 8.5a.
Data supportive of directional interaction with Treatment 1 always being superior to Treatment 2.

Figure 8.5b.
Data supportive of directional interaction with Treatment 1 not always being superior to Treatment 2.

Figure 8.5.
Diagrams of interaction between the treatments and covariate X1.

In Stage 1 directional interaction was statistically tested. It should be noted that if the researchers had tested for homogeneity of regression slopes with the "hope" of verifying that it existed, a nondirectional interaction would have been tested. Researchers recommending such a test (i.e., of homogeneity of slopes) are doing so with the intent of *wanting not to find significance* and therefore hope the data support the Restricted Model over the Full Model. If this is the researchers' intent, then we suggest they make sure that the Restricted Model is the preferred model by setting a high alpha level, at a level of, say, .60 or .70. When the Research Hypothesis describes the expected state of affairs, the researchers generally set the alpha level at somewhere around .01 or .05. The situation is conceptually inverted when the Statistical Hypothesis describes the expected state of affairs, so the appropriate alpha level would be closer to .60 or .70 (Cohen, 1970).

In Stage 1, as presented (a directional hypothesis for interaction), the researchers' expectation was that Treatment 1 would have a greater slope than Treatment 2. If the data indicated, on the other hand, that Treatment 2 has a much greater slope than Treatment 1, the Research Hypothesis of Stage 1 cannot be accepted, but one would *not* want to act as though the slopes were homogeneous. Whenever the results turn out to be in the direction opposite to that hypothesized, the researchers must go back to the theory "drawing board" to attempt to discover the anomaly. Therefore, if the slopes were quite discrepant in the direction opposite to that hypothesized in Stage 1, it would not be appropriate to conduct Stage 2 of the analysis.

Stage 2: Testing the Stage 2 Research Hypothesis. If in Stage 1 it was found that there was no interaction, the researchers would then proceed to conduct Stage 2 with the following Stage 2 directional Research Hypothesis:

Directional Research Hypothesis—Stage 2: Treatment 1 is more effective than Treatment 2 in producing higher posttest scores when the influence of pretest ability is covaried.

The Statistical Hypothesis for Stage 2 would be as follows:

Directional Statistical Hypothesis—Stage 2: Treatment 1 is not more effective than Treatment 2 in producing higher posttest scores when the influence of pretest ability is covaried.

Recall that the results obtained from Stage 1 lead the researchers to conclude that the regression lines had a common slope. Thus to test the Stage 2 Research Hypothesis, the Full Model must have two coefficients associated with the treatments (a_1 and a_2) but a common coefficient (b_3) for the covariate X1 (i.e., a common slope for the two regression lines). Thus, the Full Model for Stage 2 is:

$$Y1 = a_1 T1 + a_2 T2 + b_3 X1 + E_2 \qquad \text{(Model 8.15)}$$

The reader should note that this Full Model is the same as the Restricted Model associated with the Statistical Hypothesis in Stage 1, which is why it is labeled Model 8.15.

To reflect the Statistical Hypothesis for Stage 2, which requires the posttest means of the groups to be equal, the coefficients for the treatments must be set equal (i.e., $a_1 = a_2$). This common posttest mean, which is equal to the common Y intercept of the two regression lines, is labeled a_3. Since T1 and T2 are mutually exclusive vectors, which when added yield the unit vector, the Restricted Model for Stage 2 can be expressed as:

$$Y1 = a_3 U + b_3 X1 + E_3 \qquad \text{(Model 8.16)}$$

where a_3 and b_3 are regression coefficients calculated to minimize the sum of the squared elements in the new error vector, E_3.

In Stage 2 the Full Model (Model 8.15) allows for two parallel lines with different Y intercepts, and the Restricted Model (Model 8.16) forces the two lines into one common line for both treatments (i.e., two lines with the same slopes and same Y intercepts). If statistical significance is obtained when the difference between the R^2 values of Model 8.15 and Model 8.16 is obtained with df_n and df_d equal to (3 - 2) and (N - 3), respectively; and if a_1 is greater than a_2, then the Stage 2 Research Hypothesis is supported. The two lines of best fit will be parallel, and Treatment 1 will be more effective than Treatment 2, specifically (a_1 - a_2) units more effective, as estimated by the Full Model. If $a_1 = a_2$ or $a_1 < a_2$, the Research Hypothesis is untenable and Treatment 1 is not more effective than Treatment 2 when pretest scores are covaried.

Homogeneity of Regression Slopes with More than One Covariate

When researchers deal with complexities such as the first covariate example in this chapter, several covariates may be used and tests for homogeneity of regression slopes might not be made. If the researchers suspect interaction with any of the covariates, then that interaction should be tested because ignoring that interaction inflates the unknown variance. On the other hand, if the researchers have good reason not to expect interaction, parallel slopes can be assumed (or they could be tested for homogeneity) across the several covariates. Some group bias on the covariates of interest will be adjusted by using the over and above test. If treatment is interacting with at least one of the covariates, as shown in Figure 8.5a, the assumed parallel lines in the analysis of covariance will give a larger error term, which *may* or *may not* mask the superiority of the treatment. The major risk is that the researchers may discard a new treatment that is very good beyond a certain point on the covariate but not much better (and possibly inferior) at other levels. In other situations, the researchers would accept one treatment as better when the Statistical Hypothesis is rejected because of a difference in covariate means. The approach presented here is the best procedure at this time for dealing with groups that are not initially equivalent. The procedure is not entirely satisfactory, but it is better than assuming that the groups are equivalent.

Given the interaction illustrated in Figure 8.5b, forcing parallel lines (i.e., automatically assuming homogeneity of regression lines) will almost completely mask the fact that Treatment 1 is better for individuals who have pretest scores beyond 4 and inferior to Treatment 2 for individuals who score below 4 on the pretest. Whether one tests for homogeneity in an ANCOVA is a judgmental matter and depends upon expectations of how the treatments will function for individuals at different points across the covariate. It must be stressed that it is important for researchers to not *automatically* assume that homogeneity of regression slopes exists.

The ANCOVA models can be thought of in terms of lines, as depicted in Figures 8.1 through 8.5. If there is an interaction between the groups and the covariate, as in Figure 8.4a, then a line is needed for each group (two lines in Figure 8.4a). Each line needs a Y-intercept and a slope; thus if two lines are needed, four parameters must be determined (requiring four pieces of information in the Full Model). If there is no interaction (the lines are homogeneous) then they all have the same slope—but possibly different intercepts. The parallel-line model would be called for here. With two groups (Figure 8.4b), one would need two intercepts, but only one slope because the two lines are parallel. With four groups (Figure 8.3), one would need four intercepts but only one slope because the four lines are parallel.

The traditional ANCOVA Research Hypothesis compares the k-parallel-line model with the single-line model. Figure 8.1 contains two parallel lines, and Figure 8.2 contains only one line. The data can be modeled both ways; the question is: Which way is best in terms of high R^2 and few pieces of information? As indicated in point 2 of Table 8.1, the Full Model contains k + 1 pieces of information and the Restricted Model has two pieces of information. If there are two groups, as in Figure 8.1 and Figure 8.2, then k is equal to 2 and pieces of information in the Full Model is 3, and the pieces of information in the Restricted Model is 2.

A careful review of Table 8.1 reveals two important points regarding the test(s) for homogeneity of regression slopes and the test for differences among group means. First, the Restricted Model for the test of homogeneity of variance is the Full Model for the ANCOVA test. Likewise, the Restricted Model for the ANCOVA test is the Full Model for the correlation test.

Table 8.1
Relationships between the ANCOVA Models

1. Interaction between covariate and groups—test for homogeneity
   Full Model—k-lines model
     • line for each of k groups (k intercepts, k slopes)
   Restricted Model—parallel-lines model
     • k parallel lines (k intercepts, 1 slope)
2. Difference between groups, over and above covariate—traditional ANCOVA test
   Full Model—parallel-lines model
     • k parallel lines (k intercepts, 1 slope)
   Restricted Model—single-line model
     • one line (1 intercept, 1 slope)
3. Covariate related to criterion—test for correlation
   Full Model—single line model
     • one line (1 intercept, 1 slope)
   Restricted Model—unit vector model
     • one horizontal line (1 intercept, no slope)

Second, any model can be either a Restricted Model or a Full Model; it all depends upon the Research Hypothesis. For any Full Model, one must realize that there are "fuller" models, but assumes that the fuller models are not better models. What is usually the case is that some of these fuller models are better. If the R^2 is less than 1.00 for the Full Model, then better fuller models exist. With this Full Model, the researcher just has not obtained the necessary predictor information or has not expressed the data with the right functional relationship (the topic of the entire next chapter).

ANCOVA AND INCREASED POWER

Previously in this chapter we presented a study that served as a basis of discussion regarding the use of a covariate. Recall that in that study the researchers wanted to test the influence of an innovative reading program on student reading scores. In the study 20 male students and 20 female students were randomly divided into the two treatments and the researchers used an analysis that utilized the students' pre-treatment reading scores as a covariate. Based on the results of the analysis, the researchers were unable to reject the Statistical Hypothesis that stated: Treatment 1 is *not* superior to Treatment 2 with respect to post-treatment reading scores, over and above the pre-treatment scores.

It was further stated that the researchers might still maintain a belief, appropriately, that the Research Hypothesis is true—but the relatively small treatment effect examined with a *sample of only modest size* led to a relatively small F value and, therefore, a high one-tailed probability value. Indeed, the Research Hypothesis may yet be true; but based upon the study's findings the Statistical Hypothesis cannot be refuted. To understand the researchers concern, we should review the concepts of (a) an effect size, (b) a Type I error and (c) a Type II error, which were first introduced in chapter 2, and introduce the concept of the power of a statistical test.

Effect Size

With respect to the study designed to determine whether Treatment 1 is superior to Treatment 2, the researchers must be aware that their ability to determine if this state is true in the population is partially dependent on how much the mean of Treatment 1 exceeds the mean of Treatment 2. This difference, which is expressed in various forms, is referred to as the effect size. The effect size can be thought of as how far apart the means of two groups are in terms of standard deviation units (e.g., 1.2 standard deviations or .50 standard deviations). Another way of expressing the effect size is in terms of the proportion of variance accounted for in the criterion variable by the treatment variable. In correlational analyses the r^2 would provide this information; while in a one-way ANOVA and a two-way ANOVA designs the eta square value and the partial eta square value, respectively, would serve as a measure of effect size.

Type I Error

When conducting a statistical test of a Research Hypothesis, researchers will determine whether the Statistical Hypothesis should be rejected (e.g., they believe that Treatment 1 is superior to Treatment 2) or it should not be rejected (e.g., they believe Treatment 1 is not superior to Treatment 2). If the researchers reject the Statistical Hypothesis and Treatment 1 is truly not superior to Treatment 2 for the population, they are in error. Such an error is referred to as a Type I error. Alpha (α) is the probability of making a Type I error. Thus, when the researchers set alpha for this study, they were establishing the chance of committing a Type I error.

A legitimate question can be raised regarding the setting of the alpha level: If the researchers set the alpha level and it is the chance of committing a Type I error, why do they not always set it very high (say .20 rather than .05)? The answer to this question lies in the fact that the researchers could commit another type of error and the chance of committing this error is inversely related to the alpha level. This other type or error is referred to as a Type II error.

Type II Error

If the researchers fail to reject the Statistical Hypothesis and Treatment 1 is truly superior to Treatment 2 for the population, their decision is wrong. Such an error is referred to as a Type II error and beta (β) is the chance of committing this type of error. Although the researchers often do not reflect on the level of beta for their statistical test, test, say from .10 to .01, their decision regarding the level at which they will set alpha, however, directly influences the level of beta. As the level of alpha is decreased, the beta level increases, and, of course as the level of alpha is increased, the level of beta is decreased.

Power

A third statistical test concept, which is related to the concepts previously discussed (i.e., a Type I error, alpha, a Type II error, and beta) is the power of a statistical test. The power of a statistical test is the probability of detecting a population fact when that population fact truly exists (e.g., the post-treatment mean of Treatment 1 is higher than the post-treatment mean of Treatment 2). Since the probability of not detecting a population fact is defined as the probability of making a Type II error, power is then one minus the probability of making a Type II error. Researchers would obviously want the power level to be high for a statistical test they are conducting.

How to Increase Power

The researchers involved in the study designed to determine whether Treatment 1 is superior to Treatment 2 indirectly expressed a concern about the power level when they stated that the Research Hypothesis may be true even though they failed to reject the Statistical Hypothesis. Of course the time for the researchers to address their concern regarding the power level is at the initial stage of planning for the study. A number of factors could lead to a low power level, including:
1. a small sample size;
2. a stringent (low) alpha level;
3. the need to detect a small effect size;
4. a large error term used in the analysis of the sample data, which could be caused by any of the following:
   a. measurement instruments with low reliability and validity;
   b. not including individual differences if they are in the design;
   c. not blocking on known, relevant variables;
   d. not including nonlinear and interaction predictor variables if thought relevant; and
   e. not adjusting for initial difference between the groups (i.e., not using covariates).

Note that each of these criteria has an associated cost. The researchers could increase the sample size, however, that would require more time and financial resources. They could change the alpha level from .05 to .10, but that would increase the chance of a Type I error. The researchers could decide to detect only large effect sizes, however, this would dictate that smaller effect sizes are of no interest. They could reduce the error term used in the analysis by changing one or more of the sub-points listed above under the error term factor (point 4).

It is the realization that researchers can increase the power of a statistical test by adjusting for initial differences between the groups that we want to stress. The use of covariates will generally increase the R^2 value of the Full Model, and thus increase the power of the statistical test being conducted. Thus, it is important for researchers to understand that one benefit of utilizing covariates is the potential increase in the power levels of the statistical tests conducted through such an analysis.

ARGUMENT FOR THE USE OF ANCOVA

Researchers are urged to use ANCOVA even when random assignment is made and even when the group means of the covariates are nearly identical. Mueller (1990) discussed this use of ANCOVA and stated concerns for using the technique when groups have not been randomly assigned. We, in this text, take the position that the statistical technique is "unaware" of whether the subjects have been randomly assigned. Making a covariance adjustment is better than not making one.

In cases where pretest ability is known, the inclusion of these data in an over and above analysis will usually provide a better estimate of within-group variance ($\hat{v}_w$). Usually people who score high on the pretest will score relatively high on the posttest, and those who score low on the pretest will usually score fairly low on the posttest. The correlation between pretest and posttest is often greater than zero. Therefore, the R^2 of a model containing knowledge of both treatment and pretest scores will be larger than an R^2 of a model using only knowledge of treatment. The over and above analytic procedure still statistically tests the unique contribution of the independent variable (e.g., treatment), but the proportional estimate of the population variance within, as measured by $(1 - R_F^2) / (N - m1)$, will be smaller at the expense of having only one less degree of freedom (due to the inclusion of the covariate) and at the cost of collecting the covariate score. Essentially, the reasoning is: Why throw away knowledge regarding the sample when one has it? If the task of researchers is to strive to account for as much of the variance as possible (R^2 as close to 1.0 as possible), then it may be appropriate to go further and recommend that the researchers include many covariates that account for nonrandom criterion variance. Chapter 1 was written with this viewpoint.

A word of caution is in order here. The ideal covariate is one that is *not* correlated with the variance in the criterion variable accounted for by the predictor variable or variables, but *is correlated* with the criterion variable. If the covariate is correlated with the variance in the criterion variable accounted for by the predictor variable or variables, the predictor variables' effects are confounded with the covariate. In this case, the amount of variance in the criterion variable accounted for by the predictor variables will be attributed to the covariate, and what might have originally been an effective predictor or predictors may go undetected. Such are the trials and tribulations of researchers in the behavioral sciences.

In summary, unfortunately researchers must conduct studies that manipulate treatments with *intact* groups (i.e., the participants have not been randomly assigned). The use of ANCOVA by behavioral science and educational researchers can provide useful information in such situations, and we encourage researchers to conduct such analyses until future developments in statistical techniques provide a better analytic procedure.

CHAPTER 9

NONLINEAR RELATIONSHIPS

Some researchers assume the GLM deals only with straight lines. In this chapter we illustrate various types of non-rectilinear (also called nonlinear or curvilinear) relationships that researchers may encounter in their research and analyze with the GLM. Observed nonlinear sample data may be due to at least two conditions. First, the theoretical expectation is reasonable regarding a nonlinear function existing among the predictor variable(s) and the criterion variable (e.g., the curvilinear relationship in Newton's Law, $d = 1/2gt^2$). Second, the scaling of the X and Y variables is arbitrary, so departures from a rectilinear relationship may be a scaling artifact. The two major sections of this chapter deal with these two conditions.

Before discussing these two conditions, the concepts of homoscedasticity and heteroscedasticity are examined. The readers who are tempted to skip this section should instead at least skim the material because a basic understanding of these concepts may be needed to communicate with statisticians. In addition we treat the violations of assumptions, such as heteroscedasticity, as the basis for good information for use in model development rather than as conditions that preclude statistical analysis.

HOMOSCEDASTICITY AND HETEROSCEDASTICITY

In chapter 2 the assumption of homogeneity of variance, that is, equal group variances, was discussed in relation to ANOVA. Given two samples that receive different treatments, least squares procedures assume that each sample is from a common population, and therefore these samples come from "populations" with equal variance. Violation of this assumption usually does *not* upset the inferences made when using the F distribution. When deriving a line of best fit, homogeneity of variance is reflected by equal variance about the line of best fit for each scale point on the X axis. This equal variance about the line of best fit is what is called *homoscedasticity*. The scattergram in Figure 9.1 illustrates a case in which homogeneity of variance is present in the sample and is thus a reasonable expectation for the population. Note that at scale point 10 on the X axis the observed Y scores are distributed in about the same way as they are for observations at scale point 20 on the X axis. Given such a data plot, that is, one in which the distribution of points around the line of best fit is symmetrical, the researcher can assume that homoscedasticity exists.

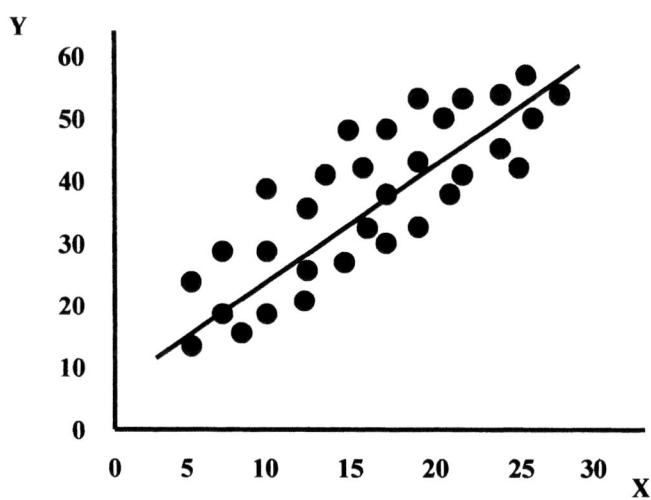

Figure 9.1.
A representation of a line in which the Y observations are homogeneously distributed about the line for all values of X.

When scattergrams depart from the ideal and distributions are not symmetric about the hypothesized line of best fit, that is, heteroscedasticity is present; most statisticians will admonish the researcher to be cautious in interpreting that line because the observed heteroscedasticity suggests the line has different errors across scale values on the X axis. Figure 9.2 contains scattergrams that illustrate three heteroscedasticity cases.

If one views the scattergrams in Figure 9.2 from a straight-line point of view, one may be concerned because an assumption is violated, and the R^2 is small due to heteroscedasticity. From the position of a researcher, systematic departures around the line of best fit as illustrated ought to be seen as a starting point for inquiry—it is very likely that in all three cases the straight-line model constructed to fit the data is not appropriate. Given the systematic departures from homoscedasticity in three scattergrams contained in Figure 9.2, one might suspect that a theoretically unexpected interaction (see chapter 7) between the X variable and another unknown variable(s) yields the odd distributions.

For the sake of simplicity, suppose all three cases illustrated are due to what is called a *treatment-by-aptitude interaction,* and that the researchers were unaware of the "treatment." Consider Figure 9.2a. If the researchers were interested in the relationship of ability (X) to criterion performance (Y), they may ask: Why is there so much variability at the extremes of the ability continuum?

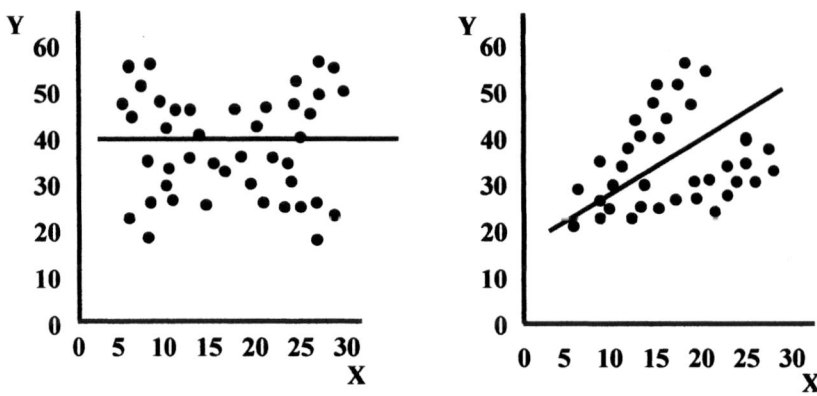

Figure 9.2a.
Heteroscedasticity in which $R^2 = 0$.

Figure 9.2b.
Heteroscedasticity in which $R^2 > 0$ and the error increases as X increases.

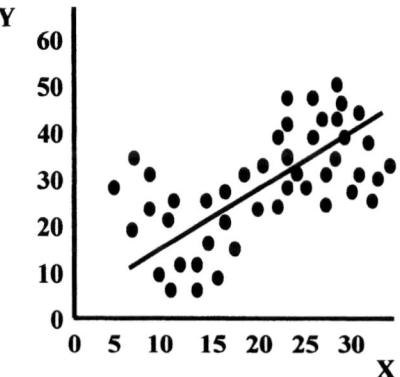

Figure 9.2c.
Heteroscedasticity in which $R^2 > 0$ and the greater error exists at low and high levels of X.

Figure 9.2.
Three hypothetical cases of heteroscedasticity.

In the hypothetical case under consideration, the sample may be drawn from two different classes of students with two teachers. Can the results be due to teacher differences? For example, perhaps Teacher A is an effective instructor for students who are below average, and Teacher B is an effective instructor for students who are above average. If the researchers ignored teacher effects, the one line in Figure 9.2a would be the outcome. If the researchers expanded the simple linear equation (i.e., $Y = a_0U + b_1X + E_1$) to include two lines, one for students of Teacher A and one for students of Teacher B, the scattergram in Figure 9.3 might be the outcome. As depicted in the scattergram contained in Figure 9.3, the heteroscedasticity is removed because the systematic teacher effects are accounted for, and the R^2 will be dramatically increased from zero to an R^2 substantially greater than zero.

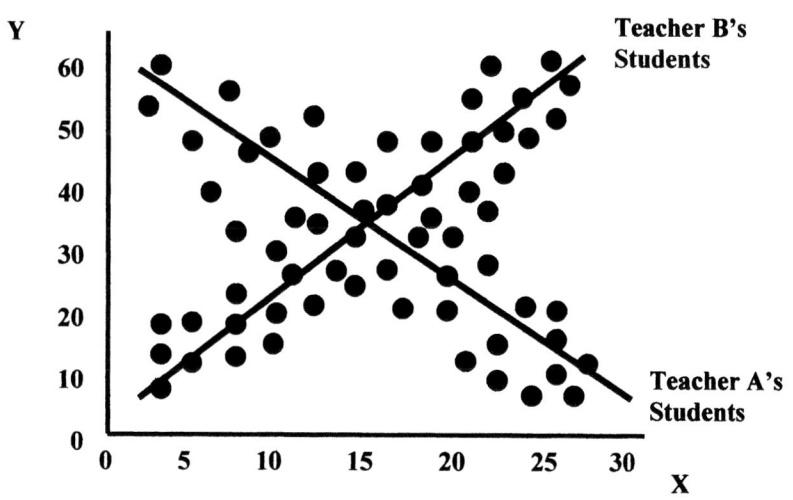

Figure 9.3.
Data from Figure 9.2a in which the observed variability in extremes is explained by teacher effects (disordinal interaction)—The scatter of points around the two lines reflect homoscedasticity.

The apparent heteroscedasticity in Figure 9.2b might be due to an ordinal interaction (two or more lines that are *not* parallel, that do not cross in the range of interest). The heteroscedasticity in Figure 9.2c may be due to a strong relationship between variables Y and X for one treatment but no relationship for another treatment. The two diagrams in Figure 9.4 illustrate these two possible explanations for the observed heteroscedasticity in Figures 9.2b and 9.2c.

So far this discussion of heteroscedasticity has dealt with straight lines of best fit. Consider the conditions observed in the scattergrams contained in Figures 9.5 and 9.6. A review of the scattergrams presented in Figures 9.5a and 9.6a suggest that unequal variability across X might be due to a nonlinear relationship between X and Y.

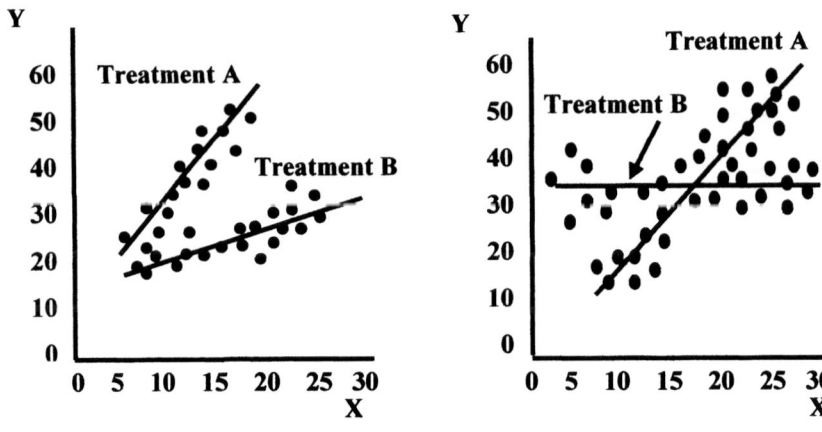

Figure 9.4a.
Explanation of heteroscedasticity in Figure 9.2b with an ordinal interaction effect.

Figure 9.4b.
Explanation of the heteroscedasticity in Figure 9.2c with a differential degree of relationship between Y and X for Treatments A and B.

Figure 9.4.
Possible explanations for the heteroscedasticity in Figures 9.2b and 9.2c.

The scattergrams in Figures 9.5b and 9.6b illustrate a situation in which Treatments A and B have two different nonlinear relationships between X and Y. In both instances, the apparent heteroscedasticity is due to the nonlinear relationship between X and Y. Whether the curves represented in Figure 9.6 are due to a theoretical expectation or are due to poor scaling of the variables is undetermined at this point. The rest of this chapter deals with the two concepts of nonlinear relationships and inappropriate scaling of the variables.

In summary, homoscedasticity (i.e., equal variability) is statistically desirable because the R^2 is enhanced. Furthermore, the homoscedasticity assumption implicitly assumes that the proper form of regression equation has been fitted. When equal variability about the line of best fit is lacking, the best research strategy may be to (a) search for unknown variables that may account for the variability and include those variables in the analysis and/or (b) identify the correct functional form of the relationship between X and Y, which may include rescaling one or more of the variables. Ideally, if one has all the relevant predictor variables and the appropriate functional form, there will be no variability about the line of best fit, an R^2 of 1.0 will be observed, and this whole section would be superfluous.

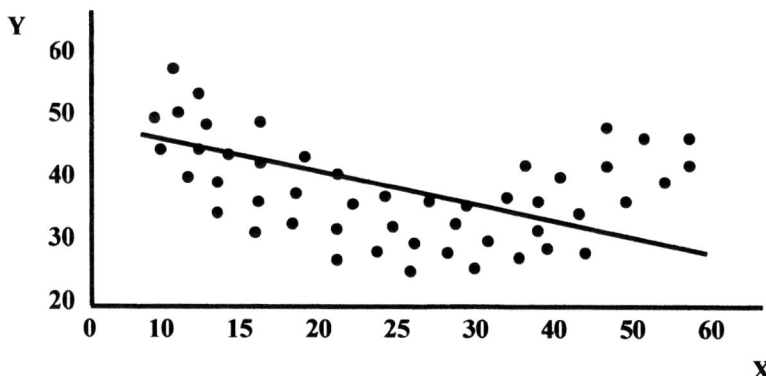

Figure 9.5a.
Scattergram 1 with heteroscedasticity.

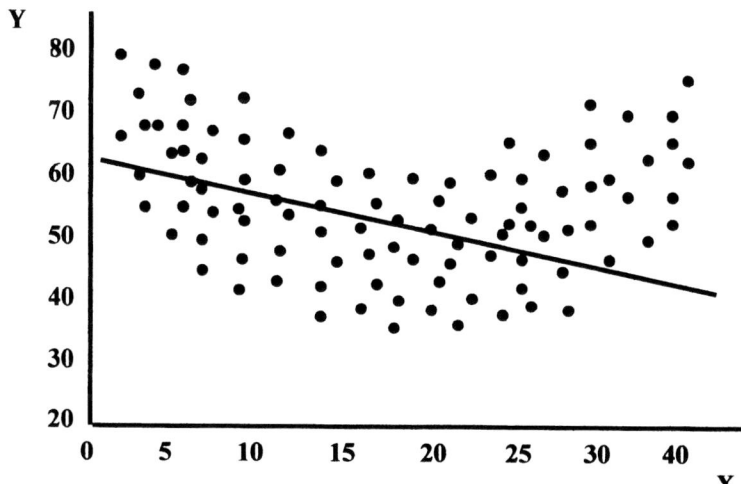

Figure 9.5b.
Scattergram 2 with heteroscedasticity.

Figure 9.5.
Two illustrations of heteroscedasticity that suggest straight-line relationships are not the best functional relationships between X and Y.

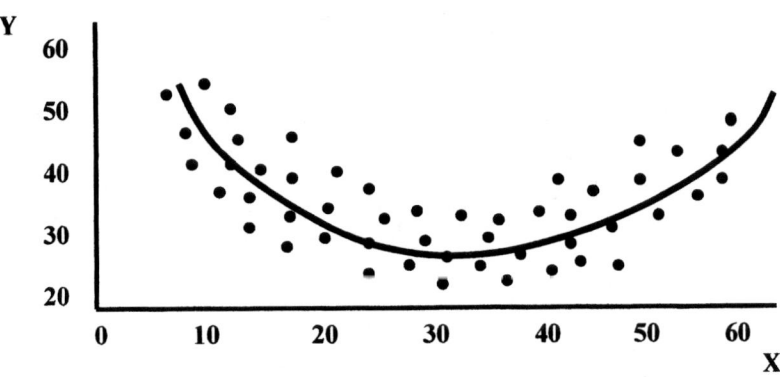

Figure 9.6a.
The removal of unequal variability about the line of best fit given in
Figure 9.5a by fitting a nonlinear function to the data.

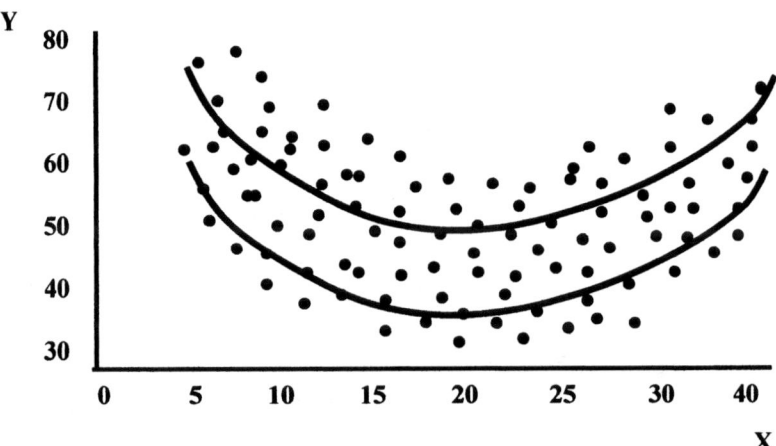

Figure 9.6b.
The removal of unequal variability about the line of best fit given in
Figure 9.5b by fitting two nonlinear functions to the data.

Figure 9.6.
Elimination of heteroscedasticity in Figures 9.5a and 9.5b.

FITTING EXPECTED NONLINEAR
FUNCTIONAL RELATIONSHIPS

The Second-Degree Relationship

In many areas of research, the relationship between X and Y is expected to be a specific curvilinear relationship. The general linear model allowing for a second-degree relationship is as follows:

$$Y = a_0U + b_1X + s_1X^2 + E_1 \hspace{3cm} \text{(Model 9.1)}$$

The four diagrams contained in Figures 9.7 illustrate four relationships that may be fit with the second-degree model. A review of these curves leads to a few general observations:
1. The numerical value of a_0 is the point where the curve passes through the Y axis at $X = 0$.
2. The numerical value of b_1 is the slope of the curve at $X = 0$. Notice that Figures 9.7a and 9.7d have a positive slope when crossing the Y axis.
3. When s_1 is negative, the open end of the curve is down; and when s_1 is positive, the open end is up.
4. When s_1 and b_1 are of the same sign (i.e., both are positive or both are negative) the maximum (or minimum) is located at a negative value on the X axis. When s_1 and b_1 are of opposite signs (i.e., one is negative and one is positive) the maximum (or minimum) is located at a positive value on the Y axis.

A U-Shaped Relationship: The Expected Relationship

In many psychomotor skill learning conditions, the theory postulates a positively accelerating curvilinear relationship. For example, for computer keyboard data entry of records, hours of practice and data-entry production (measured in single-digit numbers entered per hour) might take a form such that little gain in data-entry production is observed for the first few hours of practice, and then a spurt is observed. Figure 9.8 represents this expectation; the solid line represents the curvilinear fit. As shown, the straight line may do a fairly decent job of representing the curved data; but if the curved line is the best fit, then at the extremes (1-2 and 9-10 hours of practice) errors of under prediction with the straight-line model will be observed (the solid line is above the dashed); and the middle hours of practice will yield errors of over prediction (the dashed line is above the line of best fit).

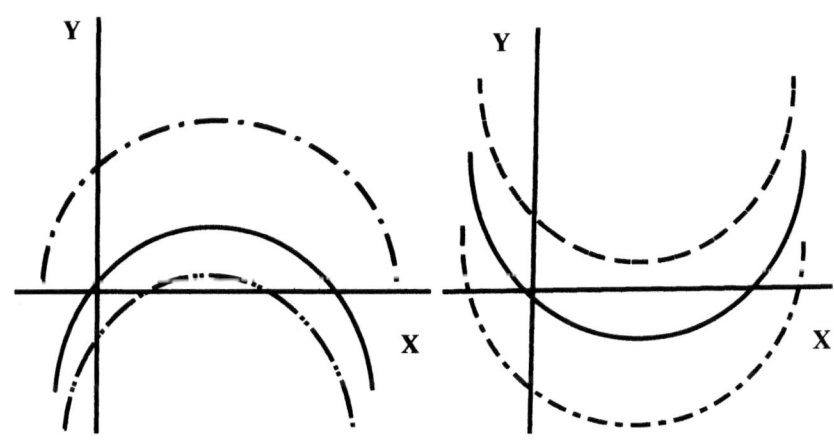

Figure 9.7a.
Coefficients b_1 and s_1 are positive
and negative, respectively.

Figure 9.7b.
Coefficients b_1 and s_1 are negative
and positive, respectively.

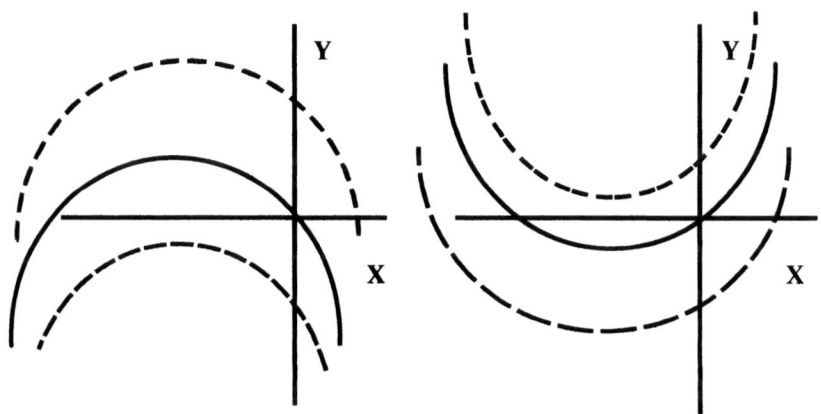

Figure 9.7c.
Coefficients b_1 and s_1 are negative.

Figure 9.7d. Coefficients b_1 and
s_1 are positive.

Figure 9.7.
Changes in the second-degree curve for various combinations of signs for coefficients a_0, b_1, and s_1 (the solid line represents $a_0 = 0$, the upper dashed line represents $a_0 > 0$, and the lower dashed line represents $a_0 < 0$).

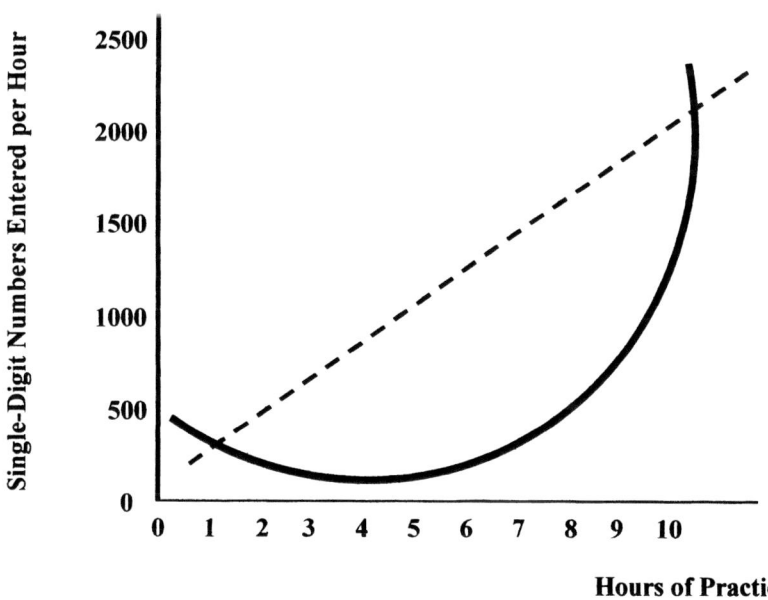

Figure 9.8.
The function of best fit as represented by a nonlinear relationship (solid line) and a linear relationship (dashed line).

A linear regression equation that includes a second-order polynomial of hours of practice vector X will fit the solid (curved) line in Figure 9.8 fairly well. The linear model for this second-order polynomial is as follows:

$$Y_1 = a_0U + b_1X + s_1X^2 + E_1 \qquad \text{(Model 9.2)}$$

where:
1. Y_1 = single-digit numbers entered per hour;
2. U = unit vector;
3. X = number of hours of practice; and
4. X^2 = the square of the elements in vector X.

Note that in this and succeeding models, the unit vector (U), the weighting coefficients (here, a_0, b_1, and s_1) and the error vector (here, E_1) will generally not be defined.

The vectors, which contain hypothetical data, would appear as follows:

$$\text{Person} \quad Y \quad = \quad a_0 U \; + \; b_1 X \; + \; s_1 X^2 \; + \; E_1$$

$$
\begin{bmatrix} 1 \\ 2 \\ 3 \\ 4 \\ \cdot \\ \cdot \\ \cdot \\ 60 \end{bmatrix}
\begin{bmatrix} 500 \\ 510 \\ 750 \\ 1700 \\ \cdot \\ \cdot \\ \cdot \\ 2500 \end{bmatrix}
= \; a
\begin{bmatrix} 1 \\ 1 \\ 1 \\ 1 \\ \cdot \\ \cdot \\ \cdot \\ 1 \end{bmatrix}
+ b_1
\begin{bmatrix} 1 \\ 2 \\ 5 \\ 9 \\ \cdot \\ \cdot \\ \cdot \\ 10 \end{bmatrix}
+ s_1
\begin{bmatrix} 1 \\ 4 \\ 25 \\ 81 \\ \cdot \\ \cdot \\ \cdot \\ 100 \end{bmatrix}
+
\begin{bmatrix} ? \\ ? \\ ? \\ ? \\ \cdot \\ \cdot \\ \cdot \\ ? \end{bmatrix}
$$

Note that the first score represented in Y is 500 single-digit numbers entered in an hour with one hour's practice ($X = 1$; $X^2 = 1$). The 60th score in Y is 2500 and $X = 10$; thus, 2500 single-digit numbers were entered in an hour for the person with 10 hours of practice ($X^2 = 10 * 10 = 100$).

Suppose a researcher had available 60 naive students in a data-entry class and wanted to determine if the surge in performance represented a likely non-chance event. One could monitor each student's performance after each hour of practice and then use a repeated-measures test for a curvilinear relationship. To remain conceptually simple, the present example will not include the repeated-measures design (chapter 10 discusses the repeated-measures design).

Recall that the researchers expect the relationship between the students' number of hours practiced to be related in a nonlinear fashion to their number of single-digit numbers entered accurately, (i.e., student productivity). Specifically, they viewed the relationship to be one in which little gain or even a slight loss in productivity will be observed for the first few hours of practice as the students familiarize themselves with the number keypad, and then it will increase at a rapid rate. Thus, they expect a positive second-degree component of the relationship in addition to the positive linear component.

To test these two expectations, the researcher may randomly assign six students to each of the 10 practice conditions (6 for one hour, 6 for two hours, etc.). The researcher would then give the practice and record the number of single-digit numbers entered for each student.

The reader may note that 10 groups were formed (one for each of the whole hours of practice), yet these 10 groups were cast into one continuous vector with values 1 through 10. One would expect that 5.5 hours of practice would yield production somewhere between the production observed by the 5- and 6-hour practice groups, and so on for all half-hour periods; therefore, it is reasonable to consider practice to be represented on a continuum.

The researchers would state the Research Hypothesis, which is labeled the Overall Research Hypothesis, as follows:

Research Hypothesis: For the specified population, the curvilinear relationship between hours of practice and the number of single-digit numbers entered exhibits a negative linear component and a positive second-degree component with the left-hand side of the function abbreviated.

The corresponding Statistical Hypothesis, which is labeled the Overall Statistical Hypothesis, is as follows:

Statistical Hypothesis: For the specified population, the curvilinear relationship between hours of practice and the number of single-digit numbers entered does not exhibit a negative linear component or a positive second-degree component with the left-hand side of the function abbreviated.

The appropriate Full Model for the Research Hypothesis is Model 9.2, and three conditions must be met by this model in order for the researchers to reject the corresponding Overall Statistical Hypothesis. These conditions are:

1. Condition 1: The coefficient associated with the linear component (b_1) must be negative.
2. Condition 2: The coefficient associated with the second-degree component (s_1) must be positive.
3. Condition 3: If Conditions 1 and 2 are met, a plot of the line of best fit must show that productivity at first decreases as low levels of training increase, but increase at an increasing rate as training levels continue to increase beyond a minimum productivity level, which is reached at the lower end of the training range.

To meet the first two conditions, two Research Hypotheses and two corresponding Statistical Hypotheses must be stated and tested. If both conditions are met, the researchers would expect a function to be formed such as the one contained in Figure 9.7b—with the left-hand side of the function somewhat abbreviated.

Testing Condition 1. The directional Research Hypothesis and its corresponding Statistical Hypothesis for testing Condition 1 are as follows:

Research Hypothesis for Condition 1: For the specified population, the curvilinear relationship between hours of practice and the number of single-digit numbers entered exhibits a negative linear component over and above the second-degree component of the relationship.

Statistical Hypothesis for Condition 1: For the specified population, the curvilinear relationship between hours of practice and the number of single-digit numbers entered does not exhibit a negative linear component over and above the second-degree component of the relationship.

The restriction that must be placed on the Full Model (Model 9.2) to obtain the appropriate Restricted Model for the Statistical Hypothesis for Condition 1 is:

$b_1 = 0$ (1 restriction)

Placing this restriction on the Full Model (Model 9.2) results in the following Restricted Model:

$$Y_1 = a_0U + s_1X^2 + E_2 \qquad \text{(Model 9.3)}$$

The researchers would first examine the sign of the coefficient for the linear component (b_1) in the Full Model to verify if, indeed, it was negative. Next they would conduct an F test of the difference between the R^2 values of the Full Model (Model 9.2) and the Restricted Model (Model 9.3) to determine if the amount of unique variation in the single-digit numbers entered per hour accounted for by the negative linear component is statistically significant. This F test would be calculated using Equation 4.1, which is as follows:

$$F(df_n, df_d) = \frac{(R_F^2 - R_R^2)/(df_n)}{(1 - R_F^2)/(df_d)} \qquad \text{(Equation 4.1)}$$

If the coefficient for the linear component of the relationship was negative and the one-tailed probability of the F test is less than the established alpha level, the researchers would reject the Statistical Hypothesis for Condition 1 and state that the data support the claim that a negative linear component exists in the second-degree polynomial relationship.

Testing Condition 2. The directional Research Hypothesis and its corresponding Statistical Hypothesis for testing Condition 2 are as follows:

Research Hypothesis for Condition 2: For the specified population, the curvilinear relationship between hours of practice and the number of single-digit numbers entered exhibits a positive second-degree component over and above the linear component of the relationship.

Statistical Hypothesis for Condition 2: For the specified population, the curvilinear relationship between hours of practice and the number of single-digit numbers entered does not exhibit a positive second-degree component over and above the linear component of the relationship.

The restriction placed on the Full Model (Model 9.2) to obtain the appropriate Restricted Model for the Statistical Hypothesis for Condition 2 is:

$$s_1 = 0 \quad \text{(1 restriction)}$$

Placing this restriction on the Full Model (Model 9.2) results in the following Restricted Model:

$$Y_1 = a_0U + b_1X + E_3 \qquad \text{(Model 9.4)}$$

Once again, the researchers would first examine the sign of the coefficient for the second-degree component (s_1) in the Full Model (Model 9.2) to verify that it is positive. Next, using Equation 4.1, they would conduct an F test of the difference between the R^2 values of the Full Model (Model 9.2) and the Restricted Model (Model 9.4) to determine if the amount of unique variation in the single-digit numbers entered per hour accounted for by the positive second-degree component is statistically significant. If the coefficient for the second-degree component of the relationship was positive and the one-tailed probability of the F test was less than the established alpha level, the researchers would reject the Statistical Hypothesis for Condition 2 and state that the data support the claim that a positive second-degree component exists in the second-degree polynomial relationship.

Condition 3—Plotting the function. As stated in Condition 3, a plot of the line of best fit must show productivity at first decreases as low levels of training increase, but increases at an increasing rate as training levels continue to increase beyond a minimum productivity level, which is reached at the lower end of the training range. To obtain a plot of the relationship between single-digit numbers entered per hour (Y_1) and the number of hours of practice (X) the researchers would use the coefficient values estimated for the Full Model (Model 9.2), that is, the values for a_0, b_1, and s_1, and approximately six selected number of training-hours values. The researchers would select low, medium, and high values from the range of training hours in the sample data. The calculated single-digit numbers entered per hour for each training-hours value is obtained by substituting the selected training-hours value into Model 9.2. To illustrate, assume one of the number of training-hours value was 2. The value of 2 is substituted into Model 9.2 as follows:

$$Y'_1 = a_0(1) + b_1(2) + s_1(2)^2$$

where Y'_1 is the predicted Y_1 value. Note that when the researchers substitute the training hours value of 2 into Model 9.2, the unit vector value is set equal to 1 (remember, regardless of which training-hours value is used, the value for the unit vector will always be 1). In addition, the error term is set equal to 0 because the single-digit numbers entered per hour (Y'_1) is calculated directly from the function estimated by Model 9.2.

The single-digit numbers entered per hour is calculated by first multiplying the linear component coefficient (b_1) by 2; and by multiplying the second-degree component coefficient (s_1) by 4 (i.e., 2^2). Next the researchers would add these two products to the value of the model's constant term (i.e., the value for a_0). The other single-digit numbers entered per hour would be calculated in the same manner. Once these six single-digit numbers entered per hour are obtained, they would be plotted along with their corresponding number of training hours on a graph in which the Y and X axes were labeled "Single-Digit Numbers Entered per Hour" and "Number of Training Hours," respectively. The curvilinear function plotted in Figure 9.8 is assumed to be the one produced by the function estimated by the Full Model (Model 9.2).

A review of the curvilinear function contained in Figure 9.8 reveals a number of important characteristics. First, the single-digit numbers entered per hour decrease slightly for low levels of training. Second, a minimum level of the number of entries per hour is reached for a fairly low level of training—approximately 4 hours. (A method of calculating the exact predictor value that corresponds to the minimum criterion value is presented in the section of this chapter entitled Applied Research Hypothesis 9.1.) Third, once this minimum productivity level is reached, additional hours of training are associated with increasing productivity levels that increase at an increasing rate. These characteristics match those stipulated in Condition 3.

Overall test results. If the researchers rejected both Statistical Hypotheses for the two conditions, it would indicate that the linear and second-degree components of the relationship between single-digit numbers entered per hour and the number of hours of practice were negative and positive, respectively, and statistically significant. Based on these test results the researchers would review the plot of the curvilinear function estimated by the Full Model (Model 9.2) to determine if its characteristics matched those stated in Condition 3. If all three conditions are met, the researchers would reject the Overall Statistical Hypothesis and support the Overall Research Hypothesis that stated for the specified population, the curvilinear relationship between hours of practice and the number of single-digit numbers entered exhibits a negative linear component and a positive second-degree component with the left-hand side of the function abbreviated.

It is important to note that the coefficient s_1 in the Full Model needs to be positive because the researchers expected a *positive* second-degree component in the relationship between hours of practice and the number of single-digit numbers entered. If s_1 was negative, however, the curve would eventually descend, an inverted U would be formed, which would be perplexing to the researchers. Such a finding would suggest that continued practice would yield less production beyond a certain point. Such a finding would go contrary to all expectations regarding skill learning and if observed should call for an immediate replication so that a strong empirical base can be established.

Inverted U-Shaped Curve: The Expected Relationship

Researchers in the behavioral sciences and education may encounter relationships that reflect the type of relationships depicted in Figures 9.7a and 9.7c, that is, relationships that take the form of an inverted U-shape curve. For example, arousal (central neural excitement) has been found to yield an inverted U-shaped relationship to cognitive performance on moderately difficult tasks. That is, those who have either low or high arousal levels score poorly, and those who are moderately aroused produce the highest responses. Figure 9.9 represents the inverted U relationship in which individuals who score 1 or 10 on an arousal measurement (X) score 5 on the cognitive performance task of moderate difficulty. Subjects with moderate arousal levels (say, 5 and 6) will score above 20 on the cognitive performance task. The theory claims that low-aroused and high-aroused subjects do poorly on the cognitive performance task for different reasons. The person who is barely aroused has insufficient excitation for focusing on the task. On the other hand, the highly aroused individual has such a high level of central excitation that trivial stimuli distract the person from the primary cognitive task. Moderate levels provide sufficient central excitation to focus on the task but not so much that it yields distraction.

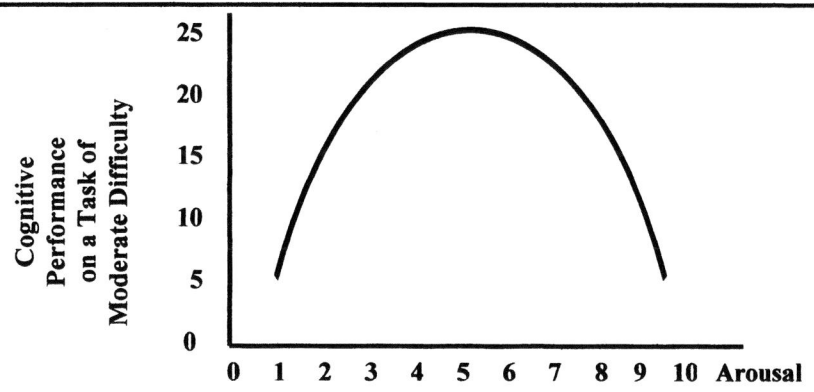

Figure 9.9.
The expected inverted U relationship between arousal and cognitive performance.

Given such theoretical expectations, in an experimental situation the researcher might pose the following research question:

For a given population, does an inverted U relationship between measured arousal level and performance on a cognitive task of moderate difficulty exist?

The researcher might then select a sample of subjects who represent a specific population (e.g., 16-year-old males in high school) and provide the moderately difficult task while monitoring the individuals' arousal levels. The research expectation might provide the following Overall Research Hypothesis:

Research Hypothesis: Among 16-year-old high school boys, an inverted U relationship between arousal level and cognitive performance on a task of moderate difficulty does exist with the maximum cognitive performance occurring near the middle of the range of the arousal scores.

As illustrated in Figures 9.7a and 9.7c, the inverted U relationship is a very specific expectation.

To obtain the inverted U in Figure 9.7a, the coefficient associated with the arousal vector must be positive, and the coefficient associated with the vector containing squared arousal scores must be negative. The sign of the Y intercept, however, may be positive, negative, or zero. A plot of the data must be made to determine if the change in direction is within the range of observed X scores.

The following vector display provides an example to illustrate the coefficients generated for a second-degree polynomial model that perfectly fits the data for an inverted U relationship:

$$\text{Subject} \quad Y \quad = \quad a_0 U \quad + \quad b_1 X \quad + \quad s_1 X^2 \quad + \quad E_1$$

$$
\begin{matrix}
1 \\ 2 \\ 3 \\ 4 \\ 5 \\ 6
\end{matrix}
\begin{bmatrix} 0 \\ 4 \\ 6 \\ 6 \\ 4 \\ 0 \end{bmatrix}
= 0
\begin{bmatrix} 1 \\ 1 \\ 1 \\ 1 \\ 1 \\ 1 \end{bmatrix}
+ 5
\begin{bmatrix} 0 \\ 1 \\ 2 \\ 3 \\ 4 \\ 5 \end{bmatrix}
+ (-1)
\begin{bmatrix} 0 \\ 1 \\ 4 \\ 9 \\ 16 \\ 25 \end{bmatrix}
+
\begin{bmatrix} 0 \\ 0 \\ 0 \\ 0 \\ 0 \\ 0 \end{bmatrix}
$$

If the three elements for subject 1 in vectors U, X, and X^2 are multiplied by the appropriate coefficients (weights), the predicted score is zero and the observed score is zero. If the same procedure is done for all six subjects, the predicted score increases and then decreases as X increases. Inspection of the values for Y in the vector display confirms that the change in direction is within the range of observed X scores.

When dealing with inverted U-shape relationships, as illustrated in the four diagrams contained in Figure 9.7, it is important to consider how much of the U-shaped curve is present in the sample data. In most research in the behavioral sciences and education, X and Y values are positive; therefore, most researchers would be concerned solely with the upper right-hand quadrant. Figure 9.7a represents an inverted U outcome state. If the values for X range from zero to the right-hand side of the figure, then the inverted U relationship is obtained. However, if the X values range from zero to a small value of X, what can one say? Figure 9.10 is a representation of Figure 9.7a over the range of scores from zero to one such small value, 4, where a positive b_1 and a negative s_1 do not yield an inverted U. If one observed such a plot in the arousal study, one might conclude that the population the sample represents does *not* have individuals with high levels of arousal. Indeed, if a larger population is defined, one that included arousal levels up to the right-hand scale value as in Figure 9.7a, one might expect to find that the inverted U-shaped relationship does, in fact, exist.

Recall that the Overall Research Hypothesis currently being considered for testing is as follows:

Overall Research Hypothesis: Among 16-year-old high school boys, an inverted U relationship does exist between arousal level and cognitive performance on a task of moderate difficulty with the maximum cognitive performance occurring near the middle of the range of the arousal scores.

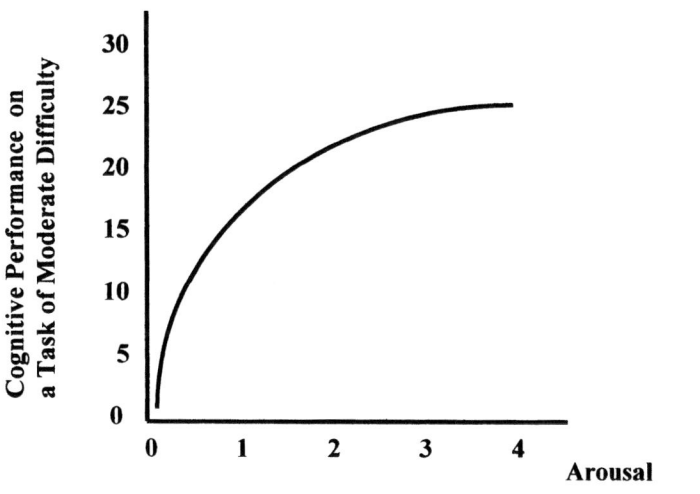

Figure 9.10.
Data from Figure 9.7a with the range of observed scores limited from 0 to 4—A second-degree curve with coefficients $a_0 > 0$, $b_1 > 0$, and $s < 0$.

The corresponding Overall Statistical Hypothesis is:

Overall Statistical Hypothesis: Among 16-year-old high school boys, an inverted U relationship does not exist between arousal level and cognitive performance on a task of moderate difficulty with the maximum cognitive performance occurring near the middle of the range of the arousal scores.

The Full Model that reflects this Research Hypothesis is as follows:

$$Y_2 = a_0U + b_1X + s_1X^2 + E_1 \qquad\qquad \text{(Model 9.5)}$$

where:
1. Y_2 = cognitive performance on the task of moderate difficulty;
2. U = unit vector;
3. X = the continuous predictor variable (arousal level); and
4. X^2 = a vector containing the square of the elements in vector X.

Three conditions must be met in order for the researchers to reject the Statistical Hypothesis and support the Research Hypothesis. These conditions are:

1. Condition 1: The coefficient associated with the linear component (b_1) must be positive.
2. Condition 2: The coefficient associated with the second-degree component (s_1) must be negative.
3. Condition 3: If Conditions 1 and 2 are met, a plot of the line of best fit must show that those individuals with the lowest and highest arousal scores (X) have somewhat the same (minimum) score on cognitive performance (Y_2), and the maximum cognitive performance score is approximately midway between the lowest and highest arousal scores.

To meet Conditions 1 and 2, two Research Hypotheses and two Statistical Hypotheses must be stated and tested.

Testing Condition 1. The directional Research Hypothesis for testing Condition 1 is as follows:

Research Hypothesis for Condition 1: Among 16-year-old high school males, a positive linear component between arousal and cognitive performance on a task of moderate difficulty does exist, over and above the second-degree component of the relationship.

Given the Full Model (Model 9.5), the Research Hypothesis implies that b_1 must be greater than 0. The associated Statistical Hypothesis for Condition 1 is as follows:

Statistical Hypothesis for Condition 1: Among 16-year-old high school males, a positive linear component between arousal and cognitive performance on a task of moderate difficulty does not exist, over and above the second-degree component of the relationship.

The Statistical Hypothesis requires that $b_1 = 0$. Recall that in the previous section (the study that examined the relationship between student productivity and training hours) this type of restriction placed in the Full Model to produce the Restricted Model, and an F test of the differences between the R^2 values of the Full Model and the Restricted Model was used to determine if the coefficient differed significantly from 0. Since only one restriction is placed on the Full Model, the researchers could use a t test of the b_1 coefficient in the Full Model to produce an equivalent test (a concept presented in previous chapters).

The researchers would first examine the sign of the coefficient for the linear component (b_1) in the Full Model (Model 9.5) to verify that it is positive. If the coefficient is positive and the one-tailed probability of the t test of the b_1 is less than the established alpha level, the Statistical Hypothesis for Condition 1 is rejected, and the researchers would state that a positive linear component of the polynomial relationship between arousal and cognitive performance on a task of moderate difficulty does exist, over and above the second-degree component of the relationship.

Testing Condition 2. The directional Research Hypothesis for testing Condition 2 is as follows:

> Research Hypothesis for Condition 2: Among 16-year-old high school males, a negative second-degree component of the relationship between arousal and cognitive performance on a task of moderate difficulty does exist, over and above the linear component of the relationship.

Note that the Full Model that reflects this Research Hypothesis for Condition 2 is Model 9.5, and this Research Hypothesis requires s_1 to be less than 0 in the Full Model.

The corresponding Statistical Hypothesis for Condition 2 is as follows:

> Statistical Hypothesis for Condition 2: Among 16-year-old high school males, a negative second-degree component of the relationship between arousal and cognitive performance on a task of moderate difficulty does not exist, over and above the linear component of the relationship.

The restriction required by this Statistical Hypothesis is $s_1 = 0$. Again, since only one restriction is required, the researchers can test this Statistical Hypothesis by conducting a *t* test of the s_1 coefficient in the Full Model.

The researchers would first examine the sign of the coefficient for the second-degree component (s_1) in the Full Model (Model 9.5) to verify that it is negative. If the coefficient is negative and the one-tailed probability of the *t* test of the s_1 is less than the established alpha level, the Statistical Hypothesis for Condition 2 is rejected, and the researchers would state that a negative second-degree component of the polynomial relationship between arousal and cognitive performance on a task of moderate difficulty does exist, over and above the linear component of the relationship.

Condition 3—Plotting the function. As stated in Condition 3, a plot of the line of best fit must show that those individuals with the lowest and highest scores on the X variable (arousal) have somewhat the same (minimum) score on Y_2 (cognitive performance), and the maximum score on Y_2 is approximately midway between the lowest and highest scores on X.

To obtain a plot of the relationship between arousal (X) and cognitive performance on a task of moderate difficulty (Y_2), the researchers would use the coefficient values estimated for the Full Model (a_0, b_1, and s_1), and a number of selected arousal scores, which would include low, medium, and high scores. As discussed in the previous example, the calculated cognitive performance score for each arousal score is obtained by substituting the arousal score into the Full Model (Model 9.5). A numerical example of this procedure is presented in the Applied Research Hypothesis 9.1 section of this chapter.

A plot of the line of best fit could yield many relationships, but all will be some form of the three lines depicted in Figure 9.11. Figure 9.11a illustrates a case in which the inverted U-shape relationship exists, but the upper end of the arousal continuum is not present in the population. The extrapolated dashed line does yield the inverted U. Given such a finding, the researcher may wish to seek a population that includes the higher end of the arousal scale; or the researcher may conclude that for the specified population an inverted J relationship exists (no highly aroused subjects are observed in the population of 16-year-old high school males).

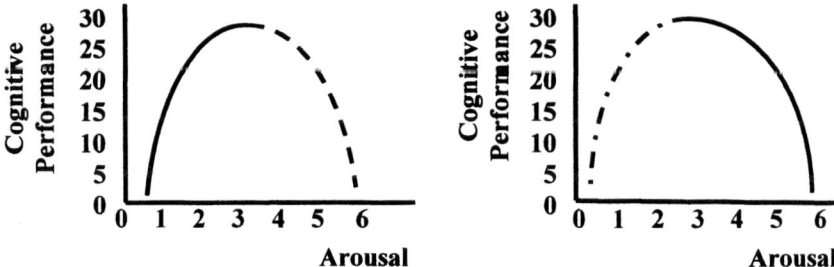

Figure 9.11a.
The upper end of the arousal continuum is not present in the population.

Figure 9.11b.
The lower end of the arousal continuum is not present in the population.

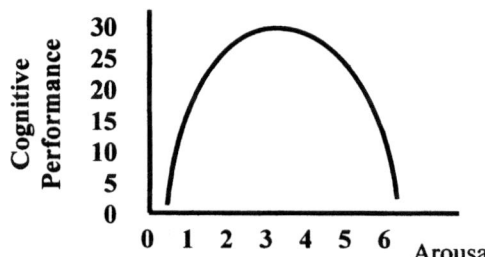

Figure 9.11c.
Entire arousal continuum is present in the population.

Figure 9.11.
Three possible outcome states for Model 9.5—b_1 is positive and s_1 is negative.

Figure 9.11b is the same as Figure 9.11a, except that no lowly aroused 16-year-old high school males are observed. Figure 9.11c represents the idealized inverted U-shaped relationship. Given a regression line that produces a diagram similar to the one presented in Figure 9.11c, and Conditions 1 and 2 are met, the researchers would conclude that: Among 16-year-old high school males, an inverted U-shaped relationship exists between arousal level and cognitive performance on a moderately difficult task.

Assume that Figure 9.9 contains the plot of the function estimated by the Full Model (Model 9.5), a review of this plot reveals that the function is indeed an inverted U shape with the maximum cognitive performance score occurring near the middle of the range of the arousal scores included in the sample. These characteristics match those stipulated in Condition 3.

Overall test results. If the researchers rejected both the Statistical Hypotheses for Conditions 1 and 2, it would indicate that the linear and second-degree components of the relationship between cognitive performance scores and arousal scores are positive and negative, respectively, and statistically significant. If a review of the plot of the curvilinear function estimated by the Full Model (Model 9.5) revealed that its characteristics matched those stated in Condition 3, the researchers would reject the Overall Statistical Hypothesis. Based on this decision, the researchers would state that among 16-year-old high school boys, an inverted U relationship between arousal level and cognitive performance on a task of moderate difficulty does exist with the maximum cognitive performance occurring near the middle of the range of the arousal scores.

Notes of caution. When conducting this type of model building and statistical testing of curvilinear relationships, researchers need to be aware of two issues. First, testing for the signs of the regression coefficients must be done within the context of making one restriction at a time upon the same Full Model (Model 9.5 in this example) because the sign of each coefficient was posited within the context of the other variables in the Full Model. An inverted U shape requires both a linear component and a second-degree component. (Whether a nonzero Y-intercept is necessary depends on the Research Hypothesis and was of no interest in this application.)

The two diagrams contained in Figure 9.12 show two possible lines using Model 9.5. The diagram in Figure 9.12a represents a curvilinear function in which the second-degree component is positive; while Figure 9.12b represents a curvilinear function in which the second-degree component is negative. When b_1 is *not* included in the Full Model for either function (i.e., a U-shape function or an inverted U-shape function), the curve is symmetrical about the Y axis and thus does not provide a good fit when the maximum point departs markedly from the Y axis (e.g., the function diagramed in Figure 9.11c).

Second, because the linear component and the second-degree component are highly correlated, one should be careful when interpreting the magnitude of the regression coefficients. The coefficients could be used in a predictive situation, however, as the predicted criterion would be approximately the same from one set of coefficients to another set of coefficients (i.e., one sample to another sample). The magnitude of the R^2 will remain somewhat constant. McNeil and Spaner (1971) presented a more generally applicable argument for including correlated predictor variables in regression models.

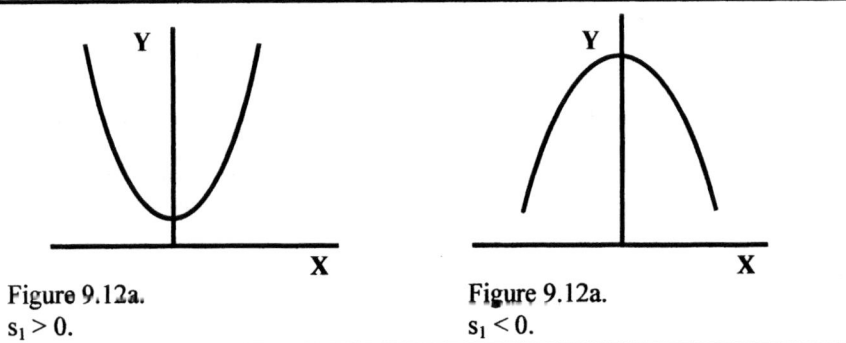

Figure 9.12a.   Figure 9.12a.
$s_1 > 0$. $s_1 < 0$.

Figure 9.12.
Curves must be symmetric about the Y axis if the linear component is not included in the model.

GENERAL RESEARCH HYPOTHESIS 9.1

General Research Hypothesis 9.1 (GRH 9.1) requires an analysis of a research question that questions the existence of a U-shaped relationship between
a criterion variable (Anxiety) and a predictor variable (DrugLevel). That is, the
researchers believe that as the drug level is increased anxiety levels will decrease, but after a certain drug level, an increase in the drug dosage is associated
with an increase in anxiety.

The General Overall Research Hypothesis 9.1 (GORH 9.1), which reflects
the researchers' expectations regarding the relationship between anxiety and
drug level, is as follows:

GORH 9.1: For a given population, the relationship between the subjects'
anxiety levels and the prescribed drug levels reflects a negative linear component and a positive second-degree component for which the minimum
level of anxiety is located near the mid-range of the drug levels.

The corresponding General Overall Statistical Hypothesis 9.1 (GOSH 9.1) is as
follows:

GOSH 9.1: For a given population, the relationship between the subjects'
anxiety levels and the prescribed drug levels reflects neither a negative linear component or a positive second-degree component for which the minimum level of anxiety is located near the mid-range of the drug levels.

The Full Model reflects the conditions regarding the linear and second-
degree components of the relationship between anxiety and drug level stipulated
in GORH 9.1 as follows:

Anxiety $= a_0U + b_1(\text{DrugLevel}) + s_1(\text{DrugLevel})^2 + E_1$ (Model 9.6)

The relationship described in GORH 9.1 contains the following three conditions:

1. Condition 1: The coefficient associated with the linear component (b_1) must be negative.
2. Condition 2: The coefficient associated with the second-degree component (s_1) must be positive.
3. Condition 3: If Conditions 1 and 2 are met, a plot of the line of best fit must show that those individuals with the lowest and highest drug levels have somewhat the same maximum anxiety levels, and the minimum anxiety level is located approximately midway between the lowest and highest drug levels.

The following two General Research Hypotheses and their corresponding General Statistical Hypotheses are constructed by the researchers to reflect Conditions 1 and 2:

General Research Hypothesis for Condition 1 (GRH for C1): For a given population, the relationship between the subjects' anxiety levels and the prescribed drug levels reflects a negative linear component over and above a positive second-degree component.

General Statistical Hypothesis for Condition 1 (GSH for C1): For a given population, the relationship between the subjects' anxiety levels and the prescribed drug levels does not reflect a negative linear component over and above a positive second-degree component.

General Research Hypothesis for Condition 2 (GRH for C2): For a given population, the relationship between the subjects' anxiety levels and the prescribed drug levels reflect a positive second-degree component over and above a negative linear component.

General Statistical Hypothesis for Condition 2 (GSH for C2): For a given population, the relationship between the subjects' anxiety levels and the prescribed drug levels does not reflect a positive second-degree component over and above a negative linear component.

The restrictions required by these Research Hypotheses are as follows:

$b_1 = 0$ (GSH for C1 restriction)
$s_1 = 0$ (GSH for C2 restriction)

The researchers would construct the following Restricted Models to reflect these two restrictions:

$$\text{Anxiety} = a_0 + s_1(\text{DrugLevel})^2 + E_2 \qquad\qquad \text{(Model 9.7)}$$

$$\text{Anxiety} = a_0 + b_1(\text{DrugLevel}) + E_3 \qquad\qquad \text{(Model 9.8)}$$

An F test of the difference between the R^2 values of the Full Model (Model 9.6) and the Restricted Model (Model 9.7) would determine if the linear component is statistically significant. If the one-tailed probability of this F test is less than the established alpha level *and* the coefficient for the DrugLevel variable in the Full Model (b_1) is negative, the researchers would reject GSH for Condition 1. Such a test result would support the condition that required a negative liner component of the relationship between anxiety and drug level.

In the same manner, an F test of the difference between the R^2 values of the Full Model (Model 9.6) and the Restricted Model (Model 9.8) would determine if the second-degree component is statistically significant. If the one-tailed probability of this F test is less than the established alpha level *and* the coefficient for the DrugLevel variable in the Full Model (s_1) is positive, the researchers would reject GSH for Condition 2. Such a test result would support the condition that required a positive second-degree component of the relationship between anxiety and drug level.

If both Condition 1 and Condition 2 are supported by the statistical testing of the linear and second-degree components of the relationship between anxiety and drug level, the researchers would plot the function estimated by the Full Model (Model 9.6). To plot this function, the researchers would select two low drug levels, two moderate drug levels, and two high drug levels. Each of these drug levels would be substituted into the Full Model along with the estimated coefficient values (i.e., the values for a_0, b_1, and s_1) to calculate the corresponding anxiety level. The researchers would plot the six pairs of values on a graph in which the Y and X axes are labeled "Anxiety" and "Drug Level," respectively. If an examination of this graph reveals that the minimum drug level is near the mid-point of the drug levels, Condition 3 is supported.

If all three conditions are met, the researchers would reject GOSH 9.1. They would be willing to state the data support that for a given population, the relationship between the subjects' anxiety levels and the prescribed drug levels reflects a negative linear component and a positive second-degree component for which the minimum level of anxiety is located near the mid-range of the drug levels. If any one or more of the three conditions are not met, the researchers would note how the relationship between anxiety and drug level deviated from the hypothesized relationship and call for further research to investigate this revealed deviation.

APPLIED RESEARCH HYPOTHESIS 9.1

Applied Research Hypothesis 9.1 (ARH 9.1) illustrates the type of analysis presented in the previous sections, that is, an investigation of a potential U-shaped relationship, in this case an abbreviated inverted U-shaped relationship between employees' job satisfaction scores (X2) and scores that measure their degree of compulsive behavior (X5). Specifically, the researchers expected the employees' job satisfaction scores would increase at a decreasing rate as their compulsive behavior scores increased, but once the compulsive behavior scores approached the upper end of the range of scores, job satisfaction scores would decrease. Thus, the researchers expect an inverted J-shaped curve with the hook of the curve turning down.

To estimate the second-degree component of this relationship, the researchers need to generate one new variable. This new variable, which is labeled X28, is formed by squaring the X5 variable (the compulsive behavior scores).

Based on the type of relationship the researchers expect to exist between the employees' job satisfaction scores and their compulsive behavior scores, the Applied Overall Research Hypothesis 9.1 (AORH 9.1) is stated as follows:

AORH 9.1: For a given population, the relationship between the employees' job satisfaction scores and their compulsive behavior scores reflects a positive linear component and a negative second-degree component for which the maximum job satisfaction score is located toward the upper end of the range of compulsive behavior scores.

The corresponding Applied Overall Statistical Hypothesis is as follows:

AOSH 9.1: For a given population, the relationship between the employees' job satisfaction scores and their compulsive behavior scores do not reflect a positive linear component or a negative second-degree component for which the maximum job satisfaction score is located toward the upper end of the range of compulsive behavior scores.

Full Model, Restricted Model, and Required Conditions

The Full Model that reflects the relationship between the employees' job satisfaction scores and their compulsive behavior scores stipulated by AORH 9.1 is as follows:

$$X_2 = a_0U + a_5X5 + a_{28}X28 + E_1 \qquad \text{(Model 9.9)}$$

Three conditions must be met in order for the researchers to reject the AOSH 9.1. These conditions are:

Condition 1—The coefficient associated with the linear component (a_5) must be positive.

Condition 2—The coefficient associated with the second-degree component (a_{28}) must be negative.

Condition 3—If Conditions 1 and 2 are met, a plot of the line of best fit must show a J-shaped curve in which the hook of the curve turns down at the upper end of the range of compulsive behavior scores.

The following two Research Hypotheses and their corresponding Statistical Hypotheses are constructed by the researchers to reflect Conditions 1 and 2:

Applied Research Hypothesis 9.1 for Condition 1 (ARH 9.1 for C1): For a given population, the relationship between the employees' job satisfaction scores and their compulsive behavior scores reflects a positive linear component, over and above the second-degree component.

Applied Statistical Hypothesis 9.1 for Condition 1 (ASH 9.1 for C1): For a given population, the relationship between the employees' job satisfaction scores and their compulsive behavior scores does not reflect a positive linear component, over and above the second-degree component.

Applied Research Hypothesis 9.1 for Condition 2 (ARH 9.1 for C2): For a given population, the relationship between the employees' job satisfaction scores and their compulsive behavior scores reflects a negative second-degree component, over and above the linear component.

Applied Statistical Hypothesis 9.1 for Condition 2 (ASH 9.1 for C2): For a given population, the relationship between the employees' job satisfaction scores and their compulsive behavior scores does not reflect a negative second-degree component, over and above the linear component.

These Statistical Hypotheses will be tested by using the t tests of the linear coefficient (a_5) and the second-degree coefficient (a_{28}) in the Full Model (Model 9.9). The Full Model will be estimated with the Microsoft Excel computer software.

Analysis of ARH 9.1 with Microsoft Excel

The data file entitled "GLM DATA EXCEL FORMAT" is used in conjunction with ARH 9.1, and is listed in the internet site listed in Appendix A. After this file as been accessed with the Excel computer software, the X28 variable must be generated. As noted in the previous section, the variable X28 is formed by squaring the values in variable X5.

Before the Excel Regression menu is accessed, the predictor variables contained in the Full Model (Model 9.11) must be placed next to each other in the data file. Thus the researchers would copy and paste the X5 and X28 variables in the data set so that they are adjacent to each other in the data file. Once these predictor variables are placed next to each other in the data file and the Microsoft Excel Regression menu has been displayed (refer to chapter 4 for a discussion of the steps used to access the Regression menu), the following four steps are completed in the Regression menu to statistically test ASH 9.1 for Condition 1 and Condition 2 (see Exhibit 9.1):

1. Click on the box next to "Input Y Range" (see Oval 1). Next, click and drag on the cells located in the column in which variable X2 is located—including the name (label) of the variable.

2. Click on the box next to "Input X Range" (see Oval 2). Next, click and drag on the cells containing the predictor variables X5 and X28—including their names (labels). Remember these variables must be located next to each other in the data file.

3. Click on the box in front of "Labels" (see Oval 3).

4. Click on the "OK" button (see Oval 4).

After these four steps are completed, the output window for the Full Model (Model 9.9) will appear as displayed in Exhibit 9.2.

Exhibit 9.1
Microsoft Excel Regression menu—Full Model (Model 9.9) for ARH 9.1 regarding Condition 1 and Condition 2.

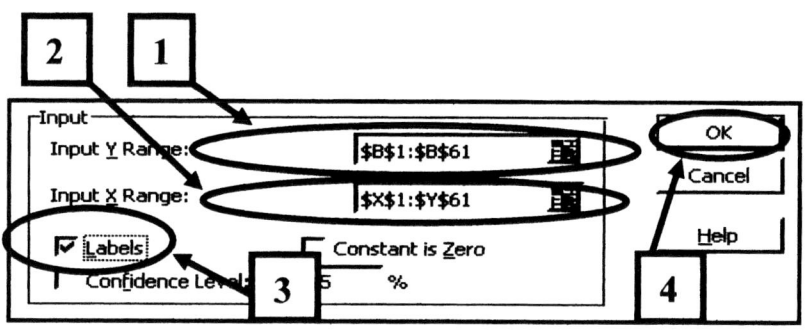

The key output information listed in Exhibit 9.2 is as follows:

1. The R^2 value for the Full Model (Model 9.9) is .957 (see Oval 1).

2. The coefficient for the linear component, which is listed under the heading "Coefficients" in the row entitled X5, is 2.99 (see Oval 2). Its corresponding t test and probability values, which are located in Oval 3, are 27.48 and 1.427E-34 (i.e., $p < .001$).

3. The coefficient for the second-degree component, which is listed under the heading "Coefficients" in the row entitled X28, is -0.0386 (see Oval 2). Its corresponding t test and probability values, which are located in Oval 3, are -23.15 and 1.139E-30 (i.e., $p < .001$).

Test results for Condition 1. Since ASH 9.1 for Condition 1 is directional, the probability of the t test for the linear component is divided by 2, which produces a value that is less than .001. Due to the fact that the linear component coefficient (2.99) is positive and its one-tailed t test probability ($p / 2 < .001$) is less than the established alpha level of .05, ASH 9.1 for Condition 1 is rejected. Thus, ARH 9.1 for Condition 9.1 is supported and the stipulated condition is met.

Exhibit 9.2.
Microsoft Excel regression output—Full Model (Model 9.9) for ARH 9.1 regarding Condition 1 and Condition 2.

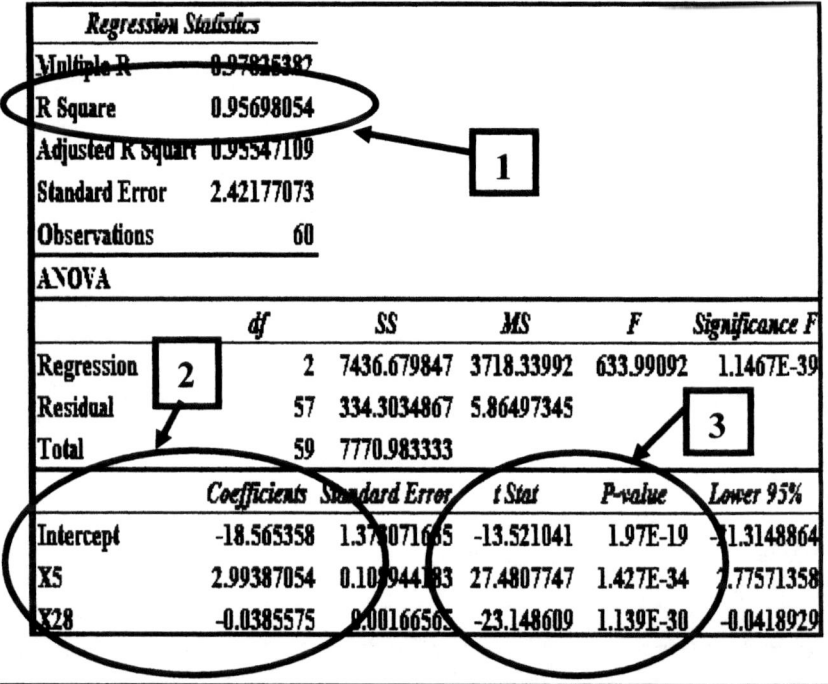

Regression Statistics					
Multiple R	0.97825382				
R Square	0.95698054				
Adjusted R Square	0.95547109				
Standard Error	2.42177073				
Observations	60				

ANOVA					
	df	SS	MS	F	Significance F
Regression	2	7436.679847	3718.33992	633.99092	1.1467E-39
Residual	57	334.3034867	5.86497345		
Total	59	7770.983333			

	Coefficients	Standard Error	t Stat	P-value	Lower 95%
Intercept	-18.565358	1.37307165	-13.521041	1.97E-19	-1.3148864
X5	2.99387054	0.10994483	27.4807747	1.427E-34	2.77571358
X28	-0.0385575	0.00166565	-23.148609	1.139E-30	-0.0418929

Test results for Condition 2. The coefficient for the second-degree component (-0.039) was negative as required by Condition 2. Once again, ASH 9.1 for Condition 2 is directional; thus the probability of the t test for the second-degree component is divided by 2, which produces a value that is less than .001. Since the second-degree component coefficient (-0.039) is negative and its one-tailed t test probability ($p / 2 < .001$) is less than the established alpha level of .05, ASH 9.1 for Condition 2 is rejected. Thus, ARH 9.1 for Condition 2 is supported and the stipulated condition is met.

Test results for Condition 3. To determine whether Condition 3 is supported by the estimated Full Model (Model 9.9), six compulsive behavior scores (i.e., 7, 17, 27, 37, 47, 57) were selected from the range of scores (i.e., 7 to 57). The selected values were typed into column A of a Microsoft Excel data file as shown in Exhibit 9.3 (Oval 1).

Exhibit 9.3.
Calculation of the estimated job satisfaction scores and steps used to plot the function estimated by the Full Model (Model 9.9).

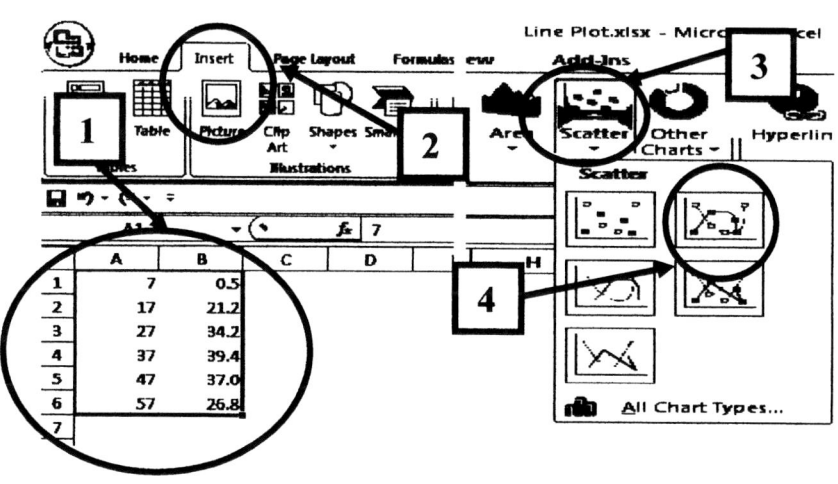

The corresponding job satisfaction score for the first compulsive behavior score (i.e., 7) was calculated through the use of Microsoft Excel by entering the following command in column B row 1:

$$= -18.5654 + 2.9939*A1 - 0.03856*A1*A1$$

This command requires the Y intercept (-18.5654) of the Full Model (Model 9.9) to be added to the products of the coefficient for variable X5 and the compulsive behavior score of 7 (2.9939 * 7) and the coefficient for variable X28 and behavior score of 7 squared (-.03856 * 49). The resulting job satisfaction score is 0.5 (see Oval 1 of Exhibit 9.3). The command typed in column B row 1 is copied and pasted in column B rows 2 through 6. The six calculated job satisfaction scores that correspond to the six selected compulsive behavior scores are listed in column B of Exhibit 9.3.

The following steps are completed to plot the function estimated by the Full Model (see Exhibit 9.3):
1. Highlight the values listed in columns A and B (Oval 1).
2. Click on "Inset" (Oval 2).
3. Click on "Scatter" (Oval 3).

4. Click on the box located in the top-right corner of the "Scatter" display (Oval 4).

Once these steps are completed, the plot of the function estimated by the Full Model (Model 9.9) will be produced (see Exhibit 9.4). A review of the function estimated by the Full Model (Model 9.9) reveals that the job satisfaction scores increase at a decreasing rate as compulsive behavior scores increase. However, once compulsive behavior scores reach approximately 39, further increases in compulsive behavior scores are associated with decreases in job satisfaction scores.

Exhibit 9.4.
Plot of the function estimated by the Full Model (Model 9.9).

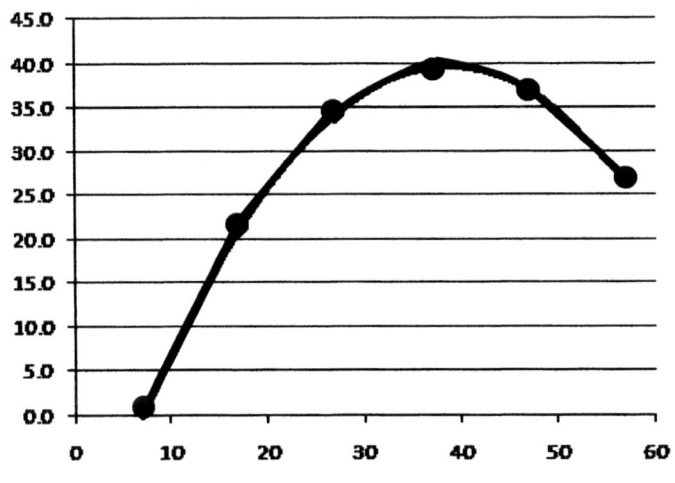

Calculation of the maximum job satisfaction score. As previously discussed, a visual examination of the function estimated by the Full Model (Model 9.9) plotted in Exhibit 9.4 reveals that the highest job satisfaction score corresponds to a compulsive behavior score of approximately 39. The researchers can determine the maximum compulsive behavior score by setting the first derivative of the function estimated by the Full Model, with respect to X5, equal to 0. The first derivative set equal to 0 is as follows:

$$0 = 2.994 + 2*(-0.0386)X5 \qquad \text{(Equation 9.1)}$$

If one is not familiar with the procedure used to calculate the first derivative with respect to X for second-degree polynomial functions such as those functions presented in this chapter, it is simply the coefficient for the linear component plus twice the value of the second-degree coefficient, which serves as the as the coefficient for the X variable (i.e., X5 in this example). Solving Equation 9.1 for X5 produces a compulsive behavior score of 38.8.

Overall test results. A review of the statistical test results of Conditions 1 and 2 reveals that the positive linear component and the negative second degree component of the polynomial function were significant. In addition the function estimated by the Full Model (see Exhibit 9.4) matches the type of inverted J-shaped function stipulated in AORH 9.1. Thus, AOSH 9.1 is rejected and the type of function described in AORH 9.1 is supported by the data.

Writing the Results for ARH 9.1 in APA Style

The results for AORH 9.1 could be written in APA style as follows:

It was hypothesized that the employees' job satisfaction scores would increase at a decreasing rate as their compulsive behavior scores increased, but once the compulsive behavior scores approached the upper end of the range of scores, job satisfaction scores would decrease. The analysis of the data revealed that 96% of the variation in job satisfaction scores was accounted for by the variation in compulsive behavior scores when the relationship was expressed as a second-degree polynomial function. The estimated coefficients for the linear component and the second-degree component of the polynomial function were 2.994 and -0.0386, respectively. Both of the one-tailed t tests of estimated positive linear component (t (57) = 27.48, $p / 2 < .01$) and the negative second-degree component (t (57) = -23.15, $p / 2 < .01$) were statistically significant at the .05 alpha level. A review of the estimated function revealed that the job satisfaction scores increase at a decreasing rate as compulsive behavior scores increase. However, once compulsive behavior scores reach approximately 39, further increases in compulsive behavior scores were associated with decreases in job satisfaction scores. Thus, the hypothesized second-degree polynomial relationship between job satisfaction and compulsive behavior scores was supported by the data.

THIRD-DEGREE POLYNOMIAL LINE FITTING

Curved lines of best fit may take many forms in addition to the second-degree polynomial presented in the previous sections. In many psychomotor studies, an S-shaped curve is expected. For example, consider the problem in this chapter regarding the relationship between data-entry production and hours of practice. After several hours of practice, a surge in production might be observed. Given further practice, improvement in production will eventually level off due to the limits of the method, the person entering the data, and the data-entry procedure. Figure 9.13 illustrates a possible shape of the relationship between data-entry production and hours of practice.

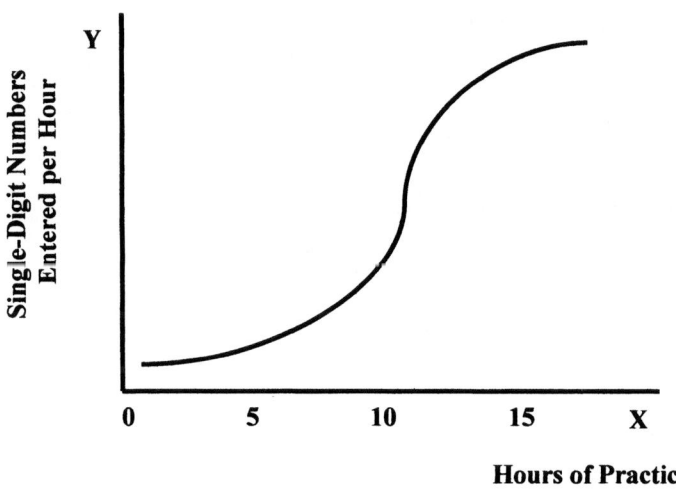

Figure 9.13.
S-Shaped curve resulting from the functional relationship between hours of practice and single-digit numbers entered per hour.

To fit the data for the first 10 hours of practice (see Figure 9.13), the simple second-degree polynomial represented by Model 9.1 will be adequate. Model 9.1 is once again as follows:

$$Y = a_0U + b_1P + s_1P^2 + E_1 \hspace{2cm} \text{(Model 9.1)}$$

Beyond 10 hours of practice, the curve is expected to change direction. To investigate such a situation, the researcher who conducted the previous study could use that data and could select 50 more subjects, randomly assign these to five practice groups (hours of practice 11, 12, 13, 14, and 15), and test the Research Hypothesis: For naive data-entry students, the number of single-digit numbers entered per hour slowly increases, increases rapidly, and then levels off with increased hours of practice.

A linear regression model that includes the unit vector (U), hours of practice (P), hours of practice squared (P^2), and hours of practice cubed (P^3) will allow the line of best fit to reflect the hypothesized S-shaped curve:

$$Y = a_0U + b_1P + s_1P^2 + c_1P^3 + E_2 \hspace{2cm} \text{(Model 9.12)}$$

where:
1. Y = the number of single-digit numbers entered per hour;
2. U = unit vector

3. P = Hours practiced;
4. P^2 = squared elements of P;
5. P^3 = cubed elements of P; and
6. a_0, b_1, s_1, and c_1 represent the estimated coefficients for the unit vector, linear component, second-degree component, and third-degree component.

For the particular S-shaped curve in Figure 9.13 to be obtained, c_1 must be negative and s_1 positive because the criterion scores are lower for large values of P. Hence the highest-powered term, P^3, must have a negative coefficient (c_1). To get the inflection point, the second-degree term must be opposite to that of the third-degree term. If the curve is horizontal at the Y axis, then the linear coefficient (b_1) will be zero. If the curve is slanted downward from left to right, b_1 will be negative, and if slanted upward, b_1 will be positive. The sign and magnitude of the linear term and the Y-intercept are not relevant to the determination of the S-shaped curve (analogous to the sign and magnitude of the intercept being irrelevant to whether there was a second-degree term, as in Figure 9.7). One may test each of the coefficients by restricting (restricting out) one coefficient at a time from the Full Model. If each of the two variables adds to the predictability and each of the weights has the appropriate sign, then the Full Model is the preferred model to fit the data. If $c_1 = 0$, then the expected leveling has not yet occurred, and the researcher might wish to extend the study to figure out when practice should stop.

When dealing with higher-order relationships, the signs of the weights become difficult to specify without resort to analytic geometry. If one has a mastery of analytic geometry, one may be able to specify the expected signs of weights and go about testing expectations. Without such training, the researcher may use empirical procedures to develop models and test expectations. Indeed, it can be difficult to ascertain the curve of best fit. Finding the *best* curve usually requires a data-snooping exercise. Curve fitting is complicated even more by the prospect that powers other than whole numbers might be operating. Suppose, for instance, that the exponent is unknown in the following equation:

$$Y = aX^m \hspace{3cm} \text{(Equation 9.2)}$$

Taking the logarithm of both sides of Equation 9.2 results in Equation 9.3, which is as follows:

$$\log_e Y = \log_e a + m(\log_e X) \hspace{2cm} \text{(Equation 9.3)}$$

If $\log_e X$ and the unit vector are used to predict $\log_e Y$, then the following model results:

$$\log_e Y = a_0 U + b_1 \log_e X + E_3 \hspace{2cm} \text{(Model 9.11)}$$

The unknowns in Equation 9.2 can be found empirically by Model 9.11; b_1 is the numerical value of the exponent m, and a_0 is the value for $\log_e a$.

Many researchers have been mistakenly given some misconceptions regarding curve fitting. First, not all the lower powers need to be included. That is to say, the data may be well fit by the second-degree term only. Second, the lower-powered terms are not the "simplest" variables. Complexity should be viewed as the number of predictor variables needed to account satisfactorily for criterion variance (McNeil & McShane, 1974). Researchers have for too long first investigated linear relationships because they thought those were the simplest. Third, the functional fit over the range of sample values is applicable to that range. Generalizations beyond that range are good guesses, but are based on the assumption that the same functional relationship occurs as in the range originally studied (an empirical question, which additional data gathering can answer). The third-degree function in Figure 9.13 is essentially linear from 7 to 12 hours of practice. And it is a second-degree function from 0 to 10 hours of practice.

Hypothesis Seeking Through the Descriptive Use of Multiple Regression Models

As stated in chapter 1, the task of the researcher is to build theoretical models and cast linear equations that will encompass the variables and functional relationships that account for a large proportion of observed criterion variance. Ultimately one hopes to approximate an R^2 of 1.00. In the quest to increase predictability, one may use linear equations descriptively to find the most parsimonious predictive model. Using the 15 groups of hours of practice that were used as the predictor variable in the function presented in Figure 9.13, one could develop a regression model with 15 linearly independent vectors:

$$Y = a_0 U + a_1 X1 + a_2 X2 + \ldots a_{15} X15 + E_4 \qquad \text{(Model 9.12)}$$

where:
1. Y = the criterion;
2. U = unit vector;
3. X1 = 1 if the criterion is from a person with one hour of practice, 0 otherwise;
4. X2 = 1 if the criterion is from a person with two hours of practice, 0 otherwise;

.

.

.

15. X15 = 1 if the criterion is from a person with 15 hours of practice, 0 otherwise; and
16. $a_0, a_1, a_2, \ldots, a_{15}$ represent the coefficients for the unit vector and predictor variables 1 through 15.

Model 9.12 will be the best fitting model (in terms of R^2) for any set of data points because each X-axis value is allowed to have its own mean. The model yields maximum curvilinearity and is referred to as the *eta coefficient model*. Though the eta coefficient model yields maximum curvilinearity, it has two significant drawbacks. First, the model requires many coefficients to be calculated, and therefore the likelihood of being able to replicate those coefficients on successive samples is low. Second, the eta coefficient model does not allow for inferences about Y values for values between the specified X-axis values, whereas using continuous predictor variables does allow for such generalizations (McNeil, 1970b).

If the S-shaped curve is the relationship the data take, the third-degree model (Model 9.12) will account for almost as much criterion variance as the eta coefficient model, but the third-degree model will do so in a more parsimonious fashion because only four pieces of information are needed, whereas 15 are needed in the eta coefficient model. Furthermore, the elimination of the unit vector or the linear component (i.e., a_0 or b_1 can be set equal to 0) may yield a model, which contains only three pieces of information, as predictive a model as Model 9.12 but is a more parsimonious model. If the restriction $a_0 = 0$ is placed on Model 9.12, the following model is produced:

$$Y = b_1P + s_1P^2 + c_1P^3 + E_5 \qquad \text{(Model 9.13)}$$

If the restriction $b_1 = 0$ is placed on Model 9.12, the following model is produced:

$$Y = a_0U + s_1P^2 + c_1P^3 + E_6 \qquad \text{(Model 9.14)}$$

Through this process, the researchers will find the most predictive and parsimonious model. The resulting model should be used as the basis for a cross-validation study. While it is the best model for the sample data, the results it produces may be restrictive with respect to the population to which the results can be generalized. This restriction is due to the fact that the relationships reflected in the model were not hypothesized beforehand, and they were discovered by searching through the sample data.

To further illustrate the use of regression models to seek hypotheses for future testing, consider a case in which the researchers have knowledge of gender, treatment, and some continuous covariate, but little expectation about how these variables interrelate in the prediction of criterion performance. To facilitate this hypothesis seeking procedure, we assume the study under consideration involves three treatment conditions, two sexes, and an ability measure, plus a criterion measure. Bivariate plots could be constructed between the ability measure and the criterion for each of the possible groups [males in Treatment 1 (MT1), males in Treatment 2 (MT2), males in Treatment 3 (MT3), females in Treatment 1

(FT1), females in Treatment 2 (FT2), and females in Treatment 3 (FT3)]. The plots might appear as presented in Figures 9.14a through 9.14f.

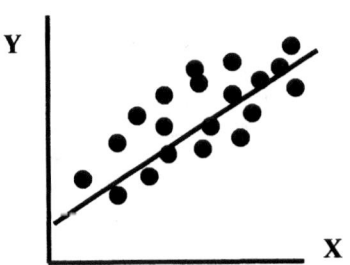

Figure 9.14a.
$Y = a_1MT1 + b_1X1 + E_1$ males in Treatment 1.

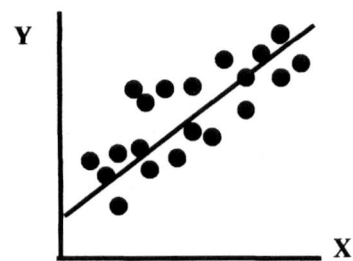

Figure 9.14b.
$Y = a_2FT1 + b_2X2 + E_2$ females in Treatment 1.

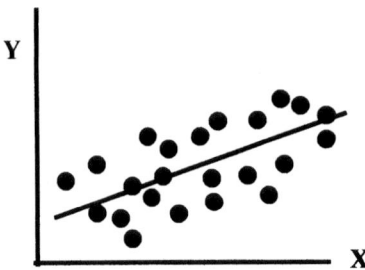

Figure 9.14c.
$Y = a_3MT2 + b_3X3 + E_3$ males in Treatment 2.

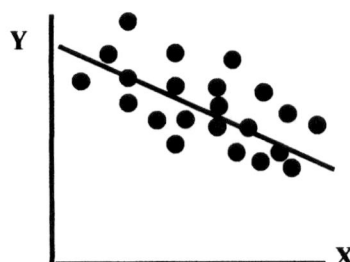

Figure 9.14d.
$Y = a_4FT2 + b_4X4 + E_4$ females in Treatment 2.

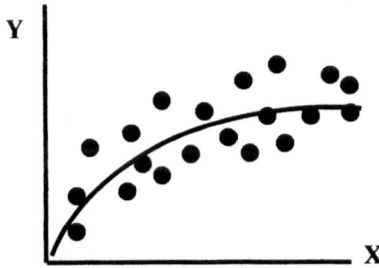

Figure 9.14e.
$Y = a_5MT3 + b_5X5 + s_5(X5)^2 + E_5$ males in Treatment 3 ($a_5 = 0$).

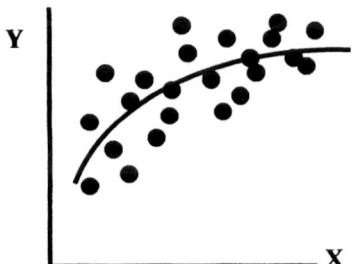

Figure 9.14f.
$Y = a_6FT3 + b_6X6 + s_6(X6)^2 + E_6$ females in Treatment 3.

Figure 9.14.
Six groups and six lines that might reflect the lines of best fit—See Model 9.15 for the definition of vectors.

Each of the lines of best fit is the most parsimonious when considering only that combination of sex and treatment; however, an inspection of the figures might lead to a more parsimonious model for the entire sample. There are 13 linearly independent vectors that describe the six lines (note that for Figure 9.14e a second-degree polynomial is included, but the Y-intercept is 0, thus $a_5 = 0$; and Figure 9.14f has three linearly independent vectors since $a_6 \neq 0$). The males and females who had Treatment 1 look like they have the same slope, but have different Y-intercepts. Thus $b_1 = b_2$ seems like a reasonable restriction. Furthermore, the curves for males and females who had Treatment 3 seem the same, with the exception of having different Y intercepts; thus $b_5 = b_6$ and $s_5 = s_6$ seem like reasonable restrictions. The curve for males appears to go through the origin, so setting the Y intercept equal to 0 ($a_5 = 0$) seems reasonable. The curve for females, though, needs a non-zero Y intercept.

Model 9.15 is one possible model reflecting all six lines in one linear equation. Model 9.15 is as follows:

$$Y = a_1 MT1 + b_1 X1 + a_2 FT1 + b_2 X2 + a_3 MT2 + b_3 X3 + a_4 FT2 + b_4 X4 + b_5 X5 + s_5 X5^2 + a_6 FT3 + b_6 X6 + s_6 X6^2 + E_7 \qquad \text{(Model 9.15)}$$

where:
1. Y = the criterion;
2. MT1 = 1 if Y comes from a male in Treatment 1, 0 otherwise;
3. X = ability if Y comes from a male in Treatment 1, 0 otherwise;
4. FT1 = 1 if Y comes from a female in Treatment 1, 0 otherwise;
5. X2 = ability if Y comes from a female in Treatment 1, 0 otherwise;
6. MT2 = 1 if Y comes from a male in Treatment 2, 0 otherwise;
7. X3 = ability if Y comes from a male in Treatment 2, 0 otherwise;
8. FT2 = 1 if Y comes from a female in Treatment 2, 0 otherwise;
9. X4 = ability if Y comes from a female in Treatment 2, 0 otherwise;
10. X5 = ability if Y comes from a male in Treatment 3, 0 otherwise;
11. $X5^2$ = ability squared if Y comes from a male in Treatment 3, 0 otherwise;
12. FT3 = 1 if Y comes from a female in Treatment 3, 0 otherwise;
13. X6 = ability if Y comes from a female in Treatment 3, 0 otherwise; and
14. $X6^2$ = ability squared if Y comes from a female in Treatment 3, 0 otherwise.

The researchers derived Model 9.15 from an inspection of the data rather than from a prior Research Hypothesis. The expected equalities presented in the previous paragraph can be verified by imposing the appropriate restrictions on Model 9.15 and then comparing the resulting R^2 values for each model. Because empirical line fitting as seen in this context is an attempt to describe parsimoniously the data, no Research Hypothesis is necessary. One may successively cast models that reflect $b_1 = b_2$, $b_5 = b_6$, and $s_5 = s_6$ and compare the R^2 values of those models with Model 9.15.

In the context of Model 9.15, one may investigate the possibility that $b_1 = b_2$, both equal to b_7, a common slope. If vectors X1 and X2 are added to form X7, then b_7 is a common slope for males and females under Treatment 1. The resulting model is as follows:

$$Y = a_1MT1 + a_2FT1 + b_7X7 + a_3MT2 + b_3X3 + a_4FT2 + b_4X4 + b_5X5 + s_5X5^2 + a_6FT3 + b_6X6 + s_6X6^2 + E_8 \qquad \text{(Model 9.16)}$$

There are 13 linearly independent vectors in Model 9.15 and 12 in Model 9.16; thus Model 9.16 is more parsimonious. If Model 9.16 is judged by the researcher to be as predictive as Model 9.15 (i.e., the difference between the R^2 values of the two models is not statistically significant), the researcher would prefer Model 9.16 to Model 9.15. In such a case Model 9.16 becomes the preferred model to which additional models will be compared.

The researcher may then wish to investigate the possibility that $b_5 = b_6$, both equal to b_8, a common weight; and $s_5 = s_6$, both equal to s_8, a common weight. These two restrictions can be made one at a time, sequentially, or simultaneously. If done together, vectors X5 and X6 are added to get vector X8 and vectors $X5^2$ and $X6^2$ are added to get $X8^2$. Assuming the researchers selected Model 9.16 over Model 9.15, placing these two restrictions in Model 9.16 produces the following model:

$$Y = a_1MT1 + a_2FT1 + b_7X7 + a_3MT2 + b_3X3 + a_4FT2 + b_4X4 + b_8X8 + s_8X8^2 + a_6FT3 + E_9 \qquad \text{(Model 9.17)}$$

The coefficients b_8 and s_8 will give the common linear and second-degree slopes for males and females given Treatment 3. Note that Model 9.16 has 12 linearly independent vectors, while Model 9.17 has 10 linearly independent vectors. Thus, Model 9.17 is a more parsimonious model.

If the R^2 loss between Model 9.17 and Model 9.16 is large, it would not reveal to the researcher whether one or the other restriction caused the loss in R^2, but rather both of the restrictions lead to the loss in the R^2 value. On the other hand, if the R^2 loss from Model 9.16 to Model 9.17 is not statistically significant, the researchers could conclude that $b_5 = b_6$ and $s_5 = s_6$. In this case the researchers would conclude that even though Model 9.16 is equally predictive as Model 9.17, the researchers would select Model 9.17 as a better description of the data than Model 9.16 because it is more parsimonious. The researchers could then formulate a hypothesis based on Model 9.17 and then test this hypothesis on new, "unsnooped" data.

When using the empirical line-fitting procedure just described, researchers should be aware that they are data-snooping and hypothesis seeking. Under such circumstances, the obtained R^2 values must be seen as *descriptive* statistics, which can be used for future hypothesis testing. *One should not use the snooped data inferentially.* A replication with new subjects is necessary before generalizing to the population. (A more extensive discussion of data-snooping appears in chapters 6 and 12.)

RESCALING OBSERVED NONLINEAR RELATIONSHIPS

In the preceding section of this chapter, we discussed transforming (squaring, cubing, etc.) variables to reflect the hypothesized nonlinear functional fit between the predictor set and the criterion. In those instances, it was expected that the actual constructs (those underlying phenomena that the predictors and the criteria were measuring, however imperfectly) were related to one another in a nonlinear manner. When a nonlinear relationship between variables is observed, however, it may not be due to the constructs being related nonlinearly; it could be due instead to either variable being a poor measure (or map) of the construct it represents.

Our concern in the final section of this chapter is with transforming a variable such that the numbers more accurately reflect the construct under consideration. Because constructs, by definition, cannot be observed, researchers must realize that the measurements they use to indicate those constructs are arbitrary numbering systems that someone has decided are "good."

The "goodness" of the arbitrary scoring could be ascertained by observing the overlap between the numbers and the construct. The Venn diagram in Figure 9.15 represents what researchers usually assume to be the state of affairs—a complete overlap between the measured variable and the construct.

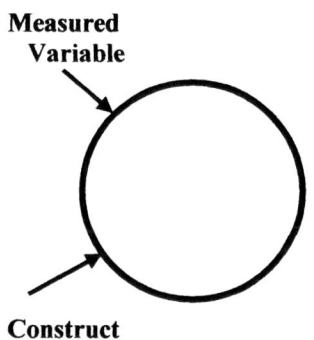

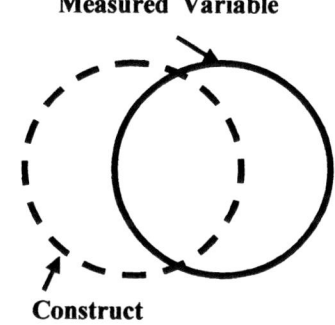

Figure 9.15.
The assumed overlap between the measured variable and the construct that the variable measures—100% overlap.

Figure 9.16.
The assumed overlap between the measured variable and the construct that the variable measures—less than 100% overlap.

The Venn diagram in Figure 9.16 represents what may be a more likely state of affairs—only a partial overlap between the measured variable and the construct. The process of transforming variables is an attempt to reach as closely as possible the state of affairs depicted in Figure 9.15.

There is no conventional or obvious way to transform a variable such that the numbers will better match the construct. A thorough knowledge of one's area of research will certainly help. A well-thought-out nomological net also would be of some assistance. But because the transformations must ultimately be tested by successive empirical verifications, perhaps the most efficient method would be through data-snooping. Data-snooping was briefly discussed earlier (see the section entitled "Hypothesis and Data Snooping" in chapter 6 and it is discussed later in chapter 12). We first give several examples to illustrate the need for transformation of variables—a process we refer to as *rescaling*.

Difficulty of Items

The numbers assigned to measure a construct should reflect that the scale points are in order (e.g., a score of 5 reflects more of the construct than does a score of 4 and less than does a score of 6) and should reflect the distance between scale points (e.g., the construct distance between scores of 4 and 6 is the same as the construct distance between 8 and 10).

The arbitrarily scored scale in Figure 9.17 has seven items, but it covers eight units on the construct. (Note: It must be emphasized here that attainment of actual construct scores—the criterion in this part of chapter 9—would most likely not be possible.) If a unit weight is assigned to each item on the scale, it will not accurately map the construct. Persons who get scores of 4, 5, 6, or 7 on the arbitrarily scored scale really have 5, 6, 7, or 8 units of the hypothetical construct. In order for the arbitrarily scored scale to map the construct, one unit would have to be added to the arbitrarily scored scale if the arbitrarily scored scale score is greater than three. Another possibility would be to add another item to the seven-item test.

To further illustrate this point suppose that the scale developer had done a relatively good job in writing items and assigning numbers to them, with the exception of the very difficult items. Only a few difficult items were included, and they were very difficult. Figure 9.18 might be the result of this scale.

Items 1 through 5 each represent one unit on the construct, but items 6 through 10 cover a distance from more than 5 to 25 on the construct. Assigning a unit weight to each item will not map the construct accurately; the arbitrarily scored scale must be modified. Either additional items would have to be added to the instrument, or the scale would have to be transformed. Since the "right" items would have to be added, transforming the existing scale is easier.

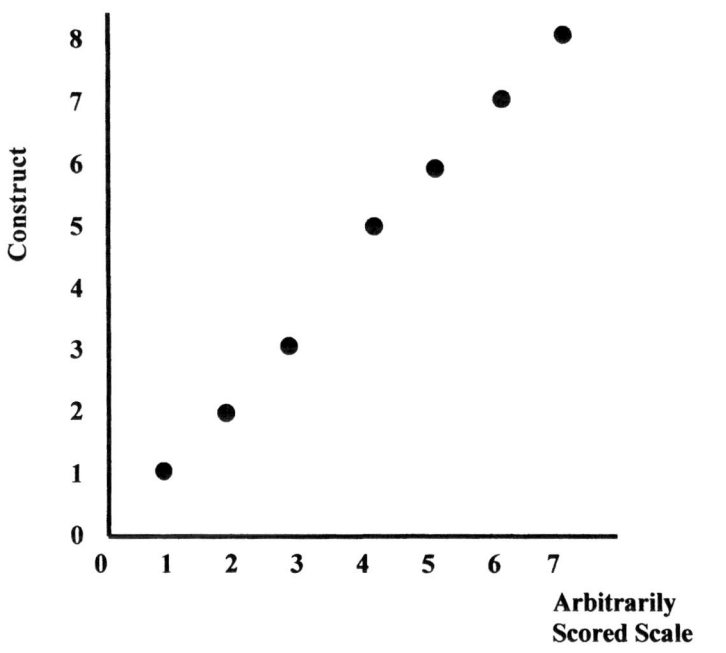

Figure 9.17.
An arbitrarily scored scale that does not accurately reflect construct distances between scale points.

If the arbitrarily scored value is five or below, it is not necessary to transform the value; while an arbitrarily scored value that is greater than five, must be rescaled to map it onto the construct. Assuming the line of best fit over the range of arbitrarily scored values from 5 to 10 had a slope of 4.0 and a Y-intercept value of -15, an arbitrarily scored value of 5 or greater (indentified as X) would be transformed through Equation 9.4, which is as follows:

Transformed Score = (4.0) * (X) – 15   (Equation 9.4)

The reader should note in the previous two examples that the rescaling suggested would simply result in all data points falling on a straight line. The rescaled arbitrarily measured scale rectilinearly maps the criterion. The resultant scoring system is also arbitrary but has an added feature—a linear fit with the construct. That perfect mapping with the construct is the goal of this discussion.

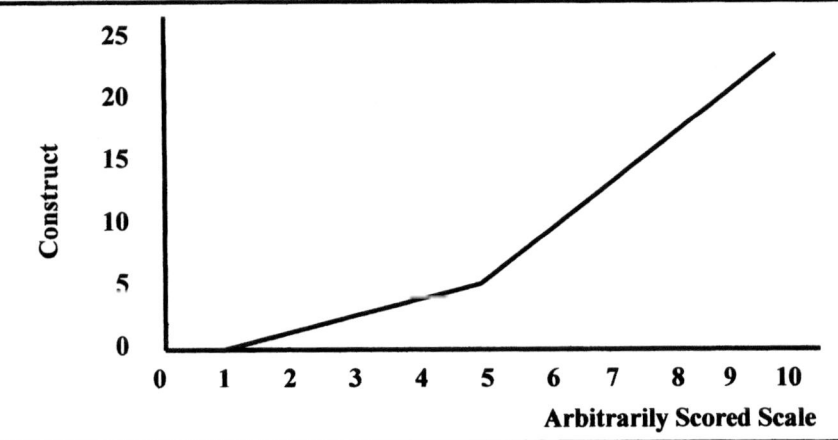

Figure 9.18.
Relationship between an arbitrarily scored scale and the construct when only a relatively few difficult items are included.

Ceiling Effect

The term *ceiling effect* implies that there is not enough discrimination at the high end of the arbitrarily scored scale, as in Figure 9.19. One could overcome this problem by including additional difficult items or by rescaling the existing arbitrarily scored scale.

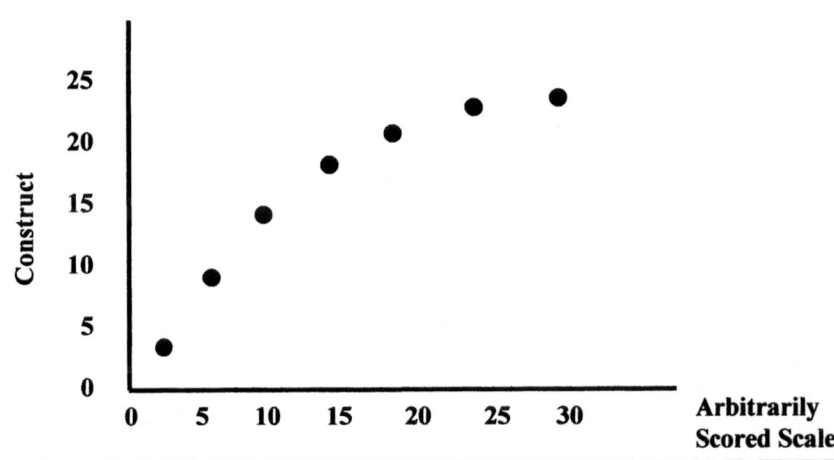

Figure 9.19.
Data depicting a ceiling effect.

The construct can be better mapped by taking the square root of the arbitrary scaled value (X) and then multiplying by 4 as indicated by Equation 9.5, which is as follows:

$$\text{Transformed Score} = (\sqrt{X}) * (4) \qquad \text{(Equation 9.5)}$$

The multiplication by 4 is not a necessary step, since the first step (that of taking the square root) results in the scores falling on a straight line. The multiplication by 4 simply results in the rescaled scores numerically matching the (arbitrarily scored) construct.

Guessing Effect

If a 100-item, 4-choice, multiple-choice test is developed some 25% of the items can be correctly answered solely by guessing. One could therefore argue that any score of 25 or below reflects a construct score of zero. The remaining scores also could be adjusted for the guessing effect to maintain a scale from 0 to 100. Figure 9.20 reflects this construct.

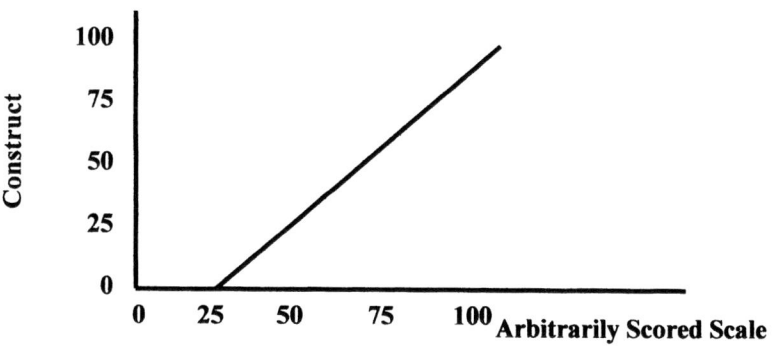

Figure 9.20.
The possible relationship between a 100-item multiple-choice test and the construct.

The required adjustments to transform each arbitrarily scaled score (X) are as follows:
1. If the arbitrarily scaled score (X) is less than or equal to 25, X is set equal to 0.
2. If the arbitrarily scaled score (X) is greater than 25, the score is transformed by using Equation 9.6, which is as follows:

$$\text{Transformed Score} = (4/3) * (X - 25) \qquad \text{(Equation 9.6)}$$

These data transformations set the rescaled score equal to zero if the original score was less than or equal to 25. If the original score was greater than 25, the rescaled score is adjusted for guessing.

Conceptual Rescaling

Haupt (1993) investigated drug use of 4th, 5th, and 6th graders. One question the students were asked to respond to was: How often do you take part in any school clubs, sports groups, or other activities? The answers were originally scored as never = 1, sometimes = 2, I used to = 3, and a lot = 4. Because club participation was used to infer the amount of time spent in structured activities, the scoring for responses "sometimes" and "I used to" were reversed to reflect the continuum of time in structured activities depicted in Figure 9.21.

Another example of conceptual rescaling was provided by Presley and Huberty (1988). They make the point that, when scores are to be weighted for the purposes of summing together, the standard deviations must be considered. The best way to do that is to transform all scores so that they have a common mean and common standard deviation. So if you wanted to construct a course grade by giving 40% weight to the midterm exam and 60% to the final, then the rescaling would be completed by substituting the standardized scores for the midterm (ZMIDTERM) and final (ZFINAL) into Equation 9.7 as follows:

$$\text{GRADE} = (.40)(\text{ZMIDTERM}) + (.60)(\text{ZFINAL}) \qquad \text{(Equation 9.7)}$$

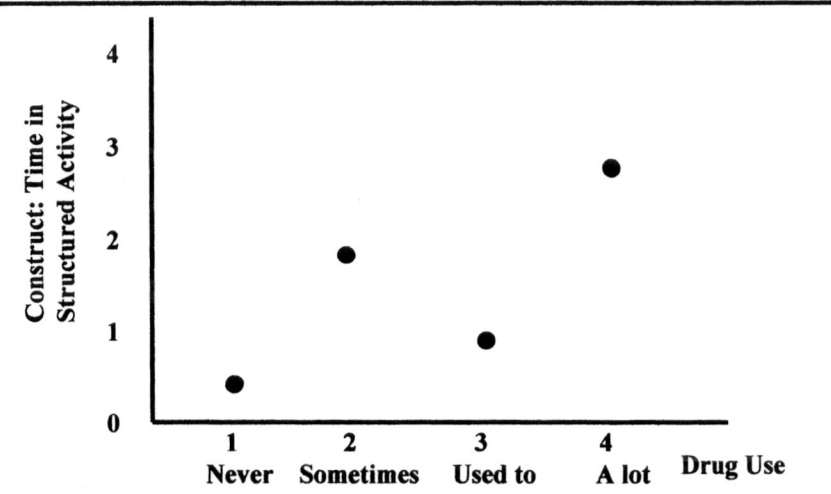

Figure 9.21.
An arbitrarily scored scale that needs rescaling based on conceptual notions about the construct.

The Spearman Rank Correlation

Researchers are often encouraged to rank their data when they have no faith in the interval property of their data and then use the Spearman Rank Correlation if an index of relationship is desired. The lowest score is assigned a value of 1, the next lowest a value of 2, and so on. If one wanted to find the rank correlation between, say, X1 and X2, one could determine the rank of each score on each variable and then obtain the R^2 for the following model:

$$R1 = a_0U + b_1R2 + E_1 \qquad \text{(Model 9.18)}$$

where:
1. R1 = the ranked X1 values;
2. U = unit vector;
3. R2 = the ranked X2 values; and
4. a_0 and b_1 are the coefficients for the unit vector and the X2 ranked values.

The square root of the R^2 of Model 9.18 would be the desired numerical value for the Spearman Rank Correlation. The important point to realize is that assigning the ranks of 1, 2, and so on is one way of rescaling.

Square Root Transformation

Researchers are sometimes admonished, when they discover unequal variances in several groups, to make a square root transformation. Depending upon the nature of the data, this procedure may serve the purpose of equalizing the (sample) variances. Two points should be made. First, the statistical assumption is about the population variances not about the sample variances. Second, blindly applying a square root transformation will not always produce the desired results. The kind of transformation required on the criterion to obtain homogeneous variances depends on the data and will not always be a square root function. Indeed, one should realize from previous chapters that the unequal variances might be due not to scaling considerations but to inherent interactions.

Nonlinear Transformations of the Criterion

The Spearman Rank Correlation and square root transformation actually make transformations on the criterion variable. They are both routine approaches that are applied somewhat mechanically to all variables—predictors and criterion. One could, though, make a specific transformation of the criterion, either for the purpose of rescaling or to reflect the functional fit between the predictor variable(s) and the criterion variable.

The Pythagorean Theorem is a good example of the utility of transforming the criterion. The reader may remember that the Pythagorean Theorem is:

$$C^2 = A^2 + B^2.$$

In terms of the GLM it is expressed as follows:

$$C^2 = a_1A^2 + b_1B^2 + E_1 \qquad \text{(Model 9.19)}$$

where:

1. C^2 = square of the hypotenuse of a right triangle;
2. A^2 = square of the length of one side of the triangle;
3. B^2 = square of the length of the other side of the triangle; and
4. a_1 and b_1 are both equal to 1.

McNeil, Evans, and McNeil (1979) pointed out several interesting aspects about this regression solution:

1. The R^2 value is 1.00, and thus the error components are 0 for each subject (each triangle).
2. The linear component of A, the linear component of B, and the unit vector are all absent from this best model.
3. An R^2 value of 1.00 could not have been achieved if the criterion had not been squared.
4. The transformations on the criterion and the predictors are not for rescaling purposes (as the lengths were measured by ruler, a very good interval way to measure distance.) The transformations were made to match the functional fit.

Functional Fit or Rescaling?

When transformations on the arbitrarily scored variables are made, the researcher is either rescaling variables or allowing for a given functional fit. Which of the two is being done is not important from a predictive point of view—the crucial concern is whether the R^2 has been increased by the transformation. A simple example may be of value here. Suppose that Model 9.20, which contains a linear term, yields an R^2 value of .60, but that Model 9.21, which contains only a second-degree term, yields an R^2 value of 1.00.

$$Y = a_0U + b_1X + E_1 \qquad \text{(Model 9.20)}$$

$$Y = a_0U + c_1X^2 + E_2 \qquad \text{(Model 9.21)}$$

These two models cannot be compared by the general F test because one is not a restriction of the other; indeed, both have two predictor pieces of information. These two models can be compared, however, on the basis of degree of predictability. Based upon the goal of predictability, Model 9.21 would be the preferred model because it yields a higher R^2 value. If the researcher felt that X was a good nonlinear mapping of the construct, then Model 9.21 would reflect a functional fit, as discussed in the early part of this chapter. If the squaring of X was a result of acknowledging the arbitrariness of the scaling of the original measure, then Model 9.21 would reflect the rescaling notions discussed in the latter part of this chapter. Whether the transformation represents a functional fit or rescaling is important from an "understanding" point of view.

Suppose that a given criterion is predicted quite well by a given predictor, as in Figure 9.22. The researchers' theory expects that the same functional fit ought to be the case for higher values of the predictor variable. As shown in Figure 9.23, though, the linear fit observed over the lower values of the predictor does not generalize to higher values. Indeed, the relationship over the entire range of values in Figure 9.23 is better depicted by a second-degree curve. If higher values of the predictor variable were investigated, perhaps the functional fit would be other than a second-degree one. The functional fit that is claimed for predictor values outside the range originally investigated must be empirically verified with those new values of the predictor variable.

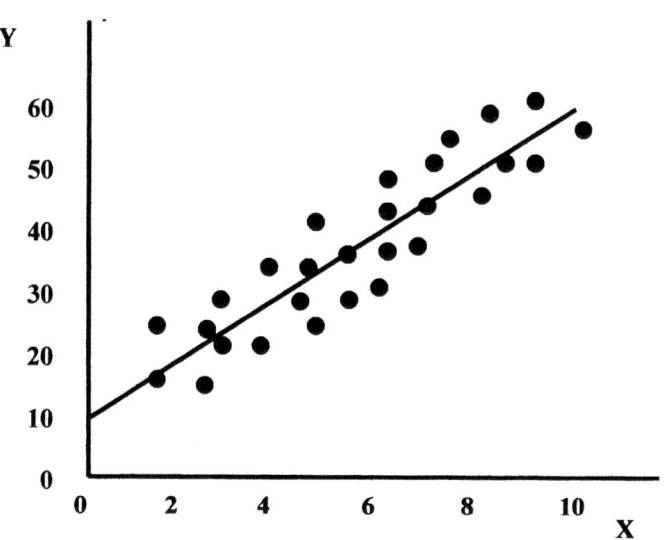

Figure 9.22.
A linear fit over the range of X values 0 through 10.

The important concern of this chapter is to emphasize the notion that numbers put on measurements are arbitrary. Initial attempts at numbering most likely do not do a good job of mapping the construct. Given that the researchers have rescaled the variable and believe the rescaled variable does in fact map the construct, the concern then is to reflect (with regression models) the hypothesized functional fit. Using the X variable only allows for a linear fit, whereas other types of fits may be required. Given that the scaling is arbitrary, there is no more reason to expect a linear fit than a nonlinear fit. Variable X is actually X^1, and realizing that should make it clear that there are an infinite number of other exponents that can be investigated (e.g., X^2, X^3, $X^{1.5}$, $X^{.3}$, $X^{.313}$).

Researchers and students often ask us to justify our choice when we use, say, $X^{1.3}$. We respond: Why do you use $X^{1.000}$? The linear component is a default value that in the age of computers should be tested for its value not automatically assumed to be of value. If the linear component is used, then some justification should be provided—just as justification should be provided for any other exponent.

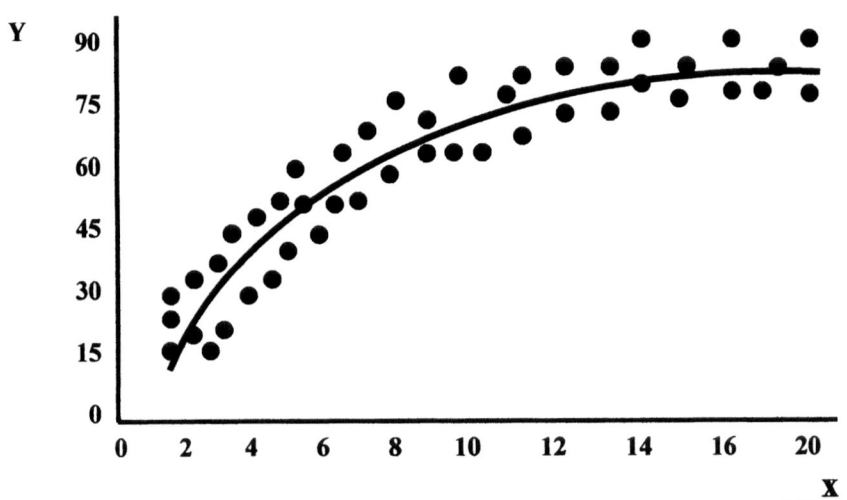

Figure 9.23.
A nonlinear fit over the range of X values 0 through 20.

Transformation of the Criterion Variable

As discussed earlier, the criterion can be transformed. In Model 9.22 each value in the criterion variable (Y) is transformed to the square root of the original value. Model 9.22 is as follows:

$$\sqrt{Y} = a_0U + b_1X + E_3 \qquad\qquad \text{(Model 9.22)}$$

It is important for researchers to understand that when the criterion variable has been transformed in a model, such as in Model 9.22, its R^2 value cannot be compared to a model in which the criterion variable has not be transformed (or transformed in a different manner). The reason the R^2 values cannot be compared is because the variance in the untransformed criterion variable and the transformed criterion variable are not equal.

Reliability and Validity

Reliability is the extent to which a variable yields the same value on repeated measures. *Validity* is the extent to which a variable measures what it is supposed to measure. The regression approach can be used to determine the magnitude of both.

An illustration of a procedure used to estimate and statistically test a test-retest reliability coefficient. Test-retest reliability is simply the correlation between two administrations of the same test. To illustrate how a GLM can be used to estimate a test-retest reliability value and statistically test this value, consider a case in which the researchers conducted a study to determine if the estimated test-retest reliability value exceeds .80. The variables FirstScore and SecondScore contained the test scores obtained from the first and second administrations of the test, respectively, for a group of students. The model required to estimate the test-retest reliability of the test is as follows:

$$\text{SecondScore} = a_0U + b_1\text{FirstScore} + E_1 \hspace{2cm} \text{(Model 9.23)}$$

It should be noted that this type of investigation is basically a correlational analysis. Thus, it does not matter which variable is used as the criterion variable and which variable is identified as the predictor variable. That is, regardless of which variable is used as the criterion variable, the same Multiple R value (correlation value), which is the square root value of the model's R^2 value, is produced. Researchers should note that the R^2 value and the Multiple R value will always be positive even if the relationship between the criterion and predictor variables is negative. Thus researchers need to verify that the relationship between the criterion and predictor variables is positive by checking the sign of the predictor variable coefficient, which will be positive when the relationship between the criterion and predictor variables is positive

For this illustration, we are assuming the R^2 value for Model 9.23 is .72 and the study involved 100 students. The test-retest coefficient is equal to the multiple R value, which is the square root of the model's R^2 value. Thus, the test-retest value is equal to .85 (assuming the coefficient for the FirstScore variable, b_1, is positive). To verify the value of .85, the researchers can execute the "F Test Calculation" program, which can be found in the internet site for this text (see Appendix A for the internet site address and Appendix D for a list of the commands needed to construct this Excel program). Once this program has been accessed with the Microsoft Excel program, the window listed in Exhibit 9.7 will be displayed.

The researchers must complete the following steps to obtain the F test of the test-retest coefficient (see Oval 1 in Exhibit 9.5):

1. Model 9.25 is viewed as the Full Model. Thus, the R^2 value for Model 9.25 (.72) is entered in column B in the row entitled "R-Squared Full Model."
2. Since the researchers hoped to find that the test-retest coefficient exceeded .80, the R^2 value of .64, which is obtained by squaring the .80 value, is entered as the R^2 value of the Restricted Model. (Thus, .64 is entered in column B in the row entitled "R-Squared Restricted Model."
3. Since the Full Model (Model 9.25) contains two pieces of information (i.e., the unit vector and the FirstScore variable), the value of 2 is entered in column B in the row entitled "Regression DF Full Model."
4. Since the R^2 value that the R^2 value of the Full Model (Model 9.25) is to be compared with has been set equal to .64, which is one piece of information, the value of 1 is entered in column B in the row entitled "Regression DF Restricted Model."
5. Since the study used test scores from 100 students and the Full Model (Model 9.25) contained two pieces of information, the value of 98 (i.e., 100 − 2) is entered in column B in the row entitled "Residual DF Full Model."

Exhibit 9.5.
Microsoft Excel F test of the test-retest coefficient.

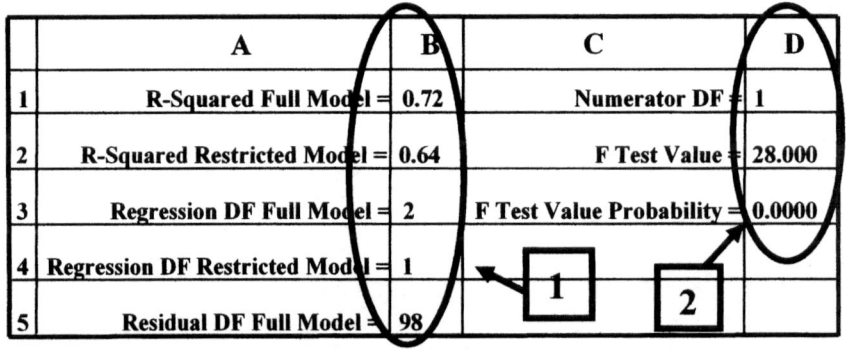

	A	B	C	D
1	R-Squared Full Model =	0.72	Numerator DF =	1
2	R-Squared Restricted Model =	0.64	F Test Value =	28.000
3	Regression DF Full Model =	2	F Test Value Probability =	0.0000
4	Regression DF Restricted Model =	1	1	
5	Residual DF Full Model =	98	2	

Once these values are entered into the Microsoft Excel program, the F test value of the difference between the two R^2 values of .72 and .64, and the corresponding probability value are listed in Oval 2 of Exhibit 9.5. The F test value is 28.00 and its two-tailed probability value is less than .001. Recall that the researchers are conducting a statistical test to determine whether the test-retest coefficient exceeded .80, which requires the R^2 value of the Full Model (Model 9.25) to exceed .64, that is, a one-tailed test is being conducted. Since the one-tailed probability ($p / 2 < .001$), which is the two-tailed probability value divided by 2, is less than the established alpha level of .05, the researchers would be willing to conclude that the R^2 value of the Full Model (Model 9.25) exceeds

.64. Thus, the analysis indicates that the estimated test-retest coefficient of .85, which is the square root value for the R^2 value of .72, is statistically higher than .80.

Validity. Reliability is of minor concern, though, as validity is the ultimate test of any variable. The purpose of rescaling is to get the measure to be a valid indicator of the construct. The aim of functional curve fitting is to get the predictor set to be a valid indicator of the criterion. Most measurement texts correctly say that a variable can be highly reliable yet have no validity for a particular construct. For example, shoe size can be reliably measured, yet it is not considered a valid measure of IQ.

Furthermore, many constructs meaningfully fluctuate over a short period. Breathing rate changes from one hour to the next; anxiety fluctuates from one task to another; and reading interest fluctuates from one comic strip to another. The concern in each case should be: Does the measure accurately reflect the criterion?

It is our position that too much attention has been placed on the reliability of measures, and not enough attention has been placed on the validity. Although some form of reliability is a necessary but not sufficient condition for validity, and because validity is the ultimate goal, more attention should be placed on validity.

The Criterion as an Approximation of the Construct

In the past, a cause-and-effect relationship has been interpreted for data that, although highly significant, yielded small R^2 values. One of the positions we take in this text is that researchers can more fruitfully spend their time by finding variables that will increase the R^2, rather than trying to understand how that small proportion of criterion variance has been predicted. Related to this problem, though, is the task of determining whether an adequate overlap exists between the construct and the criterion measure of that construct. Too often the criterion variable is equated with the construct, when in reality it is important for researchers to realize that *the criterion variable is only an approximation of the construct.* Unfortunately, there is no way at the present to know what overlap exists between the construct and the criterion variable.

Figure 9.24 presents Venn diagrams that represent four possible relationships between criterion, construct, and predictor variables. The Venn diagram contained in Figure 9.24b is the state of affairs assumed by most researchers, that is, the criterion variable accounts for 100% of the variation in the construct (horizontal lines area). The Venn diagram presented in Figure 9.24a, however, is more likely the state of affairs. Only a partial overlap exists between the criterion and the construct (areas marked by vertical and horizontal lines). One may do well to continue trying to account for the criterion variance beyond that accounted for by predictor variables Q, X, and Z (vertical lines) in Figure 9.24a because of the remaining overlap between construct and criterion variable that has yet to be accounted for by these predictor variables (horizontal lines).

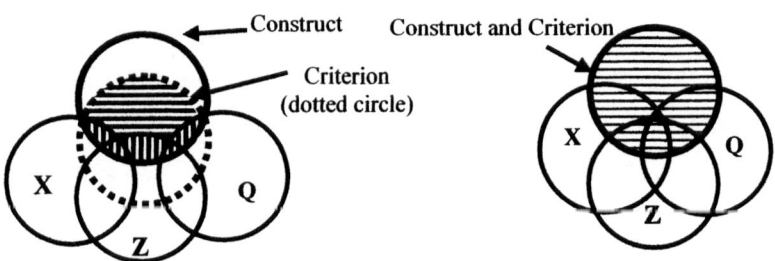

Figure 9.24a.
The criterion variable accounts for some of the variation in the construct (horizontal lines).

Figure 9.24b.
The criterion variable accounts for all of the variation in the construct (horizontal lines area).

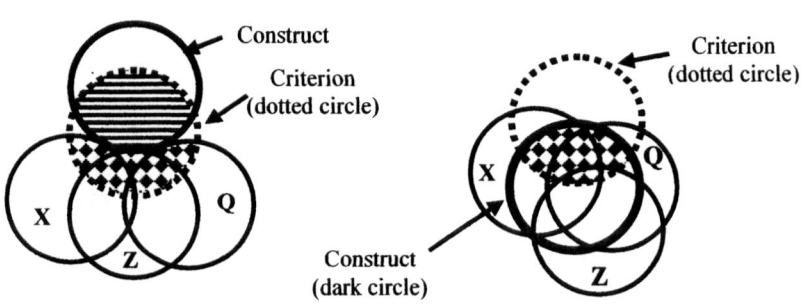

Figure 9.24c.
Predictor variables account for none of the common variation between the criterion variable and the construct (horizontal lines area)—No overlap by X, Q, or Z.

Figure 9.24d.
All of the common variation between the criterion variable and the construct is accounted for by the predictor variables (diamond-marked area).

Figure 9.24.
Venn diagrams that represent possible relationships between criterion, construct, and predictor variables (X, Q, and Z).

The Venn diagrams presented in Figures 9.24c and 9.24d also may depict the existing state of affairs. In the Venn diagrams contained in Figure 9.24c, the common overlap between the criterion variable and the construct (horizontal lines area) has yet to be accounted for by the three predictor variables. That portion of the criterion that is not predictive of the construct has, unfortunately, been well accounted for by the predictor variables (diamond-marked area). Any interpretation regarding the predictor variables that explain, let alone cause, the construct in this situation will be grossly mistaken. If the predictor variables are expected to map the construct, according to researchers' theory, then the predictor variables probably should be rescaled.

In Figure 9.24d all of the common variation between the criterion variable and construct is accounted for by the predictor variables (diamond-marked area). This situation may also call for the rescaling of the criterion measure. If various rescaling measures do not map the construct better, then the researcher should either turn to another criterion measure or realize that the construct is multidimensional. In this latter case, multiple criterion measures would be necessary, an extension of multiple regression notions. Many multivariate procedures do exist but are extensions beyond the scope of this text. The interested reader may refer to Bray and Maxwell (1985); Hair, Black, Babin, Anderson, and Tatham (2006); Hand and Taylor (1987); Harris (1985); Stevens (2002); and Tatsuoka (1988).

Nonlinear Interaction

The models and figures to be discussed in this section deal with those instances in which the researcher expects a second-degree relationship between one of the predictor variables and the criterion variable. Model 9.24 allows for a first-degree relationship and a second-degree relationship between variables Q and Y, while the relationship between variables P and Y is linear.

$$Y = a_0U + b_1P + c_1Q + d_1Q^2 + E_1 \qquad \text{(Model 9.24)}$$

This relationship is depicted in Figure 9.25; the predicted Y scores form a curved plane, with curved edges along the Q sides and straight edges along the P sides.

When an interaction between variables P and Q is added to Model 9.24, the following model is formed:

$$Y = a_0U + b_1P + c_1Q + d_1Q^2 + e_1(P * Q) + E_2 \qquad \text{(Model 9.25)}$$

The graphic representation of Model 9.25 is shown in Figure 9.26. The curved plane formed by the predicted criterion scores has curve edges along the Q sides and straight lines with differing slopes along the P sides.

Suppose that two continuous predictor variables (P and Q) each have a U-shaped relationship with the criterion variable Y. The model used to reflect these relationships would require second-degree terms for both the P and Q variables. The model, which includes two second-degree terms, would be constructed as follows:

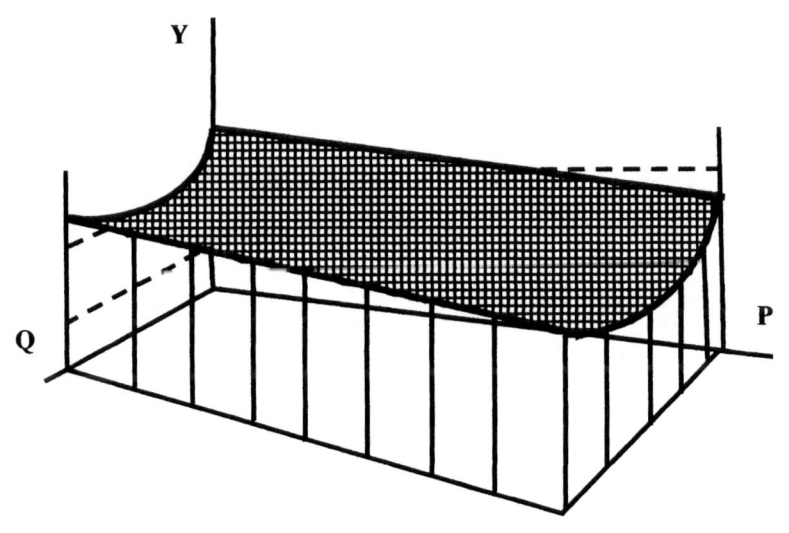

Figure 9.25.
A plane curved on one axis fit by Model 9.24.

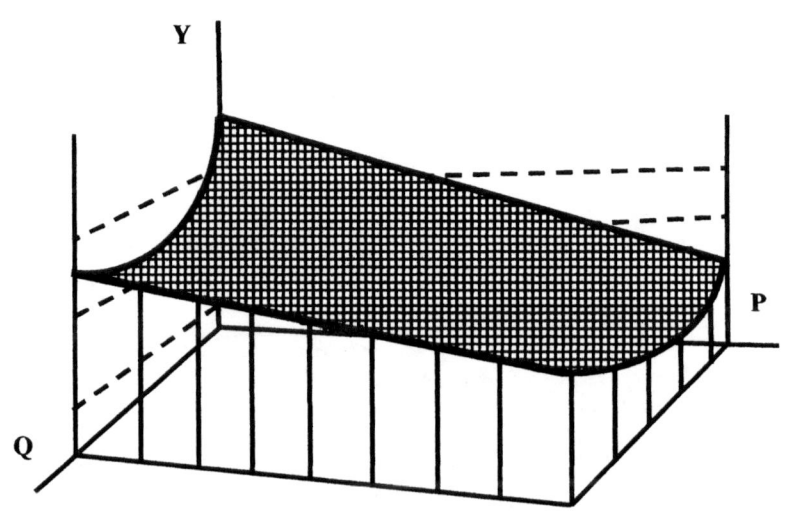

Figure 9.26.
A twisted plane curved on one axis fit by Model 9.25

$$Y = a_0U + b_1P + s_1P^2 + c_1Q + d_1Q^2 + E_3 \qquad \text{(Model 9.26)}$$

The graph of Model 9.26, which is presented in Figure 9.27, has U shaped curves for both the P and Q sides.

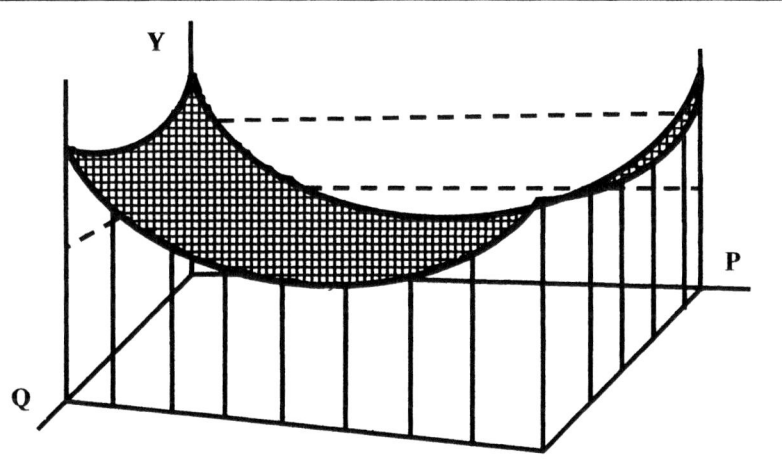

Figure 9.27.
A plane curved on both axes fit by Model 9.26.

These models and figures are only examples of the infinite types of relationships one may expect to find and are presented as a means of helping the researcher to visualize those relationships. We must again emphasize, though, that being able to draw a figure of a model is not a necessary step. If a model yields a high degree of predictability, that is sufficient. Indeed, figures in more than three dimensions are a little difficult to communicate—let alone comprehend.

Interpretation of Complex Interactions versus Predictability

Whenever researchers find a significant interaction, they should try to graph the interaction and interpret it. If the researchers have hypothesized a directional interaction before the fact, then the interpretation of the significant interaction should not be difficult. If the researchers did not suspect interaction, then a replication study should substantiate the results before they begin to have faith in actual existence of that interaction. (Actually, this requirement applies not just to interaction. Any non-hypothesized result needs replication before being accepted.)

Researchers may continually find an interaction term that accounts for a large proportion of the criterion variance; that is, the goal of predictability has been satisfied in achieving a large R^2. But why this particular interaction exists may not be explainable at the time it is found. To throw away this high degree of predictability because of a failure to comprehend it seems a bit egotistical. On the other hand, obtaining premature closure on interpretations of data may be one of the most drastic mistakes researchers make. Researchers could argue that insightful interpretations are made as a function of the data "telling" us something new, rather than as a function of our preconceived notions twisting the data so that it makes sense to us.

Interacting Variables: One Way to View all Variables

A strong case was made in chapter 7 and again in this chapter for the consideration of interaction variables. All variables can be considered as interacting variables if one remembers that any number to the zero power is equal to one. Any variable that is a multiplication of two or more other variables, say $P * Q$, is an interaction variable. By convention, the powered superscript is dropped when it is equal to one; therefore $P * Q$ also could be represented as $P^1 * Q^1$. Also, $P^2 * Q^1$ is an interaction variable, and $P^0 * Q^1$ can be considered as an interaction variable. But any number to the zero power is equal to one, therefore $P^0 * Q^1$ is simply the variable Q. Thus Q can be conceptualized as the interaction of P to the zero power and Q to the first power. Similarly, the unit vector (U) can be conceptualized as Q^0, $P^0 Q^0$, or $P^0 Q^0 R^0$, and so on.

Mendenhall (1968, p. 94) presents the regression model allowing for interaction in a two-by-three ANOVA, which is as follows:

$$Y = a_0 U + b_1 P + c_1 Q + d_1 Q^2 + e_1 (P * Q) + f_1 (P * Q^2) + E_1 \qquad \text{(Model 9.27)}$$

where:
1. P = two distinct scores representing the two levels of the first independent variable; and
2. Q = three distinct scores representing the three levels of the second independent variable.

Model 9.27 can be further conceptualized as interaction variables as illustrated in the following model:

$$Y = a_0 (P^0 * Q^0) + b_1 (P^1 * Q^0) + c_1 (P^0 * Q^1) + d_1 (P^0 * Q^2) + e_1 (P^1 * Q^1) + f_1 (P^1 * Q^2) + E_1 \qquad \text{(Model 9.27a)}$$

Each term in the above model is represented as the interaction between one independent variable and the other independent variable. Models 9.27 and 9.27a are conceptually and mathematically equivalent.

One could extend this notion further by conceptualizing polynomial terms as interactions. For instance, P^2 is equivalent to $P * P$. That is to say, the second-degree term is an interaction between the two linear terms. (Jack Byrne, in a personal communication, first brought this to our attention.) We do not view this way of conceptualizing variables as necessary for all variables, but there might be some value in it for some readers.

These notions at least explain to us why the weight for the unit vector has conventionally been depicted as a_0. If one examines, say, a third-degree model, such as the following model, one can detect a pattern:

$$Y = a_0U + a_1X + a_2X^2 + a_3X^3 + E_2 \qquad \text{(Model 9.28)}$$

If one realizes that variable X and the unit vector U can be represented as X^1 and X^0, respectively, then Model 9.28 could be rewritten as follows:

$$Y = a_0X^0 + a_1X^1 + a_2X^2 + a_3X^3 + E_2 \qquad \text{(Model 9.28a)}$$

The subscript for each weighting coefficient for each vector is the same as the power of that vector. The subscript associated with the coefficient for the unit vector is related to the power to which the predictor variable has been taken—namely zero.

Trend and Time Series

Economists for many years have been concerned with time-series analyses, on which data (usually representing some entity) are repeatedly measured over several months or years. The amount of time separating the repeated measures is, of course, up to the researcher. The time span could be days, hours, or even smaller units such as minutes or seconds. One usually does *not* consider time by itself as a causative factor; but given the pattern or trend of the criterion scores over an extended period, one *may* be able to predict the criterion value at a given later time period. With data of this sort the concern is not so much with finding causal factors, as it is with ascertaining trends that exist in the data. The only variables usually measured are the criterion variable and time at which the criterion was observed.

Overall linear trend and a sub-time period second-degree component. Assume that the director of a counseling agency intends to predict the future number of clients who are counseled. The director believes that the relationship between the number of people counseled and time divided into monthly units reflects an overall positive linear trend and a second-degree component within yearly time periods (specifically a U-shaped second-degree component). See Figure 9.28 for an illustration of this relationship. The data used to generate the predictive model consists of the following:

1. The criterion variable (Y) consists of the number of clients counseled each month for a four-year period (thus N is equal to 48).
2. The predictor variable represented by X consists of numbers 1 through 12 for each of the 4 years.

3. The predictor variable represented by T consists of numbers 1 through 48 for each of the months in the four-year period.

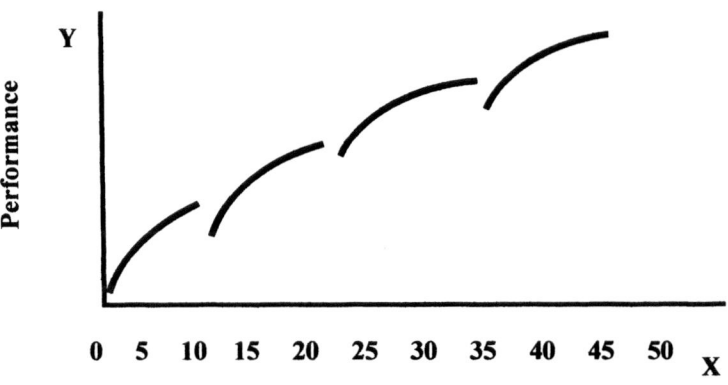

Figure 9.28.
Similar yearly pattern superimposed on an overall positive linear trend.

The following Full Model allows for an overall linear trend in the data, as well as a common second-degree function within each year:

$$Y = a_0U + b_1X + c_1X^2 + d_1T + E_1 \qquad \text{(Model 9.29)}$$

If an overall linear trend does not exist, the following restriction is placed on the Full Model (Model 9.29):

$d_1 = 0$ (1 restriction)

Placing this restriction on the Full Model produces the following Restricted Model:

$$Y = a_0U + b_1X + c_1X^2 + E_2 \qquad \text{(Model 9.30)}$$

An F test of the difference between the R^2 values of the Full Model (9.29) and the Restricted Model (9.30) will indicate if the overall linear trend in the number of clients is statistically significant. Note that since the director expected a positive overall trend, the one-tailed probability of the F test would be used in this testing procedure. If the director did not hypothesize that the trend would be positive, the two-tailed probability would be used. If the coefficient for the overall linear trend (d_1) was positive, and the one-tailed probability of the F test is less than the established alpha level, the director would be willing to state that the numbers of clients reflect a positive overall linear trend.

If a second-degree component (U shape) within each year does not exist, the following restrictions are placed on the Full Model (Model 9.29) one at a time:

$b_1 = 0$ (1 restriction on the linear component)
$c_1 = 0$ (1 restriction on the yearly second-degree component)

Placing these restrictions on the Full Model one at a time produces the following Restricted Models:

$$Y = a_0U + c_1X^2 + d_1T + E_3 \hspace{3cm} \text{(Model 9.31)}$$

$$Y = a_0U + b_1X + d_1T + E_3 \hspace{3cm} \text{(Model 9.32)}$$

An F test of the difference between the R^2 values of the Full Model (9.29) and the Restricted Model (9.31) will determine if the yearly linear component is statistically significant; while an F test of the difference between the R^2 values of the Full Model (9.29) and the Restricted Model (9.32) will indicate if the yearly second-degree component is statistically significant. Once again, note that since the director expected a U-shaped yearly relationship with a positive linear yearly trend and a negative second-degree component, one-tailed probability values of the F test would be used in this testing procedure. When the specific signs of these components are not established a priori, two-tailed probability values are used. If the signs of the linear component of the yearly trend and the second-degree component are positive and negative, respectively, and the one-tailed probabilities of the F tests are less than the established alpha level, the director has support for the position that the numbers of clients reflect a U-shaped yearly second-degree component.

If both the overall trend and the yearly second-degree component of the relationship between the number of clients and time are statistically significant, the director would use the Full Model (Model 9.29) to forecast the number of clients for future time periods. For example, when using the Full Model to predict the number of clients for the first month of the fifth year, the director would substitute the value of 1 for X and the value of 49 for T into the model. The predicted value of the model is assessed by comparing the accuracy of this predicted number of clients to the actual—once that figure is known. Since the director's goal was to forecast the number of clients the agency counseled in future time periods, the accuracy of the model's predictions is the key method of assessing the utility of the model.

Sinusoidal variation of values around an overall linear trend. The variation of values around a linear trend can reflect various types of functions. The type of fluctuations around a linear trend may best be represented by a specific trigonometric function. In the previous section the researchers anticipated that the relationship between the number of people counseled and time divided into monthly units reflects an overall positive linear trend and a second-degree component within yearly time periods. In this section the researchers expect the number of clients counseled by a counseling agency will reflect variation around an overall positive trend that follows a specific trigonometric function—namely a yearly sinusoidal function.

The scattergram contained in Figure 9.29 illustrates the variation of the number of clients over time for this type of function. A review of this scattergram reveals that the numbers of clients increase in a linear manner over time. However, the numbers of clients appear to vary in a systematic fashion around this positive linear trend. That is, for some months the observed criterion values are much higher than the linearly predicted values. For other months, the observed criterion values are much less than the linearly predicted values.

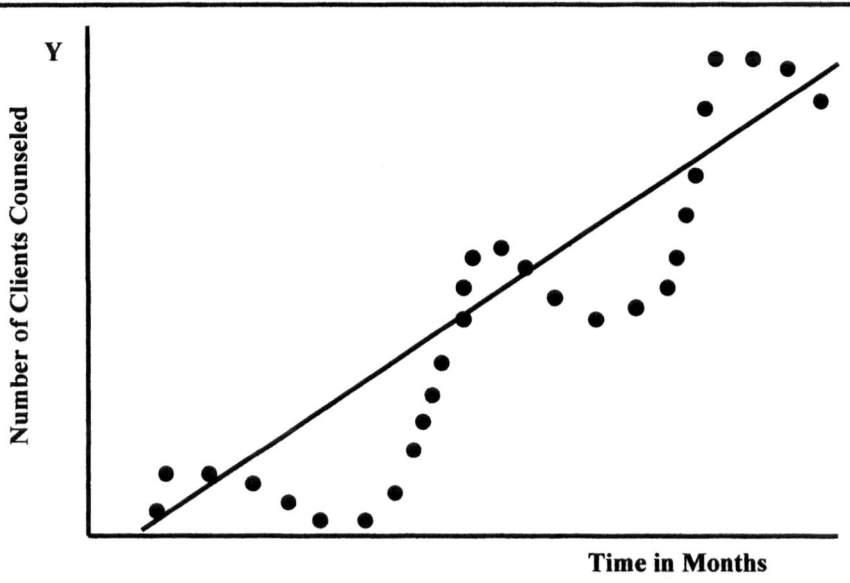

Figure 9.29.
Sinusoidal relationship between month and the number of clients counseled superimposed on the linear relationship.

Since every point in the scattergram contained in Figure 9.29 is not positioned on the linear trend line, the R^2 must be less than 1.00. This suggests that only a linear fit is not the best fit for the data. The systematic values of the errors in prediction suggest that there are some other fits to be found—that the criterion is not fluctuating solely in a linear fashion over time. One way of attempting to discover the other trend(s) in the data is to rotate the figure such that the straight line going through the data points is horizontal as illustrated in Figure 9.30. The problem now is to discover the remaining trend, after the linear trend has been extracted (or to discover the trend in the data, over and above the linear trend).

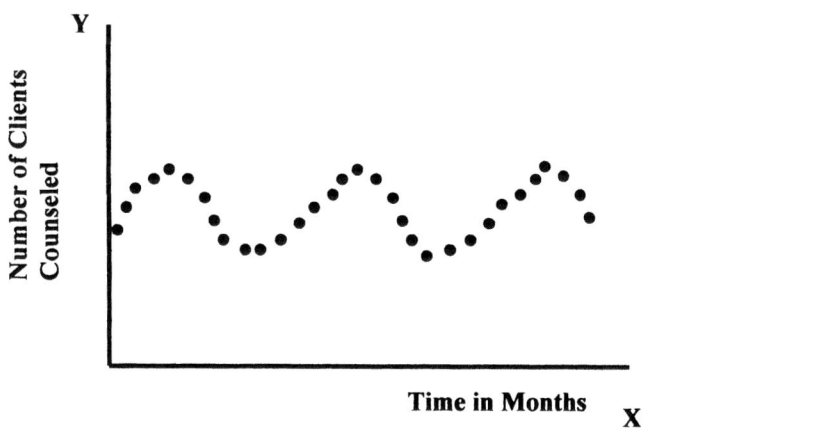

Figure 9.30.
Sinusoidal relationship between month and number of clients counseled after removing the linear trend in the number of clients counseled.

Recall that the director expects the number of clients counseled by a counseling agency reflect variation around an overall positive trend that follows a yearly sinusoidal function. To design a Full Model used to forecast future numbers of clients counseled by the agency, the following variables are formed:

1. The criterion variable (Y) consists of the number of clients counseled each month for a four-year period (thus N is equal to 48).
2. The predictor variable represented by T consists of numbers 1 through 48 for each of the months in the four-year period.
3. The predictor variable represented by X consists of numbers equal to SIN(T)(6.283/12), where 6.283 is equal to twice the value of pi and 12 is the expected timeframe for the sinusoidal function (i.e., the 12 months in a year).

The following Full Model allows for an overall linear trend in the data, as well as a common yearly sinusoidal function:

$$Y = a_0U + b_1X + d_1T + E_4 \qquad \text{(Model 9.33)}$$

If an overall linear trend and the yearly sinusoidal function do not exist, the following restrictions are placed on this Full Model:

$b_1 = 0$ (1 restriction for the overall linear trend)
$d_1 = 0$ (1 restriction for the yearly sinusoidal function)

Placing these restrictions on the Full Model one at a time produces two Restricted Models. The Restricted Model formed when b_1 is set equal to 0 is as follows:

$$Y = a_0U + d_1T + E_5 \qquad\qquad\qquad\qquad \text{(Model 9.34)}$$

The Restricted Model formed when d_1 is set equal to 0 is as follows:

$$Y = a_0U + b_1X + E_6 \qquad\qquad\qquad\qquad \text{(Model 9.35)}$$

Note that the Full Model (Model 9.33) contains three pieces of information and both Restricted Models contain two pieces of information. As discussed in previous chapters, when an F test contains 1 degree of freedom in the numerator, which is the case when the Full Model contains only one more piece of information than the Restricted Model, a t test of the appropriate Full-Model coefficient, can be used. Since both Restricted Models contain just one less piece of information than the Full Model, t tests of the b_1 and d_1 coefficients in the Full Model can be used rather than F tests of the differences between the R^2 values of the Full Model and the Restricted Models to determine whether the positive overall linear trend and the yearly sinusoidal function are significant.

If the t test of the b_1 coefficient indicates that the coefficient value is statistically significant and the d_1 coefficient is positive and statistically significant, the director can claim that the relationship between the numbers of clients counseled by the agency and time reflects a yearly sinusoidal function and a positive overall trend. Thus, the Full Model (9.33) can be used to forecast future numbers of clients counseled by the agency. Once again, the real value of the model is assessed by comparing the accuracy of its predictions to the actual numbers of clients—once those figures are known.

One caution regarding time-series models. When predicting with time-series models it is assumed that the relevant influences on the criterion variable remain the same as in the empirically analyzed time periods. Unforeseen events (such as wars, changes in policy, and changes in consumer habits) will have their (unpredictable) effect on the magnitude of the criterion. Predictions based on the estimated model (the only reason for doing the study in the first place) will be accurate as long as all the relationships between all influences on the criterion variable remain constant over future time periods.

CHAPTER 10

DETECTION OF CHANGE

The purpose of this chapter is to present models and testing procedures for Research Hypotheses that behavioral science and education researchers may pose when using various research designs. Specifically, we discuss the analysis of data obtained from the following: (a) a one-group design, (b) a one-group—pretest and posttest design, (c) a two-groups—posttest only design, (d) a two-group—pretest and posttest design, (e) a matched pairs—design, (f) a design used to identify a point change in time-series data, and (g) individual organism design. Included in the discussions of the analyses of these various designs is our initial discussions of three important analytic techniques: (a) the use of person variables in repeated measures designs (b) the use of a surrogate person variable and (b) the modeling and testing of trend lines with different Y-intercept points and slopes.

The models and the statistical testing procedures presented for each of these designs will indicate if a change has occurred, but detected changes to not necessarily indicate that a causal factor has been identified; the degree of internal validity of the design (i.e., the design's ability to isolate the impact of the treatment) is the determining factor with regards to identifying a causal factor. If the sampling procedure is considered as part of a study's research design, the design also determines, at least in part, the degree to which the results can be generalized. Before we discuss the hypotheses and models used with the aforementioned designs, the concept of ascertaining causality is presented.

ASCERTAINING CAUSALITY

One of the major goals of scientific inquiry is to identify causal relationships because one aim of the human endeavor is to "make things better" (by one's own definition) by upsetting the prediction that exists and causing a more desirable outcome. Some researchers suggest they are *not* trying to find out causal relationships, but a little questioning about their purposes often reveals a causative purpose.

Researchers may be able to predict that students with certain characteristics will learn more slowly than others, or that certain people will contract cancer, or that certain agricultural fields will yield fewer crops. But these researchers want to upset those predictions (i.e., have the students who are expected to learn less learn more; those people who are expected to get cancer would not contract the disease; and those fields that are expected to yield few crops would yield more.

Random Assignment

To upset predictions, researchers follow a multi-step process. First, the researchers formulate ideas regarding possible factors that cause changes in the outcome of interest (i.e., the criterion variable).

Second, the researchers formulate directional hypotheses regarding the identified factors *that might cause* changes in the criterion variable, and as previously stated we believe that hypotheses that are based on theory and observations should be directional.

Third, relying on intuition, theory, and logical tenets, the researchers design the study in a manner that: (a) eliminates competing explanations for changes in the criterion variable and (b) allows the hypothesized causal factors to be manipulated. If the study's design poses both of these characteristics, it will allow the researchers to determine if changes in those causal factors bring about the expected change in the criterion variable. Campbell and Stanley (1966) have a well-organized categorization of possible competing causal factors. Possible competing causal factors must be eliminated, either by the design of the study or by logical analysis. The design of the study cannot rule out all the possible causal factors, so logical analysis must always be used to ascertain causality.

The major tool researchers have used in their studies that significantly reduced the number of possible competing causal factors is random assignment. One should note, though, that random assignment does not guarantee that the samples are equivalent; random assignment only maximizes the probability that the samples will be equivalent. Seldom are two random samples equal on a given variable, let alone equal on all variables.

Fourth, the researchers test the directional hypotheses. If the directional hypotheses are supported and competing causal factors are eliminated, the researchers will be able to identify a causal factor.

Again, causality can only be ascertained by manipulating a variable and observing the predicted result. Ascertaining the impact of changing one variable is not an easy task for researchers. Variables in the real world are highly correlated, and a change in one predictor variable, therefore, *may* change one or more of the other predictor variables. In addition, simply because two variables are correlated does not mean that a change in one of the correlated variables causes a change in the other variable. Changes in both variables could be caused by changes in one or more variables not included in the analysis. Only a tight logical analysis can tease out the causative variables. Manipulation of the proposed causative variables is a necessary step in determining causality. The ultimate test for having discovered causal factors for the hypothesized changes in the criterion variables, however, rests upon successive replications of those manipulated causal factors (see Newman, McNeil, and Fraas, 2004).

Random Sampling

Another issue important to researchers is the degree to which the findings can be generalized (i.e., the external validity of the design). One technique used by researchers to increase the generalizability of the findings is random sampling. However, behavioral science and education researchers often find it difficult to randomly sample from the populations to which they would like to generalize their results because the populations are often so large or entirely theoretical that a random sample of that population is physically impossible. Most often the subjects are sampled from some accessible population, sometimes randomly but more often than not nonrandomly, which may differ from the population to which the researchers wish to generalize. Such a situation restricts the generalizability of the findings. It is the responsibility of the researchers to determine to the degree of this restriction. Once again, one method of establishing the degree of generalizability of the results is to replicate the findings in various settings and identify key elements that must exist for the findings to be replicated (see Newman, McNeil, and Fraas, 2004).

Detecting a Change versus Detecting a Causal Relationship

As previously discussed, when researchers use designs that include random assignment and random sampling they are in a better position to identify causal factors and generalize the results. Often behavioral science and educational researchers are unable to use designs that possess these characteristics. Should such research not be conducted? We believe that such research should be conducted, but the limitations of such studies should be well documented by the researchers. The Research Hypotheses and models used to analyze the data obtained from the various designs presented in the following sections will detect changes. Whether the researchers are or are not willing to identify a causal factor based on a detected change depends on the "strength" of the design. Readers should keep this point in mind when reading the following sections.

ONE-GROUP DESIGNS

Change hypotheses, that is, hypotheses that deal with a change in a single population mean, can be investigated using only one group. Such studies usually are not as conclusive as when a second group (i.e., a comparison group) is included, but one-group studies can be of some benefit, and the GLM approach can be used to answer the types of questions posed in such studies.

One-Group Posttest-Only Design

Consider a project using a counseling method intended to reduce alienation of children from their parents. One of the project objectives might be that after six weeks of participation the alienation mean score of the children in the project will be less than five on an instrument designed to measure the degree of child alienation. If the project director is interested only in how the project works for the few children in the project, all that needs to be done is to visually compare the sample mean to the value of 5 to determine if it is less than five. But often the project director's goal is to assess the adequacy of the project with the intent of adopting it in other settings. If this is the goal of the project director, a more involved assessment process is needed including the use of a design referred to as a One Group Posttest Only Design, which requires the subjects to be post-tested after the implementation of the treatment.

The first step in this process would be to formulate a Research Hypothesis, which could be stated as follows:

Research Hypothesis: After six weeks of counseling, the mean alienation score will be less than five.

The corresponding Statistical Hypothesis is as follows:

Statistical Hypothesis: After six weeks of counseling, the mean alienation score will not be less than five.

The Full Model that reflects this Research Hypothesis, which must calculate the alienation mean score, is as follows:

$$Y = a_0 U + E_1 \qquad \text{(Model 10.1)}$$

where:
1. Y = alienation scores; and
2. U = the unit vector.

Readers by now should recognize this model as the Null Model, which contains only the unit vector. Thus this Full Model has only one piece of information ($m1 = 1$). It is important to note that the value of the coefficient for the unit vector is equal to the sample mean.

Since the coefficient for the unit vector is equal to the mean alienation score for the sample data and the Statistical Hypothesis requires that the alienation mean equal 5, the restriction can be stated as follows:

$a_0 = 5$ (1 restriction)

Placing this restriction on the Full Model results in the following Restricted Model:

$$Y = 5U + E_2 \qquad \text{(Model 10.2)}$$

Since U is equal to 1 for all subjects, 5U is a constant, and subtracting that constant from both sides yields the following Restricted Model:

$$(Y - 5) = E_2 \qquad \text{(Model 10.2a)}$$

This Restricted Model contains a piece of information (i.e., m2 = 0). The F test as calculated with Equation 4.1, which was first presented in chapter 4, is applicable providing that the unit vector is in both the Full and Restricted Models. If the unit vector is not included in the models, and in this analysis it is not, then the F test must be calculated with an alternate equation, which was provided by Bottenberg and Ward (1963):

$$F = \frac{(ESS_R - ESS_F)/(m1 - m2)}{(ESS_F)/(N - m1)} \qquad \text{(Equation 10.1)}$$

where:
1. ESS_R = error sum of squares in the Restricted Model (ESS_R is equal to the squared values in E_2);
2. ESS_F = error sum of squares in the Full Model (ESS_F is equal to the squared values in E_1);
3. m1 = number of linearly independent vectors in the Full Model (the full number of pieces of information); and
4. m2 = number of linearly independent vectors in the Restricted Model (the restricted number of pieces of information).

Equation 10.1 is a more general one than Equation 4.1, although the latter is more conceptually appealing. Notice that the degrees of freedom are calculated in exactly the same manner in the two equations. If one were to apply Equation 10.1 to any of the previous examples in this text, an identical F value would result.

For the F test used to statistically test the Statistical Hypothesis in this section, the numerator degrees of freedom value is equal to 1 (i.e., m1 - m2 = 1 - 0), while the denominator degrees of freedom value is equal to $N - 1$ (i.e., $N - m1$). Once the F test is calculated with Equation 10.1, the researchers would reject the directional Statistical Hypothesis if two conditions are met: (a) the sample mean is less than 5, which requires the a_0 coefficient in the Full Model to be less than 5 and (b) the one-tailed F test probability value is less than the established alpha level.

The actual application of the process discussed in this section is quite arduous even with the use of the SPSS for Windows and Microsoft Excel computer programs—the reason being that the regression programs are not capable of estimating the required Full and Restricted Models. An equivalent test, which has its underpinnings based on the GLM, is the one-sample t test. This type of t test can quite easily be conducted with the "One-Sample t test" program in the SPSS for Windows program; and although the Microsoft Excel program does not have a program that calculates a one-sample t test, it can be calculated, along with its corresponding probability value with the execution of a few commands. The point of presenting the GLM approach for this type of statistical test is to expose readers to the GLM basis of the one-sample t test.

One-Group Pretest-Posttest Design

The researchers are often interested in the gain in the criterion variable from the beginning of the treatment to the end of the treatment. Therefore, not only are posttest scores recorded but the pretest scores on the same instrument are also recorded. Such a design is referred to as a One-Group Pretest-Posttest Design. Researchers can analyze the data obtained from this type of research design using any of three types of analyses.

Analysis of the criterion variable that consists of gain scores. As previously stated, in a One-Group Pretest-Posttest Design each student has a pretest score and a posttest score. The criterion variable can be formed by subtracting the pretest score from the posttest score for each subject, yielding a score that is referred to as (a) a gain score, (b) a difference score, or (c) a change score. This type of criterion variable (i.e., one that contains gain scores) is used in the models that reflect the following directional Research and Statistical Hypotheses:

Directional Research Hypothesis: For a given population, the mean gain score is greater than zero (or any specified value).

Directional Statistical Hypothesis: For a given population, the mean gain score is not greater than zero (or any specified value).

The Full Model that reflects this Research Hypothesis is as follows:

$$Y = a_0U + E_1 \qquad\qquad\qquad\qquad\qquad\qquad\text{(Model 10.3)}$$

where Y and U represent the gain scores the unit vector, respectively. Since the coefficient for the unit vector is equal to the mean gain score for the sample data and the Statistical Hypothesis requires that the sample mean gain score is equal to 0 (i.e., the mean posttest and pretest scores do not differ), the restriction can be stated as follows:

$$a_0 = 0 \quad \text{(1 restriction)}$$

Placing this restriction on the Full Model (Model 10.3) results in the following Restricted Model:

$$Y - 0U \mid E_2 \qquad\qquad \text{(Model 10.4)}$$

This Restricted Model can be expressed as follows:

$$Y = E_2 \qquad\qquad \text{(Model 10.4a)}$$

The F test used to statistically test the Statistical Hypothesis is calculated by using Equation 10.1. For the F test used to statistically test the Statistical Hypothesis in this section, the numerator degrees of freedom value is equal to 1 (i.e., m1 - m2 = 1 - 0), while the denominator degrees of freedom value is equal to N - 1(i.e., N - m1). The researchers would reject the directional Statistical Hypothesis if two conditions are met: (a) the mean gain score is greater than 0, which requires the a_0 coefficient in the Full Model to be positive and (b) the one-tailed F test probability value is less than the established alpha level.

We do not recommend that this approach be used to test the Statistical Hypothesis stated in this section because, similar to the concern expressed in the testing procedure presented in the previous section, the actual application of this produce is quite arduous. The same results can be obtained by conducting a one-sample t test. Again, the point of presenting the GLM approach for this type of statistical test is to expose readers to the fact the GLM is the basis of the one-sample t test in which the criterion variable consists of change scores.

Analysis of a criterion variable consisting of pretest and posttest scores. Similar to the type of analysis discussed in the previous section, the type of analysis presented in this section utilizes both the pretest and posttest scores. In this analytic procedure, however, the criterion variable consists of both of the pre-test and posttest scores rather than the change scores.

Typical directional Research and Statistical Hypotheses for this type of design would be as follows:

Directional Research Hypothesis: For a given population, the posttest mean is higher than the pretest mean.

Directional Statistical Hypothesis: For a given population, the posttest mean is not higher than the pretest mean.

The criterion variable (Y) consists of each subject's pretest and posttest scores. Due to the fact that both the pretest and posttest scores are placed in the criterion vector, the number of scores in the criterion variable will be twice the number of subjects. When this type of criterion variable is used in the analysis the predictor variable (X), which is referred to as a time-period variable, consists of 0 and 1 values, with the value of 0 being used when the criterion score is a pretest score and the value of 1 being used when the criterion score is a posttest score.

The Full Model that reflects the condition stated in the Research Hypothesis is as follows:

$$Y = a_0 U + a_1 X + E_1 \qquad \text{(Model 10.5)}$$

It is important to understand four characteristics of this model. First, the criterion variable (Y) contains both the pretest and posttest scores. Second, the predictor variable (X) is a dichotomous variable consisting of the time-period vector composed of 0 and 1 values. Third, the model contains two pieces of information, which are the unit vector and the time-period variable. Fourth, the coefficient for the treatment variable (a_1) is equal to the mean posttest score minus the mean pretest score.

Since the Statistical Hypothesis requires the mean posttest and pretest scores to be equal, the restriction placed on the Full Model is as follows:

$$a_1 = 0 \quad \text{(1 restriction)}$$

Placing this restriction on the Full Model results in the following Restricted Model:

$$Y = a_0 U + E_2 \qquad \text{(Model 10.6)}$$

Note that this Restricted Model is the Null Model, that is, it contains only the unit vector. Thus this Restricted Model contains only one piece of information.

To statistically test the Statistical Hypothesis, the researchers would test the difference between the R^2 values of the Full Model (Model 10.5) and the Restricted Model (Model 10.6) with Equation 4.1, which was previously presented in chapter 4 and is as follows:

$$F(df_n, df_d) = \frac{(R_F^2 - R_R^2)/(df_n)}{(1 - R_F^2)/(df_d)} \qquad \text{(Equation 4.1)}$$

Since one restriction was placed on the Full Model, the numerator degrees of freedom (df_n) value is equal to 1. As previously mentioned the number of scores recorded in the criterion variable is twice the number of subjects because the vector contains both the pretest and posttest scores. Thus, the denominator degrees of freedom (df_d) value is equal to $2N - 1$, where N is the number of subjects. The researchers would reject the directional Statistical Hypothesis if two conditions are met: (a) the mean posttest score exceeds the mean pretest score, which requires the a_1 coefficient in the Full Model to be positive and (b) the one-tailed F test probability value is less than the established alpha level.

It is important for researchers to understand that the analytic procedure described in this section regarding the statistical test of the difference between the mean pretest and posttest scores has one significant deficiency. The models used in the analysis ignore an important piece of information, which is that every person is tested twice. The exclusion of this type of information from the models creates two concerns for the researchers.

First, one assumption of statistical analysis—independence of observations—is violated. Second, the analysis is not as powerful as it might be because individual differences are large in most areas. That is to say, if a subject scores low on the pretest, that same subject is expected to score relatively low on the posttest. Conversely, if a person scores high on the pretest, that subject is expected to score high on the posttest. The inclusion of the information regarding which person the set of scores belonged to can be done with a set of person variables, which will be discussed in the next section.

The inclusion of the information regarding the person from which the scores were obtained through the use of person variables will cause two opposite forces to impact the power level of the statistical test of the difference between the pretest and posttest mean scores. First, the inclusion of person variables will reduce the denominator degrees of freedom, which will decrease the power level. Second, the increase in the amount of variance accounted for in the criterion variable by the inclusion of the person variables will increase the power level. Usually the increase in power caused by the increased amount of variation accounted for by the inclusion of the person variables will far outweigh the decrease in power caused by the decrease in the denominator degrees of freedom value. Thus, the inclusion of person variables will generally increase the power of the statistical test.

Analysis of a criterion variable consisting of pretest and posttest scores with the use of a pretest covariate. When statistically testing the difference between the means of pretest and posttest scores with the GLM approach, it is essential that they include information regarding the person for which the given pair of scores was recorded. Such information can be included in the Full and Restricted Models by constructing a set of person variables.

In a set of person variables each variable represents one of the N subjects, and each variable is a dichotomous variable consisting of the values of 0 and 1. Person variable 1, which represents subject 1, contains a value of 1 when the pretest and posttest scores contained in the criterion variable are recorded for subject1, but 0 otherwise. The other person variables are formed in the same manner. To illustrate the construction of person variables, the following variables, which are listed in Table 10.1, have been formed for seven students:

1. The criterion variable Y contains the 7 pretest scores followed by the 7 posttest scores recorded for the 7 students.
2. The time-period dichotomous predictor variable T consists of 0 and 1 values. The value for variable T is 0 when the corresponding criterion

value is a pretest score and 1 when the corresponding criterion value is a posttest score.

3. A person variable is constructed for each of the 7 subjects in the study (P1 through P7). Each person variable contains 0 and 1 values. For a given person variable the value of 1 indicates that the corresponding value in the criterion variable is a pretest score or posttest score for this subject, while the value of 0 reveals the score is another subject's score.

To further illustrate the structure of the person variables, consider the person variable for Sam, which is labeled P1. Only when the criterion variable represented Sam's pretest score or posttest score, the P1 person variable contains a value of 1. The values are assigned in a similar manner for the other six students.

Note that if the researchers horizontally sum these seven person variables, the unit vector is formed, that is, every value in the vector is 1. Thus, the set of person variables are linearly dependent. Due to this fact, when the Full and Restricted Models contain the unit vector, one of the person variables cannot be placed in the models.

Typical directional Research and Statistical Hypotheses for this type of design would be the same as the ones posed in the previous section, which are as follows:

Directional Research Hypothesis: For a given population, the posttest mean is higher than the pretest mean.

Directional Statistical Hypothesis: For a given population, the posttest mean is not higher than the pretest mean.

It is important for researchers to understand, however, that the directional Research and Statistical Hypotheses could also be stated as follows:

Directional Research Hypothesis: For a given population, the time-period variable accounts for some of the variation in the test scores over and above the amount of variation accounted for by the person variables.

Directional Statistical Hypothesis: For a given population, the time-period variable does not account for some of the variation in the test scores over and above the amount of variation accounted for by the person variables.

This version of the Research and Statistical Hypotheses stresses the need for the analysis to include information regarding for which person a given score was recorded (i.e., the need to include person variables in the models).

Table 10.1
Criterion Variable, Time-Period Predictor Variable, and Person Variable

Student	Y	T	P1	P2	P3	P4	P5	P6	P7
Sam's Pre	14	0	1	0	0	0	0	0	0
Joe's Pre	15	0	0	1	0	0	0	0	0
Sue's Pre	15	0	0	0	1	0	0	0	0
Pat's Pre	18	0	0	0	0	1	0	0	0
Sal's Pre	20	0	0	0	0	0	1	0	0
Hal's Pre	21	0	0	0	0	0	0	1	0
Tom's Pre	23	0	0	0	0	0	0	0	1
Sam's Post	19	1	1	0	0	0	0	0	0
Joe's Post	21	1	0	1	0	0	0	0	0
Sue's Post	18	1	0	0	1	0	0	0	0
Pat's Post	25	1	0	0	0	1	0	0	0
Sal's Post	28	1	0	0	0	0	1	0	0
Hal's Post	26	1	0	0	0	0	0	1	0
Tom's Post	30	1	0	0	0	0	0	0	1

The Full Model that reflects the condition stated in the Research Hypothesis is as follows:

$$Y = a_0U + a_1T + p_1P1 + p_2P2 + p_3P3 + p_4P4 + p_5P5 + p_6P6 + E_1$$
$$\text{(Model 10.7)}$$

Three points should be noted regarding this Full Model. First, the person variable for Tom (P7) is not included in this Full Model. Since this model includes the unit vector (U) and the seven person variables are linearly dependent, one of the person variables must be excluded from the model. Second, this model contains eight pieces of information (i.e., the unit vector, the time-period variable, and six person variables). Third, the coefficient for the time-period variable (a_1) will equal the mean posttest score minus the mean pretest score.

Since the Statistical Hypothesis requires the mean posttest and pretest scores to be equal, the restriction is as follows:

$$a_1 = 0 \quad \text{(1 restriction)}$$

Placing this restriction in the Full Model results in the following Restricted Model:

$$Y = a_0U + p_1P1 + p_2P2 + p_3P3 + p_4P4 + p_5P5 + p_6P6 + E_1 \qquad \text{(Model 10.8)}$$

This Restricted Model contains seven pieces of information.

To statistically test the Statistical Hypothesis, the researchers would test the difference between the R^2 values of the Full Model (Model 10.7) and the Restricted Model (Model 10.8) with the calculation of an F test through Equation 4.1. Assuming the data contained in Table 10.1 are the data used in the analysis, the numerator and denominator degrees of freedom values for the F test are 1 (i.e., 8 - 7) and 6 (i.e., 14 - 8), respectively. The researchers would reject the directional Statistical Hypothesis if two conditions are met: (a) the posttest mean is greater than the pretest mean and (b) the one-tailed F test probability value is less than the established alpha level.

Of the three types of analyses discussed for one-group studies, the analysis that includes the person variables makes the best use of the information obtained from the study. Another way of saying this is that, in most cases, the inclusion of the person variables will yield a greater level of statistical power.

Researchers may encounter certain research designs that incorporate additional repeated measures, which necessitate the use of multiple time-period variables in the Full Model. The use of these multiple time-period variables is analogous to the use of multiple groups in a one-way ANOVA. If the researchers were interested in testing whether the means of a test administered in three testing periods differed, they could place a restriction on the Full Model that would require those differences to be equal in the resulting Restricted Model. If the F test of the difference between the R^2 values of the Full and Restricted Models was statistically significant, the researchers would conclude that at least one of the differences between the mean scores recorded for the three time periods was statistically significant. The researchers determine which of the multiple time-period variable means differed by applying the multiple comparison techniques discussed in chapter 4.

GENERAL RESEARCH HYPOTHESIS 10.1

The discussion of General Research Hypothesis 10.1 (GRH 10.1) provides a template for constructing Full and Restricted Models used to test hypotheses that involve repeated measures of a criterion variable. In this hypothetical study the criterion variable (Y) consists of both the pretest and posttest scores, that is, two measures. The time-period variable (T) is a dichotomous variable that consists of 0 and 1 values. Each posttest score in the criterion variable has a corresponding value of 1 in the time-period variable, while pretest scores have corresponding values of 0. It should be noted that if the study involved an additional measure of the criterion variable, say, a retention test, the criterion would include not only the pretest and posttest scores but also the retention test scores. In addition, one additional time-period variable would be created to indicate when the score was a retention score.

If the number of individuals in the study is k, a dichotomous person variable is constructed for each of these k individuals. The person variables consist of 0 and 1 values and they are labeled P1 through Pk. The P1 variable, which is the person variable for person 1, will contain a value of 1 only when a criterion value is a score for person 1 (regardless of whether it is a pretest or posttest score); while a 0 is assigned to P1 when a criterion value is a score for another person. The other person variables are formed in the same manner.

For repeated-measures analyses, there will be as many elements in each vector as there are subjects, multiplied by the number of times they were measured. Thus, in this hypothetical study, each vector contains $2N$ values because each person had a pretest and posttest score.

Hypotheses and Models

If the researchers are willing to indicate which of the two criterion measures, say, the posttest mean is higher than the pretest mean; they would state GRH 10.1 as follows:

Directional GRH 10.1: For a given population, the posttest mean is higher than the pretest mean, over and above expected individual differences.

The corresponding directional General Statistical Hypothesis (GSH 10.1) is as follows:

Directional GSH 10.1: For a given population, the posttest mean is not higher than the pretest mean, over and above expected individual differences.

If the researchers are not willing to indicate which of the two criterion measures is higher, they would state the nondirectional GRH 10.1 as follows:

Nondirectional GRH 10.1: For a given population, the posttest mean is different from the pretest mean, over and above expected individual differences.

The corresponding nondirectional General Statistical Hypothesis (GSH 10.1) is as follows:

Nondirectional GSH 10.1: For a given population, the posttest mean is not different from the pretest mean, over and above expected individual differences.

The Full Model that reflects GRH 10.1 is as follows:

$$Y = a_0U + a_1T + p_1P1 + p_2P2 + \ldots + p_{k-1}P(k-1) + E_1 \qquad \text{(Model 10.9)}$$

This Full Model will have $[1 + 1 + (k - 1)]$ pieces of information (i.e., the unit vector, the time-period vector, and the k minus 1 person vectors).

Since the coefficient for the treatment variable (a_1) is equal to the posttest mean of the treatment group minus the posttest mean of the comparison group and the Statistical Hypothesis stipulates that the adjusted posttest means of the two groups do not differ, the restriction is stated as follows:

$a_1 = 0$ (1 restriction)

Placing this restriction on the Full Model (Model 10.7) produces the following Restricted Model:

$$Y = a_0U + p_1P1 + p_2P2 + \ldots + p_{k-1}P(k-1) + E_2 \qquad \text{(Model 10.10)}$$

This Restricted Model has k pieces of information (i.e., the unit vector and k minus 1 person vectors).

F Test of the Difference between the R^2 Values

To statistically test the Statistical Hypothesis, the researchers would test the difference between the R^2 values of the Full Model (Model 10.9) and the Restricted Model (Model 10.10) with the calculation of an F test through the use of Equation 4.1. The numerator and denominator degrees of freedom values for this test are 1 and k -1, respectively. If the researchers used the nondirectional GRH 10.1, they would reject nondirectional GSH 10.1 if the two-tailed F test probability value was less than the established alpha level. If the researchers used GSH 10.1, which is directional, two conditions must be met before it could be rejected: (a) the posttest mean must be greater than the pretest mean, which requires the coefficient of the treatment variable (a_1) to be positive and (b) the one-tailed F test probability value must be less than the established alpha level.

Use of a Surrogate Person Variable

When the Full and Restricted Models must contain person variables to accurately reflect the hypotheses researchers may encounter situations in which the number of individuals included in the analysis is large, which would require a large number of person variables. If the number of required person variables is large, the researchers may find that the formation of the person variables is quite time consuming. Williams (1974a) describes a procedure researchers could use in such cases that requires the formation of just one variable, which we will refer to as a surrogate person variable, rather than the large set of person variables.

If the study involves two measures of the criterion variable, the value entered for person 1 in the surrogate variable is the total of the pretest and posttest score (the average of the two scores could also be used). This total score is entered in the vector of the surrogate variable when the criterion variable is either the pretest score or posttest score for person 1. The other values for the surrogate variable are formed in the same manner for the other individuals in the study.

When surrogate person variable labeled SPV is used rather than the k -1 person variables the Full Model represented by Model 10.9 would be modified as follows:

$$Y = a_0U + a_1T + a_2SPV + E_1 \qquad \text{(Model 10.9a)}$$

This Full Model (Model 10.9a) will produce the same R^2 value as the Full Model represented by Model 10.9, which contained k -1 person variables. It is very important for the researchers to realize, however, that the SPV variable is representing not one piece of information but rather $k - 1$ pieces of information. Thus Model 10.9a contains $k + 1$ pieces of information (i.e., the unit vector, the time-period variable, and the k minus 1 person variables as represented by the surrogate person variable).

The Restricted Model that contained the k minus 1 person variables (Model 10.10) would also be modified to form the following Restricted Model:

$$Y = a_0U + a_2SPV + E_2 \qquad \text{(Model 10.10a)}$$

Again, this Restricted Model (Model 10.10a) will produce the same R^2 value as the Restricted Model represented by Model 10.10, which contained k minus 1 person variables. Thus, Model 10.10a contains k pieces of information of information (i.e., the unit vector and the k minus 1 person variables as represented by the surrogate person variable).

The difference between the R^2 values of these two models would be tested with an F test calculated with Equation 4.1. When calculating this F test value it is, of course, important to use the appropriate degrees of freedom values. When a surrogate person variable is included in the models, the F test of the difference between the R^2 values of the two models calculated by the standard statistical programs, such as SPSS for Windows, will be incorrect because the program will not use the appropriate number of degrees of freedom values. The t tests of the variable coefficients will also be incorrect for the same reason.

As previously noted, the SPV represents k -1 pieces of information. Thus, the Full Model (Model 10.9a) contains $k + 1$ pieces of information and the Restricted Model (Model 10.10a) has k pieces of information, which produces numerator and denominator degrees of freedom values for the F test of 1 and $N - (k + 1)$, respectively.

We believe SPV is a very useful tool for researchers to have in their arsenal of analytic tools. The application of a surrogate person variable to a hypothetical set of data is presented in the next section—also refer to Fraas and McDougall (1983) for an analysis that utilized a surrogate person variable.

APPLIED RESEARCH HYPOTHESIS 10.1

This section provides a numerical example of the GLM testing procedure of Applied Research Hypothesis 10.1 (ARH 10.1), which involves a repeated measurement of the criterion variable. The data used in the analysis are listed in Table 10.1. The students were exposed to a treatment that was designed to increase the students' self-confidence levels. The criterion variable contains the pretest and posttest self-confidence scores of the seven students exposed to the treatment in a study that used a One-Group Pretest-Posttest Design. One of the predictor variables was a dichotomous time-period variable in which the 0 and 1 values indicate whether the corresponding self-confidence score was a pretest score or a posttest score, respectively. The other predictor variables formed a set of dichotomous person variables. The researchers posed ARH 10.1 as a directional hypothesis as follows:

Directional ARH 10.1: For a given population, the posttest self-confidence mean is higher than the pretest self-confidence mean, over and above individual differences.

The corresponding Applied Statistical Hypothesis 10.1 (ASH 10.1) is as follows:

Directional ASH 10.1: For a given population, the posttest mean is not higher than the pretest mean, over and above individual differences.

Full and Restricted Models

The Full Model that reflects the condition stated in ARH 10.1 is as follows:

$$Y = a_0U + a_1T + p_1P1 + p_2P2 + p_3P3 + p_4P4 + p_5P5 + p_6P6 + E_1$$
$$\text{(Model 10.11)}$$

Since this Full Model contained the unit vector, only six of the seven person variables were included in the model. In addition to the unit vector and the six person variables, this Full Model included the time-period variable. Thus the model has eight pieces of information (i.e., $m1 = 8$).

Due to the fact that the a_1 coefficient in the Full Model is equal to the mean posttest minus the mean pretest and ASH 10.1 requires the two means to be equal, the restriction can be stated as follows:

$a_1 = 0$ (1 restriction)

Placing this restriction in the Full Model results in the following Restricted Model:

$$Y = a_0U + p_1P1 + p_2P2 + p_3P3 + p_4P4 + p_5P5 + p_6P6 + E_1 \quad \text{(Model 10.12)}$$

This Restricted Model contained the unit vector and six person vectors (the time-period variable has been eliminated). Thus the model has 7 pieces of information (i.e., $m2 = 7$).

To statistically test the Statistical Hypothesis, the researchers would test the difference between the R^2 values of the Full Model (Model 10.11) and the Restricted Model (Model 10.12) with the calculation of an F test through the use of Equation 4.1. The numerator and denominator degrees of freedom values for the F test are 1 (i.e., 8 - 7) and 6 (i.e., 14 - 8), respectively. The researchers would reject the ASH 10.1 if two conditions are met: (a) the posttest self-confidence mean is greater than the pretest self-confidence mean, which requires the coefficient for the treatment variable (a_1) to be positive and (b) the one-tailed F test probability value is less than the established alpha level.

Analysis of ARH 10.1 with Microsoft Excel—Models Containing Person Variables

The data listed in Table 10.1 have been stored in a file entitled "TABLE 10.1 DATA," which is located in the internet site for this text (see Appendix A for the address). Once this file and the Excel Regression menu are accessed (see chapter 4), the following four steps are completed to obtain the output for the Full Model (Model 10.11) used to statistically test ASH 10.1 (see Exhibit 10.1):

1. Click on the box next to "Input $\underline{Y}$ Range" (see Oval 1). Note that the criterion variable (Y) is located in column A. Next, click and drag on the cells located in the column in A, including the name of the variable.
2. Click on the box next to "Input $\underline{X}$ Range" (see Oval 2). Note that the predictor variables contained in the Full Model (Model 10.11) are located in columns B through H. Next, click and drag on the cells containing the predictor variables T, P1, P2, P3, P4, P5, and P6, including their names.
3. Click on the box in front of "Labels" (see Oval 3).
4. Click on the "OK" button (see Oval 4).

After these four steps are completed, the output window for the Full Model (Model 10.11) will appear as displayed in Exhibit 10.2.

Three values contained in the Full Model output window displayed in Exhibit 10.2 are used in the calculation of the F test. The first piece of information is the R^2 value of .974 (see Oval 1). The second and third key values are the regression degrees of freedom value of 7 and the residual degrees of freedom value of 6 (see Oval 2). The two other pieces of information needed to calculate the F test are obtained from the Restricted Model output.

Exhibit 10.1.
Microsoft Excel Regression menu—Full Model (Model 10.11) for ARH 10.1.

Input

Input Y Range: $A$1:$A$15

Input X Range: $B$1:$H$15

☑ Labels   ☐ Constant is Zero

☐ Confidence Level: 3   %

1   2   3   4

OK

Cancel

Exhibit 10.2.
Microsoft Excel regression output—Full Model (Model 10.11) for ARH 10.1.

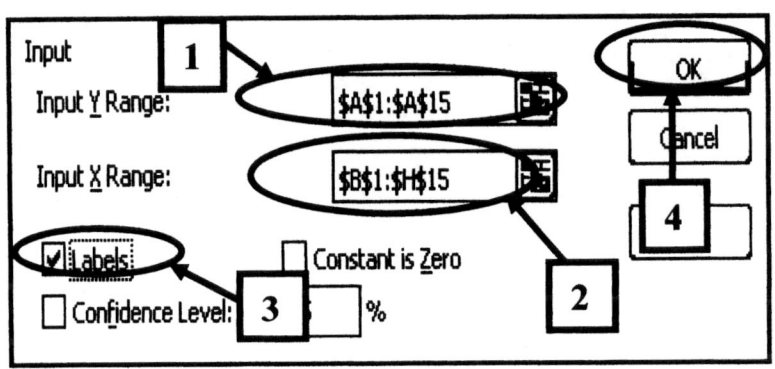

Regression Statistics					
Multiple R	0.0867				
R Square	0.9736				
Adjusted R Square	0.9427				
Standard Error	1.1852				
Observations	14				

ANOVA					
	df	*SS*	*MS*	*F*	*Significance F*
Regression	7	310.5000	44.3571	31.5763	0.0003
Residual	6	8.4286	1.4		
Total	13	318.9286			

	Coefficients	*Standard Error*	*t Stat*	*P-value*	*Lower 95%*
Intercept	23.5714	0.8959	26.3090	0.0000	21.3791
T	5.8571	0.6335	9.2452	01	4.3070
P1	-10.00	1.1852	-8.4372	02	-12.9001
P2	-8.50	1.1852	-7.1716	0.0004	-11.4001
P3	-10.00	1.1852	-8.4372	0.0002	-12.9001
P4	-5.00	1.1852	-4.2186	0.0056	-7.9001
P5	-2.50	1.1852	-2.1093	0.0794	-5.4001
P6	-3.00	1.1852	-2.5312	0.0446	-5.9001

The output for the Restricted Model is obtained by completing the follow-ing steps in the Regression menu (see Exhibit 10.3):

1. Click on the box next to "Input $\underline{Y}$ Range" (see Oval 1). Next, click and drag on the cells in the column in which variable Y is located, which is column A.
2. Click on the box next to "Input $\underline{X}$ Range" (see Oval 2). Next, click and drag on the cells in the columns in which the P1, P2, P3, P4, P5, and P6 variables are located, which are columns C through H.
3. Click on the box in front of "Labels" (see Oval 3).
4. Click on the "OK" button (see Oval 4).

After these four steps are completed, the output window for the Restricted Mod-el (Model 10.12) will be produced as displayed in Exhibit 10.4.

Exhibit 10.3.
Microsoft Excel Regression menu—Restricted Model (Model 10.12) for ASH 10.1.

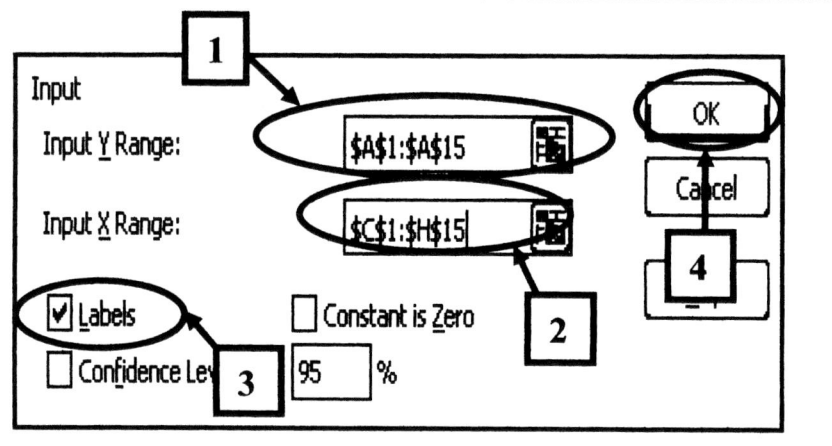

The two values obtained from the Restricted Model (Model 10.12) output, which is displayed in Exhibit 10.4, are (a) the R^2 value of .597 (see Oval 1) and (b) the regression degrees of freedom value of 6 (see Oval 2). Remember, the Microsoft Excel output windows for the Full and Restricted Models do not pro-vide the appropriate F test for the difference between the R^2 values of these two models. However, the appropriate F test value can be calculated with the use of the Microsoft Excel file entitled "F Test Calculation," which is contained in internet site listed in Appendix A. Once this file is accessed, the following val-ues are entered into the file (see Oval 1 in Exhibit 10.5):

1. The R^2 value of the Full Model (.994) is entered into Row 1 of Column B.
2. The R^2 value of the Restricted Model (.597) is entered into Row 2 of Column B.
3. The regression degrees of freedom value for the Full Model (7) is entered into Row 3 of Column B.
4. The regression degrees of freedom value for the Restricted Model (6) is entered into Row 4 of Column B.
5. The residual degrees of freedom value for the Full Model (6) is entered into Row 5 of Column B.

Exhibit 10.4.
Microsoft Excel regression output—Restricted Model (Model 10.12) for ASH 10.1.

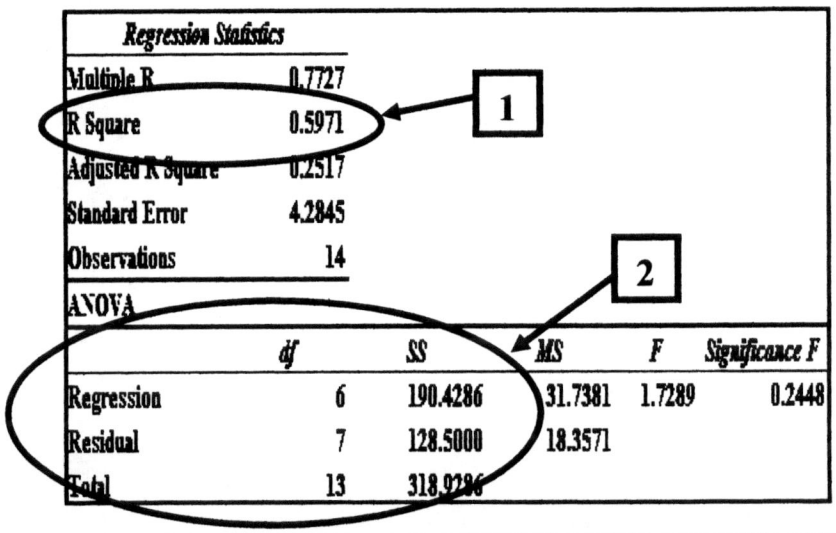

As previously stated, the researchers will reject ASH 10.1 if two conditions are met: (a) the posttest self-confidence mean is greater than the pretest self-confidence mean and (b) the one-tailed F test probability value is less than the established alpha level. The researchers can determine if the first condition is met by examining the coefficient value for the time-period variable (T) in the Full Model (Model 10.11). Recall that the posttest and pretest scores were assigned values of 1 and 0, respectively, in the T variable. Thus, the T variable coefficient is equal to the sample mean posttest score minus the sample mean pretest score. Since the T variable coefficient value is a positive 5.86 (see Oval 3 in Exhibit 10.2), the positive mean posttest self-confidence score exceeds the mean pretest self-confidence score.

Exhibit 10.5.
Microsoft Excel *F*-test calculation for ASH 10.1

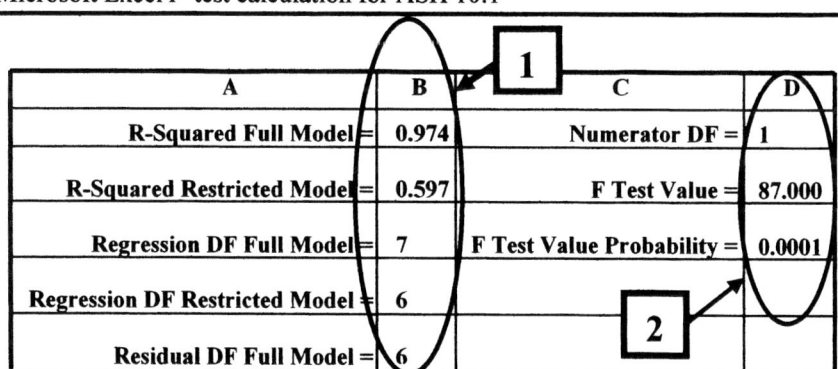

With respect to the second condition, the one-tailed *F* test probability value
($p / 2 < .001$), which is equal to one half of the *F* test probability value list in
Oval 2 in Exhibit 10.5. Since this one-tailed probability value is less than the
established alpha value of .05 and the mean posttest self-confidence score exceeds the mean pretest self-confidence score, ASH 10.1 is rejected. Thus, the
researchers have evidence that for a given population, the posttest self-
confidence mean is higher than the pretest self-confidence mean, over and above
individual differences.

Analysis of ARH 10.1 with Microsoft Excel—Models Containing a Surrogate Person Variable

As discussed in the previous section (General Hypothesis 10.1), we discussed the concept of a surrogate person variable. In this section we present the
printouts of the Full Model and Restricted Model that used a surrogate person
variable rather than the set of person variables. Before these models can be constructed and analyzed, the surrogate person variable (SPR) must be formed.
Once the researchers have the Microsoft Excel data file displayed as presented
in Exhibit 10.6, they would construct the SPR value as follows:
1. The variable label "SPV" is typed in row 1 of column J, which is the
   first blank column.
2. The command "=A2+A9" is typed in row 2 of column J. This command
   adds the pretest (located in A2) and posttest scores (located in A9) for
   the first student.
3. The command typed in row 2 of column J is copied and pasted in rows 3
   through 8 of column J.
4. The command "=A2+A9" is typed in row 9 of column J.

5. The command typed in row 9 of column J is copied and pasted in rows 10 through 15 of column J.

It should be noted that these commands are the appropriate commands when the set of pretest scores precede the set of posttest scores in the criterion vector (Y). Once the steps are completed, each SPV value is equal to the sum of the given student's pretest and posttest scores (see Oval 1 Exhibit 10.6).

Exhibit 10.6.
Calculation of surrogate person variable (SPV) values for variable (SPV) for ASH 10.1 and ARH 10.1.

Y	T	P1	P2	P3	P4	P5	1	P7	SPV
14	0	1	0	0	0	0	0	0	33
15	0	0	1	0	0	0	0	0	36
15	0	0	0	1	0	0	0	0	33
18	0	0	0	0	1	0	0	0	43
20	0	0	0	0	0	1	0	0	48
21	0	0	0	0	0	0	1	0	47
23	0	0	0	0	0	0	0	1	53
19	1	1	0	0	0	0	0	0	33
21	1	0	1	0	0	0	0	0	36
18	1	0	0	1	0	0	0	0	33
25	1	0	0	0	1	0	0	0	43
28	1	0	0	0	0	1	0	0	48
26	1	0	0	0	0	0	1	0	47
30	1	0	0	0	0	0	0	1	53

The Full Model that used the surrogate variable is constructed as follows:

$$Y = a_0U + a_1T + a_2SPV + E_1 \qquad \text{(Model 10.13)}$$

Complete the following steps to obtain the analysis for this Full Model (Model 10.13)—see Exhibit 10.7:

1. Click on the box next to "Input Y Range" (see Oval 1). Next, click and drag on the cells in the column in which variable Y is located, which is column A.
2. Move the variable T (column B) to the column next to the SPV variable, which is column K, and click on the box next to "Input X Range" (see Oval 2). Next, click and drag on the cells in the columns in which the variables SPV (column J) and T (column K) are located.

3. Click on the box in front of "Labels" (see Oval 3).
4. Click on the "OK" button (see Oval 4).

The output for the Full Model (Model 10.13) is listed in Exhibit 10.8. The key piece of information listed in the output is the R^2 value of .974 (see Oval 1).

Exhibit 10.7.
Microsoft Excel Regression menu—Full Model (Model 10.13) for ARH 10.1.

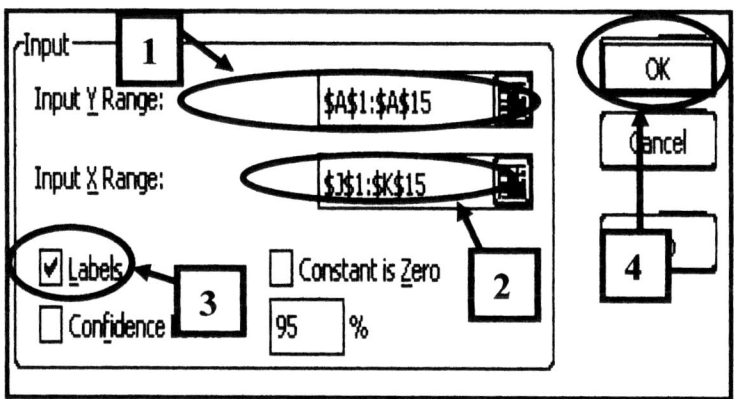

Note that this R^2 value is the same as the R^2 value of the Full Model (Model 10.11) that contained the six person variables (see Oval 1 of Exhibit 10.2). For the Full Model that contains the SPV variable (Model 10.13), however, the regression and residual degrees of freedom values of 2 and 11 are *not* correct (see Oval 2 of Exhibit 10.8). Thus, it is important for the researchers to understand that since the SRV variable represents six person variables, the F test of the model's R^2 value and t tests of the variables' coefficients listed on the output for Model 10.13 are incorrect. Also note that the treatment variable (T) coefficient for Model 10.13 is 5.86 (see Oval 2 in Exhibit 10.8), which matches the coefficient for the T variable in Model 10.11 (see Oval 3 in Exhibit 10.2).

When calculating the correct F test for ASH 10.1 the researchers must identify the correct degrees of freedom values. The researchers must remember that SRV in Models 10.13 and 10.14 represents k -1 person variables. Thus, for the Full Model (Model 10.13) the correct regression degrees of freedom value is 7, which is equal to the number of predictor variables in the model (i.e., the time-period variable and the six person variables that SPV represents). The correct residual degrees of freedom value for this Full Model is 6 (i.e., 14 – 8), which is equal to the number of scores minus the number of pieces of information in the model. (Recall that the number of pieces of information in the model includes the unit vector, the time-period variable, and the six person variables represented by SPV.)

Exhibit 10.8.
Microsoft Excel regression output—Full Model (Model 10.13) with the surrogate person variable (SPV) for ARH 10.1

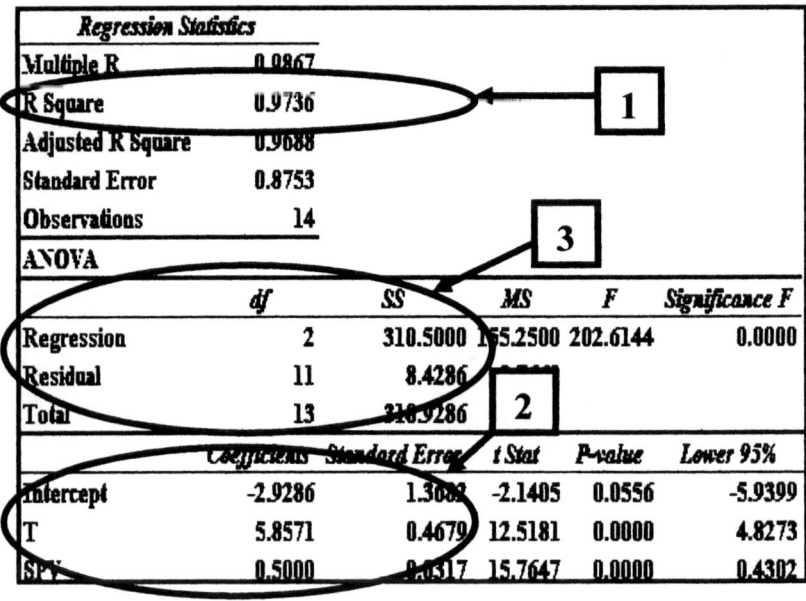

Regression Statistics					
Multiple R	0.9867				
R Square	0.9736				
Adjusted R Square	0.9688				
Standard Error	0.8753				
Observations	14				

ANOVA					
	df	SS	MS	F	Significance F
Regression	2	310.5000	155.2500	202.6144	0.0000
Residual	11	8.4286			
Total	13	318.9286			

	Coefficients	Standard Error	t Stat	P-value	Lower 95%
Intercept	-2.9286	1.3682	-2.1405	0.0556	-5.9399
T	5.8571	0.4679	12.5181	0.0000	4.8273
SPV	0.5000	0.0317	15.7647	0.0000	0.4302

The corresponding Restricted Model is as follows:

$$Y = a_0U + a_2SPV + E_1 \qquad \text{(Model 10.14)}$$

To obtain the Excel output for this Restricted Model (Model 10.14), complete the following steps (see Exhibit 10.9):
1. Click on the box next to "Input $\underline{Y}$ Range" (see Oval 1). Next, click and drag on the cells in the column in which variable Y is located, which is column B.
2. Click on the box next to "Input $\underline{X}$ Range" (see Oval 2). Next, click and drag on the cells in the column in which the variable SPV is located (column K).
3. Click on the box in front of "Labels" (see Oval 3).
4. Click on the "OK" button (see Oval 4).
Once these steps are completed, the output for the Restricted Model (Model 10.14 will be displayed (see Exhibit 10.10).

Exhibit 10.9.
Microsoft Excel Regression menu—Restricted Model (Model 10.14) for ASH
10.1.

Exhibit 10.10.
Microsoft Excel regression output—Restricted Model (Model 10.14) with the
surrogate person variable (SPV) for ASH 10.1

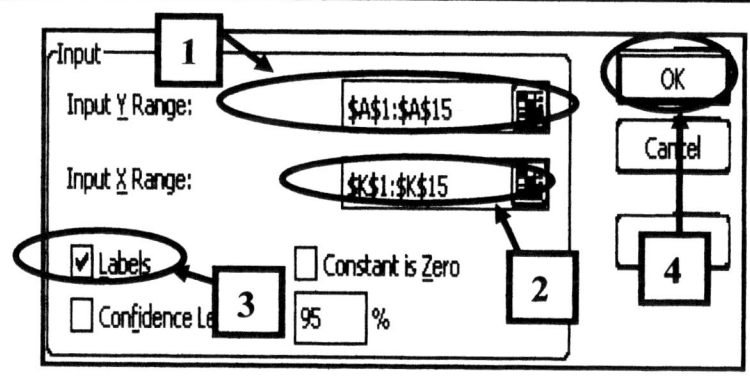

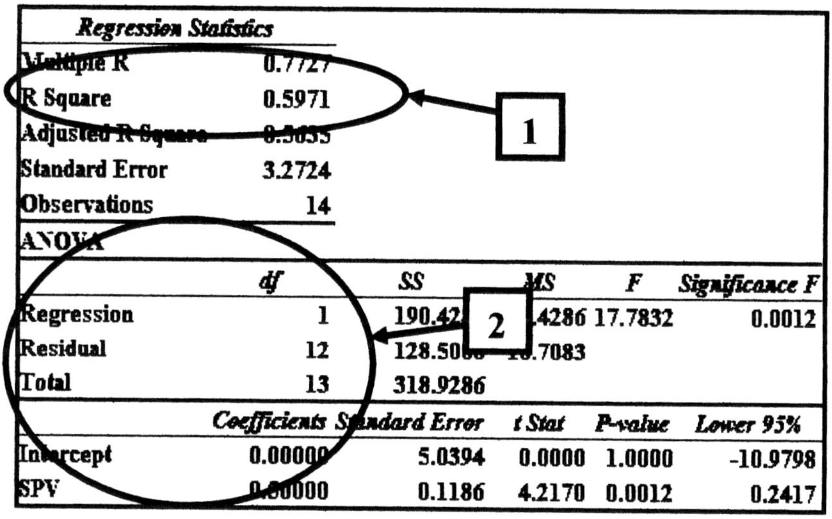

The key piece of information listed in the output for this Restricted Model
(Model 10.14), which is presented in Exhibit 10.10 is the R^2 value of .597 (see
Oval 1). Again, note that this R^2 value is the same as the R^2 value of the
Restricted Model (Model 10.12) that contained the six person variables (see

Oval 1 of Exhibit 10.4). But as with the Full Model containing the SPV variable (Model 10.13), the residual degrees of freedom value listed in the output for the Restricted Model containing the SPV variable (Model 10.14) is incorrect. The correct residual degrees of freedom value for Model 10.14 is six, which is equal to the number of person variables represented by SPV.

To calculate the appropriate F test used to statistically test ASH 10.1, the researchers would enter the R^2 values and the corrected degrees of freedom values for the models that contained SRV (i.e., Models 10.13 and 10.14) into the Microsoft Excel program entitled "F Test Calculation," which are the same values as those listed in Exhibit 10.5. Of course, with the corrected degrees of freedom values, the models that contained SRV (i.e., Models 10.13 and 10.14) will produce the same F test and probability values as those calculated when using the values obtained from the models that included the six person variables (i.e., Models 10.11 and 10.12).

Writing the Results for ARH 10.1 in APA Style

The researchers may present their findings regarding the statistical test of ASH 10.1, which is the Statistical Hypothesis corresponding to the ARH 10.1, as follows:

> As indicated by the coefficient for the time-period variable in the Full Model (5.86), the mean posttest self-confidence score exceeded the mean pretest self-confidence score by 5.86 points. The one-tailed probability value of the F test used to statistically test the difference between the R^2 values of the Full Model (.97) and the Restricted Model (.60) indicated that this difference between the pretest and posttest scores is statistically significant at the .05 level [$F(1, 6) = 87.00$ ($p / 2 < .01$)]. The effect size of the difference between the mean posttest self-confidence score and the mean pretest self-confidence score, which is measured by the difference between the R^2 values of the Full and Restricted Models, is .37. This value is classified as a large effect size according to the criteria established by Cohen (1988).

TWO-GROUP DESIGNS

The goal of researchers who use two-group models is to have a strict comparison group, but usually subjects cannot be randomly assigned to a treatment and a comparison group. Therefore, researchers in the behavioral sciences and education often have to be satisfied with comparison groups that are somewhat similar to the treatment group.

Two-Group Posttest-Only Design

The introduction of a comparison group often allows the elimination of many competing explanatory variables. The simplest design entailing both a comparison group and an experimental group is one in which the groups are not pretested but rather just posttested. One reason for not including a pretest is to avoid the external validity issue Stanley and Campbell (1966) referred to as a pretest-treatment interaction effect, which exists when the pretest sensitizes the subjects to the treatment, and thus the results of the study can be generalized only when a pretest is administered along with the treatment.

The major problem with this design is the lack of knowledge the researchers have regarding the equivalency of the two groups (i.e., the treatment and comparison groups). Even if the subjects are random assigned to the groups does not guarantee that the characteristics of the subjects in the groups are initially equal, and difference between the mean posttest scores of the groups could be due to this lack of group equivalency. Such a lack of equivalency will decrease the researchers' ability to isolate the effect of the treatment on the posttest, that is, the study's internal validity is decreased. If researchers are more concerned about this possible decrease in the study's internal validity than the pretest-treatment interaction external validity issue, they will general apply the research design presented in the following section.

Two-Group Pretest-Posttest Design

In the Two-Group Pretest-Posttest Design the subjects in the treatment and comparison groups are pretested and posttested. The pretest measurements provide the researchers some knowledge of initial standing of the equivalency of the two groups with respect to the subjects' pretest scores. The researchers can utilize three different methods to analyze the data obtained from such a design with the first of the three methods being slightly easier to handle conceptually and operationally and the last method being more powerful statistically.

Readers should note that the statistical tests of the Statistical Hypotheses in the following sections are conducted with F tests. Since each of the Statistical Hypotheses places only one restriction on the Full Model, the statistical test of the Statistical Hypothesis could also be conducted with a t test of the treatment variable coefficient estimated by the Full Model. Readers should refer to the sections in chapter 8 entitled "General Research Hypothesis 8.1" and Applied Research Hypothesis 8.1" for additional discussions of the analytic technique that used an F test, and the section in chapter 8 entitled "A t test of the Group Coefficient in the Full Model" for a discussion of the analytic technique that used a t test.

Analysis of the criterion variable that consists of gain scores. One analytic method that can be used with a Two-Group Pretest-Posttest Design forms the values in the criterion variable by subtracting each subject's posttest score from the corresponding pretest score. These resulting scores are referred to as gain scores, difference scores, or change scores.

To illustrate this analytic method we will assume the criterion variable, which consists of change scores, is labeled Y. In addition, the predictor dichotomous treatment variable (X) consists of the values of 0 (comparison group) and 1 (treatment group). The directional Research and Statistical Hypotheses are stated as follows:

> Directional Research Hypothesis: For a given population the mean gain scores of the treatment method is higher than the mean gain scores of the comparison method.

> Directional Statistical Hypothesis: For a given population the mean gain scores of the treatment method is not higher than the mean gain scores of the comparison method.

The Full Model that reflects the Research Hypothesis is as follows:

$$Y = a_0 U + a_1 X + E_1 \qquad\qquad \text{(Model 10.15)}$$

It is important for researchers to understand that the coefficient for the treatment variable (a_1) is equal to the sample change-score mean of the treatment group minus the sample change-score mean of the comparison group.

Since the Statistical Hypothesis stipulates that the change-score means of the two groups do not differ, the restriction can be stated as follows:

$a_1 = 0$ (1 restriction)

Placing this restriction into the Full Model produces the following Restricted Model:

$$Y = a_0 U + E_2 \qquad\qquad \text{(Model 10.16)}$$

Using Equation 4.1, the researchers would conduct an F test of the difference between these R^2 values of these Full and Restricted Models. Since the hypotheses are directional hypothesis, the Statistical Hypothesis would be rejected if two conditions are met: (a) the mean gain score of the treatment group must exceed the mean gain score of the control group and (b) the difference in group gain scores must be statistically significant. These conditions will be met if: (a) in the Full Model (Model 10.15) the treatment variable coefficient (a_1) is positive and (b) the one-tailed F test probability value is less than the established alpha value.

Analysis of the criterion variable that consists of posttest scores with the use of a pretest covariate. If the researchers are willing to assume that the correlation between pretest and posttest is not 1.0, which is a realistic assumption in behavioral science or educational research, then the pretest scores can be used as a covariate. In this analysis the criterion variable (Y) consists of the subjects' posttest scores and the two predictor variables are the treatment variable (X1) and the pretest variable (X2). The dichotomous treatment variable consists of the values of 0 (comparison group) and 1(treatment group), and the continuous pretest variable consists of the subjects' pretest scores.

The directional Research and Statistical Hypotheses would be stated as follows:

Directional Research Hypothesis: For a given population the posttest mean of the treatment group is higher than the posttest mean of the comparison group, over and above pretest score differences.

Directional Statistical Hypothesis: For a given population the posttest mean is treatment group is not higher than the posttest mean of the comparison group, over and above pretest score differences.

The Full Model that reflects the Research Hypothesis is as follows:

$$Y = a_0U + a_1X1 + a_2X2 + E_1 \qquad \text{(Model 10.17)}$$

Researchers should note that the coefficient for the treatment variable (a_1) is equal to the estimated *adjusted* posttest mean of the treatment group minus the *adjusted* posttest mean of the comparison group.

Since the Statistical Hypothesis stipulates that the adjusted posttest means of the two groups do not differ, the restriction can be stated as follows:

$a_1 = 0$ (1 restriction)

Placing this restriction into the Full Model results in the following Restricted Model:

$$Y = a_0U + a_2X2 + E_2 \qquad \text{(Model 10.18)}$$

The researchers would conduct an *F* test of the difference between these R^2 values of these Full and Restricted Models by using Equation 4.1. Since the hypotheses are directional hypothesis, the Statistical Hypothesis would be rejected if two conditions are met: (a) the mean posttest score of the treatment group must exceed the mean posttest score of the control group and (b) the difference between the mean posttest scores of the treatment and comparison groups must be statistically significant. These conditions will be met if: (a) the treatment variable coefficient (a_1) is positive and (b) the one-tailed F test probability value is less than the established alpha value.

Analysis of a criterion variable consisting of posttest scores with the use of pretest scores from different instruments. In some research situations it is unwise to administer the same test before treatment as the one administered after treatment. In this situation, other behaviors related to the posttest can be assessed before the treatment and used as covariates. The researcher can assess as many behaviors as are suspected might be different between groups.

To illustrate this analytic procedure the criterion variable (Y) is assumed to consist of posttest scores and the dichotomous treatment variable (T) consists of the values of 0 (comparison group) and 1 (treatment group). In addition to this treatment variable, the researchers collect data on three behaviors of the subjects that are represented by the continuous predictor treatment variable (T). The three predictor variables, which are labeled PreA, PreB, and PreC, represent behavior measures A, B, and C, respectively.

The directional Research Hypothesis would be stated as follows:

Directional Research Hypothesis: The mean posttest score of the treatment group is higher than the mean posttest score of the comparison group, over and above the differences in the behaviors A, B, C.

The corresponding directional Statistical Hypothesis would be stated as follows:

Directional Statistical Hypothesis: The mean posttest score of the treatment group is not higher than the mean posttest score of the comparison group, over and above the differences in the behaviors A, B, C.

The Full Model that reflects the Research Hypothesis is as follows:

$$Y = a_0 U + a_1 T + b_1 PreA + b_2 PreB + b_3 PreC + E_1 \qquad \text{(Model 10.19)}$$

Since the coefficient for the treatment variable (a_1) is equal to the adjusted mean of the treatment group minus the adjusted mean of the comparison group and the Statistical Hypothesis requires the two adjusted means to be equal, the restriction can be stated as follows:

$$a_1 = 0 \quad \text{(1 restriction)}$$

Placing this restriction into the Full Model produces the following Restricted Model:

$$Y = a_0 U + b_1 PreA + b_2 PreB + b_3 PreC + E_1 \qquad \text{(Model 10.20)}$$

The researchers would conduct an F test of the difference between the R^2 values of these Full and Restricted Models by using Equation 4.1. Since the hypotheses were directional, the Statistical Hypothesis would be rejected if two conditions are met: (a) the mean posttest score of the treatment group must exceed the mean posttest score of the control group and (b) the difference in group posttest scores must be statistically significant. These conditions will be met if: (a) the treatment variable coefficient (a_1) is positive and (b) the one-tailed F test probability value is less than the established alpha value.

Analysis of a criterion variable consisting of posttest scores using person variables and the interaction effect between treatment and time. The lengthy and probably confusing heading simply indicates that this analytic method combines the notions of interaction and repeated measures (with directionality, of course). In this analytic technique the interaction effect between the treatment and the time is tested to determine if the mean posttest score exceeds the mean pretest score. Figure 10.1 contains the type of interaction effect between treatment and time that the researchers may expect to find in the data.

In the diagram contained in Figure 10.1a the pretest means of the treatment (Tr) and comparison (C) groups are equal, but in the other two diagrams the pretest means of the groups are not equal. In Figure 10.1b the pretest mean of the treatment group exceeds the pretest mean of the comparison group, but in Figure 10.c the pretest mean of the comparison group exceeds the pretest mean of the treatment group.

The directional Research Hypothesis that describes the type of interaction effect between treatment and time that the researchers expect to find can be stated as follows:

Research Hypothesis: The relative effectiveness of the treatment method as compared to the comparison method is greater at post-period than at pre-period, over and above individual differences.

The corresponding directional Statistical Hypothesis is stated as follows:

Statistical Hypothesis: The relative effectiveness of the treatment method as compared to the comparison method is not greater at post-period than at pre-period, over and above individual differences.

The Full Model reflecting the Research Hypothesis is as follows:

$$Y = a_0U + a_1(T1 * Tr) + a_2(T1 * C) + a_3(T2 * Tr) + p_1P1 + p_2P2 + \ldots +$$
$$p_{N-1}(PN\text{-}1) + E1 \qquad \text{(Model 10.21)}$$

The variables used in the models designed to statistically test this Statistical Research Hypothesis are as follows:

1. Y = the criterion variable containing both pretest and posttest scores,

2. T1 = 1 if score from pre, 0 otherwise;
3. T2 = 1 if score from post, 0 otherwise;
4. Tr = 1 if score from treatment, 0 otherwise;
5. C = 1 if score from comparison, 0 otherwise;
6. P1 = 1 if score from person 1, 0 otherwise;
7. P2 = 1 if score from person 2, 0 otherwise;
8. .
9. .
10. .
11. P_N = 1 if score from person N, 0 otherwise.

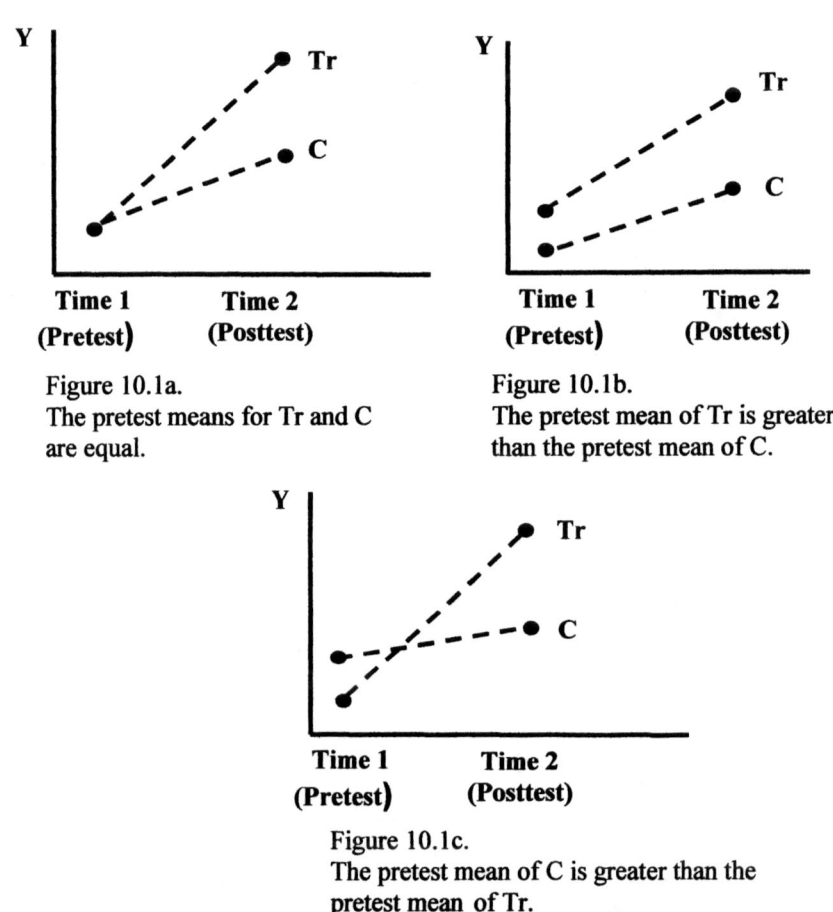

Figure 10.1a.
The pretest means for Tr and C
are equal.

Figure 10.1b.
The pretest mean of Tr is greater
than the pretest mean of C.

Figure 10.1c.
The pretest mean of C is greater than the
pretest mean of Tr.

Figure 10.1.
Three possible time-by-treatment interaction effects in which the treatment (Tr)
increases more than the comparison (C).

The restriction stated in the Statistical Hypothesis requires the two regression lines estimated by the Full Model to be parallel, that is, there is no interaction effect between time and treatment. It is important to note that in the Full Model (Model 10.21) the distance between the pretest mean for the treatment group and the pretest mean for the comparison group is estimated by the difference between coefficient a_1 and a_2. In addition, distance between the posttest mean for the treatment group and the posttest mean for the comparison group is estimated by the a_3 coefficient—remember that the variable (T2 * C) is not in Model 10.21. If the two lines are parallel, the restriction requires the following:

$(a_1 - a_2) = a_3$ (1 restriction)

Placing this restriction into the Full Model results in the following Restricted Model:

$$Y = a_0U + a_5T1 + a_7X + p_1P1 + p_2P2 + \ldots + p_{N-1}(PN\text{-}1) + E_2 \quad \text{(Model 10.22)}$$

For an in-depth discussion of these types of models designed to test for an interaction effect of two dichotomous variables (i.e., treatment and time) refer to the General Hypothesis 7.1 section of chapter 7.

The researchers would conduct an F test of the difference between the R^2 values of these Full and Restricted Models by using Equation 4.1. Since the hypotheses were directional, the Statistical Hypothesis would be rejected if two conditions are met: (a) the positive difference between the mean posttest score of the treatment group and the comparison group must exceed the difference between the mean pretest score of the treatment group and the comparison group and (b) the amount of variation accounted for by the interaction between the time and treatment must be statistically significant. These conditions will be met if: (a) in the Full Model (Model 10.21) coefficient a_3 is positive and greater than the difference between coefficients a_1 and a_2 and (b) the one-tailed F test probability value is less than the established alpha value.

THE PROCESS OF MATCHING

The purpose of this section is to familiarize students with the concept of matching more thoroughly than it was presented in chapter 8. Researches can use two different matching methods: (a) individual pairs can be matched or (b) groups are matched on indices (e.g., means and standard).

As mentioned in the Two-Group Design section, readers should note that the statistical tests of the Statistical Hypotheses in the following sections are conducted with F tests. Since each of the Statistical Hypotheses places only one restriction on the Full Model, the statistical test of the Statistical Hypothesis could also be conducted with a t test of the treatment variable coefficient estimated by the Full Model. Once again, readers should refer to the sections in chapter 8 entitled "General Research Hypothesis 8.1" and Applied Research Hypothesis 8.1" for additional discussions of the analytic technique that used an F test, and the section in chapter 8 entitled "A t test of the Group Coefficient in the Full Model" for a discussion of the analytic technique that used a t test.

Matching Groups

When researchers match groups, subjects are measured on several suppo- sedly relevant variables prior to the formation of the groups. The study's sub- jects are then assigned to groups such that the resultant groups are similar (not necessarily equivalent) on the several supposedly relevant variables. The sole requirement of similarity is often that of equal means. An additional requirement that is sometimes used is that of similar standard deviations. Unfortunately, only visual comparisons of the group means and standard deviations are usually em- ployed to ascertain similarity. A totally indefensible procedure that is sometimes used in place of visual comparisons of the mean values and standard deviation values of the groups is the visual comparison of the extreme scores of the vari- ous groups. But as most researchers know, extreme scores tell little about the bulk of the scores.

Matching variables not used in the analysis. After the two matched groups are obtained, the researchers may not use the matching variables in the analysis of the data; sometimes the matching variables are only used for the purpose of delineating the two groups. When the matching variables are not used in the data analysis—regardless of whether a pretest is or is not administered as part of the study—the format of the directional Research Hypothesis and its corresponding directional Statistical Hypothesis could be stated as follows:

> Directional Research Hypothesis: For a given population the mean posttest score of the treatment group is higher than the mean posttest scores of the comparison method.

> Directional Statistical Hypothesis: For a given population the mean posttest score of the treatment group is not higher than the mean posttest scores of the comparison method.

If the posttest scores are contained in the criterion variable labeled Y and the predictor dichotomous treatment variable (X), in which the values of 1 and 0 represent the treatment group and comparison group, respectively, the Full Model would be constructed as follows:

$$Y = a_0U + a_1X + E_1 \hspace{4cm} \text{(Model 10.23)}$$

Since the a_1 coefficient is equal to the mean posttest score for the treatment group minus the mean posttest score for the comparison group and the Statistical

Hypothesis requires the two posttest means to be equal, the restriction can be stated as follows:

$a_1 = 0$ (1 restriction)

The Restricted Model formed by the placement of this restriction on the Full Model is as follows:

$$Y = a_0U + E_2$$ (Model 10.24)

Note that this Restricted Model is the Null Model.

The researchers would conduct an F test of the difference between the R^2 values of these Full and Restricted Models by using Equation 4.1. Since the hypotheses were directional, the Statistical Hypothesis would be rejected if two conditions are met: (a) the mean posttest score of the treatment group must exceed the mean posttest score of the comparison group and (b) the difference in the mean group posttest scores must be statistically significant. These conditions will be met if: (a) the treatment variable coefficient (a_1) in the Full Model is positive and (b) the one-tailed F test probability value is less than the established alpha value.

It is important to note that although the matching variables were not used in the data analysis, the matching process does allow the researchers to establish a higher degree of internal validity of their research design. We now turn our attention to the type of analysis in which the matching variable is used in the data analysis process.

Using the matching variables as covariates. In the previous section, the matching variables were not used in the data analysis procedure. Researchers could, however, use the matching variable as covariates. That is, the matching variable could be used as a means of reducing the within variability, thus resulting in a smaller error component.

To illustrate how researchers would conduct such an analysis, assume the criterion variable (Y) consists of posttest scores. The treatment variable (X1) is dichotomous in which the values of 1 and 0 represent the treatment group and comparison group, respectively; and the variables used to match the study's subjects, which will serve as covariates, are labeled X2 and X3.

The directional Research Hypothesis and its corresponding Statistical Hypothesis for such an analysis are as follows:

Directional Research Hypothesis: For a given population the mean posttest score of the treatment group is higher than the mean posttest score of the comparison method over and above the covariates.

Directional Statistical Hypothesis: For a given population the mean posttest score of the treatment group is not higher than the mean posttest score of the comparison method over and above the covariates.

The Full Model would be constructed as follows:

$$Y = a_0U + a_1X1 + a_2X2 + a_3X3 + E_1 \qquad \text{(Model 10.25)}$$

Due to the fact that the Statistical Hypothesis requires the posttest means of the treatment and comparison groups to be equal and the a_1 coefficient in this Full Model is equal to the mean posttest score for the treatment group minus the mean posttest score for the comparison group, the restriction can be stated as follows:

$$a_1 = 0 \quad \text{(1 restriction)}$$

The Restricted Model formed by the placement of this restriction on the Full Model is as follows:

$$Y = a_0U + a_2X2 + a_3X3 + E_2 \qquad \text{(Model 10.26)}$$

Again, the researchers would conduct an F test of the difference between the R^2 values of these Full and Restricted Models by using Equation 4.1. Since the hypotheses were directional, the Statistical Hypothesis would be rejected if two conditions are met: (a) the mean posttest score of the treatment group must exceed the mean posttest score of the comparison group and (b) the difference in the mean group posttest scores must be statistically significant. These conditions will be met if: (a) the treatment variable coefficient (a_1) in the Full Model is positive and (b) the one-tailed F test probability value is less than the established alpha value.

Problems with matching groups. When researchers match groups two important problems may exist—regardless of whether the matching variables are used or not used in the data analysis. First, the relevancy of the matching variables is often of questionable value. If there is no relationship between the matching variable and the criterion, then the results may be generalized to subjects that differ on the matching variable, although the researchers who have needlessly matched do not know that this generalization can be made.

Second, it is difficult to equate the groups on the matching variables. It is very difficult to come up with two groups having equal means on several matching variables. Of course, the more matching variables one is interested in, the more unlikely it is that one can come up with equivalent groups. This problem will be further delineated in the following discussion concerning matched pairs of individuals.

Matched Pairs of Individuals

A more thorough technique for conducting the matching process involves matching individuals. Thus, for every person in the treatment group, there is a like person in the comparison group. The extent to which the comparison subject is like the treatment subject is again a function of the number and relevancy of the matching variables. Gender may be an important matching variable in one study, whereas in another study it may not be relevant. Whether a variable should be considered as a matching variable is an empirical question that can only be estimated ahead of time on theoretical grounds by the researcher.

Analyzing the data for matched pairs. Since for every person in the experimental group there is a like person in the comparison group, person vectors can be used. The method used to analyze match-pairs data is similar to the method presented in General Research Hypothesis 10.1 section of this chapter that was used to analyze the difference between the subjects' mean pretest and posttest scores.

When the data set contains matched-pairs data the criterion variable (Y) consists of the posttest scores for each pair of subjects. Instead of a dichotomous time-period variable used in the General Research Hypothesis 10.1 section, the researchers would construct a dichotomous treatment variable (Tr). In this treatment variable a value of 0 is assigned to the variable if the subject is a member of the comparison group, while a value of 1 is assigned to the variable if the subject is a member of the treatment group. The set of person variables used in the General Research Hypothesis 10.1 section are replaced with a set of matched-pairs variables (labeled MP1, MP2, . . ., MPk) which have a similar structure. The matched-pairs variable for a given pair of subjects is assigned a value of 1 if a given criterion posttest score belongs to either one of the subjects in the pair, and for all other criterion posttest values a value of 0 is assigned.

Two characteristics of the variables generated for the analysis of the matched-pairs data should be noted. First, if the study contains k pair of subjects, the study will include k match-pair variables. In addition, if the study includes N subjects, each variable (i.e., the criterion variable, the treatment variable, and the match-pairs variables) will contain 2N elements.

The directional Research Hypothesis and its corresponding Statistical Hypothesis that deal with the comparison of the posttest scores of the treatment and comparison groups are stated as follows:

Directional Research Hypothesis: For a given population the mean posttest score of the treatment group is higher than the mean posttest score of the comparison method over and above the person variables.

Directional Statistical Hypothesis: For a given population the mean posttest score of the treatment group is not higher than the mean posttest score of the comparison method over and above the person variables.

The Full Model that reflects the Research Hypothesis is as follows:

$$Y = a_0U + a_1Tr + m_1MP1 + m_2MP2 + \ldots + m_{k-1}MP(k-1) + E_1$$
$$\text{(Model 10.27)}$$

Since the Statistical Hypothesis requires the posttest means of the treatment and comparison groups to be equal and the a_1 coefficient in this Full Model is equal to the mean posttest score for the treatment group minus the mean posttest score for the comparison group, the restriction can be stated as follows:

$$a_1 = 0 \quad \text{(1 restriction)}$$

When this restriction is placed on the Full Model (Model 10.27), the following Restricted Model results:

$$Y = a_0U + m_1MP1 + m_2MP2 + \ldots + m_{k-1}MP(k-1) + E_2 \quad \text{(Model 10.28)}$$

The Statistical Hypothesis will be rejected if two conditions are met. First, the posttest mean of the treatment group is greater than the posttest mean of the comparison group, which requires the treatment variable coefficient in the Full Model (a_1) to be positive. Second, the one-tailed probability value of the F test used to statistically test the difference between the R^2 values of the Full and Restricted Models is less than the established alpha level.

Problems with the use of matched pairs. Researchers should be aware of three major problems that they may encounter when analyzing matched-pairs data. First, the degrees of freedom are drastically reduced. For an "uncorrelated" or "independent" t test, the degrees of freedom is $N - 2$, where N is the total number of people in the two groups. For the matched pairs analysis (referred to as *correlated* or *dependent t test* by some), the degrees of freedom is the number of pairs minus one. Thus, if one decides to match, the gain that one obtains in the amount of variation accounted for in the criterion variable by the inclusion of the person variables must be enough to override the loss in degrees of freedom (exactly half as many degrees of freedom). If the matching variables are chosen judiciously, the loss in degrees of freedom will be of little consequence.

Second, a more important problem with matched pairs concerns the availability of identical subjects. With individual differences as large and variable as they typically are, it is often difficult to find an adequate match. Empirical studies using this procedure often demand an initial sampling pool of 300 or 400 to get even a crude match for 30 pairs of subjects.

Third, the search for matching individuals often eliminates the extreme subjects, those who are deviant on only one measure. Conversely, those deviant matches that are found are often a function of a large error component, and on a second testing this good match may indeed turn out to be a bad match after all.

The elimination of those subjects will decrease the generalizability of the study's findings.

POINT CHANGE IN MULTIPLE TIME PERIODS

Some research questions involve the detection of a change at a certain point in time as a function of some stimulus change or input. An ideal situation would be to measure several subjects in several testing periods both before and after the introduction of a stimulus change. In some cases, however, only one data point is generated for each time period (e.g., a study conducted by Hill (2008) that investigated the equity of educational funding as measured by the Gini coefficient). In these types of studies, which are referred to as time-series studies, the number of time periods before the hypothesized change need not equal the number of time periods after the hypothesized change, although the example presented in this section does have equal numbers in the two sets of time periods.

The three diagrams contained in Figure 10.2 depict three possible situations for a set of 16 observations in which the change in values is hypothesized to have occurred at the end of period 8. In Figure 10.2a a change in the values contained in the pre-stimulus time periods versus the values contained in the post-stimulus time periods did not occur, that is, any measured change is attributed to random variation. A review of Figure 10.2b reveals the values in the post-stimulus time periods reflect a positive trend that was greater than the trend established for the values in the pre-stimulus time periods. In Figure 10.2c the trend of the values in the pre-stimulus time periods does not differ from the trend of the values in the post-stimulus time period, however, the level of the trend line for the post-stimulus time periods is higher than the level of the trend line for pre-stimulus time periods, as measured by their Y-intercept points.

In the hypothetical example presented in this section, we are assuming that a one criterion value is recorded for each of 16 time periods. In addition, a stimulus is introduced after time period 8.

The issue that researchers must address in this time-series study is: Can any change in the values after the introduction of a stimulus be attributed to more than just random variation? To determine if a change has occurred in the values contained in the two sets of time periods, the researchers need to construct a Full Model that fits a line to each set of time periods.

A regression model that allows two such lines is as follows:

$$Y = a_0U + a_1T1 + a_2X1 + a_3X2 + E_1 \qquad \text{(Model 10.29)}$$

where:

1. Y = the criterion to be predicted;
2. U = unit vector;
3. $T1 = 1$ if observation from time period 9 through 16, 0 otherwise;

4. X1 = the time period if observation is from time period 1 through 8, 0 otherwise;
5. X2 = the time period if observation from time period 9 through 16, 0 otherwise.

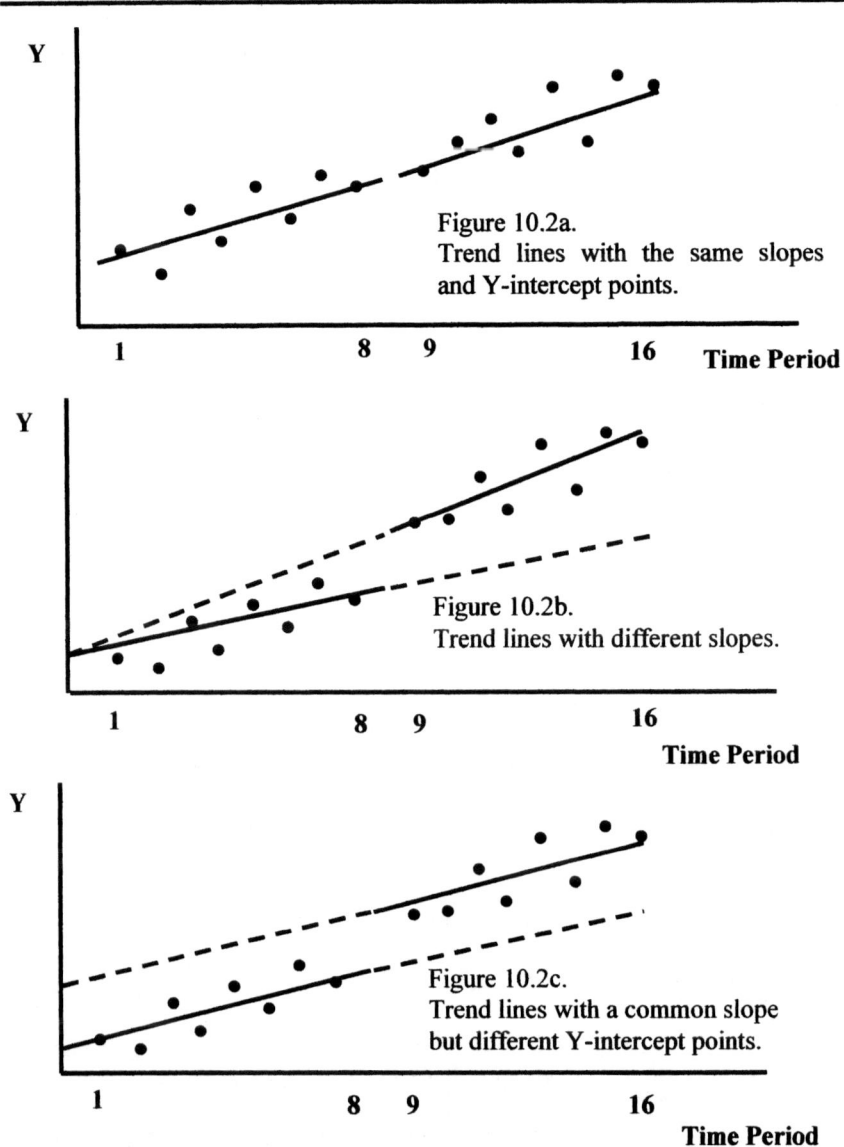

Figure 10.2a.
Trend lines with the same slopes and Y-intercept points.

Figure 10.2b.
Trend lines with different slopes.

Figure 10.2c.
Trend lines with a common slope but different Y-intercept points.

Figure 10.2.
Three possible conditions for two trend lines.

In this Full Model, the slope of the trend line for periods 1 through 8 and the slope of the trend line for periods 9 through 16 are estimated by the coefficients a_2 and a_3, respectively. In addition, the Y-intercept point of trend line for periods 1 through 8 is estimated by the a_0 coefficient, while the a_1 coefficient estimates the difference between the Y-intercept point of the trend line for periods 9 through 16 and the Y-intercept point of the trend line for periods 1 through 8. As depicted in Figures 10.2b and 10.2c, a change in the values between the two time periods can manifest itself in one of two ways regarding the two lines estimated by the Full Model (Model 10.29): (a) different slopes for the two trend lines and/or (b) different Y-intercept points for the two trend lines.

Test of the slopes. The Research and Statistical Hypotheses posed to determine if the positive slope for second trend line is greater than the slope of the first trend line are stated as follows:

Directional Research Hypothesis: The positive slope of the trend line for the values contained in the post-stimulus time periods is greater than the positive slope of the trend line for the values contained in the pre-stimulus time periods.

Directional Statistical Hypothesis: The positive slope of the trend line for the values contained in the post-stimulus time periods is not greater than the positive slope of the trend line for the values contained in the pre-stimulus time periods.

For this set of hypotheses, Model 10.29 would serve as the Full Model. As previously stated, the slopes of the first and second trend lines are estimated by the a_2 and a_3 coefficients in the Full Model (Model 10.29), respectively. Thus, the restriction placed on this Full Model by the Statistical Hypothesis (i.e., the slopes of the trend lines do not differ) is as follows:

$a_2 = a_3$ (1 restriction)

Placing this restriction into the Full Model (Model 10.22) produces the following Restricted Model:

$$Y = a_0U + a_1T1 + a_4X4 + E_2 \qquad \text{(Model 10.30)}$$

In this Restricted Model the variable X4 is the sum of variables X1 and X2, which produces values 1 through 16—the time period values. In addition, the coefficient for this X4 variable (a_4) is an estimate of the single (common) trend line (i.e., the trend line fit to periods 1 through 16).

The researchers would conduct an F test of the difference between the R^2 values of these Full and Restricted Models by using Equation 4.1. Since the hypotheses were directional, the Statistical Hypothesis would be rejected if two conditions are met: (a) the positive slope of the second trend line exceeds the positive slope of the first trend line and (b) this difference between the slopes is statistically significant. These conditions will be met if: (a) the coefficient values for a_2 and a_3 in the Full Model are positive and the value for a_3 exceeds the value for a_2 and (b) the one-tailed F test probability value is less than the established alpha value. If these two conditions are met, the researchers would state that the positive slope of the trend line for the values contained in the post-stimulus time periods is significantly greater than the positive slope of the trend line for the values contained in the pre-stimulus time periods. That is, the difference between the slopes cannot be attributed to random variation.

Test of the Y-intercept points. If the researchers are willing to conclude that the slopes of the two trend lines do not differ, they must next determine if the levels of the two sets of values differ. That is, the researchers need to test the difference between the two Y-intercept points of the two trend lines. The Research and Statistical Hypotheses posed to determine if the Y-intercept point for second trend line is greater than the Y-intercept point of the first trend line are stated as follows:

> Directional Research Hypothesis: The Y-intercept point of the trend line for the values contained in the post-stimulus time periods is greater than the Y-intercept point of the trend line for the values contained in the pre-stimulus time periods.

> Directional Statistical Hypothesis: The Y-intercept point of the trend line for the values contained in the post-stimulus time periods is not greater than the Y-intercept point of the trend line for the values contained in the pre-stimulus time periods.

Assuming that the test of the difference between the slopes of the two trend lines revealed that the difference could be attributed to random variation in the values, the researchers will assume the slopes of the two trend lines are equal. Thus, the Full Model used to test the difference between the Y-intercept points is Model 10.30, which requires one common slope for the two trend lines.

Due to the fact that the a_1 coefficient in the Full Model (Model 10.23) estimates the difference between the Y-intercept point of the trend line for the post-stimulus time periods and the Y-intercept point of the trend line for the pre-stimulus time periods, and the restriction required by the Statistical Hypothesis requires the two Y-intercept points to be equal, the restriction can be stated as follows:

$a_1 = 0$ (1 restriction)

Placing this restriction into the Full Model (Model 10.23) produces the following Restricted Model:

$$Y = a_0U + a_4X4 + E_3 \qquad \text{(Model 10.31)}$$

The researchers would conduct an F test of the difference between the R^2 values of these Full and Restricted Models by using Equation 4.1. Since the hypotheses were directional, the Statistical Hypothesis would be rejected if two conditions are met: (a) the Y-intercept point of the post-stimulus trend line exceeds the Y-intercept point of the pre-stimulus trend line and (b) this difference between the two Y-intercept points is statistically significant. These conditions will be met if: (a) the coefficient value of a_1 in the Full Model positive and (b) the one-tailed F test probability value is less than the established alpha value. If these two conditions are met, the researchers would state that the Y-intercept point of the post-stimulus trend line is significantly greater than the Y-intercept point of the pre-stimulus trend line. That is, the difference between the Y-intercept points cannot be attributed to random variation.

CURVILINEAR RELATIONSHIPS THAT ACCOUNT FOR THE DATA BETTER THAN STIMULUS CHANGE

It is important for researchers to understand that the analytic technique presented in the previous section assumes the values in the pre-stimulus and post-stimulus trend lines reflect *linear* trends. We suspect that there are quite a few curvilinear relationships in the real world, but these relationships have not been discovered because researchers have not tested for such relationships. If there is an underlying curvilinear relationship for each of the two sets of values, one may be making a costly mistake by trying to analyze the data with models that estimate two straight lines.

Figure 10.3 depicts a set of data where the performance scores after the stimulus change continue to follow the curvilinear trend that started before the stimulus change. Visually, it appears that a single second-degree curved line can account for the relationship between time and the performance scores, rather two straight lines.

The two dashed straight lines in Figure 10.3 might lead to the conclusion that the stimulus change after Time Period 20 had an effect (i.e., reflected an increase in the rate of change in Y). However, if a single second-degree curve fits all the data well (yielding a high R^2), then what was happening systematically before the introduction of the stimulus is still happening after its introduction (see the solid curve in Figure 10.3). The researchers would be hard-pressed to argue that the introduction of the stimulus led to a change in the performance scores.

The Full Model needed to test the existence of two separate second-degree curved lines is as follows:

$$Y = a_0U + b_1Z2 + s_1Z3 + a_2Z4 + b_2Z5 + s_2Z6 + E_1 \qquad \text{(Model 10.31)}$$

where:
  1. Y = the criterion performance score;
  2. U = the unit vector;
  3. Z1 = 1 if observation is from one of the time periods 1 through 20, 0 otherwise;
  4. Z2 = the time period if observation is one of the 1 through 20 time periods, 0 otherwise;
  5. Z3 = the squared value of the element in Z2;
  6. Z4 = 1 if observation is from one of the time periods 21 through 40, 0 otherwise;
  7. Z5 = the time period if observation is one of the 21 through 40 time periods, 0 otherwise; and
  8. Z6 = the squared value of the element in Z5.

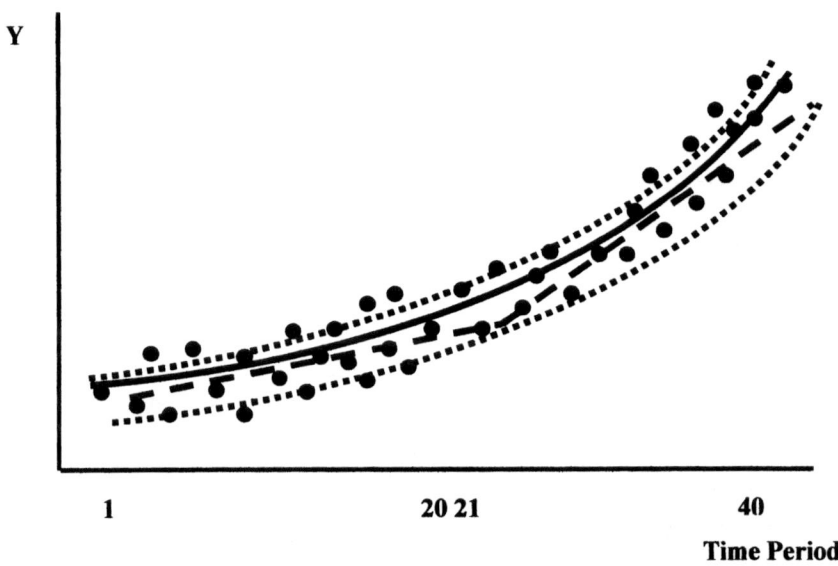

Figure 10.3.
A Case in which a single second-degree curved line (solid curve) depicts the data as well as two straight lines (dashed lines) and two second-degree curved lines (dotted curves).

Researchers should understand what each coefficient in this Full Model estimates: (a) the a_0 estimates the Y-intercept point of the first trend line; (b) the a_2 estimates the difference between the Y-intercept points of the two trend lines; (c) b_1 and b_2 estimate the linear component of the first and second trend lines, respectively; and (d) s_1 and s_2 estimate the second-degree components of the first and second trend lines, respectively. If both of the second-degree polynomials estimated by this Full Model in fact estimate a common second-degree polynomial, then the Y-intercept point, the linear component, and the second-degree component of one second-degree polynomial should equal the corresponding values in the other second-degree polynomial. Thus, the following restrictions must be tested:

$$a_2 = 0; \quad b_1 = b_2; \quad \text{and} \quad s_1 = s_2 \quad \text{(3 restrictions)}$$

Placing these restrictions into the Full Model (Model 10.31) produces the following Restricted Model:

$$Y = a_c U + b_c Z7 + s_c Z8 + E_2 \qquad \text{(Model 10.32)}$$

where:
1. U = the unit vector,
2. $Z7$ = the continuous vector of time periods, and
3. $Z8$ = the squared value of the corresponding element in $Z7$.

Once again it is important for the researchers to understand what each coefficient in this Restricted Model estimates: (a) a_c estimates the common Y-intercept point, (b) b_c estimates the common linear component, and (c) s_c estimates the common second-degree component. The researchers would calculate an F test of the difference between the R^2 values of the Full Model (Model 10.31) and the Restricted Model (Model 10.32). Since three restrictions were placed on the Full Model, the researchers must use the two-tailed probability of the F test. If the two-tailed probability value is less than the established alpha value, the researchers would conclude that the data should be modeled with two second-degree polynomial curves. If, however, the two-tailed probability value is not less than the established alpha value, the researchers would conclude that only one second-degree polynomial curve is needed and state that the stimulus had no impact on the performance scores.

Whether the researchers choose a curvilinear or linear fit to the data their decision should be a function of both theoretical notions and empirical data. A statistician cannot tell the researchers when to investigate curvilinear relationships. What we have done is to show the applicability of the GLM procedure to curvilinear types of problems (and to encourage such application whenever the researchers think that such an analysis might prove to be fruitful).

FUNCTIONAL CHANGE IN
INDIVIDUAL-ORGANISM DESIGNS

The discussion in the preceding section has many ramifications for individual-organism research. Kelly, McNeil, and Newman (1973) have documented a more extensive application of the GLM approach to repeated measures of a single organism. The following paragraphs should only serve to whet the appetites of empirically oriented single-organism researchers.

A visual inspection of the data that emanate from standard individual-organism designs can be bolstered with probability statements. Suppose that some functional relationship is established under a set of conditions, which is labeled A. Next a new set of conditions, which is labeled B, is introduced to the organism. Finally, the A conditions are reintroduced, which is labeled A'. Data points depict the functional relationship between time period and performance level for each condition. Figure 10.4 illustrates one possible outcome in which performance in condition A reflects a slightly positive trend, an accelerated increase in performance in condition B, and when condition A was reintroduced (i.e., condition A') a reversion back to the rate of increase in performance experienced in condition A.

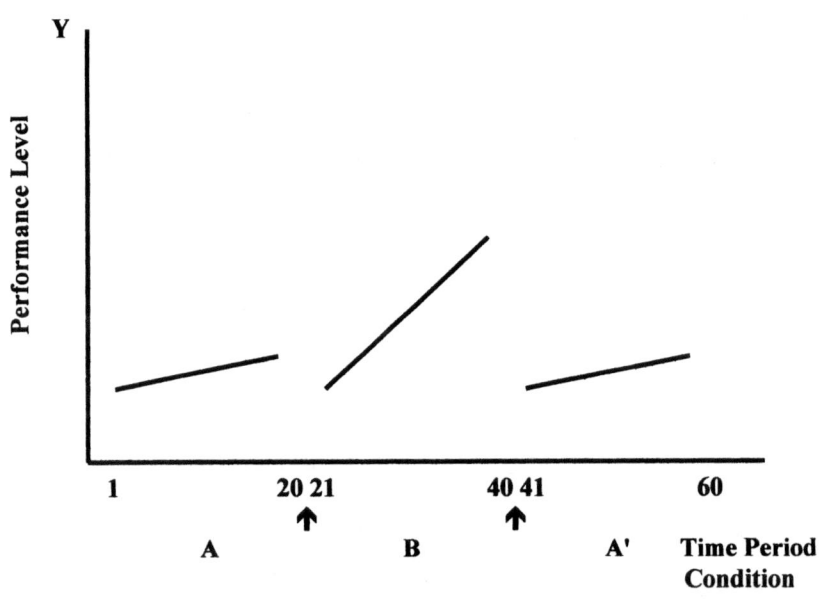

Figure 10.4.
One possible data set emanating from single-organism research.

Test of the Trend Lines for Conditions A and A'

As with any collection of data, several research questions can be tested. For instance, the researchers might ask if the A baseline function has been achieved during the A' condition. This question is addressed by, first, designing a Full

Model and a Restricted Model that are used to analyze only the data from A and A'. The Full Model is constructed as follows:

$$Y = a_0U + b_1X1 + a_3T3 + b_3X3 + E_1 \qquad \text{(Model 10.33)}$$

where:
1. Y = criterion of performance level;
2. U = unit vector;
3. X1 = time period if the observation is from one of the 1 through 20 time periods (Condition A), 0 otherwise;
4. T3 = 1 if observation is from 41 through 60 time periods, 0 otherwise (Condition A'); and
5. X3 = time period minus 40 if the observation is from one of the 41 through 60 time periods, 0 otherwise.

Note that the adjustment made on the time scores in vector X4 has the effect of superimposing the data obtained during the A' condition onto the data obtained during the A condition.

It is important for the researchers to understand what each coefficient in this Full Model estimates: (a) a_0 estimates the Y-intercept point of the Condition A line, (b) b_1 estimates the slope of the Condition A line, (c) a_3 estimates value equal to the Y-intercept point of the Condition A line minus the Y-intercept point of the Condition A' line, and (c) b_3 estimates the slope of the Condition A' line.

If the trend lines for Conditions A and A' do not differ, their Y-intercept points and slopes must be equal. Thus the restrictions placed on the Full Model (Model 10.33) are as follows:

$$a_3 = 0 \quad \text{and} \quad b_1 = b_3 \quad \text{(2 restrictions)}$$

Placing these restrictions on the Full Model (Model 10.33) results in the following Restricted Model:

$$Y = a_0U + b_cX5 + E_2 \qquad \text{(Model 10.34)}$$

where X5 = the time periods numbered 1 through 20 at which performance level was observed (recall the time periods 41 through 60 were transposed to 1 through 20). In this Restricted Model the a_0 and b_c coefficients estimate the common Y-intercept point and common slope for the two trend lines (i.e., the trend lines for Conditions A and A').

The researchers would conduct an F test of the difference between these R^2 values of these Full and Restricted Models by using Equation 4.1. If the probability of the F test is less than the established alpha level, the researchers cannot assume that the same (linear) functional relationship exists under the Conditions A and A'. If, however, the probability of the F test is greater than the established alpha level, the researchers would conclude that the difference between the two trend lines is no more than what would be expected through random fluctuations.

Test of the Trend Line for Condition B and the Common Trend Line for Conditions A and A'

A possibly more interesting change question involves the comparison of the functional relationship under Condition B with the (average) functional relationship under Conditions A and A', assuming the trend lines of those two conditions do not differ. If the Condition B is effective, then the rate of increase ought to be faster than under baseline conditions (i.e., Conditions A and A'). The superimposed starting point for all conditions might reasonably be expected to be the same (an empirically testable situation, of course).

The directional Research Hypothesis of interest could be stated as follows:

Directional Research Hypothesis: Assuming a common starting point for all conditions, the Condition B yields a faster rate of increase on the criterion Y than does the average rate of increase of Conditions A and A'.

The corresponding directional Statistical Hypothesis is as follows:

Directional Statistical Hypothesis: Assuming a common starting point for all conditions, the Condition B does not yield a faster rate of increase on the criterion Y than does the average rate of increase of Conditions A and A'.

The Full Model that reflects this Research Hypothesis is as follows:

$$Y = a_0U + b_2X2 + b_4X4 + E_3 \qquad \text{(Model 10.35)}$$

where:
1. Y = criterion of performance level;
2. U = unit vector;
3. $X2$ = time period minus 20 if the observation is from Condition B, 0 otherwise; and
4. $X4$ = time period if the observation is from Condition A and time period minus 40 if the observation is from Condition A', 0 otherwise.

Note that this Full Model contains a common Y-intercept point, which was required by the Research Hypothesis stipulation that the conditions must have a common starting point. This common Y-intercept point is estimated by the a_0 coefficient. The b_2 and b_4 coefficients estimate the slopes of the Condition B trend line and the slope of the trend line for Conditions A and A'.

Since the Statistical Hypothesis requires the slopes of the two trend lines to be equal, the following restriction will be placed on the Full Model (Model 10.35).

$$b_2 = b_4 \quad \text{(1 restriction)}$$

Placing this restriction into the Full Model (Model 10.35) yields the following Restricted Model:

$$Y = a_0U + b_cX5 + E_4 \qquad \text{(Model 10.36)}$$

where X5 = the time periods numbered 1 through 20 at which performance level was observed (recall the time periods 41 through 60 were transposed to 1 through 20).

The researchers would conduct an F test of the difference between the R^2 values of these Full and Restricted Models by using Equation 4.1. Since the hypotheses were directional, the Statistical Hypothesis would be rejected if two conditions are met: (a) the slope of the positive trend line for Condition B must exceed the positive slope of the trend line for Conditions A and A' and (b) the one-tailed F test probability value is less than the established alpha value. If these two conditions are met, the researchers would state that assuming a common starting point for all conditions, the Condition B yields a faster rate of increase on the criterion Y than does the average rate of increase of Conditions A and A'. If either of the two conditions is not met, the researchers would not be able to support such a statement.

Kelly et al. (1973) illustrate some ways GLM can be used to answer meaningful questions with single-organism research designs. Although some of the material is redundant with the material presented in this section, we encourage interested researchers to read the Kelly et al. article.

A Note Regarding Time Series

A statistical analysis of time measures often has been a special and troublesome problem. Traditional methods, when applied to time designs, can yield results that are incorrect. The basic reason for this is that such data as time measures tend to fluctuate. If it is possible to assume that the variability is uniform over the whole criterion, there is no longer a statistical problem (Boneau, 1960; Kerlinger & Pedhazur, 1973). In operant conditioning designs, the legitimacy of making the assumption that the variability is stable over the area of interest is more likely to be correct due to the procedures used in calculating baseline. The procedures generally require collection of enough data so that one can determine if the responses are stable over the area of interest. The number of observations necessary depends upon the stability and adequacy of the function being fit. At the lower limit, only one more observation than predictor pieces of information is needed to calculate the statistical index. In essence, one needs only a sufficient number of observations to provide a stable indicator of the functional relationship, and this is primarily a non-statistical decision.

CHAPTER 11

DICHOTOMOUS CRITERION VARIABLE

Although the statistical and computer solutions to dichotomous variables are the same as with continuous variables, researchers may formulate their Research Hypotheses differently depending on whether their criterion variable is continuous or dichotomous. Statisticians have developed alternate computational formulae depending on whether the criterion variable is continuous or dichotomous. Fortunately, the computer neither knows nor cares whether the criterion variable is dichotomous or continuous. This chapter demonstrates how the GLM can be used to answer questions dealing with a dichotomous criterion variable (i.e., one that contains two categories) and one or more predictor variables, which may be either continuous or dichotomous. When there are more than two categories on each of the predictor and criterion sides, then the GLM cannot be used and one must turn to a multivariate procedure, such as discriminant analysis or logistic regression with a multinomial criterion variable. These techniques will not be addressed in this text.

The first portion of this chapter discusses the technique used to analyze a dichotomous criterion variable and a dichotomous predictor variable. The next portion presents models in which the dichotomous criterion variable is predicted from a set of predictor variables that may consist of either continuous or dichotomous variables. The final portion of this chapter examines some advantages and potential problems with the use of GLM and dichotomous criterion variables.

USING GLM WITH A
DICHOTOMOUS CRITERION VARIABLE

Researchers can use the GLM approach presented in this text to statistically test a Research Hypothesis that involves a dichotomous criterion variable and a dichotomous predictor variable. As we stated throughout this text we encourage researchers to construct models that consider as many relevant variables in the analysis as possible. Although we believe it will be a rather rare case that one predictor variable will be identified as the only relevant variable, we begin our discussion of the analysis of a dichotomous criterion variable with one such hypothetical example because it will serve as the basis of more complex analyses.

A Hypothetical Study that Involves a Dichotomous Criterion Variable and a Dichotomous Predictor Variable

Assume researchers are interested in determining whether the students who completed one instructional program (Program B) passed a math proficiency test in a higher proportion than students who completed another instructional program (Program A). The data obtained from the study the researchers conducted are listed in Table 11.1.

Table 11.1
Data for the Study that Involved a Dichotomous Criterion Variable (Test) and a Dichotomous Predictor Variable (Program)

	Test Results		
	Failed	Passed	Totals
Program A	36	24	60
Program B	18	42	60
TOTALS	54	66	120

The researchers created two variables (i.e., a criterion variable and a predictor variable) for the data listed in Table 11.1. The dichotomous criterion variable, which is labeled Test, indicates whether a student failed the math proficiency test (0) or passed the math proficiency test (1); while the dichotomous predictor variable, which is labeled Program, indicates whether a student received instruction through Program A (0) or Program B (1). Note that the researchers assigned a value of 1 to the instructional program they thought would have a higher proportion of students who passed the test (i.e., Program B) because it makes interpretation of results easier.

The research question posed by the researchers is as follows:

Is the proportion of students who passed the math proficiency test higher for the students who received instruction through Program B than it is for the students who received instruction through Program A?

The directional Research Hypothesis that reflects the condition posed in this research question can take a number of forms, including the following:

Directional Research Hypothesis—Version 1 (ARH 11.1a): For the population, the proportion of students who passed the math proficiency test is higher for the students who received instruction through Program B than it is for the students who received instruction through Program A.

Directional Research Hypothesis—Version 2 (ARH 11.1b): For the population, a positive relationship exists between instructional method and the math proficiency test results.

Directional Research Hypothesis—Version 3 (ARH 11.1c): For the population, participation in Program B is associated with an increase in the probability of passing the math proficiency test.

For the second version of the Research Hypothesis it is essential that the researchers assign the value of 1 to the instructional program they believe will have a higher proportion of students who pass the test (assuming the students who pass the test are assigned the value of 1).

The corresponding Statistical Hypothesis for each version of the directional Research Hypothesis is stated as follows:

Directional Statistical Hypothesis—Version 1 (ASH 11.1a): For the population, the proportion of students who passed the math proficiency test is not higher for the students who received instruction through Program B than it is for the students who received instruction through Program A.

Directional Statistical Hypothesis—Version 2 (ASH 11.1a): For the population, a positive relationship does not exist between instructional method and the math proficiency test results.

Directional Statistical Hypothesis—Version 3 (ASH 11.1a): For the population, participation in Program B is not associated with an increase in the probability of passing the math proficiency test.

Of course these research and statistical hypotheses could be stated as nondirectional hypotheses. Once again, however, we encourage researchers to use directional hypotheses whenever possible.

As previously noted, both the predictor variable (program) and criterion variable (test) are dichotomous. Using the data listed in Table 11.1, the vectors for the variables would appear as listed in Table 11.2 for the 20 of the 120 students in the data set. Note that the Assessment variable contained in the data file is not used in this analysis. It is used in the analysis conducted later in this chapter.

Table 11.2
The Vectors for the Test, Program, and Assessment Variables for the First 20
Students in the Data Set contained in Table 11.1

Student	Test [a]	Program [b]	Assessment
1	0	0	43
2	0	0	24
3	0	0	15
4	0	0	36
5	0	0	28
6	0	0	48
7	0	1	20
8	0	1	47
9	1	0	28
10	1	0	42
11	1	0	30
12	1	0	55
13	1	1	48
14	1	1	31
15	1	1	22
16	0	1	17
17	1	1	40
18	1	1	33
19	1	1	45
20	1	1	22

[a] A student who failed the math proficiency test is assigned a value of 0, while a
student who passed the math proficiency test is assigned a value of 1.
[b] A student who was placed in Program A is assigned a value of 0, while a student who was placed in Program B was assigned a value of 1.

The Full Model designed to reflect the condition stated in the Directional
Research Hypothesis (ARH 11.1a) is as follows:

$$\text{Test} = a_0 U + a_1 \text{Program} + E_1 \qquad \text{(Model 11.1)}$$

The coefficient for the Program variable in the Full Model (a_1) is equal to the
proportion of the students who passed the test in Program B minus the proportion of the students who passed the test in Program A. Since the Statistical Hypothesis requires these two proportions to be equal, the restriction that must be
placed on the Full Model to meet this condition is as follows:

$$a_1 = 0 \qquad \text{(1 restriction)}$$

Placing this restriction on the Full Model (Model 11.1) produces the following Restricted Model:

$$\text{Test} = a_0 U + E_2 \qquad \text{(Model 11.2)}$$

Note that this Restricted Model is the Null Model.

The difference between the R^2 values of the Full and Restricted Models is statistically tested with an F test calculated with Equation 4.1 (refer to chapter 4 to review this equation). The numerator degrees of freedom value for this F test is equal to 1, which is equal to the number of restrictions placed on the Full Model, while the denominator degrees of freedom value is equal to 118, which is equal to the sample size of 120 minus the number of pieces of information contained in the Full Model (i.e., the unit vector and the program variable).

Since the Research Hypothesis is directional, two conditions must be met before the researchers can reject it. First, the proportion of the students who passed the math proficiency test in the Program B group must exceed the proportion of the students who passed the math proficiency test in the Program A group. Since the coefficient for the Program variable in the Full Model (a_1) is equal to the proportion of the students who passed the math proficiency test in the Program B group minus the proportion of the students who passed the math proficiency test in the Program A group, the value of the a_1 coefficient must be positive. Second, the one-tailed F test probability value, which is equal to one half of the two-tailed F test probability value, must be less than the established alpha level, say, .05.

If the researchers use Microsoft Excel to estimate the Full Model (Model 11.1) for the data contained in Table 11.1, the output contained in Exhibit 11.1 would be produced. The value for the program variable coefficient (i.e., the a_1 value in the Full Model) is .30 (see Oval 1). This value indicates that the proportion of the students who passed the math proficiency test in the Program B group (the group assigned the value of 1 in the data set) is .30 higher than the proportion of the students who passed the test in the Program A group (the group assigned the value of 0 in the data set). Readers should note that the coefficient for the unit vector (the intercept value) in the Full Model (a_0) is equal to the proportion of the students who passed the math proficiency test in the Program A group. This value is equal to .40, which is located in the row entitled "Intercept" for the column entitled "Coefficients" in the Microsoft Excel output sheet (see Oval 1).

Figure 11.1 contains a plot of Model 11.1. The proportion of students who passed the test in Program A, which is represented by the star, is located at the Y level of .40. This value is equal to the coefficient for the unit vector (a_0). The proportion of students who passed the test in Program B, which is represented by the dot, is equal to .70. This value is equal to the sum of the coefficient for the unit vector (a_0) and the coefficient for the program variable (a_1)—the a_1 is located in Oval 1 of Exhibit 11.1.

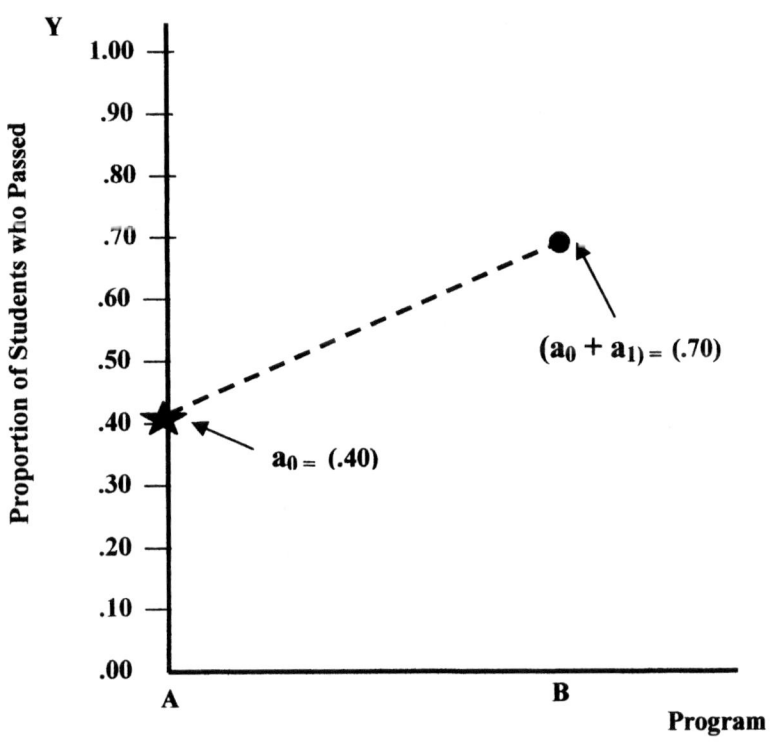

Figure 11.1.
A plot of Model 11.1 ($Y = a_0 + a_1 Program + E_1$).

Once again refer to Exhibit 11.1. The R^2 value of the Full Model is .091 (see Oval 2). Since the Restricted Model (Model 11.2) is the Null Model, which, of course, has an R^2 value equal to .00, the F test contained in the ANOVA Table is the appropriate statistical test of the Statistical Hypothesis. The F test value of 11.8 has a corresponding two-tailed probability value of .0008 (see Oval 3) or a one-tailed probability of .0004 (i.e., .0008 / 2 = .0004).

As previously stated, the Statistical Hypothesis (ASH 11.1a) can be rejected if the coefficient for the program variable (a_1) is positive and the one-tailed F test probability is less than .05. Since the a_1 coefficient is a positive .30 and the one-tailed F test probability of .0004 is less than the .05 alpha level, the researchers would reject (ASH11.1a). Thus the researchers would be willing to state that the proportion of the students who passed the math proficiency test in the Program B group is significantly higher than the proportion of the students who passed the test in the Program A group.

Exhibit 11.1.
Microsoft Excel regression output—Full Model (Model 11.1) for ARH 11.1a.

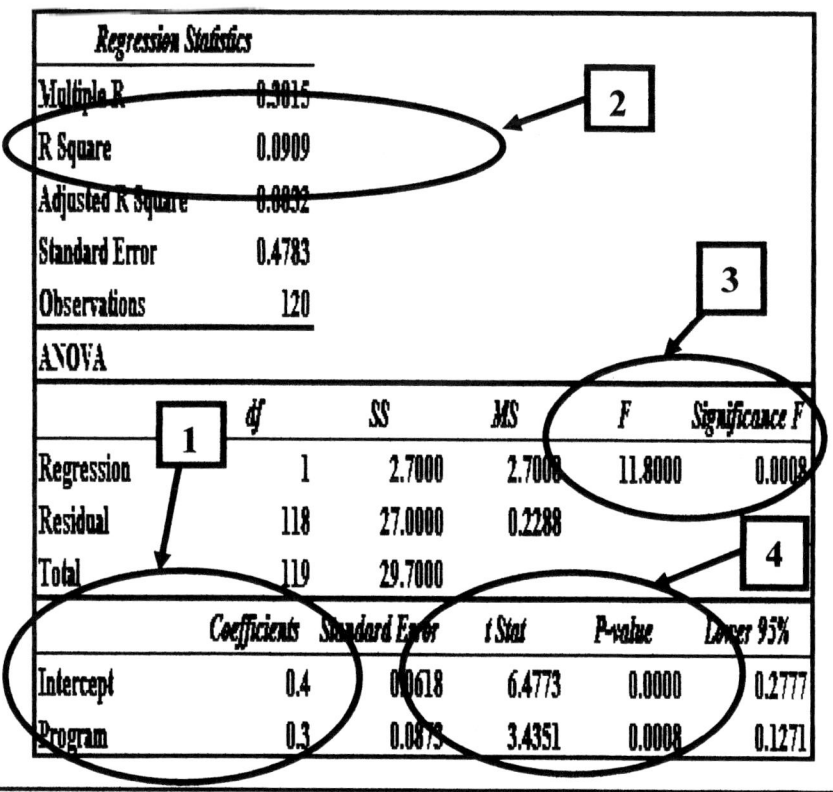

Readers should note that the statistical test of the Statistical Hypothesis required only one restriction be placed on the Full Model. Thus, the researchers could have used a t test of the program variable coefficient instead of the F test of the difference between the R^2 values of the Full and Restricted models. As listed in Oval 4 in Figure 11.1 the t test value for the program variable coefficient was 3.435 and its corresponding two-tailed probably was .0008, which is the same probability value as the F test probability. Since the Research Hypothesis (ARH11.1a) was directional, this two-tailed probability would be divided by 2 to produce a one-tailed probability value of .0004. Due to the fact that the program coefficient is positive and the one-tailed t-test probability value is less than the alpha level of .05, the Statistical Hypothesis (ASH11.1a) would be rejected, as was the case when the F test was used. Again, we see that the F test applied to a case with one restriction produces the same result as a t test. The resulting probabilities are exactly the same.

USING GLM WITH A DICHOTOMOUS CRITERION VARIABLE AND MULTIPLE PREDICTOR VARIABLES

As mentioned in chapter 6, researchers often believe that the criterion variable is influenced by more than one variable, which requires their research questions and corresponding models to include more than one predictor variable. The use of GLM allows researchers to include multiple predictor variables in an analysis of a dichotomous criterion variable, and such a model is often referred to as a linear probability model.

Applied Research Hypothesis 11.2

The purpose of this section is to present an empirical example of an analysis of Applied Research Hypothesis (ARH 11.2), which involves a dichotomous criterion variable, with a linear probability model. The researchers are interested in determining whether the students who complete one instructional program are more likely to pass a math proficiency test than students who completed another instructional program over and above the influence of the students' pre-treatment math ability scores. The variables included in the analysis are as follows:

1. The criterion variable, which is labeled "test," is a dichotomous variable in which the values of 0 and 1 are assigned to the students who do not pass the test and those who do pass the test, respectively.
2. The predictor variable labeled "program" is a dichotomous variable in which the values of 0 and 1 are assigned to the students instructed with Program A and Program B, respectively.
3. The continuous predictor variable labeled "assessment" is a continuous variable. These values measure the students' pretreatment math abilities.

The data corresponding to these variables are contained in the SPSS data file entitled "SPSS Data for a Dichotomous Criterion Variable," which is located in the internet site for this text (see Appendix A). To conduct the analyses presented in this section access this data file.

The research question posed is as follows:

Are the students who complete Program B more likely to pass a math proficiency test than students who complete Program A over and above their pre-treatment math ability scores?

The directional Research Hypothesis and associated Statistical Hypothesis constructed to reflect this directional research question are as follows:

ARH 11.2: For the population, students who complete Program B are more likely to pass a math proficiency test than students who complete Program A over and above their pre-treatment math ability scores.

ASH 11.2: For the population, students who complete Program B are not more likely to pass a math proficiency test than students who complete Program A over and above their pre-treatment math ability scores.

Full and Restricted Models

The Full Model that reflects ARH 11.2 is as follows:

$$Test = a_0U + a_1Program + a_2 Assessment + E_1 \qquad \text{(Model 11.3)}$$

ASH 11.2 requires the probability values of the two groups (i.e., the Program A and B groups) to be equal, and since the coefficient for the program variable (a_1) estimates the difference between the probabilities of passing the math proficiency test when students are taught with Program B rather than Program A over and above the influence of their pre-treatment math ability scores, which are contained in the vector for the assessment variable, the restriction can be expressed as follows:

$$a_1 = 0 \qquad \text{(1 restriction)}$$

Placing this restriction on the Full Model (Model 11.3) results in the following Restricted Model:

$$Test = a_0U + a_2 Assessment + E_2 \qquad \text{(Model 11.4)}$$

The difference between the R^2 values of the Full and Restricted Models is statistically tested with an F test. Since the ASH 11.2 is directional, two conditions must be met before the researchers can reject it. First, the estimated probability of passing the math proficiency test must be higher for the students in the Program B group than for the students in the Program A group. This condition requires the coefficient for the Program variable in the Full Model (a_1) to be positive (refer to the previous section for a discussion regarding the reason why the program variable coefficient must be positive). Second, the one-tailed F test probability value must be less than the established alpha level.

Analysis of ARH 11.2 with SPSS for Windows

Once the data contained in the file entitled "DICHOTOMOUS CRITERION DATA" are accessed from the internet site (see Appendix A) and the commands needed to access the SPSS for Windows Linear Regression menu (see chapter 4) have been completed, the Linear Regression menu presented in Exhibit 11.2 will be displayed.

Exhibit 11.2.
SPSS Linear Regression menu—Restricted Model (Model 11.4) for ASH 11.2.

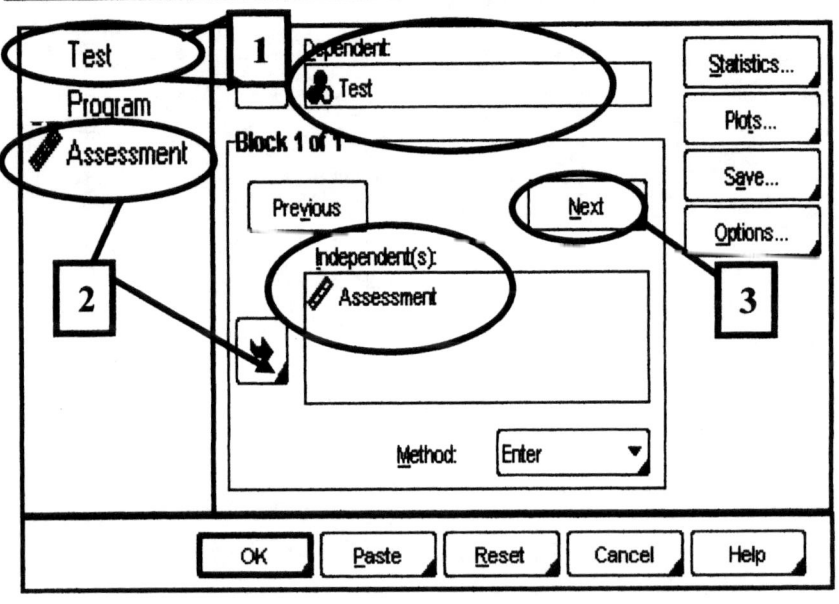

The criterion variable (Test) and the predictor variable (Assessment) contained in the Restricted Model (Model 11.4) for ARH 11.1 must be entered into the Linear Regression menu. As illustrated in Exhibit 11.2, these variables are entered as follows:

1. Click on the test variable (see Oval 1) and click on the arrow key next to the "Dependent" box. This will identify test as the criterion variable.
2. Click on the assessment variable (see Oval 2) and click on the arrow key next to the "Independent(s)" box. This will identify the assessment variable as the predictor variable in the Restricted Model.
3. Click on the "Next" key (see Oval 3).

Once these three steps are completed, the Linear Regression menu contained in Exhibit 11.3 will be displayed. To obtain the Full Model (Model 11.3) used to test ASH 11.2, the following steps are completed in this menu:

1. Click on the program variable (see Oval 1) and click on the arrow key next to the "Independent(s)" box. This will include the program variable along with the assessment variable as the predictor variables in the Full Model.
2. Click on the "Statistics" key (see Oval 2).

Exhibit 11.3.
SPSS Linear Regression menu—Full Model (Model 11.3) for ARH 11.2.

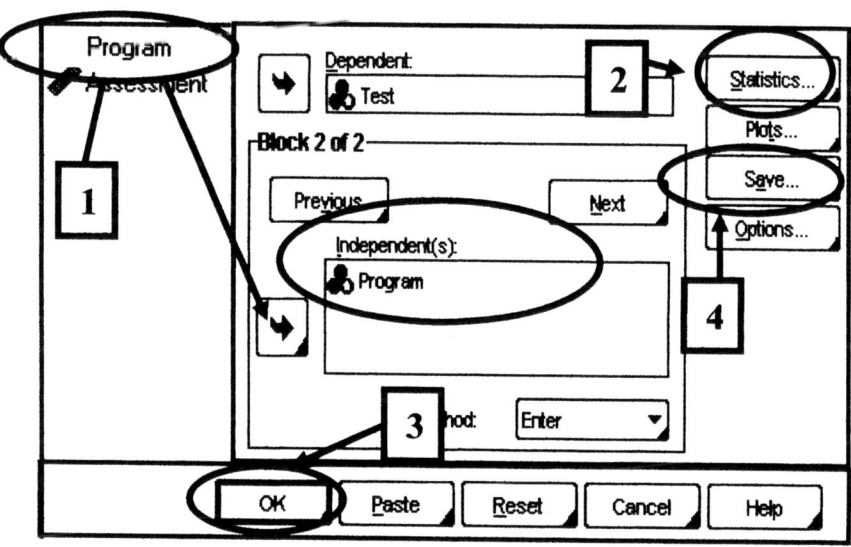

After these two steps are completed, the Statistics menu in Exhibit 11.4 will be displayed, and the following steps are completed in this menu:

1. Click on the box in front of "R square change," Model fit," and "Estimate" (see Oval 1).
2. Click on the "Continue" key (see Oval 2).

Upon completion of these steps, the Linear Regression menu will once again be displayed as shown in Exhibit 11.3. Click on the "OK" button in this menu (see Oval 3 in Exhibit 11.3). Once this task is completed the output window for the Full Model (Model 11.3) and the Restricted Model (11.4) will be displayed, as illustrated in Exhibit 11.5.

The key pieces of information contained in Exhibit 11.5 are as follows:

1. The output identifies Restricted Model (Model 11.4) and the Full Model (Model 11.3) as Model 1 and Model 2, respectively (see Oval 1).
2. The R^2 values for the Restricted Model (Model 11.4) and the Full Model (Model 11.3) are .069 and .172, respectively (see Oval 2).
3. The difference between the R^2 values of the Restricted Model (Model 11.4) and the Full Model (Model 11.3) is .103 (see Oval 3).
4. The F test of the difference in the R^2 values of the two models is 14.504 (see Oval 4). The probability of F test, which is listed as .000 to three decimal places, is less than .001(see Oval 4). Since the ARH 11.2 is di-

rectional the two-tailed probability value is divided by 2 to obtain the one-tailed probability value, which is of course is also less than .001.
5. The regression coefficient for the Program variable is .320 (see Oval 5).

Exhibit 11.4. SPSS Statistics menu for ASH 11.2 and ARH 11.2.

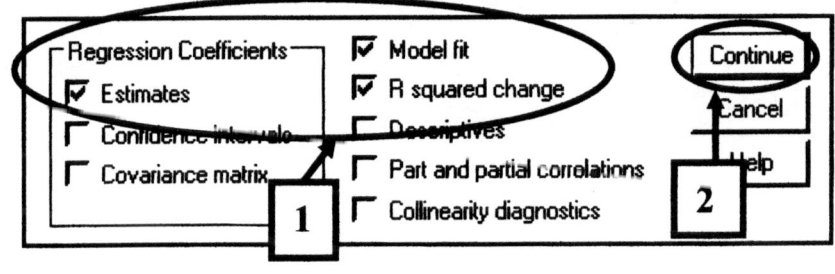

Exhibit 11.5.
SPSS regression output—Full Model (Model 11.3) and the Restricted Model (Model 11.4) for ASH 11.2 and ARH 11.2.

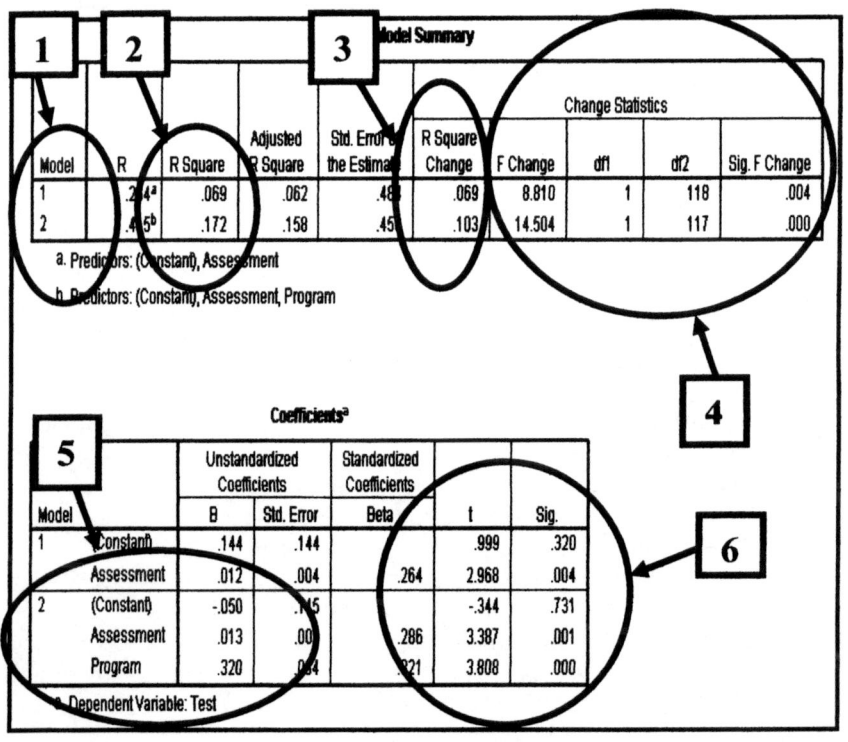

Since the regression coefficient for the program variable (.320) is positive and the one-tailed probability value ($p / 2 < .001$) of the F test is less than the alpha level of .05, the statistical hypothesis ASH 11.2 is rejected. Thus analysis supports the claim that the students who complete Program B are more likely to pass a math proficiency test than students who complete Program A over and above their pre-treatment math ability scores.

Figure 11.2 contains a plot of the Full Model (Model 11.3). The plot contains two regression lines, one for each group (i.e., Program A and Program B).

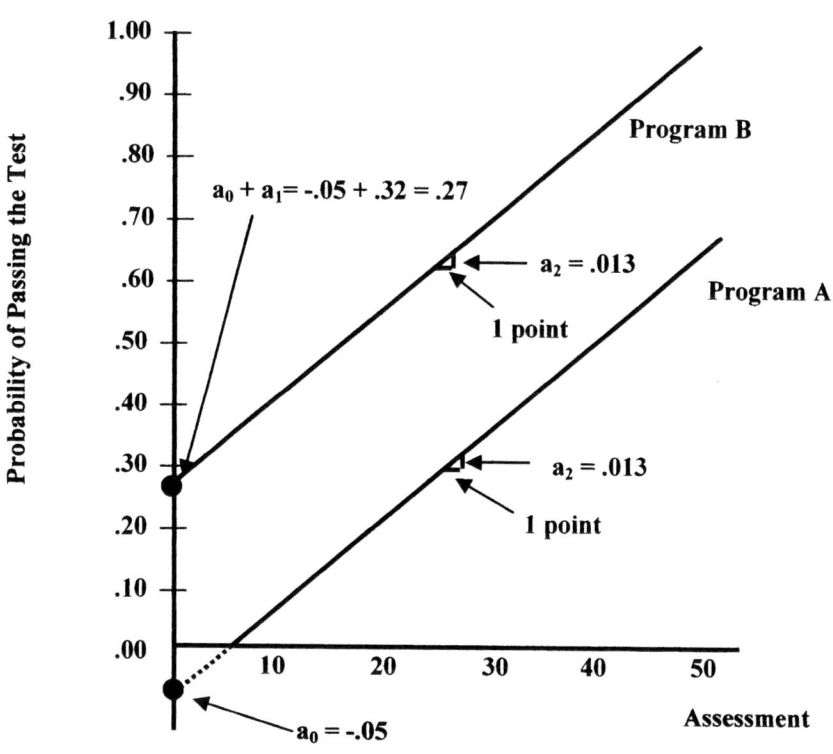

Figure 11.2.
A plot of Model 11.3 ($Y = a_0 + a_1\text{Program} + a_2\text{Assessment} + E_1$).

It is important to note that this Full Model requires the slopes of the two regression lines to be equal. The slope of each regression line is equal to the coefficient for the assessment variable ($a_2 = .013$). Since The Full Model (Model 11.3) used by the researchers does not allow the slopes of the two regression lines to differ, they are assuming that the pre-treatment math ability assessment scores do not interact with the instructional programs. The researchers could include, however, such an interaction effect, and we encourage researchers to consider and test for such interaction effects. The procedures that researchers would use to test for interaction effects when the criterion variables are dichotomous variables are exactly the same as those presented in the sections entitled "General Research Hypothesis 7.2" and "Applied Research Hypothesis 7.2" in chapter 7.

As indicated in Figure 11.2, the Y intercept for the Program A regression line is equal to the coefficient for the unit vector ($a_0 = -.05$), while the Y intercept for the Program B regression line (.27) is equal to the sum of the coefficient for the unit vector ($a_0 = -.05$) and the coefficient for the Program B variable ($a_1 = .32$). Since the two regression lines are assumed to be parallel (i.e., the distance between the lines is constant), the probability of passing the test is .32 (the coefficient value for the program variable) higher for students who received Program B than Program A for any given pre-treatment assessment score.

The t Test of the Program Variable Coefficient

Note that the statistical test of the program variable coefficient required only one restriction to be placed on the Full Model (Model 11.3). Thus, the test of the program variable coefficient in Model 11.3 could be conducted with the coefficient's t test and its corresponding probability value. If the researchers decided to use the t test of the program variable coefficient to statistically test ARH 11.2, the output for the Full Model (Model 11.3) would be used.

The t test of the program coefficient, which is listed in Oval 6 in Exhibit 11.5, is 3.808, and its corresponding two-tailed probability value is less than .001. Since ARH 11.2 was a directional Research Hypothesis, this two-tailed probability is divided by two to produce the appropriate one-tailed probability, which is also less than .001. Since the sign of the coefficient for the program variable is positive (see Oval 5 of Exhibit 11.5) and its one-tailed probability level is less than the alpha level of .05, the condition stated in ARH 11.2 is satisfied. Thus, the researchers would be willing to state that the students who complete Program B are more likely to pass a math proficiency test than students who complete Program A over and above their pre-treatment math ability scores. Again, note that the t test of the program variable coefficient and the F test of the difference between the R^2 values of the Full and Restricted Models (i.e., Model 11.3 and Model 11.4) produce exactly the same probability value and hence the same conclusion. Indeed, with 1 df in the numerator of the F test, t^2 is equal to F. In this application, the t value squared is very close to the reported F value ($3.808^2 = 14.501$—the reported F was 14.504), the discrepancy occurring because of rounding differences.

Prediction and Classification

Sometimes researchers may be interested in assessing the Full Model's ability to accurately classify the students regarding whether they will fail or pass the test. To obtain such an assessment of Model 11.3, follow the steps described in Exhibit 11.3 except request the "Save" menu by clicking on the "Save" key (see Oval 4 in Exhibit 11.3) before clicking on the "OK" key (see Oval 4).

Once the Save menu is displayed as illustrated in Exhibit 11.6, complete the following two steps:

1. Click on the box in front of "Unstandardized" located under the heading "Predicted Values" (see Oval 1).
2. Click on the "Continue" key (see Oval 2).

After these two steps are completed, the Linear Regression menu will reappear (see Exhibit 11.3). Click on the "OK" key (see Oval 3). Return to the data file and the predicted probability values for each student will be displayed in this data file (see Oval 1 in Exhibit 11.7). Note that only the first 20 of the 120 predicted probability values are listed in Exhibit 11.7.

Exhibit 11.6.
SPSS Save menu.

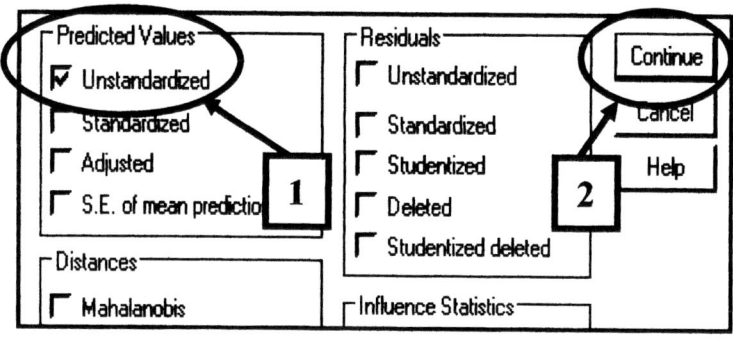

To obtain the degree of accuracy of these predicted probability values, each predicted probability value must be placed in either the fail category or the pass category by using the Compute Variable menu. To access this menu click on the following:

> Transform
> Compute Variable

The Compute Variable menu will be displayed as presented in Exhibit 11.8.

Exhibit 11.7.
SPSS data file containing the predicted probability values.

Test	Program	Assessment	PRE_1
0	0	43	.50920
0	0	24	.26213
0	0	15	.14506
0	0	36	.41821
0	0	38	.31415
0	0	48	.57429
0	1	20	.52961
0	1	47	.88080
1	0	28	.31415
1	0	42	.49625
1	0	30	.34017
1	0	55	.66534
1	1	48	.89381
1	1	31	.67269
1	1	22	.55562
0	1	7	.49059
1	1	44	.78975
1	1	33	.69870
1	1	45	.85478
1	1	22	.55562

Complete the following steps in the "Compute Variable" menu to form a variable labeled "Predicted_Category," which classifies each student as a member of the "fail" category (predicted probability is less than or equal to .50) or a member of the "pass" category (predicted probability is greater than .50):

1. Type the variable name "Predicted_Category" in the box located below the title "Target Variable" (see Oval 1).
2. In the box located below the title "Numeric Expression" type the number 0. This will assign a value of 0 to the students who meet the condition stipulated in the next step (see Oval 2).
3. Click on the "If" key (see Oval 3).

Once these steps are completed the Compute Variables: If Cases menu will be displayed as presented in Exhibit 11.9.

Complete the following steps in the Compute Variables: If Cases menu:

1. Click on the circle located in front of the line entitled "Include if case satisfies condition:" (see Oval 1).
2. Click on the pre_1 variable and move it to the right-hand box (Oval 2).
3. To the right of the pre_1 variable type "< =.50" (see Oval 3).
4. Click on the "Continue" key (see Oval 4).

Upon completion of these steps the Compute Variable menu will reappear (see Exhibit 11.8); click on the "OK" key (see Oval 4). Once this task is completed any student with a predicted probability value of less than or equal to .50 will be assigned a value of 0 in the predicted_category variable, which indicates the student is predicted to fail the test.

Exhibit 11.8.
The SPSS Compute Variable menu used to form the fail-pass categories for the predicted probability values.

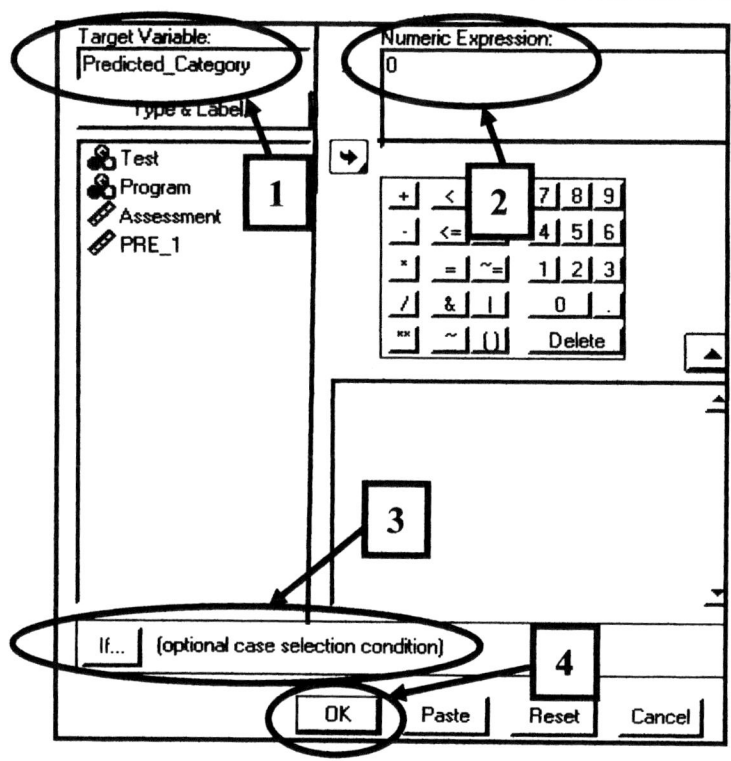

To assign a value of 1 to each student who is predicted to pass, which is a student with predicted probability value greater than .50, the same steps used to assign the value of 0 to each student who is predicted to fail the test are completed with the following exceptions:

1. In the box located below the title "Numeric Expression" located in the "Compute Variable" menu type the value of 1 rather than the value of 0 as previously done in Oval 2 of Exhibit 11.8.
2. Type ">" in the right-hand box located in the "Compute Variables: If Cases" menu following the variable name of "pre_1" rather than "< =" as previously done in Oval 3 of Exhibit 11.9.
3. Click on the "Continue" key in the "Compute Variables: If Cases" menu (Oval 4 in Exhibit 11.9).
4. This will place you back in the "Compute Variables" menu. Click on the "OK" key (Oval 4 in Exhibit 11.8).

Exhibit 11.9.
The SPSS Compute Variables: If Cases menu used to form the fail-pass categories for the predicted probability values.

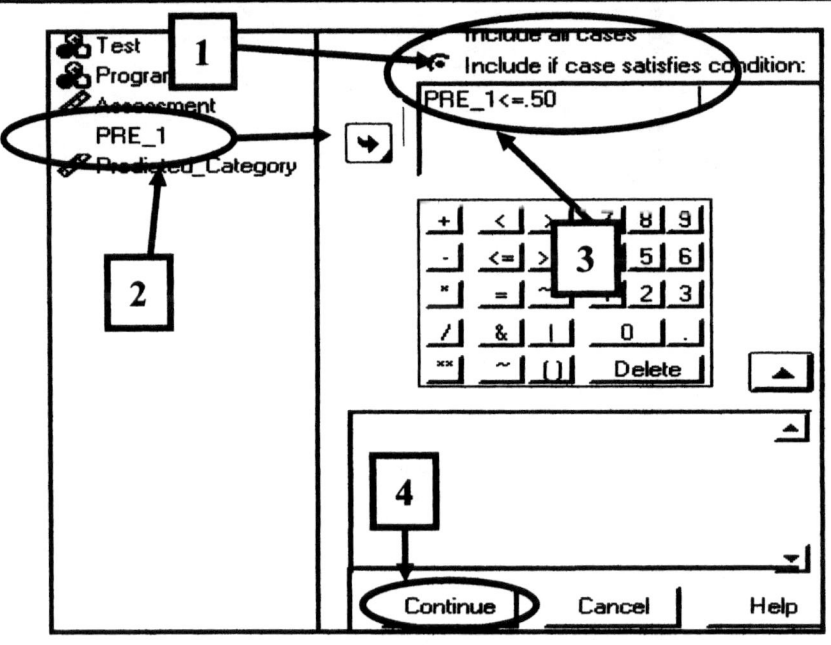

After the 0 and 1 values have been assigned to the predicted_category variable, access the Crosstabs menu by clicking on the following commands:

>   Analyze
>   Descriptive Statistics
>   Crosstabs

The Crosstabs menu in Exhibit 11.10 will appear. Complete the following steps
in this Crosstabs menu:
1. Click on the test variable and move it to the box located under the heading "Row(s)" (see Ovals 1).
2. Click on the "Predicted_Category" variable and move it to the box located under the heading "Column(s)" (see Ovals 2).
3. Click on the "OK" key (see Oval 3).
Once these steps are completed the crosstabulation output will be displayed as
presented in Exhibit 11.11.

Exhibit 11.10.
The SPSS Crosstabs menu.

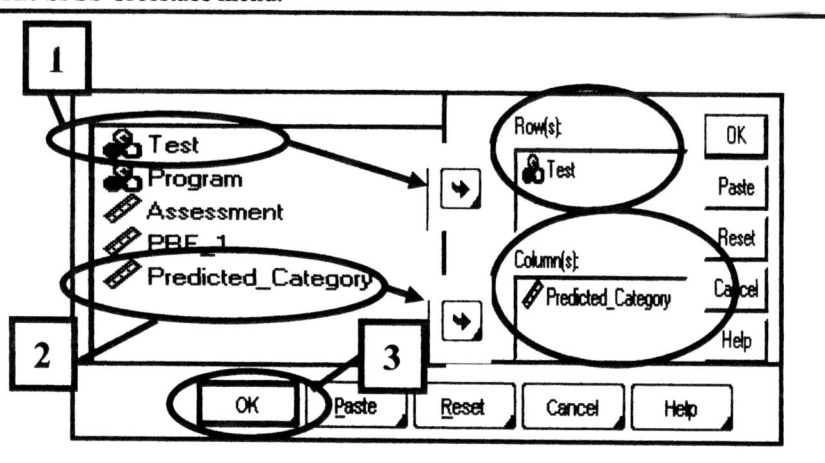

Exhibit 11.11.
The SPSS Crosstabs output.

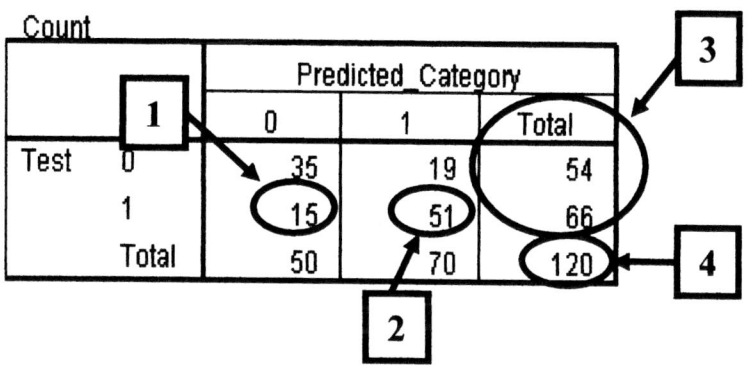

A review of the output window contained in Exhibit 11.11 reveals that 35 students who failed the test were classified as failing the test (see Oval 1). The number of students who passed the test and were classified as passing the test was 51 (see Oval 2). Thus, the sum of these two cells, which is 86 (i.e., 35 + 51), is the number of students correctly classified by the Full Model (Model 11.7). Dividing 86 by the total number of students (120), which is equal to .72, gives the proportion of students correctly classified by the model. This value is often reported by researchers as a measure of the model's adequacy.

Two points need to be mentioned regarding this measure of a model's adequacy. First, it is a better practice to use a "holdout" group when calculating the proportion of correct classifications. A holdout group is a randomly selected subset of the sample, which usually is equal to approximately 25% of the total sample not included in the estimation of the model's parameters. Once the model's parameters are estimated, which are the coefficient values for the unit vector and the predictor variables, their values are used to estimate the predicted probability of those persons in the holdout sample. These estimated probability values are used to place the holdout persons into one or the other dependent variable categories (i.e., the fail and pass categories). The proportion of the persons correctly classified by the model is then used to assess the model's adequacy with respect to classification accuracy.

Researchers tend to prefer the calculation of the model's proportion of correctly classified persons to be calculated in this manner. Why? If the persons who are used to estimate the model's parameters are also used to determine the proportion of persons correctly classified by the model, the value will tend to overestimate the accuracy of the model.

Second, the proportion of persons correctly classified by the model, regardless of whether or not a holdout group is used, can be compared to various criteria. The issue is: What is an acceptable level of classification accuracy? Is 60%, 70%, or 80% adequate or is our proportion of correctly classified persons no better than one would expect by chance? Some researchers compare the proportion of persons correctly classified to one or more criteria to judge whether the values generated by their model is adequate.

Two such criteria are known as the maximum chance criteria and the proportional chance criteria. We refer readers to Hair, Black, Babin, Anderson, and Tatham (2006) for a discussion of these criteria and their use. Hair et al. also discusses a Press' Q Statistic, which is a statistical test used to determine if the model's ability to classify is better than one would expect by chance. Readers should be aware, however, that this statistical test is sensitive to the sample size.

The issue of judging a model's level of classification accuracy can also be addressed by calculating and reporting a value known as the adjusted proportion of correctly classified persons (Long, 1997). When expressed as a percentage, this adjusted proportion value will indicate the percentage reduction in prediction errors due to the model. The calculation of this adjusted proportion value is based on the classification table contained in Exhibit 11.11, which was produced by completing the steps previously discussed in this section. The adjusted proportion correctly classified is calculated by using the following formula:

$$\text{Adjusted Proportion} = \frac{(\text{count in cell A} + \text{count in cell D}) - \text{maximum row count}}{\text{total count} - \text{maximum row count}}$$

(Equation 11.1)

where:

1. The count in cell A is the number of students who failed the test and were classified by the model as failing, which is equal to 35 (see Oval 1).
2. The count in cell D is the number of students who passed the test and were classified by the model as passing, which is equal to 51 (see Oval 2).
3. The maximum row count is the larger of the two row numbers (i.e., 54 and 66), which is 66 (see Oval 3).
4. The total count is the total number of students, which is equal to 120 (see Oval 4).

Substituting these values into Equation 11.1 produces the following Adjusted Proportion of correctly classified students:

$$\text{Adjusted Proportion} = \frac{(35 + 51) - 66}{120 - 66} = .37$$

The value of .37 indicates that the model reduces the errors in prediction by 37%.

Writing the Results for ARH 11.2 in APA Style

The results for ARH 11.2 could be written in APA style as follows:

The Full Model, which contained the program and assessment predictor variables, had an R^2 value of .17 and it reduced prediction errors by 37%. The R^2 value for the Restricted Model, which contained only the assessment predictor variable, was .07. The difference between the R^2 values of the Full and Restricted Models ($\Delta R^2 = .10$), was statistically significant at the one-tailed alpha level of .05, $F(1, 117) = 14.50$, $p / 2 < .01$. Since the coefficient for the program variable was restricted to equal zero in the Restricted Model, this significant F test indicates that the coefficient for the program variable is statistically significant. The program coefficient value of .32 indicates instruction with Program B is associated with an increase of .32 in the probability that a given student will pass the math proficiency test, holding constant the student's pre-treatment math ability score.

THE USE OF GLM WITH A
DICHOTOMOUS CRITERION VARIABLE:
ADVANTAGES AND POTENTIAL PROBLEMS

The GLM approach to hypothesis testing of a dichotomous criterion variable has a number of advantages over other statistical approaches, including the
following:

1. The GLM approach forces the researcher to state the Research Hypothesis. Unfortunately, this is not always done in chi-square and phi coefficient analyses.
2. The GLM approach is easily generalized to all other least squares hypotheses. Separate computing formulae and different rules for calculating
   degrees of freedom are not necessary.
3. The stating and testing of directional hypotheses is encouraged by the
   GLM approach, whereas directional chi-square analyses and directional
   phi coefficient analyses are at best mentioned only briefly in statistics
   books.
4. The GLM approach considers the number of cases and the number of
   categories in the predictor variable in the calculation of the denominator
   degrees of freedom. Since the chi-square test of significance assumes
   many subjects, the probability statement is more exact when it is calculated from the F test that results from the GLM approach.

It should be noted, however, that the application of GLM to a dichotomous
criterion variable poses three potential problems. The first potential problem is
related to the issue of the type of information the technique provides. The other
two potential problems are related to underlying assumptions of the technique.

First, a coefficient for a given predictor variable, which indicates the change
in the conditional probability of being classified in the group assigned the value
of one in the criterion variable for a one-unit change in the predictor variable, is
linear and unaffected by the initial conditional probability value. Thus, a one-
unit change in the predictor variable will produce the same change in the predicted probability when the initial probability is .90 as when it is .50. This linear
feature of the predictor coefficient may cause a predicted value to be less than 0
or greater than 1. As noted by Austin, Yaffee, and Hinkle (1994), predicted values that are located below 0 and above 1 are illogical and not interpretable.

Second, the assumption of normality of the error term in a GLM is not tenable for a model in which the criterion variable is a dichotomous variable, because the error term for a given set of predictor variables can take on only two
values. As noted by Gujarati (1988) "although OLS (a model estimated by the
ordinary least squares method) does not require the disturbances (error term values) to be normally distributed, we assumed them to be so distributed for the
purpose of statistical inference, that is, hypothesis testing" (p. 469).

Third, Gujarati (1988) demonstrated that variance of the error term values is
heteroscedastic. Although this condition does not result in biased OLS estimates,

such estimates are inefficient. Thus the validity of the statistical tests conducted on the OLS coefficients is questionable.

With respect to the first potential problem, we believe that although such predicted probability values (i.e., ones that are less than 0 or greater than 1) may make some researchers uncomfortable, they may not affect the predictability of the model with respect to its classification accuracy. Regarding the second and third potential problems, we agree with Wooldride (2008) who stated: "It turns out that, in many applications, the usual OLS [ordinary least squares] statistics are not far off, and it is still acceptable in applied work to present a standard OLS analysis of a linear probability model" (p. 250).

CHAPTER 12

THE STRATEGY OF RESEARCH AS VIEWED FROM THE GLM APPROACH

We discuss the General Linear Model in this chapter as it relates to several goals of research: predictability, parsimony, replication, and validity generalization. These goals are presented with the development of a well-established physical law. Our emphasis is upon the percentage of variance that can be accounted for in the criterion under investigation rather than on statistical significance from random events. Additional remarks concerning curvilinear relationships and data snooping are also presented.

MEETING THE GOALS OF RESEARCH WITH THE GLM

The material that follows in this section represents one way which Sir Isaac Newton might have developed the law of gravity. The material is adapted from McNeil (1970a) and is presented with the intent of showing the advantage of using the GLM technique as one's statistical tool. Think of it as a narrative interview with Sir Isaac Newton.

It seems that for years some of our most competent physicists have been looking for the functional relationship between the amount of time an object has been falling in space and the distance the object has fallen. Stating the problem symbolically, $d = f(t)$, and verbally, "Distance is what function of time?" I believe that I have finally discovered this functional relationship between time and distance. This means that, if I know the amount of time an object has been falling, I can tell you how far it has fallen.

One of the surprising findings that my research strategy has led me to is that I need know *nothing* about the object, if I can assume that the object is falling in a vacuum, where there is no resistance to its fall. The only additional variable that I need to know is what I call the "gravitational constant"—computed from the forces being exerted by the earth, the sun, the moon, and other heavenly bodies. For any one time and place, this variable has the same value for any object.

Goal 1: Predictability

To give a little background, let me review previous research. Galileo (1632/1953) published much data on falling objects, but he only investigated the linear relationship between time and distance. Galileo and, before him, Aristotle used the GLM procedure in testing their hypotheses, but these great thinkers did not realize the value and the flexibility of this procedure. They only used a very restricted form of the technique, the form that ascertains the rectilinear relationship between the two variables.

Galileo used the following model:

$$D = a_0U + b_1T + E_1 \qquad\qquad \text{(Model 12.1)}$$

where:
1. D = the vector containing the distances the objects have traveled;
2. T = the vector containing the amount of time the objects have traveled;
3. U = the Unit vector that allows the regression constant to be nonzero;
4. E_1 = the vector containing the differences between the actual distance and the predicted distance (that predicted from the pool of predictor variables on the right-hand side of the equation); and
5. a_0 and b_1 are weighting coefficients of best fit, determined from the sample data. "Best" is here defined in the least squares sense, minimizing the sum of the squared values in E_1.

Model 12.1 produced a significant fit of the data, but the R^2 value was discouragingly low ($R^2 = .40$). This R^2 value is the squared value of the correlation between the observed and predicted values and can be interpreted as the proportion of variance in the distance-measures criterion accounted for by the predictor variables in the sample (in Model 12.1, the regression constant and the continuous variable of time).

A Linear Model that Allows for a Curvilinear Relationship

Upon looking at a bivariate plot of the data in Figure 12.1, it becomes obvious that there is not simply a linear relationship—but a curvilinear relationship between time and distance. And, because the rate of acceleration of the curve continuously increases, there must be a second-degree curvilinear relationship. The GLM model that allows a second-degree relationship to exist is:

$$D = a_0U + b_1T + c_1T^2 + E_2 \qquad\qquad \text{(Model 12.2)}$$

All the variables in Model 12.1 appear in Model 12.2, with the addition of T^2. Each element of T^2 is the squared value of the corresponding value in T (thus, if the object took 4 seconds to fall the designated distance, then T^2 would be 4^2 or 16).

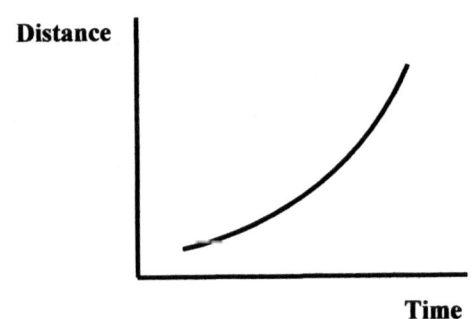

Figure 12.1. Bivariate plot of data suggesting a curvilinear relationship between time and distance.

The variable T^2 allows the second-degree curve in the data to manifest itself, if in fact there is a second-degree curve in the data. I fit Model 12.2 to the data for a 100-pound stone. What surprised me was not that I obtained a significant fit of the data but that the R^2 value was .99. Distance could be almost exactly predicted from knowledge of the regression constant, time, and time squared.

Goal 2: Parsimony

Parsimony means to explain something in the simplest way possible. With respect to the problem at hand, Model 12.2 incorporates three pieces of information in predicting the criterion of distance. The notion of parsimony demands that we investigate the predictability of the criterion by dropping out T, T^2, or U, or some combination of these three vectors. If T^2 is dropped out of the model (actually I am hypothesizing that the weight (c_1) associated with T^2 is equal to 0), I end with the model used by Galileo. I tested the predictability (R^2) of Galileo's model (Model 12.1) against the R^2 of Model 12.2 and found a significant decrease ($p < .01$) in the amount of variance being accounted for in Galileo's model. Model 12.2 was accounting for 99% of the variance, while Galileo's model was accounting for some 40%. I would like to add parenthetically here that many researchers had interpreted Galileo's results incorrectly. Galileo had tested the predictability of his model against one that contained no differential information:

$$D = a_0U + E_3 \qquad\qquad\qquad \text{(Model 12.3)}$$

The test showed that there was a significant decrease in predictability, and in particular, indicated that the rectilinear relationship between time and distance was something other than 0. As the R^2 of .99 in Model 12.2 indicated, a model can be built that more accurately predicts the criterion of distance. The point is that we should not rest on our haunches until we get close to accounting for 100% of the variance.

A second model more parsimonious than Model 12.2 involves hypothesizing that the weight associated with T is equal to 0. Again, this reduces to dropping that particular variable out of the pool of predictor variables:

$$D = a_0U + c_1T^2 + E_4 \qquad \text{(Model 12.4)}$$

Model 12.4 does not contain the time variable at all, except in its squared form, and the surprising thing is that Model 12.4 was as predictive as the original Full Model, Model 12.2. In fact, the R^2 of Model 12.4 was also .99, exactly equal to that of Model 12.2. What was also true, but fully anticipated, was that the weight for the Unit vector was equal to 0, (i.e., the Y intercept is 0). This, in effect, substantiates our impression that the distance covered by a falling stone in 0 minutes is exactly 0 feet. The final values for the more parsimonious model are $a_0 = 0$ and $c_1 = 16$.

$$D = (0 * U) + (16 * T^2) \qquad \text{(Model 12.4)}$$

The goal of parsimony has been satisfied in that the criterion has been predicted with a small amount of predictor information: Here only one bit of predictor information is needed, that of the square of time.

Goal 3: Replicability

As indicated previously, the regression weights were optimum weights for those particular sample data. I am concerned, of course, about predicting the distance measure for not just this one sample of stones but for any stone from a clearly defined population of stones. My first concern was to check the replicability of the obtained regression weights for another set of data on 100-pound stones.

One process of checking replicability simply involves taking the obtained regression weights (from Model 12.4) and applying them to the new data. This process produces predicted distances, and the predicted distances can be then correlated with the actual observed distances. In this instance, every new time score was simply squared and then correlated with the associated distance score. The resulting (Pearson Product Moment) correlation was .98, yielding an R^2 of .96, which again is quite satisfactory. The regression weights found for Model 12.4 were checked for replicability on several other samples of data with 100-pound stones, and all R^2 values were above .95.

Goal 4: Validity Generalization

There are several aspects to the replication problem. I just discussed how I replicated the functional relationship in a single population. My students were interested in generalizing this functional relationship to other populations particularly populations of lighter stones. I discovered that the stones' weights were unimportant. Here is how I obtained evidence to make this statement. I wondered about the generalizability of the results on the 100-pound stones to lighter stones. I obtained data on 50-pound stones and proceeded to apply the regression weights from the 100-pound stones in Model 12.4. To my surprise, the R^2 resulting from this replicability study was comparable to the aforementioned replicability studies. That is, the functional relationship between time and distance was the same for both 100-pound stones and 50-pound stones.

It was not a great intellectual leap to investigate the possibility that the functional relationship was similar for stones of all weights. To check this hypothesis, I collected distance and time data on stones of various weights and then added pound measures to the predictor pool of variables in Model 12.4:

$$D = a_0U + c_1T^2 + h_1P + E_5 \qquad\qquad \text{(Model 12.5)}$$

where P is the weights in pounds of the stone being measured.

Of course, to make this model viable, I had to measure objects of various weights. When Model 12.5 was applied to a set of data on stones of differing pounds, the regression weight h_1 turned out to have a numerical value of 0. Thus, when the predictive efficiency of Model 12.5 was compared against the predictive efficiency of Model 12.4, no predictive information was lost. The hypothesis—that the functional relationship between time and distance was similar for objects of any weight (more specifically, for the range of weight values I used in the analysis)—was supported. To put it another way, weight is not needed in defining the functional relationship between distance and time. The functional relationship has been generalized across all weights.

A Return to the Gravitational Constant

I still have not discussed the gravitational constant. This is the most difficult discussion of all, but it should be included because it increases the generalizability of my findings and also illustrates the flexibility of the GLM approach.

All the data that have been presented so far were obtained in my experimental labs. But objects move in places other than my lab. In particular, it had been known for quite some time that objects move faster when they are at lower altitudes. Also, the functional relationship that I found did not replicate on observations made by Galileo in outer space. Therefore, the functional relationship between time and distance may be modified by the gravitational field in which the object is measured; that is, various observations provided clues about what kinds of variables might be important to investigate.

I shall not go into the actual calculation of this gravitational constant, except to indicate that it can be quite accurately measured. The gravitational constant will be symbolized as G. I subjected data that included objects measured within various gravitational fields to Model 12.6 (an extension of Model 12.4) to check the functional relationship within the various gravitational fields:

$$D = a_0 U + c_1 T^2 + i_1 G + E_6 \qquad \text{(Model 12.6)}$$

Model 12.6 yielded an R^2 value of .50, indicating that 50% of the criterion variance was predicted by the predictor set. To be sure, this is a large amount but not as high as I had been accustomed to. I was careful not to make the same mistake that Galileo had made years before, that of stopping with an extremely parsimonious model that does only a fair job of accounting for the criterion variance. Generally, one reduces the amount of predictability when the variable pool is reduced. Conversely, the inclusion of another predictor variable will generally increase the predictability, the question being: Is the increase in predictability significant?

Curvilinear Interaction

The bivariate plot (Figure 12.2) between time and distance for the various G levels was visually inspected, and it became obvious that there was likely an interaction between the gravitational constant (G) and the square of time (T^2) (Figure 12.3). A model was constructed that allowed this hypothesized interaction to exist, the vector $(G * T^2)$ represents a single number that is the product of the gravitational field and the square of time:

$$D = a_0 U + c_1 T^2 + i_1 G + j_1 (G* T^2) + E_7 \qquad \text{(Model 12.7)}$$

In accordance with the anticipated results, R^2 was .99, indicating that Model 12.7 was indeed a good reflection of the functional relationship. I carefully noted that a_0, c_1, and i_1 all had numerical values extremely close to 0, indicating that the variables associated with these regression weights were all somewhat useless in obtaining the high degree of predictability. Thus the hypothesis that the parameter value of these weights is, in fact, equal to 0 was tested. By setting a_0, c_1, and i_1 in Model 12.7 equal to 0, the following model is derived:

$$D = j_1 (G * T^2) + E_8 \qquad \text{(Model 12.8)}$$

When the R^2 of Model 12.8 was tested against the R^2 of Model 12.7, a significant decrease in the amount of variance being accounted for was *not* found.

In fact, the R^2 of Model 12.8 was .98, well within the decrease expected from the measurement errors in our data.

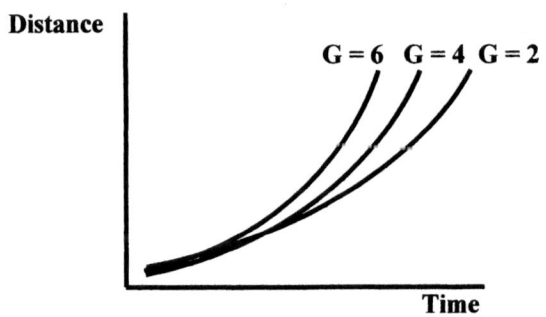

Figure 12.2.
Bivariate plot of data indicating a curvilinear interaction between time and gravity (G) in the prediction of distance.

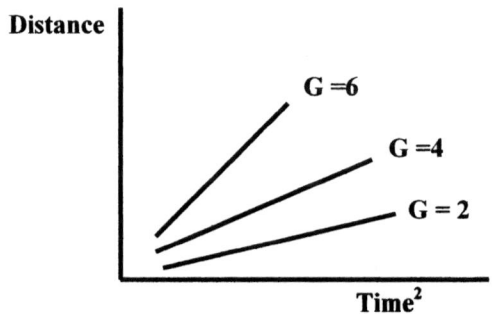

Figure 12.3.
Bivariate plot of data indicating a curvilinear interaction between time squared and gravity (G) in the prediction of distance.

The goal of predictability was met: The interaction model predicted a high percentage of the criterion variance, although a good linear relationship did not exist in the prediction of distance. The "simplified" interaction model (Model 12.8) predicted just as satisfactorily as Model 12.7; therefore, the simplified interaction model was accepted as a better model (because it simultaneously met the goals of predictability and parsimony).

After running successful replications on other sample data, Model 12.8 was deemed to be "lawful" for the phenomenon being studied (because it met the goals of predictability, parsimony, replicability, and to some extent validity generalization). The functional relationship between time and distance (for stones of various weights dropped from various altitudes) is thus that function indicated by Model 12.8.

From my GLM analyses of Model 12.8, I note that the numerical value of j is .5. Verbally then, D is equal to one-half the product of G times T^2; symbolically, $D = 1/2(G * T^2)$. If we call distance d, the gravitational constant g, and time t, we have the following symbolic equation: $d = 1/2g\ t^2$. Because it always yields an R^2 of close to 1.00 and I found it, I will call it Newton's Law.

Our Remarks

After reading Newton's discussion, we hope that behavioral science researchers will pay attention to the methodology that he presented. It seems inappropriate for these researchers to argue continually that the behaviors they look at are inherently complex but then to use only simplex models; the behaviors may indeed be complex, but how many researchers have even attempted to investigate curvilinear interaction as Newton did? It is time for us to realize that complex behavior demands complex methodology, as well as complex theory.

There is a problem about what variables one should choose, but that is a content question that cannot be answered by a methodologist. Transformations and intricate relationships should be sought rather than discarded. Interaction has traditionally been viewed as a bad effect rather than as an effect that will help in explaining the phenomenon at hand.

The notion of a curvilinear relationship is not new to methodology, but we must admit that few investigators make a habit of checking the curvilinear predictability. Indeed, some statistical books discourage the inclusion of highly correlated predictor vectors in regression models (all polynomials are highly correlated with the original vector). Whether a researcher should include a vector that allows for curvilinearity is partly a function of past research findings and partly a function of the nature of the data. Though the addition of another predictor variable cannot decrease the R^2, it is not the case that such a vector will always statistically increase the R^2 value. Whether such investigations prove rewarding is an empirical question.

Inspecting the sample data is considered a poor research strategy by many researchers. But if the goal of research is to develop scientific findings that can be generalized, then it can be argued that a researcher is essentially committing a crime when one does not "squeeze" the data for all they are worth. When such squeezing is done, though, the researcher is committed to replicating findings on an independent sample. The stability, and therefore value, of the finding must rest on the replicated study rather than on the original study alone.

THE MOST GENERAL REGRESSION MODEL

The hypotheses and their associated models presented since Chapter 1 have been of a rather restricted nature when compared with the research notions of Chapter 1. In that first chapter, we argued that behavior (whether of humans, university administrations, plots of land, etc.) is complex and that many predictor variables may be needed to account for that behavior. In accounting for such behavior, we have shown that categorical information could be used (Chapter 4), that continuous variables could be used (Chapters 5 and 6), and that transformations of originally scaled variables should be considered (Chapter 9). We must emphasize that the data and inherent functional relationships determine the variables necessary to obtain the desired R^2. Both dichotomous and continuous variables can be used in the same regression model (as discussed in Chapters 7 and 8).

Generalized Research Hypotheses were presented in the previous chapters in part to show that traditional least squares solutions are all computational simplifications for the general least squares procedure. The regression formulation was shown for such traditional solutions as the Pearson Product Moment correlation (Generalized Research Hypothesis 5.1); the t test of the point biserial correlation equals 0 (also called the t test for the difference between two independent means, as in Generalized Research Hypothesis 4.1); the t test for the difference between two dependent means (Models 10.23 and 12.24); the t test for a single population mean (Generalized Research Hypothesis 10.1); the one-way ANOVA (Generalized Research Hypothesis 4.2); the two-way ANOVA (Chapter 7); and the ANCOVA (Chapter 8). Each of these statistical procedures was computed using the same least squares procedure. To aid in the calculation of these solutions by hand, simplified formulae have been developed by statisticians over the years to fit special cases; and then these solutions became grouped according to their computational similarities (correlation, t tests, F tests, etc.).

Now that the computer is available to perform calculations, statistics should be put in their conceptual place rather than their computational place. Any hypothesis can be tested by the regression approach (as long as the researcher is using the least squares approach with a single criterion). The specific Research Hypothesis dictates the Full Model and the Restricted Model and the testing of those models. Rather than spending time searching for the "appropriate" simplified statistical test that will fit the design, the researcher should spend time explicitly stating the Research Hypothesis, so that the Full Model and the Restricted Model can be generated and tested.

As a way of organizing one's approach to stating Research Hypotheses, it may be helpful to point out that all of the Research Hypotheses discussed in previous chapters may be stated as "over and above" hypotheses. This is not to suggest that stating it that way is the best way to state *any* hypothesis but to emphasize that each is a form of the most generalized hypothesis.

All hypotheses can be either phrased or rephrased into a single structure. That structure contains a criterion variable and one or more variables that are being tested and it may contain one or more covariates (The Unit vector is nearly always a covariate, but it could be a variable being tested, a covariate, or neither. This is more fully discussed later in this chapter.) The single structure for expressing all Research Hypotheses is:

> Is knowledge of variables X1, X2, . . ., Xk valuable, over and above knowledge of variables Z1, Z2, . . ., Zg, in the prediction of the criterion Y?"

One regression model is specified by the Research Hypothesis as having full knowledge of the mentioned predictor variables:

$$Y = a_1X1 + a_2X2 + \ldots + a_kXk + c_1Z1 + c_2Z2 + \ldots + c_gZg + E_9$$
(Model 12.9)

where:
1. Y = the criterion variable;
2. X1, X2, . . ., Xk = the variables being tested;
3. Z1, Z2, . . ., Zg = the over and above variables; and
4. $a_1, a_2, \ldots, a_k, c_1, c_2, \ldots, c_g$ are least squares weighting coefficients calculated so as to minimize the squared elements of the error vector $E9_g$.

The Full Model must be compared to a Restricted Model that has as predictors only the over and above information. Each k variable of interest is restricted from the Full Model by setting its weighting coefficient equal to some numerical value. There are an infinite number of numerical values that one could use. The actual restriction(s) need to be a function of past theory, empirical findings, and expectations of the researcher. To simplify the following discussion, each of the k variables of interest will be set equal to a 0 (making k restrictions). The restrictions would be:

$$a_1 = 0; a_2 = 0; \ldots; a_k = 0$$

Forcing those restrictions onto the Full Model would result in the following Restricted Model:

$$Y = c_1Z1 + c_2Z2 + \ldots + c_gZg + E_{10}$$
(Model 12.10)

The increase in R^2 in going from the Restricted Model to the Full Model would be due to the predictive information in the X1, X2, . . ., Xk variables that is over and above the predictive information of the Z1, Z2, . . ., Zg variables in the Restricted Model.

The Research Hypothesis also could be stated as:

Research Hypothesis: Variables X1, X2, . . ., Xk are predictive of Y, holding constant the variables of Z1, Z2, . . ., Zg

or

Research Hypothesis: Variables X1, X2, . . ., Xk are predictive of Y, while covarying the effects of variables Z1, Z2, . . ., Zg

And if the variables appearing in the Full Model but not in the Restricted Model are all mutually exclusive group membership vectors, then the Research Hypothesis could be stated as:

Research Hypothesis: The k groups are different on Y, after adjusting for differences on the variables of Z1, Z2, . . ., Zg

Because there are k variables to be assessed and g covariate variables, $m1 = k + g$. (Note that it is being assumed that all linearly dependent vectors have been omitted from the Full and Restricted Models). The number of predictor pieces of information in the Restricted Model is g, therefore, $m2 = g$. The numerator degrees of freedom value for the F test then is:

$$(m1 - m2) = ((k + g) - (g)) = k$$

The R^2 of the Full Model minus the R^2 of the Restricted Model is the proportion of criterion variance accounted for by the additional k variables. Indeed, this average increase in R^2 becomes the numerator for the F test:

$$F(m1 - m2, N - m1) = \frac{(R_F^2 - R_R^2)/(m1 - m2)}{(1 - R_F^2)/(N - m1)}$$

(Equation 12.1)

For the F test (in terms of R^2) to be computed on regression models, three requirements must be met:
1. The same criterion must appear in both models.
2. All predictor variables in the Restricted Model must appear in the Full Model, either directly or as a linear combination of the Full Model predictor variables.
3. The unit vector must appear in both models.

A quick glance at Models 12.9 and 12.10 will verify that the first two requirements are met. The same criterion appears in both models, and the Z1, Z2, . . ., Zg predictor vectors in the Restricted Model appear in the Full Model. For most applications, researchers will want to include the Unit vector in both models (as one of the over and above variables, designated by Zi) as a way of taking care of the arbitrary scaling of variables. Indeed, the Unit vector so commonly appears in both models that most computerized regression programs provide it automatically. The Unit vector has been included as a covariable so often that now it is assumed to be a covariable, unless otherwise indicated. It must be remembered, though, that the Unit vector also can be tested (i.e., be one of the test variables designated by Xi) as in the single population mean question in Generalized Research Hypothesis 10.1. Furthermore, the Unit vector need not appear in either model—the variables that appear in a given model are always a function of the question the researcher is asking. (Whenever the Unit vector is not in the Full Model or is being tested and is not in the Restricted Model, one must not use the general F test with R^2 values but use Equation 10.1 with error sum of squares instead.)

Any number of Xi variables may be investigated, and any number of Zi variables may be controlled (including none if that is desired). Furthermore, the variables may be dichotomous, or they may be continuous. The criterion variable in multiple regression is limited to only one variable, but it may be either dichotomous or continuous. Combinations of each of these options result in specific kinds of Research Hypotheses, most of which were discussed in previous chapters. To reemphasize the point made earlier, though all hypotheses can be reworded in the over and above fashion, many hypotheses make more sense stated otherwise. For instance, "For a given population, Treatment A is better than Treatment B on the criterion" seems more communicative than does "For a given population, Treatment A is better than Treatment B on the criterion, over and above the overall mean."

Where is this discussion leading? Suppose Researcher A has theorized that variables X1, X2, X3, and X4 account for the variability in the criterion of interest. An obtained R^2 of .90 is used as evidence for that statement. Researcher B, though, realizes that not all the criterion variance has not been accounted for and proposes that, in addition, variable X6 is necessary. Two researchers have posited two different states of affairs, and as one is a restricted case of the other, the argument can be tested through the regression approach. Because Researcher B is proposing a new, untried variable, and because Researcher B is using more information than Researcher A, Researcher B is obligated to have a higher R^2. The magnitude of the difference in R^2 could be agreed to before data collection, or the two researchers could agree to use a test of significance for the difference in R^2, as in Generalized Research Hypothesis 12.1. With this Generalized Research Hypothesis, the text has finally arrived at the degree of complexity that was outlined in Chapter 1.

Generalized Research Hypothesis 12.1

Assume we have the following hypotheses:

Directional Research Hypothesis: For a given population, X6 is positively predictive of the criterion Y, over and above X1, X2, X3, and X4.

Nondirectional Research Hypothesis: For a given population, X6 is predictive of the criterion Y, over and above X1, X2, X3, and X4.

Statistical Hypothesis: For a given population, X6 is not predictive of the criterion Y, over and above X1, X2, X3, and X4.

The Full Model for these hypotheses is as follows:

Full Model: $Y = a_0U + a_1X1 + a_2X2 + a_3X3 + a_4X4 + a_6X6 + E_1$
(Model 12.11)

For the directional Research Hypothesis we expect $a_6 > 0$, and the restriction placed on the restricted model is $a_6 = 0$. For the nondirectional Research Hypothesis we expect $a_6 \neq 0$, and the restriction placed on the restricted model remains:

$a_6 = 0$ (1 restriction)

The Restricted Model is as follows:

$Y = a_0U + a_1X1 + a_2X2 + a_3X3 + a_4X4 + E_2$ (Model 12.12)

where:
1. Y = the criterion;
2. X1, X2, X3, X4, X6 = continuous or categorical information; and
3. a_0, a_1, a_2, a_3, a_4, and a_6 are least squares weighting coefficients calculated to minimize the sum of the squared values in the error vectors.

The degrees of freedom are as follows:
1. Degrees of freedom numerator = (m1 - m2) = (6 - 5) = 1
2. Degrees of freedom denominator = $(N - m1) = (N - 6)$.

Note that if all predictors are continuous, m1 = 6. If some reflect categorical information, the linear dependencies must be ascertained.

THE GOAL OF CONTROL IN
THE BEHAVIORAL SCIENCES

All behavioral science research is ultimately concerned with cause and effect—which is *control*. Many researchers initially react negatively to this statement, but when hard-pressed will ultimately agree that cause and effect relationships are what they are looking for. Statisticians have traditionally indicated that cause and effect relationships cannot be ascertained from correlational studies but only from ANOVA designs applied to experimental data. Familiarity with the GLM approach makes it clear that correlation and ANOVA are both subsets of the GLM and interpretations of cause and effect can only be made on logical grounds. If the logic is not defensible, then the cause and effect relationship is not tenable.

Unfortunately, many casual interpretations are based on significance levels rather than on the amount of variance accounted for (Byrne, 1974). Results that are highly significant (due to factors other than chance) may account for, say, only 1% of the variance in the criterion. A not too unusual bivariate outcome results in a correlation of .44 and an R^2 less than .20. Of course, some phenomena are more highly accounted for than others, but few studies in the literature report results leading to an R^2 greater than .50. (It would be very interesting to force all researchers to report their R^2 values and not let them hide behind statistical significance inflated by large sample sizes.

All other factors being equal, as the sample size increases, the results will become more "significant," whereas the R^2 will simply more closely approximate the population R^2. Thus, increasing the number of subjects does not artificially inflate the R^2 value, although it does deflate the probability value. Figure 12.4 depicts these relationships.

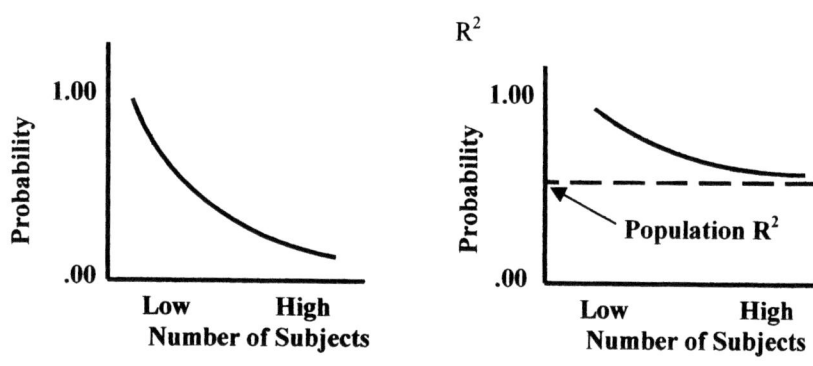

Figure 12.4.
Relationship between number of subjects, probability, and R^2.

In the past, cause and effect has been interpreted on data that, although highly significant, yielded small R^2 values. One of our positions in this text is that one should spend one's time more fruitfully by finding variables that will increase the R^2 rather than trying to understand how that small proportion of criterion variance has been predicted.

But while adding more predictor variables will increase the R^2, the question is, has a stable R^2 been obtained? Figure 12.5 contains the relationship between R^2 and the number of predictors. As illustrated, the R^2 can be increased by adding additional predictors—until there are as many predictors as there are subjects, resulting in an "overfit" model with an associated R^2 of 1.00. As Constas and Francis (1992) illustrated, with a figure similar to Figure 12.6, the interest should be on the adjusted R^2.

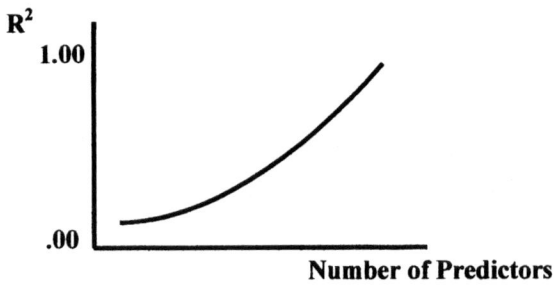

Figure 12.5.
The relationship between R^2 and the number of predictors.

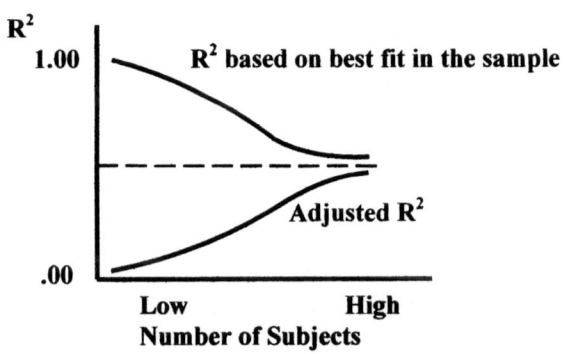

Figure 12.6.
The relationship between adjusted R^2 and R^2.

Of additional importance is the adequacy of the overlap of the criterion with the construct one wishes to consider, as discussed at the end of Chapter 9. Too often the criterion is equated with the construct, when in reality one must realize that the criterion is only an approximation of the construct, that is, there is only a partial overlap between the criterion and the construct.

If the unaccounted-for variance overlaps with the construct, then one would do well to continue trying to build up the R^2. Several consternating situations arise when the criterion only partly overlaps the construct. It may be that the criterion variance accounted for is not overlapping the construct variance. Any interpretation regarding causality in this situation would be appropriate for the criterion, but not for the construct.

On the other hand, all the common variance between the criterion and the construct may be accounted for. To attempt to build up the predictability of the criterion here would here be fruitless. Interpretation of causality may be of some value here, but one must note that complete control (bringing about desired changes or upsetting the prediction) of the construct would not be possible because of the lack of an isomorphic fit between criterion and construct. This situation calls for the selection of other criterion variables, either instead of the criterion or in addition to the criterion.

Once the R^2 has approached 1.0 and the researcher has some faith in the criterion as a good measure of the construct, then one can begin considering how to control the construct. This implies that variables must be included in the model that can be manipulated. Nonmanipulated variables that must remain in the predictive model simply indicate differences in functional relationships for various populations. That is, if gender must remain in the model, then the functional relationship is different for males than it is for females. But if all the predictor variables are ones that cannot be manipulated, then the model is simply describing a state of affairs rather than indicating some possible ways of upsetting the state of affairs.

Given a choice between greatly overlapping predictor variables, some of which are more readily manipulated than others, one would choose to include the variables that are more easily manipulated variables in the model. Some variables are easier to manipulate than are others; but, unfortunately, current statistical procedures do not allow us to incorporate these differences. These decisions must be made on subjective bases rather than on objective bases.

It may be the case that even though the R^2 is extremely high, the predictor variables may not be the causers of the criterion. The criterion and predictor variables may both be caused by some additional variable or set of variables. The technique of *path analysis* has been introduced by statisticians as a possible means of assessing the causal sequence. Those developments have not been entirely successful, but the intent is appropriate.

One real-world advantage of GLM—allowing the use of correlated predictor variables—turns out also to be one of the problems in detecting control. When one predictor variable is changed, one would expect that the values of the other (correlated) variables also might change. This is another reason for not interpreting regression weights. Indeed, the whole swarm could change as the result of changing only one variable. If the criterion is increased, then this might well be a desired goal. But when control is being considered, one usually wants to know how much effect there will be on the criterion, given a specified change in a predictor.

The GLM may give one the insight about what variable to change, but it does not generally indicate what effects to expect. If the subjects have been randomly assigned to the independent variables, then these effects are implied by the GLM approach. But these kinds of designs do not adequately reflect the real world. When subjects are sampled from the real world, the predictor variables will be correlated, and therefore the effects cannot be directly known. The effects of changing predictor variables must wait for subsequent empirical manipulation. We have already indicated that some variables are more easily manipulated than others; it also might be the case that some kinds of subjects are more easily manipulated than others. So, the assessment of control must wait for empirical verification. Ascertaining causality is always a tentative proposition. Empirical manipulation helps rule out other competing explainers.

As will be shown in the next few sections, GLM can help in the various stages of one's research. Ultimately, causers will be identified. But no statistical technique will provide an understanding of why that variable is a cause. Knowledge of one's field, theory, and common sense, combined with statistical results, nurture explanation and understanding.

A PROPOSED RESEARCH STRATEGY

In the present section, we put the processes of data snooping, hypothesis testing, replication, and manipulation into the perspective of an encompassing research strategy. The strategy involves (a) the use of *data snooping* to generate hypotheses (preferably directional hypotheses); (b) the testing of hypotheses on static variables; (c) the *replication* of the hypothesis testing; (d) the development and testing hypotheses about the effects of *manipulation* of some previously static variables; and (e) the *replication* of the manipulated-variable hypotheses. Table 12.1 displays the strategy schematically.

Data Snooping

When one is data snooping, one has little idea about what will be found. The goal of data snooping can be viewed from either of two viewpoints. The first is that the researcher is attempting to build up the R^2 to as close to 1.00 as possible, reasonable, or desirable. The second viewpoint is that the researcher is attempting to reduce the magnitude of the error vector. By identifying variables that are correlated with what was originally considered "error," the criterion can be more accurately predicted, and the magnitude of the error reduced. Once again we realize that the error vector is both error of measurement *and* error of specification. By *error of specification* we mean that not all of the correct variables were originally considered in the regression model.

Table 12.1.
Stages of a Proposed Research Strategy and their Relative Emphasis upon Probability and R^2

Stage	Emphasis on Low Probability	Emphasis on High R^2
Static Variables Stages 1, 2, & 3		
Stage 1. Data Snooping (Hypothesis Generating)	No	Yes
Stage 2. Hypothesis Testing (Static)	Yes	No
Stage 3. Replication	No	Yes
Dynamic Variables Stages 4 & 5		
Stage 4. Manipulation	Yes	Yes
Stage 5. Replication	Yes	No

When a large R^2 is observed, one may suspect that the set of variables used to obtain that R^2 is a good set of variables to investigate further. This stage of research can be very exciting because most dramatic discoveries seem to have originated in a stage of this sort; but any discoveries at this stage must be tested on successive samples before one may reasonably say that a discovery was made. Serendipitous findings at this stage can be valuable; one must simply realize that subsequent verification must be made.

Suppose that, for example, researchers have an interest in why 9th graders vary in their knowledge of and achievement in the area of biological science. The researchers have also chosen a criterion measure of this construct (realizing, of course, that it is not exactly the same as their construct); the criterion measure is a standardized biological science achievement test administered at the end of the 9th grade (after one year of biology instruction). Based on the researchers' knowledge of the research area and on discussions with biology teachers and students, the researchers decide to investigate the following variables to see if they will help to predict biology achievement:

1. gender of student;
2. socioeconomic status of student;
3. time student has spent outdoors in earlier years;
4. number of extra credits student completes;
5. (number of extra credits)^2;
6. size of biology class; and
7. gender * size of biology class

The researchers do not have enough information at this point to state hypotheses about the ways that these variables relate to biology achievement; they are data snooping.

The researchers arrange to collect data on the criterion and predictor variables from the 9th-grade students in a particular school. Just before administering the biology achievement test, the school notices that it does not have enough original copies of the test; so the school makes photocopies of the test for the final one-quarter of the students. Our researchers notice that the photocopies are not as legible as the originals. They therefore add a new variable to their list to consider—legibility of the test.

Due to the researchers' data snooping, they find that the following model, containing only some of their suspected variables, does a good job of predicting biology achievement, attaining an R^2 of .76:

$$B = a_0U + b_1S + c_1A^2 + d_1(G * X) + e_1(H * X) + f_1L + E_{11} \quad \text{(Model 12.13)}$$

where:
1. B = biology achievement test score;
2. U = unit vector;
3. S = socioeconomic status (SES) of student on 5-point scale;
4. A^2 = (Number of extra credits)^2;
5. $G * X$ = size of biology class if student is male, 0 otherwise;
6. $H * X$ = size of biology class if student is female, 0 otherwise; and
7. L = legibility of the test.

The researchers are ready to state and test their hypotheses.

Hypothesis Testing

A hypothesis cannot be made without the researcher relying on some combination of past research, theoretical orientation, and intuitive insight. Data snooping results can often aid in developing a supportable hypothesis. As discussed previously, one may in the hypothesis-testing stage compare the difference in R^2 between the Full Model and Restricted Model to some preselected amount of difference, or one may rely on the probability values as presented in this text. If only one restriction is being made on the Full Model, and if significance is obtained, one will know what is causing that significance. The goal of making an accurate causal statement is enhanced if the hypothesis is tested in a model exhibiting a high R^2. But as can be seen in Table 12.1, a high R^2 is not viewed as necessary at this stage.

To return to the example, the researchers, based on their data snooping results, state several hypotheses, of which the following is one; and they collect new data on which to test them. The researchers' Research Hypothesis is as follows:

Research Hypothesis: For 9th-grade biology students, the score obtained on the biology achievement test will be higher for boys if they are in small classes (fewer than 20) rather than in large classes, and will be higher for girls if they are in large classes (20 and more) rather than in small classes, over and above the effects of SES, number of extra credits squared, and legibility of the test.

The researchers then test their hypothesis and find that, at their chosen alpha level, the Statistical Hypothesis can be rejected; so the Research Hypothesis is accepted as tenable. Successively, the researchers find that all the variables in Model 12.13 are valuable.

Up to this point, no variables have been manipulated, although some are potentially easily manipulated. The model containing static (nonmanipulated) variables (Model 12.13) was upheld by the hypothesis testing. The static hypothesis-testing phase now needs to be replicated.

Replication

As can be seen in Table 12.1, the process of replication appears twice, once in the static phase and once in the manipulation phase. Because it is necessary to proceed from a static model to a dynamic one (containing manipulation) when hunting and testing for causal relationships, the researcher must make a decision concerning the extent and level of successful replication before manipulation. This decision is based largely on the potential cost of the manipulation intended--cost in money, time, and human resources. Rescheduling the large and small biology classes will cost money and may provoke parental, teacher, student, or taxpayer protest. Thus, one may want to replicate the static hypothesis-testing model several times before venturing to manipulate. In the process of replication (before manipulation), one should not be too concerned with probability but with obtaining a high R^2. (It would be ideal if approximately the same weighting coefficients could be obtained, but correlated predictors and small sample sizes usually keep this from happening.)

Newman, McNeil, and Fraas (2004) discussed two kinds of replication and indicated how the GLM could be used in studying them. One method of estimating the replicability of the findings deals with replication in the exact same system. The second method, which contains subjective probability values, is used to estimate the replicability of a study's findings in a system that may differ from the initial system with respect to salient variables. The incorporation of the replicability estimates delineated in this paper would provide critical information to decision makers about the likelihood that the implementation of a particular method or treatment would produce similar results in their systems.

Manipulation

The first three stages above apply to static data: that is, within-person, focal stimuli, and context variables are measured as they exist, with no effort to change them. Functional relationships are described, so that the criterion variable may be predicted. One should be aware that a variable that is predictive of a criterion in a static model may not be predictive in the same fashion as a dynamic variable. The extreme case would be a variable that simply cannot be manipulated. Another example would be a variable that, when manipulated, causes "side effects" that influence the criterion (either positively or negatively). On the other hand, variables that are found to be not predictive in a static situation may indeed be necessary causers in a dynamic situation—Hickrod, 1971, discusses this and several other interesting aspects of regression with respect to school-finance concerns. As a consequence of these realities, some statisticians take a strong stand regarding the analysis of static variables. Werts (1970) stated:

> Obviously when partial regression weights are estimated from naturalistic data it will be far from certain that these coefficients will have any resemblance to what would be found under conditions of experimental control. By asking what would happen "if," we are in effect asking the question: what would the data be like if they weren't what they are?" (p. 130)

We take a more moderate view. Researchers should be aware that what is discovered in a static situation very likely may not occur in a dynamic situation. But the researcher still may choose to investigate the static situation before the dynamic one because of the potential "cost" of creating the dynamic situation (manipulating variables under experimental control). We also see researchers getting overly enthused with their manipulations, often losing sight of the purpose of the manipulation—raising scores on the criterion.

When an R^2 of 1.0 is obtained in a dynamic situation, one can be confident that the causers have been isolated. One then needs to investigate the magnitude of the criterion scores. It may be that the causers of increasing the criterion two units may have been discovered, but a two-unit increase may not be a sufficient increase. The discovered causers may not be sufficient to obtain the desired three-unit increase. Therefore, other causers must be posited and investigated. The cycle of research has thus started again.

The manipulation step in scientific investigation can be thought of as making a change in subjects, so that predictions can be upset. Johnny is predicted to be a bad reader by the use of the static data, but certain changes can be made, resulting in Johnny becoming a good reader. A plot of land is predicted to be of low fertility, but certain changes may make that land desirable for planting of crops.

The stage of manipulation contains dynamic variables—those that have been willfully changed by the researcher. As has been emphasized previously, it is desirable to investigate *manipulatable* variables in the snooping and static hypothesis-testing stages so that the move into the manipulation phase will be facilitated.

Our biology researcher wants to manipulate and test each predictor variable in Model 12.13. This is dynamic hypothesis testing, and the researcher's hypotheses will, one hopes, be directional since, before manipulation, the researcher has investigated the variables extensively.

Are all the predictor variables in Model 12.13 manipulatable? Class size and the proportion of males to females in each class can be arranged by the school. Legibility of the test is also directly manipulatable—one needs only to select certain levels of legibility. The number of extra credits may be more difficult to manipulate; some added inducement may be necessary to get students to complete more assignments (if the hypothesized relationship is positive). Or the researcher may believe that it is the content rather than the credit of those assignments that is of benefit and may change those assignments from extra credit to required.

Socioeconomic status is not readily manipulatable—SES is generally based on income or guardian income, occupation, and education. Changing a child's family circumstances is usually impossible, so the researcher may wish to hypothesize critical elements or aspects of SES that might be varied. Nutrition, self-esteem, need for achievement, and teacher acceptance are surely related to SES and also may affect biology achievement. Indeed, biology achievement might be hampered by inadequate breakfast and a lack of desire to achieve. These may be the manipulatable variables that SES summarizes. If so, then a school breakfast program and training that would instill a need for achievement may be manipulated to increase biology achievement for students who would otherwise be predicted to perform poorly without such intervention.

Gender also is not easily manipulated, but gender is placed in the model to interact with a manipulatable variable. It thus performs a useful predictive function without being manipulated. Here, the researcher may not be concerned with gender as a manipulatable variable. The researcher may, however, wish to consider what aspects of gender (underlying dynamic variables) interact with class size to predict biology achievement. Surely simple physiology has an explanatory role—but of unknown value; socialization variables, however, are more likely to be the underlying causative factors accounted for by the gender variable. For example, in the past girls may have been informed that, "Science is not really for girls." If so, a measure of sex role identification ought to be a better predictor of biology achievement than the binary gender variable. If it is a better predictor, a program designed to break down stereotypes ought to reduce the negative influence of being a girl on biology achievement.

If one wishes to go beyond prediction to improvement, then it is important that static variables be examined carefully for their underlying dynamic causative influences on criterion measures (e.g., sex role identification may underlie the effect on achievement of the binary Gender variable).

One may not always wish to do away with all nonmanipulatable variables, however. They are valuable predictors in at least two situations: First, nonmanipulatable variables may be used until the underlying dynamic variable is found. Second, nonmanipulatable variables may be unchangeable but may interact with a dynamic variable (e.g., gender interacting with treatment). If one cannot manipulate, then one must rest on the tenuous grounds of theory. Theory always plays a large part because the variables that are included in the analysis were purposefully put there—one may not think initially of that as theory, but it certainly guided one's actions.

Replication of Manipulated Findings

Once manipulations have been found to be successful (resulting in changes in the desired direction), then replications are called for to establish the magnitude of the effect of those manipulations. Researchers should put faith in causers only when manipulations have a consistent effect.

Successful replication by the biology researcher should please the researcher. With success at each of the previous stages, the researcher can be somewhat assured that the results are valid. Additional replications on new samples of 9th-grade students will give further credence to the findings, for chance may have been operating in all the previously successful results.

What is more likely the case is that the biology researcher did not perform research at some of the other stages. Because of lack of resources or time, research is often forced to go to the manipulation stage without prior success at the other stages. In these cases, replication is a more crucial stage: But how many successful replications are needed? There is no established answer to that question, although it depends in part upon how crucial it is that the criterion of interest be under control.

Table 12.1 summarized the stages of the proposed research strategy and showed their emphases upon probability and R^2. If success on the goal of either low probability or high R^2 is not obtained at any stage, activity must revert to the initial stage of data snooping.

FURTHER READING

The GLM approach is an extremely flexible technique that opens up a host of secrets about statistics—once certain mechanical topics are mastered. New practical developments are constantly being made by those using the GLM. More mathematical treatments of the technique than presented here can be found in Maxwell and Delaney (1990), Seber (1984), Ward and Jennings (1973), and Kleinbaum and Kupper (1978). An elementary introduction for hypnotists, written by Starr (1971), contains minimal mathematical concepts.

The Special Interest Group of the American Educational Research Association has been in existence for many years, and their journal, *Multiple Linear*

Regression Viewpoints (MLRV), is a forum for new ideas. Many articles published in that journal have been referenced in this text. MLRV affords a way for the reader to keep abreast of new regression developments. Further information on MLRV, including contents of previous issues, can be obtained from the internet at http://mlrv.ua.edu/ejournal.html.

APPENDIXES

APPENDIX A
INTERNET SITE ADDRESS AND FILES

Internet Site: testingwithglm.com

Microsoft Excel Files listed on the internet site are as follows:
1. GLM DATA EXCEL FORMAT
2. Change in R-Squared Program
3. F Test Calculation Program
4. Table 10.1 Data
5. DICHOTOMOUS CRITERION VARIABLE DATA

SPSS files listed on the internet site are as follows:
1. GLM DATA SPSS FORMAT
2. DICHOTOMOUS CRITERION VARIABLE DATA

APPENDIX B
GLM DATA EXCEL FORMAT

id	X2	X3	X4	X5	X6	X7	X8	X9	X10	X11	X12	X13	X14	X15	X16	X17	X18	X19
1	4	4	26	8	1	0	82	97	0	1	0	1	80	4	1	0	0	0
2	6	9	28	9	1	0	95	101	0	1	1	0	75	6	1	0	0	0
3	6	8	41	10	0	1	103	104	0	1	0	1	82	7	1	0	0	0
4	8	10	29	10	1	0	109	106	0	1	0	1	84	7	1	0	0	0
5	10	12	40	11	1	0	84	96	0	1	0	1	85	6	1	0	0	0
6	11	13	30	11	1	0	93	100	0	1	1	0	82	6	1	0	0	0
7	12	14	39	12	0	1	98	102	0	1	1	0	84	5	1	0	0	0
8	14	15	32	13	1	0	99	103	0	1	0	1	90	5	1	0	0	0
9	16	15	37	14	0	1	87	98	0	1	0	1	92	6	1	0	0	0
10	17	16	37	15	0	1	106	105	0	1	1	0	92	6	1	0	0	0
11	18	16	36	15	0	1	113	112	0	1	1	0	90	7	1	0	0	0
12	20	17	35	17	0	1	128	125	0	1	0	1	97	11	1	0	0	0
13	22	18	34	18	1	0	89	99	0	1	0	1	100	17	1	0	0	0
14	24	18	33	58	0	1	115	118	0	1	1	0	96	18	1	0	0	0
15	26	20	32	21	0	1	125	124	0	1	1	0	98	18	1	1	0	0
16	27	20	32	22	0	1	91	100	0	1	1	0	100	18	0	1	0	0
17	28	20	31	23	0	1	116	115	0	1	1	0	101	18	0	1	0	0
18	28	21	36	55	1	0	117	117	0	1	0	1	105	18	0	1	0	0
19	29	20	30	24	0	1	123	120	0	1	0	1	106	17	0	1	0	0
20	30	21	29	53	0	1	126	120	0	1	0	1	108	17	0	1	0	0

APPENDIX B (continued)
GLM DATA EXCEL FORMAT

	A	B	C	D	E	F	G	H	I	J	K	L	M	N	O	P	Q	R	S
22	21	31	21	28	26	0	1	120	116	0	1	1	0	103	17	0	1	0	0
23	22	32	22	26	52	0	1	129	127	0	1	1	0	105	18	0	1	0	0
24	23	32	22	37	27	1	0	120	116	0	1	0	1	110	19	0	1	0	0
25	24	33	22	24	51	0	1	114	113	0	1	0	1	112	25	0	1	0	0
26	25	34	22	37	52	1	0	100	102	0	1	1	0	107	28	0	1	0	0
27	26	36	23	38	28	1	0	98	103	0	1	1	0	110	29	0	1	0	0
28	27	38	24	38	49	1	0	96	102	0	1	0	1	115	30	0	1	0	0
29	28	40	24	38	48	1	0	85	95	0	1	1	0	120	30	0	1	0	0
30	29	42	25	39	32	1	0	125	128	0	1	0	1	115	30	0	1	0	0
31	30	48	25	40	38	1	0	95	102	0	1	1	0	120	30	0	1	0	0
32	31	5	5	27	8	1	0	84	80	1	0	1	0	75	4	0	0	1	0
33	32	7	8	27	9	1	0	85	82	1	0	0	1	82	6	0	0	1	0
34	33	7	8	41	10	0	1	115	123	1	0	0	1	83	7	0	0	1	0
35	34	8	11	29	11	1	0	87	85	1	0	1	0	80	7	0	0	1	0
36	35	10	12	41	11	0	1	95	95	1	0	1	0	83	6	0	0	1	0
37	36	11	12	30	10	1	0	93	92	1	0	1	0	84	6	0	0	1	0
38	37	12	13	39	11	0	1	99	100	1	0	1	0	85	5	0	0	1	0
39	38	14	15	32	14	1	0	95	100	1	0	0	1	90	5	0	0	1	0

APPENDIX B (continued)
GLM DATA EXCEL FORMAT

	A	B	C	D	E	F	G	H	I	J	K	L	M	N	O	P	Q	R	S
40	39	16	16	37	13	0	1	118	128	1	0	0	1	92	6	0	0	1	0
41	40	17	16	37	14	0	1	88	86	1	0	0	1	93	6	0	0	1	0
42	41	18	17	36	16	0	1	94	92	1	0	1	0	90	7	0	0	1	0
43	42	20	18	35	16	0	1	102	102	1	0	1	0	92	11	0	0	1	0
44	43	21	17	34	19	1	0	89	87	1	0	0	1	98	17	0	0	1	0
45	44	23	17	33	57	0	1	116	122	1	0	0	1	100	18	0	0	1	0
46	45	25	19	32	22	0	1	97	97	1	0	0	1	103	18	0	0	0	1
47	46	27	20	32	21	0	1	103	102	1	0	0	1	105	18	0	0	0	1
48	47	28	21	31	24	0	1	90	89	1	0	1	0	100	18	0	0	0	1
49	48	29	22	36	54	1	0	96	96	1	0	1	0	102	18	0	0	0	1
50	49	30	22	30	24	0	1	108	113	1	0	1	0	103	17	0	0	0	1
51	50	30	21	39	54	1	0	100	101	1	0	1	0	104	17	0	0	0	1
52	51	31	22	28	27	0	1	91	89	1	0	0	1	108	17	0	0	0	1
53	52	32	21	26	51	0	1	112	116	1	0	0	1	110	18	0	0	0	1
54	53	32	22	37	28	1	0	104	104	1	0	1	0	105	19	0	0	0	1
55	54	33	21	24	52	0	1	92	90	1	0	0	1	112	25	0	0	0	1
56	55	33	22	37	51	1	0	120	130	1	0	0	1	110	28	0	0	0	1
57	56	35	23	38	29	1	0	122	132	1	0	0	1	113	29	0	0	0	1
58	57	37	24	38	50	1	0	128	140	1	0	1	0	110	30	0	0	0	1
59	58	39	24	38	49	1	0	124	136	1	0	1	0	113	30	0	0	0	1
60	59	41	25	39	33	1	0	126	137	1	0	0	1	114	30	0	0	0	1
61	60	46	25	40	39	1	0	130	142	1	0	0	1	124	30	0	0	0	1

APPENDIX C
Change in R-Squared Program

To construct the "Change in R-Squared Program" in a Microsoft Excel spreadsheet, complete the following steps (type the words and symbols inside " "):

Step1: Type "R-Squared" in cell A1and right justify the cell contents.

Step 2: Type "t-value=" in cell A2 and right justify the cell contents.

Step 3: Type "DFd=" in cell A3and right justify the cell contents.

Step 4: Type "R-Squared Change =" and right justify the cell contents.

Step 4: Type "=((B2*B2)*(1-B1))/(B3)" and left justify the cell contents.

When the following values (a) .934, (b) 17.316, and (c) 56 are typed in cells B1, B2, and B3, respectively, the value of .35 will be computed and placed in cell D1. The program will appear as follows:

	A	B	C	D
1	R-Squared=	0.934	R-Squared Change =	0.35
2	t Value=	17.316		
3	DFd=	56		

APPENDIX D
F Test Calculation

To construct a program in a Microsoft Excel spreadsheet that will calculate the *F* test and the corresponding probability value for a change in the R^2 values of a Full Model and a Restricted Model, complete the following steps (type the words and symbols inside " "):

Step 1: Type "R-Squared Full Model =" in cell A1and right justify the cell contents.

Step 2: Type "R-Squared Restricted Model =" in cell A2 and right justify the cell contents.

Step 3: Type "Regression DF Full Model =" in cell A3and right justify the cell contents.

Step 4: Type "Regression DF Restricted Model" in cell A4 and right justify the cell contents.

Step 5: Type "Residual DF Full Model =" in cell A5 and right justify the cell contents.

Step 6: Type "Numerator DF =" in cell C1 and right justify the cell contents.

Step 7: Type "F Test Value =" in cell C2 and right justify the cell contents.

Step 8: Type "F Test Value Probability" in cell C3 and right justify the cell contents.

Step 9: Type "=B3-B4" in cell D1 and left justify the cell contents.

Step10: Type"=((B1-B2)/D1)/((1-B1)/B5)" in cell D2 and left justify the cell contents.

Step 11: Type "FDIST(D2,D1,B5)" and left justify the cell contents. Set the decimal places to 4 for this cell.

When the following values (a) .982, (b) .877, and (c) 3, (d) 2, and (e) 56 are typed in cells B1, B2, B3, B4, and B5, respectively, the values of (a) 1, (b) 326.667, and (c) 0.0000 will be computed and placed in cells D1, D2, and D3, respectively. The Excel spreadsheet will appear as follows:

◢	A	B	C	D
1	R-Squared Full Model =	0.982	Numerator DF =	1
2	R-Squared Restricted Model =	0.877	F Test Value =	326.667
3	Regression DF Full Model =	3	F Test Value Probability =	0.0000
4	Regression DF Restricted Model =	2		
5	Residual DF Full Model =	56		

APPENDIX E
Dichotomous Criterion Variable Data

	A	B	C
1	Test	Program	Assessment
2	0	0	43
3	0	0	24
4	0	0	15
5	0	0	36
6	0	0	28
7	0	0	48
8	0	1	20
9	0	1	47
10	1	0	28
11	1	0	42
12	1	0	30
13	1	0	55
14	1	1	48
15	1	1	31
16	1	1	22
17	0	1	17
18	1	1	40
19	1	1	33
20	1	1	45
21	1	1	22
22	0	0	48
23	0	0	29
24	0	0	19
25	0	0	30
26	0	0	29
27	0	0	40
28	0	1	25
29	0	1	40
30	1	0	20

	A	B	C
31	1	0	49
32	1	0	30
33	1	0	50
34	1	1	40
35	1	1	38
36	1	1	28
37	0	1	15
38	1	1	45
39	1	1	30
40	1	1	48
41	1	1	28
42	0	0	48
43	0	0	29
44	0	0	19
45	0	0	30
46	0	0	29
47	0	0	40
48	0	1	25
49	0	1	40
50	1	0	20
51	1	0	49
52	1	0	30
53	1	0	50
54	1	1	40
55	1	1	38
56	1	1	28
57	0	1	15
58	1	1	45
59	1	1	30
60	1	1	48

Appendix E (continued)
Dichotomous Criterion Variable Data

	A	B	C
61	1	1	28
62	0	0	41
63	0	0	24
64	0	0	15
65	0	0	36
66	0	0	28
67	0	0	48
68	0	1	20
69	0	1	47
70	1	0	28
71	1	0	42
72	1	0	30
73	1	0	55
74	1	1	48
75	1	1	31
76	1	1	22
77	0	1	17
78	1	1	40
79	1	1	33
80	1	1	45
81	1	1	22
82	0	0	42
83	0	0	24
84	0	0	15
85	0	0	36
86	0	0	28
87	0	0	48
88	0	1	20
89	0	1	47
90	1	0	28

	A	B	C
91	1	0	42
92	1	0	30
93	1	0	55
94	1	1	48
95	1	1	31
96	1	1	22
97	0	1	17
98	1	1	40
99	1	1	33
100	1	1	45
101	1	1	22
102	0	0	48
103	0	0	29
104	0	0	19
105	0	0	30
106	0	0	29
107	0	0	40
108	0	1	25
109	0	1	40
110	1	0	20
111	1	0	49
112	1	0	30
113	1	0	50
114	1	1	40
115	1	1	38
116	1	1	28
117	0	1	15
118	1	1	45
119	1	1	30
120	1	1	48
121	1	1	28

References

Andrews, F., Morgan, J., & Sonquist, J. (1967). *Multiple classification analysis*. Ann Arbor: The University of Michigan, Institute for Social Research.

Austin, J. T., Yaffee, R. A., & Hinkle, D. E. (1992). Logistic regression for research in higher education. In Smart, J. C. (Ed.) *Higher Education: Handbook of Theory and Research, 8,* 379–410.

Bender, J. A., Kelly, F. J., Pierson, J. K., & Kaplan, H. (1968). Analysis of the comparative advantages of unlike exercises in relation to prior individual strength levels. *The Research Quarterly, 39,* 443–448.

Boneau, C. A. (1960). The effects of violations of assumptions underlying the *t-test. Psychological Bulletin, 57,* 49–64.

Bottenberg, R. A., & Ward, J. H. (1963). *Applied multiple linear regression*. Lackland Air Force Base, TX: Aerospace Medical Division, AD 413128.

Bray, J. H., & Maxwell, S. E. (1985). *Multivariate analysis of variance*. Beverly Hills, CA: Sage.

Byrne, J. (1974). The use of regression equations to demonstrate causality. *Multiple Linear Regression Viewpoints, 5*(1), 11–22.

Campbell, D. T., & Stanley, J. C. (1966). *Experimental and quasiexperimental designs for research*. Chicago: Rand McNally.

Castaneda, A., Palermo, D. S., & McCandless, B. (1956). Complex learning and performance as a function of anxiety in children and task difficulty. *Child Development, 27,* 328–332.

Cohen, J. (1970). Approximate power and sample size determination for common one-sample and two-sample hypothesis tests. *Educational and Psychological Measurement, 30,* 811–832.

Cohen, J. (1988). *Statistical power analysis for the behavioral sciences* (2nd ed.). New York: Academic Press.

Cohen, J., & Cohen, P. (1975). *Applied multiple regression/correlation analysis for the behavioral sciences*. New York: Halstead.

Constas, M. A., & Francis, J. D. (1992). A graphical method for selecting the best sub-set regression model. *Multiple Linear Regression Viewpoints, 19* (1), 16–25.

DuCette, J., & Wolk, S. (1972). Ability and achievement as moderating variables of student satisfaction and teacher preparation. *The Journal of Experimental Education, 41* (1), 12–17.

Fraas, J. W., & Drushal, M. E. (1987). The use of MLR models to analyze partial interaction: An educational application. *Multiple Linear Regression Viewpoints, 15*(2), 85–96.

Fraas, J. W., & McDougall, W. (1983). The use of one full MLR model to conduct multiple comparisons in a repeated measures design. *Multiple Linear Regression Viewpoints 12*(1), 42–55.

Fraas, J. W., & Newman, I. (1997). The use of the Johnson-Neyman confidence bands and multiple regression models to investigate interaction effects: Important tools for educational researchers and program evaluators. *Multiple Linear Regression Viewpoints 24*(1), 1997, 14–24.

Galileo, G. (1632/1953). *The two chief world systems* (S. Drake, Trans.). Berkeley: University of California Press. (Original work published 1632.)

Goldberg, L. R. (1972). Parameters of personality inventory construction and utilization: A comparison of prediction strategies and tactics. *Multivariate Behavioral Research Monographs, 7,*(2).

Gujarati, D. N. (1988). *Basic econometrics* (2nd ed.). New York: McGraw-Hill.

Hair, J. F., Black, W. C., Babin, B. J., Anderson, R. E., & Tatham, R. L. (2006). *Multivariate Statistics* (6th ed.). Upper Saddle River, NJ: Pearson Prentice Hall.

Hand, D. J., & Taylor, C. C. (1987). *Multivariate analysis of variance and repeated measures: A practical approach for behavioral scientists.* New York: Chapman & Hall.

Harris, R. J. (1985). *A primer of multivariate statistics* (2nd ed.). Orlando, FL: Academic Press.

Harris, R. J. (1993). "Beta" weights should be used to interpret regression variates and to assess in-context variable importance. *Mid-western Educational Research, 6*(1), 11–14.

Haupt, C. C. (1993). Facilitation effects: Determinants of drug use among Hispanic and White non-Hispanic fourth, fifth; and sixth grade public school students. *Dissertation Abstracts International, 53*(7), *530A.* (University Microfilms No. DA9232434).

Hickrod, G. A. (1971). Local demand for education: A critique of school finance and economic research circa 1959-1969. *Review of Educational Research, 41,* 35–50.

Hill, R. F. (2008). *Policy decisions and horizontal equity in Ohio: 1980-2003.* Unpublished doctoral dissertation, Ashland University, Ashland, Ohio.

Hinkle, D. E., Wiersma, W., & Jurs, S. G. (1994). *Applied statistics for the behavioral sciences* (3rd ed.). Boston: Houghton Mifflin.

Jennings, E. E. (1967). Fixed effects analysis of variance by regression analysis. *Multivariate Behavioral Research. 2*, 95–108.

Kelly, F. J., Beggs, D. L., McNeil, K. A., Eichelberger, T., & Lyon, J. (1969). *Research design in the behavioral sciences: Multiple regression approach.* Carbondale: Southern Illinois University Press.

Kelly, F. J., McNeil, K. A., & Newman, I. (1973). Suggested inferential statistical models for research in behavior modification. *The Journal of Experimental Education, 41*(4), 54–63.

Kerlinger, F. N., & Pedhazur, E. J. (1973). *Multiple regression in behavioral research.* New York: Holt, Rinehart & Winston.

Kleinbaum, D. G., & Kupper, L. L. (1978). *Applied regression analysis and other multivariable methods.* North Scituate, MA: Duxbury Press.

Kromrey, J. D., & Foster-Johnson, L. (1996). Determining the efficacy of intervention: The use of effect sizes for data analysis in single-subject research. *Journal of Experimental Education, 65*(1), 73-93.

Lewis-Beck, M. S. (1980). *Applied regression: An introduction.* Beverly Hills, CA: Sage.

Long, J. S. (1997). *Regression models for categorical and limited dependent variables: Analysis and interpretation.* Thousand Oaks, CA: Sage.

MacClelland, G. H., & Judd, C. M. (1993). Statistical difficulties of detecting interactions and moderator effects. *Psychological Bulletin, 114*(2), 376–390.

Maxwell, S. E., & Delaney, H. D. (1990). *Designing experiments and analyzing data: A model comparison perspective.* Belmont, CA: Wadsworth.

McNeil, K. A. (1970a). Meeting the goals of research with multiple regression analysis. *Multivariate Behavioral Research, 5*, 375–386.

McNeil, K. A. (1970b). The negative aspects of the eta coefficient as an index of curvilinearity. *Multiple Linear Regression Viewpoints, 1*(1), 7–17.

McNeil, K. A., & Beggs, D. L. (1969, February). *The mathematical equivalence of the test of significance of the point biserial correlation coefficient and the t-test for the difference between means.* Paper presented at the meeting of the American Educational Research Association, Los Angeles.

McNeil, K. A., Evans, J., & McNeil, J. T. (1979, April). *Non-linear transformation of the criterion.* Paper presented at the meeting of the American Educational Research Association, San Francisco.

McNeil, K. A., & Kelly, F. J. (1970). Express functional relationships among data rather than assume "intervalness." *The Journal of Experimental Education, 39*(2), 43–48.

McNeil, K. A., & McShane, M. (1974). Complexity in behavioral research, as viewed within the multiple linear regression approach. *Multiple Linear Regression Viewpoints, 4* (4), 12–15.

McNeil, K., & Newman, I. (1996). Categorical or continuous interaction? *Mid-Western Educational Researcher, 6*(4), 32-34.

McNeil, K. A., Newman, I., & Kelly, F. J. (1996). *Testing research hypotheses with the general linear model.* Southern Illinois University Press: Carbondale, IL.

McNeil, K. A., & Spaner, S. D. (1971). A defense for including highly correlated predictor variables in multiple regression models. *Multivariate Behavioral Research, 6,* 117–125.

Mendenhall, W. (1968). *Introduction to linear models and the design and analysis of experiments.* Belmont, CA: Wadsworth.

Mosteller, F., & Tukey, J. W. (1968). Data analysis, including statistics. In G. Lindzey & E. Aronson (Eds.), *The handbook of social psychology: Vol. 2* (pp. 88–203). Reading, MA: Addison Wesley.

Mueller, R. O. (1990). Teaching ANCOVA: The importance of random assignment. *Multiple Linear Regression Viewpoints. 17*(2), 1–14.

Newman, I., Benz, C., & Williams, J. D. (1990). Alternative in analyzing the Solomon four group design. *Multiple Linear Regression Viewpoints.* 17 (2), 91–103.

Newman, I., Fraas, J. W., & Herbert, A. (2002). *Testing non-nil null hypotheses with t tests of group means: A Monte Carlo study.* Paper presented at the annual meeting of the Eastern Educational Research Association, Sarasota, FL.

Newman, I., Fraas, J., & Laux, J. M. (2000). A three-step adjustment procedure for Type I error rates. *Journal of Research in Education, 10*(1), 84–90.

Newman, I., Groom, W., & Hoedt, K. (1983). Note: Suggested method of correcting alpha error build-up on orthogonal dependent variables. *Multiple Linear Regression Viewpoints. 2*(1), 56–60.

Newman, I., McNeil, K., & Fraas, J. W. (2004). Two methods of estimating a study's replicability. *Mid-Western Educational Researcher. 17*(2), pp. 36–40.

Newman, I., & Newman, C. (1999). *A discussion of low R-squares: Concerns and uses.* Paper presented at the meeting of the American Educational Research Association, Montreal, Canada.

Presley, R .J., & Huberty, C. (1988). Predicting statistics achievement: A prototypical regression analysis. *Multiple Linear Regression Viewpoints. 16*(1), 36–77.

Publication Manual of the American Psychological Association (2009). (Publication of the American Psychological Association). Washington, DC.

Rosenthal, R. (1984). *Meta-analytic procedures for social research.* Beverly Hills, CA: Sage.

Saunders, D. R. (1956). Moderator variables in prediction. *Educational and Psychological Measurement, 16,* 209–222.

Seber, G. A. F. (1984). *Multivariate observations.* New York: John Wiley & Sons.

Spaner, S. (1970). *Application of multivariate techniques to developmental data: The derivation, verification, and validation of a predictive model of twenty-four month cognitive data* (Doctoral dissertation, Southern Illinois University, 1990). *Dissertation Abstracts International, 31, (10-A).*

Starr, F. H. (1971). The remarriage of multiple regression and statistical inference: A promising approach for hypnosis researchers. *The American Journal of Clinical Hypnosis, 13,* 175–197.

Stevens, J. (2002). *Applied multivariate statistics for the social sciences* (3rd ed.). Hillsdale, NJ: Lawrence Erlbaum Associates.

Tatsuoka, M. M. (1988) *Multivariate analysis: Techniques for educational and psychological research* (2nd ed.). New York: Macmillan.

Thayer, J. D. (1990). Implementing variable selection techniques in regression. *Multiple Linear Regression Viewpoints, 17*(2), 67–90.

Thompson, B. (1999). Improving research clarity and usefulness with effect size indices as supplements to statistical significance tests. *Exceptional Children, 65*(3), 329–337.

Wackerly, D. D., Mendenhall III, W. Scheaffer, R.L. (2002). *Mathematical statistics with applications* (6th ed.). Pacific Grove, CA: Duxbury.

Ward, J., & Jennings, E. (1973). *Introduction to linear models.* Englewood Cliffs, NJ: Prentice Hall.

Werts, C. E. (1970). The partitioning of variance in school effects studies: A reconsideration. *American Educational Research Journal, 7,* 127–132.

Wilkinson, W. K., & McNeil, K. (1999). *Research for helping professions.* Pacific Grove: Brooks/Cole.

Williams, J.D. (1974a). *Regression analysis in educational research.* New York: MSS Information Corporation.

Williams, J. D. (1974b). Regression solutions to the AxBxS design. *Multiple Linear Regression Viewpoints.* 5(2), 3–9.

Williams, J. D. (1987). The use of nonsense coding with ANOVA situations. *Multiple Linear Regression Viewpoints, 15*(2), 29–39.

Williams, J. D., & Lindem, A. C. (1974). Regression computer programs for setwise regression and three related analysis of variance techniques. *Multiple Linear Regression Viewpoints.* 4(4), 30–46.

Williams, J. D., & Newman, I. (1982, March). *Using multiple linear models to simultaneously analyze a Solomon four group design.* Paper presented at the meeting of the American Educational Research Association, New York.

Winer, B. J. (1971). *Statistical principles in experimental design* (2nd ed.). New York: McGraw-Hill.

Woehlke, P., Elmore, P. B., & Spearing, D. L. (1990). Testing assumptions in multiple regression: Comparison of procedures available in SAS and SPSSX. *Multiple Linear Regression Viewpoints, 17*(2), 48–66.

Wood, D. A., & Langevin, M. J. (1972). Moderating the prediction of grades in freshman engineering. *Journal of Educational Measurement, 9*, 311–320.

Wooldride, J. M. (2008). *Introductory econometrics: A modern approach.* (2nd ed.) Cincinnati, OH: South-Western College Publishing.

Index

Authors

Keith McNeil taught statistics and research methods since 1967. He has been active in the Special Interest Group of Multiple Linear Regression in the American Educational Research Association since its inception. He has co-authored three other texts on GLM, as well as a text on research methods, and with Isadore Newman texts on survey research and writing dissertations. He is an Emeritus professor at New Mexico State University.

Isadore Newman taught statistics and research methods since 1970. He has been active in the Special Interest Group of Multiple Linear Regression in the American Educational Research Association since its inception, serving as the editor of its journal for 20 years. He has co-authored 17 books and book chapters related to research methods, three of which were co-authored with Keith McNeil on GLM, survey research, and writing dissertations. He is a Distinguished Professor Emeritus at Akron University and currently serves as visiting scholar at Florida International University.

John W. Fraas, Trustees' Distinguished Professor Emeritus, taught statistics at Ashland University in the schools of business, economics, and education for nearly four decades. While at Ashland University he was awarded numerous teaching awards including the first Teaching Excellence Award given by the Association of Collegiate Business Schools and Programs. He has published over 35 journal articles dealing with applied statistics in the fields of business management, economics, and education. In addition, he authored a text entitled "Basic Concepts in Educational Research," which was published by University Press of America.